Repair and Strengthening of Existing Reinforced Concrete Structures

Repair and Strengthening of Existing Reinforced Concrete Structures

Editor

Andreas Lampropoulos

Basel • Beijing • Wuhan • Barcelona • Belgrade • Novi Sad • Cluj • Manchester

Editor
Andreas Lampropoulos
School of Architecture,
Technology and Engineering
University of Brighton
Brighton
UK

Editorial Office
MDPI
St. Alban-Anlage 66
4052 Basel, Switzerland

This is a reprint of articles from the Special Issue published online in the open access journal *Materials* (ISSN 1996-1944) (available at: https://www.mdpi.com/journal/materials/special_issues/RSERCS).

For citation purposes, cite each article independently as indicated on the article page online and as indicated below:

Lastname, A.A.; Lastname, B.B. Article Title. *Journal Name* **Year**, *Volume Number*, Page Range.

ISBN 978-3-7258-0215-9 (Hbk)
ISBN 978-3-7258-0216-6 (PDF)
doi.org/10.3390/books978-3-7258-0216-6

© 2024 by the authors. Articles in this book are Open Access and distributed under the Creative Commons Attribution (CC BY) license. The book as a whole is distributed by MDPI under the terms and conditions of the Creative Commons Attribution-NonCommercial-NoDerivs (CC BY-NC-ND) license.

Contents

About the Editor vii

Thomaida Polydorou, Nicholas Kyriakides, Andreas Lampropoulos, Kyriacos Neocleous, Renos Votsis, Ourania Tsioulou, et al.
Concrete with a High Content of End-of-Life Tire Materials for Flexural Strengthening of Reinforced Concrete Structures
Reprinted from: *Materials* **2022**, *15*, 6150, doi:10.3390/ma15176150 1

Muhammad Ahmed, Piero Colajanni and Salvatore Pagnotta
Influence of Cross-Section Shape and FRP Reinforcement Layout on Shear Capacity of Strengthened RC Beams
Reprinted from: *Materials* **2022**, *15*, 4545, doi:10.3390/ma15134545 22

Dominik Skokandić, Anđelko Vlašić, Marija Kušter Marić, Mladen Srbić and Ana Mandić Ivanković
Seismic Assessment and Retrofitting of Existing Road Bridges: State of the Art Review
Reprinted from: *Materials* **2022**, *15*, 2523, doi:10.3390/ma15072523 42

Yong Li, Zijie Yu and Yongqian Liu
Experimental and Numerical Study of the Ultimate Flexural Capacity of a Full-Size Damaged Prestressed Concrete Box Girder Strengthened with Bonded Steel Plates
Reprinted from: *Materials* **2023**, *16*, 2476, doi:10.3390/ma16062476 65

Wenqing Wu, Hui Zhang, Zheng Liu and Yunpeng Wang
Numerical Analysis on Transverse Splicing Structure for the Widening of a Long Multi-Span Highway Concrete Continuous Box Girder Bridge
Reprinted from: *Materials* **2022**, *15*, 6805, doi:10.3390/ma15196805 93

Bi Kai, A. B. M. A. Kaish and Norhaiza Nordin
Concrete-Filled Prefabricated Cementitious Composite Tube (CFPCCT) under Axial Compression: Effect of Tube Wall Thickness
Reprinted from: *Materials* **2022**, *15*, 8119, doi:10.3390/ma15228119 111

Kyong Min Ro, Min Sook Kim and Young Hak Lee
Improved Shear Strength Equation for Reinforced Concrete Columns Retrofitted with Hybrid Concrete Jackets
Reprinted from: *Materials* **2023**, *16*, 3734, doi:10.3390/ma16103734 122

Andreas Lampropoulos, Demetris Nicolaides, Spyridon Paschalis and Ourania Tsioulou
Experimental and Numerical Investigation on the Size Effect of Ultrahigh-Performance Fibre-Reinforced Concrete (UHFRC)
Reprinted from: *Materials* **2021**, *14*, 5714, doi:10.3390/ma14195714 138

Mingzhao Chen, Xudong Deng, Rongxin Guo, Chaoshu Fu and Jiuchang Zhang
Tensile Experiments and Numerical Analysis of Textile-Reinforced Lightweight Engineered Cementitious Composites
Reprinted from: *Materials* **2022**, *15*, 5494, doi:10.3390/ma15165494 151

Giorgio Mattarollo, Norbert Randl and Margherita Pauletta
Investigation of the Failure Modes of Textile-Reinforced Concrete and Fiber/Textile-Reinforced Concrete under Uniaxial Tensile Tests
Reprinted from: *Materials* **2023**, *16*, 1999, doi:10.3390/ma16051999 173

Arash Karimi Pour, Zahra Mohajeri and Ehsan Noroozinejad Farsangi
Impact of Polypropylene Fibers on the Mechanical and Durability Characteristics of Rubber Tire Fine Aggregate Concrete
Reprinted from: *Materials* **2022**, *15*, 8043, doi:10.3390/ma15228043 . . . . **203**

Rihua Yang, Yiming Yang, Xuhui Zhang and Xinzhong Wang
Experimental Study on Secondary Anchorage Bond Performance of Residual Stress after Corrosion Fracture at Ends of Prestressed Steel Strands
Reprinted from: *Materials* **2023**, *16*, 7441, doi:10.3390/ma16237441 . . . . **229**

Paul Penzel, Maximilian May, Lars Hahn, Silke Scheerer, Harald Michler, Marko Butler, et al.
Bond Modification of Carbon Rovings through Profiling
Reprinted from: *Materials* **2022**, *15*, 5581, doi:10.3390/ma15165581 . . . . **245**

Sofija Kekez and Rafał Krzywoń
Prediction of Bonding Strength of Externally Bonded SRP Composites Using Artificial Neural Networks
Reprinted from: *Materials* **2022**, *15*, 1314, doi:10.3390/ma15041314 . . . . **266**

Huigui Zhang, Wuman Zhang and Yanfei Meng
Salt Spray Resistance of Roller-Compacted Concrete with Surface Coatings
Reprinted from: *Materials* **2023**, *16*, 7134, doi:10.3390/ma16227134 . . . . **283**

Zofia Szweda
Evaluating the Impact of Concrete Design on the Effectiveness of the Electrochemical Chloride Extraction Process
Reprinted from: *Materials* **2023**, *16*, 666, doi:10.3390/ma16020666 . . . . **299**

About the Editor

Andreas Lampropoulos

Dr Andreas Lampropoulos is a Principal Lecturer in Civil Engineering at the University of Brighton. He earned his Diploma (2003), MSc (2005) and PhD (2010) degrees in Civil Engineering (Structural Division) from the University of Patras, in Greece. His main research interests are novel construction materials and seismic strengthening/retrofitting of existing structures. He focuses on a wide range of cementitious materials such as ultra-high-performance fibre-reinforced concrete (UHPFRC), steel fibre-reinforced concrete (SFRC) and cementitious materials reinforced with nanoparticles, and he is currently working on the development and application of cement-free concretes. He has conducted extensive experimental and numerical work on the development of novel strengthening techniques for the structural improvement of reinforced concrete (RC) and unreinforced masonry (URM) structures. Dr Lampropoulos I the Chair of the International Association for Bridge and Structural Engineering (IABSE) Task Group 1.1 'Improving Seismic Resilience of Reinforced Concrete Structures' and Task Group 5.5 'Conservation and Seismic Strengthening/Retrofitting of Existing Unreinforced Masonry Structures', and he is an active member of various task groups of IABSE and the International Federation for Structural Concrete (fib).

He has published more than 80 works in international research journals, books and conference proceedings, and he is an Editorial Board Member of several international journals.

Article

Concrete with a High Content of End-of-Life Tire Materials for Flexural Strengthening of Reinforced Concrete Structures

Thomaida Polydorou [1,2,*], Nicholas Kyriakides [1,2], Andreas Lampropoulos [3], Kyriacos Neocleous [1,2], Renos Votsis [1], Ourania Tsioulou [3], Kypros Pilakoutas [4] and Diofantos G. Hadjimitsis [1,2]

1. Department of Civil Engineering and Geomatics, Cyprus University of Technology, Limassol 3036, Cyprus
2. ERATOSTHENES Centre of Excellence, Limassol 3012, Cyprus
3. School of Architecture, Technology and Engineering, University of Brighton, Brighton BN2 0JY, UK
4. Department of Civil and Structural Engineering, The University of Sheffield, Sheffield S10 2TN, UK
* Correspondence: thomaida.polydorou@cut.ac.cy

Abstract: This research investigates the performance of Steel Fiber Reinforced Rubberized Concrete (SFRRC) that incorporates high volumes of End-of-life tire materials, (i.e., both rubber particles and recycled tire steel fibers) in strengthening existing reinforced concrete (RC) beams. The mechanical and durability properties were determined for an environmentally friendly SFRRC mixture that incorporates a large volume (60% by volume aggregate replacement) of rubber particles and is solely reinforced by recycled tire steel fibers. The material was assessed experimentally under flexural, compressive and impact loading, and thus results led to the development of a numerical model using the Finite Element Method. Furthermore, a numerical study on full-scale structural members was conducted, focusing on conventional RC beams strengthened with SFRRC layers. This research presents the first study where SFRRC is examined for structural strengthening of existing RC beams, aiming to enable the use of such novel materials in structural applications. The results were compared to respective results of beams strengthened with conventional RC layers. The study reveals that incorporation of End-of-life tire materials in concrete not only serves the purpose of recycling End-of-life tire products, but can also contribute to unique properties such as energy dissipation not attained by conventional concrete and therefore leading to superior performance as flexural strengthening material. It was found that by incorporating 60% by volume rubber particles in combination with recycled steel fibers, it increased the damping ratio of concrete by 75.4%. Furthermore, SFRRC was proven effective in enhancing the energy dissipation of existing structural members.

Keywords: rubberized concrete; recycled steel fibers; rubber aggregate; End-of-life tire materials; strengthening

1. Introduction

Replacement of concrete aggregates by recycled rubber particles promotes sustainability and circular economy [1] and it particularly contributes to the elimination of End-of-life tires, an environmental and health issue recognized worldwide. Previous research [2] has identified the promising behavior of rubberized concrete, highlighting the fact that the inclusion of rubber particles as aggregate replacement can increase the energy absorption capacity of concrete significantly [3].

Even though researchers have advised against the use of high volumes of rubber particles due to the considerable compressive strength reduction, the use of larger quantities of rubber particles enhances particular concrete properties [4,5] that can improve the material behavior under impact and earthquake loading. The material flexural strength can be enhanced through the provision of adequate steel fiber reinforcement. Incorporation of two types of End-of-life tire materials (i.e., recycled tire steel fibers and recycled rubber particles) not only serves the purpose of recycling but can also increase the energy absorption of concrete [6], therefore making the material capable of absorbing energy conveyed to a structure during earthquake ground motion.

The application of novel fiber reinforced cementitious materials has been effectively used for the structural strengthening of existing Reinforced Concrete (RC) members including bridges [7] and beams [8–10]. Additional Steel Fiber Reinforced Concrete (SFRC) layers have been used with and without the presence of steel reinforcing bars [9,10], and the results have shown that the use of SFRC layers, even without steel bars, can significantly enhance the stiffness of structural elements, while the addition of steel reinforcing bars is essential for further enhancement of the ultimate load bearing capacity.

The use of materials such as SFRC provides significant benefits in terms of mechanical performance and durability, although traditional SFRC mixtures incorporate a large amount of manufactured steel fibers and are therefore costly and energy intensive. Hence, this research aims to suggest an alternative of using recycled steel fibers retrieved from End-of-life tires. Recycled tire steel fibers have been proven effective as concrete reinforcement, indicating equal or superior performance compared to manufactured steel fibers [11,12]. The mechanical properties of concrete mixtures incorporating recycled tire steel fibers and rubber particles at various contents and particle sizes has been a subject of many research studies [13–18], but the effect of using such materials to strengthen conventional RC beams, especially when including a high volume of rubber particles, has not been investigated previously.

Only one study has investigated the performance of polyvinyl alcohol (PVA) fiber reinforced rubberized cementitious composites in strengthening conventional RC beams thus far, indicating the promising potential of rubberized cementitious composites in strengthening and repairing RC structural members [19]. Thus, this research aims to evaluate for the first time the effectiveness in strengthening existing RC beams of concrete by incorporating a large volume of rubber particles (60% by volume aggregate replacement) and recycled steel fibers, both retrieved from End-of-life tires.

This study evaluates the mechanical and durability properties of environmentally friendly Steel Fiber Reinforced Rubberized Concrete (SFRRC). Experimental assessment included tests under static and cyclic compressive loading, flexural bending, impact loading as well as durability tests under chloride exposure and freeze–thaw cycles. In addition, numerical analysis has been conducted based on the obtained experimental data, which further enabled comparison to relevant conventional strengthening techniques.

2. Materials and Methods

The experimental study focused on an SFRRC mix design optimized by previous research [20], where 60% (by volume) of the mineral aggregate was replaced by rubber particles recycled from End-of-life tires. To maintain the aggregate particle size distribution in the mix, fine aggregate was replaced by finely granulated rubber of equivalent particle size, while coarse aggregate was replaced by shredded rubber of equivalent particle size.

The cement used was an EN 197-1 [21] Portland cement type CEM I 52.5 N, obtained from Vassiliko, Cyprus. In addition, we used Pulverized Fly Ash (PFA) according to EN 450-1 [22], and Microsilica or Silica Fume (MS) conforming to Silica Fume Class 1 requirements of EN13263-1:2005+A1:2009 [23]. A liquid polycarboxilic polymer-based superplasticizer conforming to EN934-2:2009 [24], with a specified mass density at $1070 \text{ kg/m}^3 \pm 30 \text{ kg/m}^3$ was also used in this study.

Diabase coarse aggregate (4–20 mm) and a blend of diabase and limestone fine aggregate (0–4 mm) from local quarries were used in this mix, at a fine to coarse aggregate ratio of 1:1.22. In replacement to 60% by volume of mineral aggregate, recycled rubber particles at equivalent sizes were used, particularly a blend of particles obtained from tire recycling plants in Croatia, Cyprus and the UK. Since no material specifications were provided, the specific gravity of a representative sample of the rubber particle blend used in this study was evaluated by EN 1097-6 [25]. The determined specific gravity was 0.8, which was taken into consideration during the mix design process, to ensure accurate aggregate replacement by volume.

Cleaned and sorted Recycled Tire Steel Fibers (RTSF) obtained from Sheffield, U.K. were used in this study. The evaluated tensile strength and Young's Modulus of the RTSF used in this study were provided by the distributor at 2560 ± 550 MPa and 200,000 MPa ± 0.5%, respectively [26].

The SFRRC mix constituents are listed in Table 1. Specimen preparation followed standard concrete mixing procedures as described by EN 206:2013+A1:2016 [27], cast in standard cubes and cylinders as well as prisms of $10 \times 10 \times 40$ cm^3, demolded 24 h after casting and cured in a water bath in standard laboratory conditions. Before testing, the specimens were taken out of the water and allowed to dry for 24 h. before hardened concrete testing commenced.

Table 1. Steel Fiber Reinforced Rubberized Concrete with High Rubber Content and Plain Concrete–Mix Design Details.

Mix Constituent	SFRRC Mix Amount (kg/m^3)	Plain Concrete Mix Amount (kg/m^3)
Cement CEM I 52,5 N	400.00	400.00
Silica Fume (micro-silica)	100.00	100.00
Fine Mineral Aggregate	310.50	573.70
Coarse Mineral Aggregate	378.00	1147.50
Fine Rubber Particles	169.70	0.00
Coarse Rubber Particles	207.00	0.00
Recycled Tire Steel Fibers	25.00	0.00
Water	225.00	225.00
Superplasticizer	3.61	3.61

To compare the response of SFRRC to Plain Concrete (PC), three standard cube PC specimens were cast as well as 6 $10 \times 10 \times 40$ cm^3 rectangular PC prisms. The plain concrete mix constituents are listed in Table 1.

2.1. Static Loading

2.1.1. Loading under Compression

The compressive strength of both cubic and cylindrical SFRRC specimens, as well as cubic PC specimens, was evaluated as prescribed by EN 12390-3:2009/AC:2011 [28]. The axial compressive load was applied on the cylinders at a displacement rate of 0.3 mm/min and on the cubes at a rate of 0.4 MPa/s, using a standard Compressive Testing Machine of 3000 kN load capacity. In addition, the static modulus of elasticity of cylindrical SFRRC specimens in compression was determined as recommended by BS 1881-121 [29].

2.1.2. Loading under Bending

Flexural strength was evaluated by loading prismatic specimens under four-point bending, using a state-of-the-art hydraulic water cooled tensile/compressive testing machine of 250 kN capacity. A custom-made yoke was used to mount two Linear Variable Difference Transducers (LVDTs), placed at midspan of each side of the prism as recommended by the JSCE-SF4 [30] method of tests for flexural strength and toughness of steel fiber reinforced concrete. Before initiating the test, a 12.5 mm long clip gauge was mounted on the bottom surface of the prism, to measure the crack mouth opening displacement (CMOD) on 5 mm wide, 15 mm deep notches, pre-sawn across the midsection of the prism. The bending test setup is shown in Figure 1.

Assessment of the post peak energy absorption of the material was conducted by studying the residual flexural strength (f_{Ri}) values obtained at the predetermined crack mouth opening displacements (CMOD) of 0.5 mm, 1.5 mm, 2.5 mm, and 3.5 mm, as recommended by RILEM TC 162-TDF [31] and the EN 14651 standard [32]. To compare the response of SFRRC to plain concrete, three plain concrete rectangular prisms were also loaded under 4-point bending, using standard procedures and identical setup as for the SFRRC prisms, shown in Figure 1.

Figure 1. Four-point bending test setup.

2.2. Cyclic Compressive Loading

The energy absorption of the material was also evaluated by loading cylindrical specimens (15 cm diameter) under cyclic compressive loading and using a universal load and displacement control testing machine. A custom-made yoke was used consisting of two circular brackets, each secured on the cylinder surface by two pins at 180-degree angle from each other and holding three LVDTs, arranged at 120-degree angles around the circumference of the specimen. The compressive loading setup is shown in Figure 2.

Figure 2. Cyclic compressive loading specimen setup.

Loading was applied by considering the average cylinder compressive strength from the respective cylinder batch and followed a specific loading command. All specimens followed identical loading commands, with the only variable being the maximum load reached. Testing was initially load controlled at a rate of 500 N/s up to a predetermined load limit of 28 kN, then followed by a 60 s pause before unloading down to 1 kN and reloading back to 28 kN twice; these cycles completed the first set of cyclic loading within the linear stress–strain behavior range. Loading continued without a pause at the same load-controlled rate up to 50 kN, before changing to displacement-controlled loading at a rate of 0.2 mm/min. The second set of loading cycles resumed at that rate by unloading down to 1 kN and loading back up to 50 kN, recurrently for a total of three cycles. Further on, specimens were loaded up to their capacity at the same displacement-controlled rate and, after that additional loading commenced to perform three cyclic loading and unloading rounds at each predetermined displacement of 5 mm, 9 mm, 12 mm, 15 mm and 20 mm, as indicated in the load-displacement plot shown in Figure 3. The displacement of the SFRRC Cylinder was determined based on the average displacement of the three mounted LVDTs.

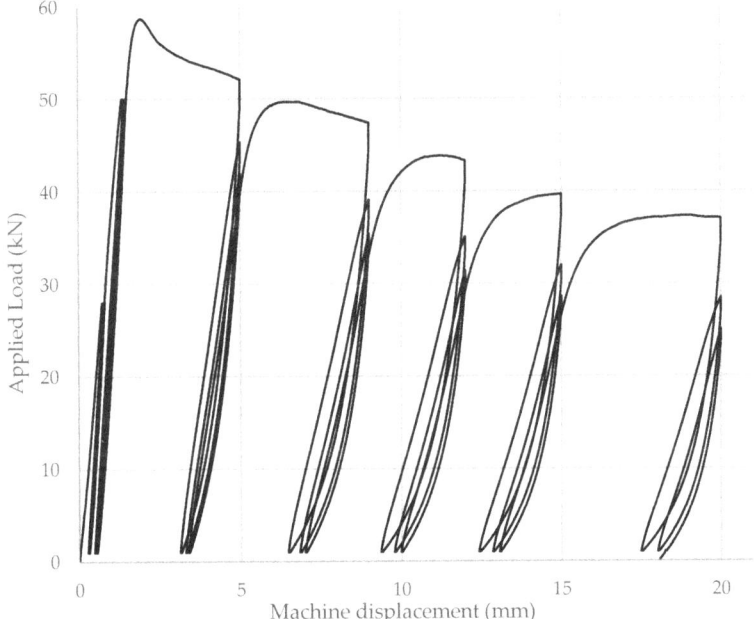

Figure 3. Cyclic loading command path.

2.3. Impact Loading

The material response to impact loading was assessed by direct impact of a steel sphere that fell free from a known height through a cast-iron, cylindrical tube onto the top surface of rectangular $10 \times 10 \times 40$ cm^3 specimens. Six SFRRC and three PC specimens were experimentally assessed under impact loading in this study. A custom-made frame was specifically machined for this study and bolted to the laboratory's strong floor to ensure stability, as shown in Figure 4; the design of this custom-made frame was based on the drop-weight impact test concept.

A stationary Kistler accelerometer with sensitivity of 1 V/g was used, mounted at the centroid of the impact surface of each prism, 150 mm from the point of impact. A Kyowa PCD-300B sensor interface and its respective DCS-100A recording software (Kyowa, Tokyo, Japan) were used for data acquisition. The sampling frequency was set to 1000 Hz to capture the first vibrational modes.

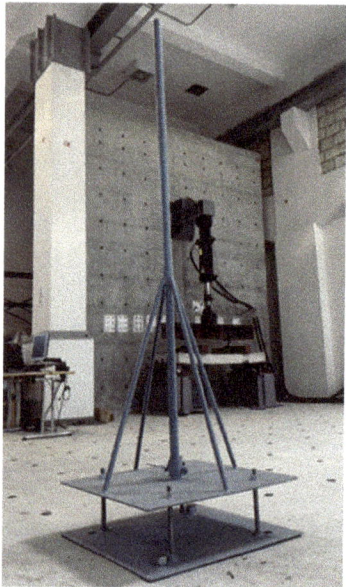

Figure 4. Impact test drop-weight frame.

The scope of this test was to measure the damping ratio ζ of both SFRRC and PC, through the calculation of the decrease in magnitude of the vibration response at the mid span of the specimen.

2.4. Durability Testing

2.4.1. Chloride Corrosion Exposure

Resistance to chloride corrosion was assessed by subjecting three sets of three rectangular, $10 \times 10 \times 40$ cm^3 prisms to wet–dry cycles. The specimens were exposed to chlorides by immersion into 3% NaCl under a schedule of 7 days in and 7 days out of solution repeatedly for 2-, 4- and 6-month periods. The mass of each prism was recorded prior to their first wet–dry cycle and upon conclusion of the respective set cycling period and compared to a fourth (control) set of prisms, which followed the same wet–dry cycling scheme but was immersed in clean water instead of NaCl solution.

2.4.2. Freeze–Thaw Resistance

The freeze–thaw durability of the material was assessed by the mass scaling loss of four cubes (15 cm side), as recommended by the CEN/TS 12390-9:2016 [33] method. Thermocouples were embedded into the cubes during casting to enable specimen temperature monitoring through the freeze–thaw cycles. The cubes were weighed at 28 days maturity before being immersed in 3% NaCl solution in stainless steel containers and placed into a freeze–thaw chamber, programmed to apply continuous cycles of freezing and thawing by alternating between -15 °C and 20 °C. To determine mass loss, the cubes were removed from the chamber during thawing at the end of each 7-, 14-, 28-, 42- and 56-day cycle, and their surfaces were thoroughly brushed, which led to material detaching from the cubes due to scaling. The detached material was collected, oven dried at 105 °C for 24 h and then weighed to the nearest 0.1 g.

3. Experimental Results and Discussion

The paper examines the properties of a SFRRC mix design that includes a high content of End-of-life tire materials such as rubber particles and steel fibers under static and cyclic compressive loading, flexural bending, impact loading and durability testing. Local materials were used to recreate an already optimized SFRRC mix with 60% by volume of its mineral aggregate replaced by a blend of rubber particles with a specific gravity of 0.8 (as determined by EN 1097-6 [25]), as expected for a blend of rubber particles obtained from End-of-life tires. Mix development was achieved through aggregate replacement by volume and was based on the specific gravity of the rubber particle sample used. The experimental behavior of cubic, cylindrical and rectangular SFRRC specimens are discussed in the following subsections.

3.1. Static Loading

3.1.1. Loading under Compression

The compressive strength of three cubic (15 cm side) and six cylindrical (15 cm diameter) SFRRC specimens were determined. An average 28-day cube compressive strength of 8.2 MPa was reached, whereas an average 28-day cylinder compressive strength of 3.5 MPa was obtained. The unexpectedly lower average compressive strength yielded by the cylindrical SFRRC specimens compared to their respective SFRRC cubic specimens is discussed as follows. To compare with PC, the compressive strength of three PC cubes (15 cm side) was also determined following standard procedures. The PC cubes reached an average 28-day cube compressive strength of 51.92 MPa, with a variance at 0.232.

The average static modulus of elasticity in compression was also determined for the high rubber content SFRRC mix by testing four cylinders (15 cm diameter), resulting in an average value of 3.263 GPa. The SFRRC static loading results summary is provided in Table 2.

Table 2. SFRRC Static loading under Compression Results per Experimental Procedure.

Experimental Procedure	Average Value	Variance
Cylinder Compressive Strength	3.499 MPa	0.274
Cube Compressive Strength	8.201 MPa	0.388
Cylinder Modulus of Elasticity	3.263 GPa	0.408

As widely accepted, aggregate replacement by rubber particles leads to a significant reduction in concrete compressive strength. In this study, a compressive strength reduction of 84.2% was observed, expected due to the high percentage of the mix aggregate being replaced by rubber particles when comparing PC cubes to SFRRC cubes.

Focusing on the SFRRC compressive test results, where both cubes and cylinders were compared, a notable 134.29% higher compressive strength was obtained for SFRRC cubes compared to SFRRC cylinders from the same batch, much greater than the discrepancy expected due to their geometry. The mix consistency is one of the reasons the cylinder compressive strengths did not reach the values expected based on their respective cube strengths. It was observed that rubber particles tend to rise to the surface soon after casting, due to their lightweight nature. Therefore, slender cylinders end up inconsistent through their height and fail in the top 1/3 of the specimen where rubber is accumulated creating a larger, weak Interfacial Transition Zone (ITZ) that leads to premature rupturing. This explains the early failure of cylinders compared to cubes in compression, thus signifying that the extremely low compressive strengths yielded by the cylinders are not realistic and could have been much higher had the mix consistency been ensured.

The uneven distribution of rubber in the cylinder results in a failure zone in the top part of the cylindrical specimen, as described in Section 3.2. It is possible that the standard procedure for mixing and placing conventional concrete is not suitable for rubberized concrete. The cylinder samples yielded significantly lower compressive strengths than their respective cubic specimens due to uneven distribution of the rubber particles in the taller cylindrical specimens. It is recommended that alternative procedures for mixing and placing rubberized concrete are developed, specifically targeting the addition of rubber particles to the mix and their accumulation to the top part of the specimen during placement, providing cohesive mixes of uniform properties. A potential solution could be to use pre-treated rubber particles, using techniques proven [6,34] to enhance mechanical properties of the material, or even placing rubber particles separately from the rest of the SFRRC mix constituents.

3.1.2. Loading under Bending

The equivalent flexural-tensile stress versus crack mouth opening displacement (CMOD) and the average of the two LVDT readings (Vertical Displacement) were recorded for each of the seven $10 \times 10 \times 40$ cm^3 rectangular SFRRC prisms tested under 4-point bending, as shown in Figures 5–7. All specimens tested under bending came from two batches of the same SFRRC mix. For comparative reasons, three PC rectangular specimens with dimensions of $10 \times 10 \times 40$ cm^3 were also tested under 4-point bending using the setup shown in Figure 1. The Equivalent Flexural-tensile Stress for PC specimen PC1 vs. Vertical Displacement is shown in Figure 5, along with the Equivalent Flexural-tensile Stress withstood by SFRRC specimen P5 vs. its Vertical Displacement and CMOD.

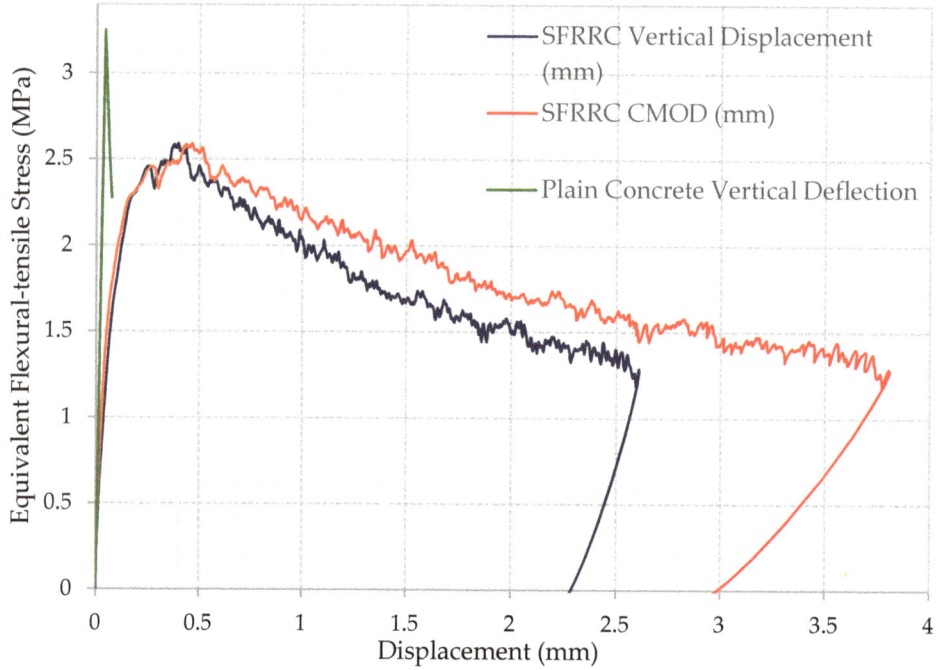

Figure 5. Equivalent Flexural-tensile stress vs. Displacement for Steel Fiber Reinforced Rubberized Concrete Specimen P5 and Plain Concrete Specimen PC1.

The residual flexural strength (f_{Ri}) of each SFFRC prism was determined at the predetermined crack mouth opening displacements (CMOD) of 0.5 mm, 1.5 mm, 2.5 mm, and 3.5 mm, as recommended by RILEM [31]. The average flexural residual strength values and coefficients of variation obtained per CMOD are reported in Table 3. The limit of proportionality [32] of the prisms tested in this study was evaluated at an average value of 2.7 MPa.

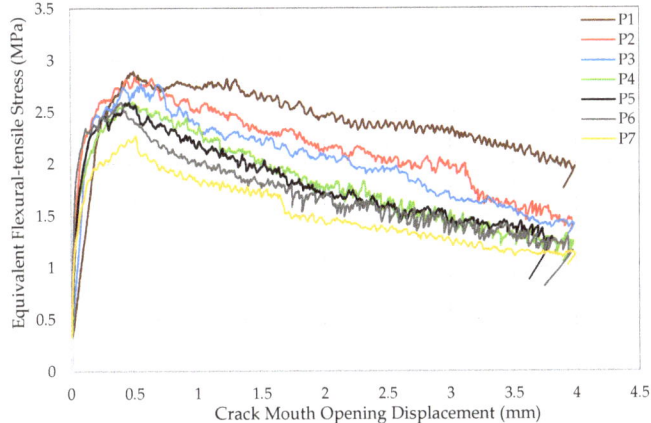

Figure 6. Equivalent Flexural-tensile stress vs. CMOD.

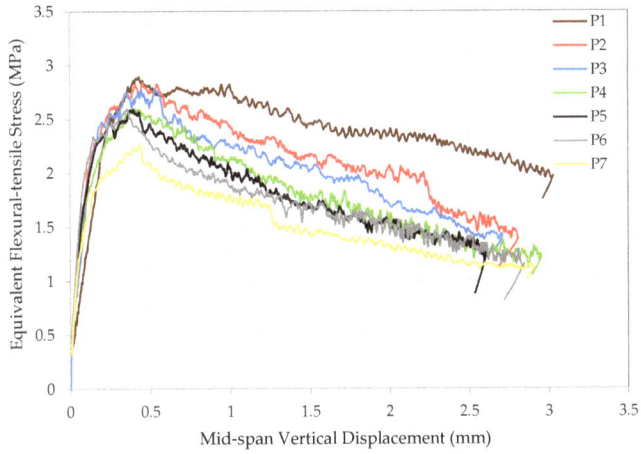

Figure 7. Equivalent Flexural-tensile stress vs. Mid-span Vertical Displacement.

Table 3. Average Residual Flexural Strength Values and Coefficients of Variation per CMOD.

	Average Flexural Residual Strength f_{Ri} (MPa)	CoV of Flexural Residual Strength f_{Ri} (MPa)	CMOD (mm)
f_{R1}	2.57	0.08	0.5
f_{R2}	2.05	0.17	1.5
f_{R3}	1.78	0.19	2.5
f_{R4}	1.45	0.25	3.5

Processing of the critical deflection values taken at points on the curves where the stress–strain behavior of the material is still linear during the static bending test rounds, in conjunction with the experimental modulus of elasticity (E) values obtained from cylinder compressive strength tests allowed the authors to determine the effective moment of inertia (I_{eff}) of the specimens. The I_{eff} was calculated using Equation (1) [35]:

$$\delta = k \frac{W L^3}{E I_{eff}} \quad (1)$$

using the experimentally determined E value, the experimental CMOD at midspan, determined at the maximum load sustained by each prism as δ, the theoretical value of k for simply supported beam setup, the total load (W) including the specimen self-weight and the measured length (L) of each prism.

The average effective moment of inertia (I_{eff}) of the beam specimens tested under 4-point bending was determined at 1.5316×10^5 mm^4. This value was used to compare to the average value of specimen stiffness (k) obtained from the impact testing applied to beam specimens made of the same material mix and attaining the same geometrical characteristics.

The equivalent flexural-tensile stress versus crack mouth opening displacement (CMOD) plot shapes indicate the significant ability of the material to absorb energy, identified by the large area under the curve. In addition to the inclusion of the recycled tire steel fibers, the post peak energy absorption is attributed to the ability of the rubber particles to undergo large deformation in tension thus promoting high energy absorption. In contrast, PC does not attain nearly as much energy absorbing capability, as indicated in the PC Equivalent Flexural-tensile stress vs. Displacement plot included in Figure 5.

In addition to the graphs, the average SFRRC residual flexural strength values and the 43.6% reduction in residual flexural strength between the CMODs of 0.5 mm and 3.5 mm, as presented in Table 3, demonstrate that the SFRRC mix is highly ductile and attains exceptional post-cracking load carrying capacity. The SFRRC mix with a 60% by volume aggregate replaced by rubber particles has demonstrated high ductility and high post-cracking load carrying capacity, also indicated by high f_{Ri} values (Table 3), compared to the obtained limit of proportionality of 2.7 MPa.

The low rate of reduction in residual strength (f_{Ri}) values observed (43.6% reduction between f_{R1} and f_{R4}) also proves the capability of the SFRRC to attain post peak energy absorption. The deformability and toughness of the material is indicated not only by the visible large area under the stress–strain curve observed in the equivalent flexural-tensile stress vs. CMOD plots Figure 6, but also by the small reduction observed in the average flexural residual strength values (f_{Ri}) in accordance with their corresponding and significant CMOD growth, as reported in Table 3.

The post cracking residual flexural-tensile strength of the SFRRC mix was classified by considering the provisions of the fib 2010 Model Code [36,37]. With an f_{R1k} at 2.25 MPa and f_{R2k} at 1.22 MPa, the material post-cracking residual strength is specified as class 2a, since the strength interval (f_{R1k}) is between 2.0 and 2.5 MPa and the residual strength ratio ($\frac{fR3k}{fR1k}$) between 0.5 and 0.7 for class a, as defined by fib Model Code 2010 5.6-1; in addition, the characteristic limit of proportionality, f_{LK} [36] was calculated at 2.33 MPa, therefore a value of 0.2964 was obtained for the $\frac{fR1k}{fLk}$ ratio. Since both relationships of the fib Model Code 2010 [36] 5.6-2 (Equation (2)) and 5.6-3 (Equation (3)) are fulfilled, fiber reinforcement in this case could substitute (also partially) conventional reinforcement at ultimate limit states.

$$\frac{fR1k}{fLk} > 0.4 \qquad (2)$$

$$\frac{fR3k}{fR1k} > 0.5 \qquad (3)$$

3.2. Cyclic Compressive Loading

Four standard SFFRC cylindrical specimens (15 cm diameter) were loaded under cyclic compressive loading to further confirm the energy absorption capabilities of the developed SFRRC mix with 60% by volume aggregate replaced by rubber particles. The stress–strain behavior of the cylinders tested in this study is represented by plotting the applied stress vs. average strain recorded by the 3 LVDTs mounted around the cylinder circumference, which were monitored throughout the application of cyclic loading and unloading rounds. The stress–strain plot obtained for cylinder B2 tested in this study is presented in Figure 8.

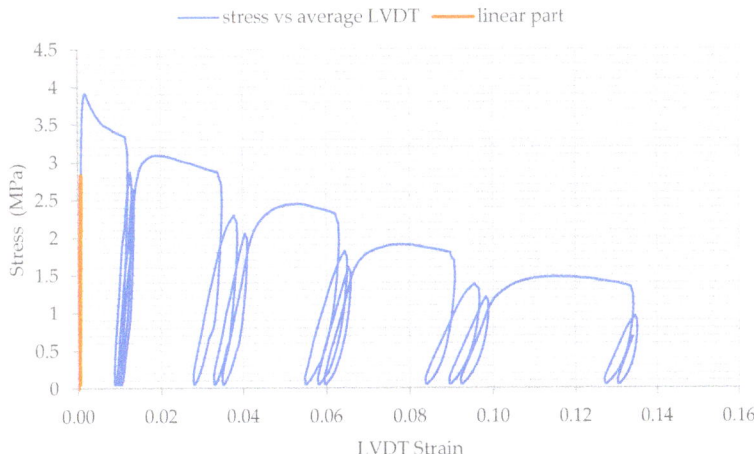

Figure 8. Stress vs. Average Strain Curve for Specimen B2 under Cyclic Compressive Loading.

The cyclic compressive loading stress–strain curve shown in Figure 8 indicates the capacity of the material to withstand continuous cycles of axial compressive loading and unloading even post-peak.

The cylinder damage initiation and development of crack formation was visually observed during the cyclic loading and unloading program. The SFRRC mix with 60% aggregate replaced by rubber was able to reach an average ultimate strain of 0.023 before crack initiation. Figure 9 displays cylinder deformation at the end of the test, after completion of the final cyclic loading/unloading rounds at 20 mm machine displacement.

Figure 9. Cyclic Compressive Loading Specimen B1 at 20 mm Displacement.

The average axial compressive strain at which significant cracking of the cylindrical specimens was observed was determined by visual inspection during the loading cycles and the results obtained by the LVDT measurements. As shown in Figure 9, the failure zone of specimen B1 is concentrated on the top part of the cylinder, indicating the inconsistency of material mix content through the cylinder specimen height. This explains the low compressive strength values obtained by the cylinders tested under static compressive loading in this study. In this case, the cylinder specimen tested had accumulation of rubber in the top 1/3 of the cylinder height.

The stress–strain behavior and shape of the curves shown in Figure 8 indicate the capacity of the SFRRC mix to absorb energy. The amount of energy absorbed by each specimen during the cyclic loading and unloading rounds is directly related to the area under their resulting stress–strain curves. The strain values before failure, corresponding to stress values that were no lower than 80% of the maximum or peak strength reached by each specimen, were considered as the ultimate value of strain reached by each specimen before crack initiation.

Visual inspection of the cylinders during the cyclic rounds of testing confirmed the material's condition at the point of 80% peak strength. The strain values noted are relatively conservative, considering the number of loading and unloading cycles applied to the specimen prior to that.

3.3. Impact Loading

Impact loading was applied onto the top surface of $10 \times 10 \times 40$ cm^3 prismatic specimens. Three identical impact hits were performed on each specimen. The displacement response time history after impact loading was plotted for every hit and were further used to determine the damping characteristics for each of the tested specimens. Figure 10 presents an indicative oscillation diagram obtained for SFRRC specimen D2, under Hit 2, from this study.

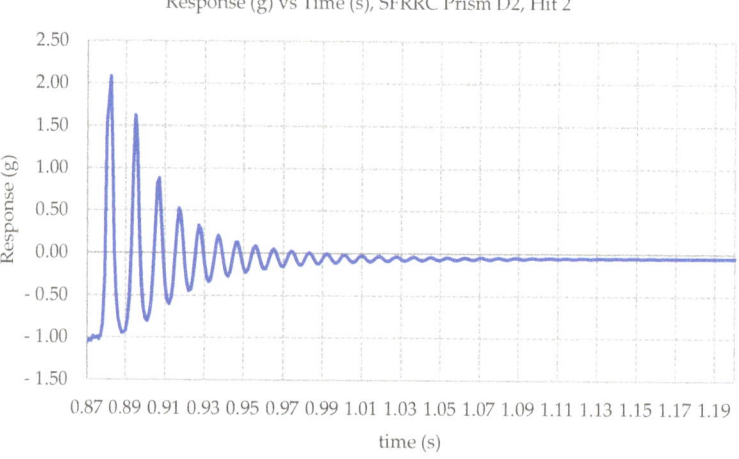

Figure 10. Impact Induced Oscillation Diagram for SFRRC Prism D2, Hit 2.

The average damping ratio, ζ, and average fundamental frequency of oscillation, f, were calculated for the set of six prismatic SFRRC specimens tested in this study, taking into consideration the average values of ζ and f obtained by the three impact hits recorded per specimen. The damping ratios were calculated using values three peaks apart from each other, as indicated in Equation (4) [38]:

$$\zeta = \frac{1}{2\pi j} \ln \frac{u_i}{u_{j+i}} \quad (4)$$

The fundamental frequency of oscillation, f, was determined in the frequency domain using Fast Fourier Transform (FFT). The average values of both ζ and f are listed in Table 4 along with their corresponding value of variance.

Table 4. SFRRC Impact specimens Damping Ratio, ζ and Frequency, f, Average and Variance.

Damping Ratio, ζ		Frequency, f (Hz)	
Average	Variance	Average	Variance
0.118	0.243	81.00	0.096

The damping ratio, ζ, and average fundamental frequency of oscillation, f, were calculated for PC specimens also, following identical procedures, for comparison purposes. The PC average damping ratio, ζ, was evaluated at 0.067 with a variance of 0.183, and the average frequency, f, for PC was determined at 52.00 Hz with a variance of 0.085, considering three impact hits on each of the three PC prisms tested in this study.

The average damping ratio (ζ = 0.118) of the SFRRC specimens tested in this study is 136% higher, relative to the widely accepted value for reinforced concrete (ζ = 0.05) [39], indicating superior energy absorbing capacity. In direct comparison to the impact performance of equivalent PC prisms tested under the same conditions during this study, there is a significant increase observed in the material damping ratio. Specifically, a 60% by volume replacement of aggregate by rubber particles led to a 75.4% increase in the value of ζ.

From the experimentally obtained frequency of oscillation, f, of the SFRRC specimens examined under impact loading during this study and reported in Table 4, the authors were able to obtain the stiffness, k, of each tested specimen, thus their moment of inertia, using the also experimentally obtained modulus of elasticity, E.

The effective moment of inertia (I = 2.9863×10^5 mm^4) of the SFRRC was calculated using the experimental parameters obtained from the impact test study. The stiffness (EI) of the specimen was calculated using the calculated value of circular frequency (ω) and the mass of the specimen was incorporated in the corresponding equation for the stiffness (K) calculation, accounting for the respective support and loading conditions. The derived I was found to be satisfactorily close to the effective moment of inertia, $I_{eff} = 1.5316 \times 10^5$ mm^4, obtained using the experimental deflection of the specimen at specific loading points in the linear section of the load-deflection curves obtained in the static bending study. More specifically, the moment at mid span from four-point bending test and the modulus of elasticity obtained from compression load testing, were used to define I_{eff}.

3.4. Durability Testing

3.4.1. Chloride Corrosion Resistance

Four groups of prisms were examined under the chloride corrosion test program. One of the three groups, the control set, followed identical wet–dry cycling schedule as the other three groups, but was instead immersed in water with no chlorides present. The average mass loss and coefficients of variation per group of prisms tested in this study are reported in Table 5.

Table 5. Average Mass Loss, Coefficients of Variation per Chloride Corrosion Test Specimen Set.

Cycle Duration (Days)	NaCl Solution % Concentration	Average Mass Loss (%)	Coefficient of Variation
56 (28 in/28 out)	3	0.98	0.067
112 (56 in/56 out)	3	1.29	0.026
168 (84 in/84 out)	No NaCl	1.23	0.043
168 (84 in/84 out)	3	1.04	0.085

In terms of durability, chloride exposure mass loss values are insignificant, and the fact that the control set of specimens which underwent the cycles in plain water with no NaCl content has experienced similar mass loss compared to the specimen sets that were immersed in 3% NaCl solutions, proves that the developed SFRRC is not susceptible to corrosion when exposed to Chloride environments. The minor mass loss that occurred in all 4 sets of specimens is attributed to the loss of small pieces of material from the corners of prisms due to handling of the specimens through the duration of the study.

3.4.2. Freeze–Thaw Resistance

Standard cube specimens were assessed for freeze–thaw durability as prescribed by CEN/TS 12390-9:2016 [33]. The mass loss due to scaling was recorded after each freeze–thaw cycle and the mass loss due to scaling vs. the number of freeze–thaw cycles the set underwent are provided in Figure 11. The average scaling level at 56 cycles was calculated at 1.56% by mass, a value that is lower than the relative scaling level threshold of 3% by mass, as indicated in the CEN/TS 12390-09 [33] standard. In addition, the total mass of scaled materials at 56 freeze–thaw cycles, over the specimen surface area was also calculated and determined to be 0.98 kg/m^2, a value just under the threshold value of 1 kg/m^2 defined by the slab-test method in CEN/TS 12390-09 [33] based on the Swedish Standard SS 13 72 44 [40]. Due to a higher absorption rate initially, as well as the external surface roughness of the specimens, mass loss due to scaling is greater after the first 8 cycles than after 14 cycles. Mass loss due to scaling continues to rise after that, as expected with subsequent freeze–thaw cycles.

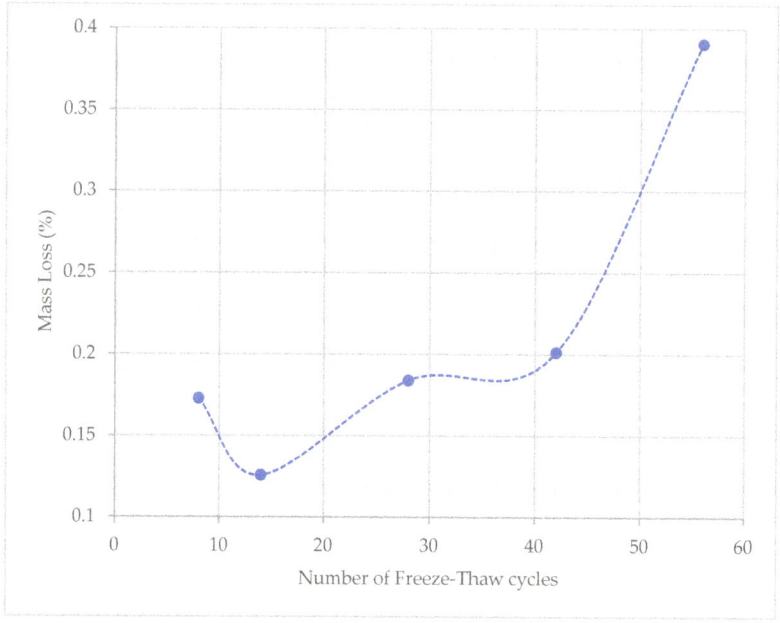

Figure 11. Mass loss (%) vs. Number of Freeze–Thaw cycles.

The freeze–thaw durability test results indicate that the SFRRC mix with 60% by volume of mineral aggregate replaced by rubber can withstand cycles of extreme temperatures successfully, with minor mass loss due to scaling.

4. Numerical Study on the Efficiency of SFRRC as a Strengthening Material
4.1. SFRCC Numerical Modeling

Numerical simulations were conducted using Finite Element Analysis (FEA) software ATENA [41]. For the simulation of SFRCC performance, the compressive strength and modulus of elasticity obtained experimentally (Sections 3.1.1 and 3.2, respectively), were used. More specifically compressive strength equal to 8.2 MPa and modulus of elasticity equal to 32.6 GPa were used. Regarding the behavior in tension, a previously developed constitutive model for UHPFRC [9] was adopted and inverse analysis was conducted to calculate the characteristic points of the model as presented in Figure 12. From the beginning of the loading until the end of the linear point (S0) there is the uncracked stage, where there is a linear stress–strain behavior. The crack formation takes places after the elastic point (S0), when the crack initiates and is followed by stain hardening until S1 (Figure 12), due to the bridging effect of the fibers. After S1 (Figure 12), the ultimate strength is reached and is followed by decreasing stresses (S2 and S3), and finally there is a complete release of stresses when the crack opens without any stress contribution.

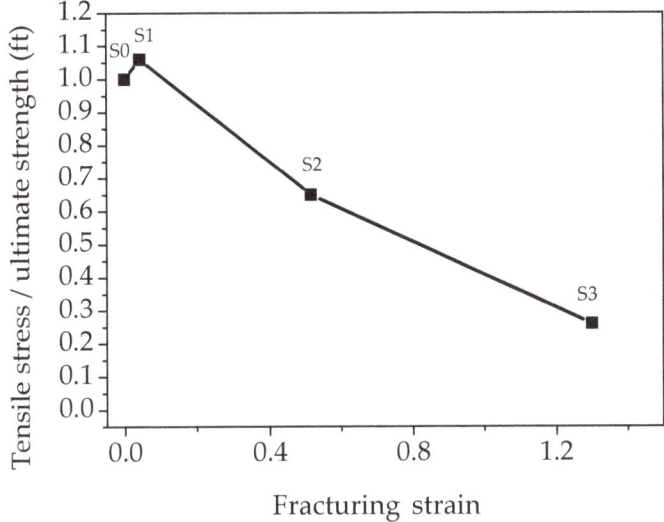

Figure 12. Constitutive model for the tensile stress–strain model of SFRCC. Adapted from [9].

For the inverse analysis, the flexural tests results presented in Section 3.1.2 were used. Numerical models were developed to simulate the response of the prisms and analyses were conducted using the model presented in Figure 13 considering different tensile strength values (0.4, 0.6, 0.8, 1.0 MPa). Indicative strain distribution at the maximum load stage for the prism with 0.8 MPa tensile strength is presented in Figure 13. The black lines indicate the crack development while the strain contours represent the strain distribution along the prism span. The maximum compressive strain at the maximum load stage was found at the top of the prism and it was equal to 1%, while the respective maximum tensile strain at the bottom of the prism was 12%.

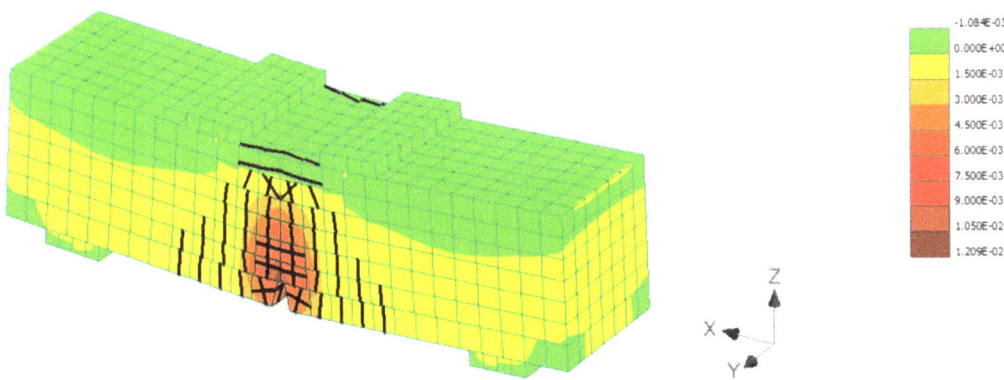

Figure 13. Indicative strain distribution at the maximum load.

The load deflection results of the numerical analyses are compared to the experimental and the results are illustrated in Figure 14. The thinner, colored lines resemble the experimental load deflection data obtained per specimen loaded under bending.

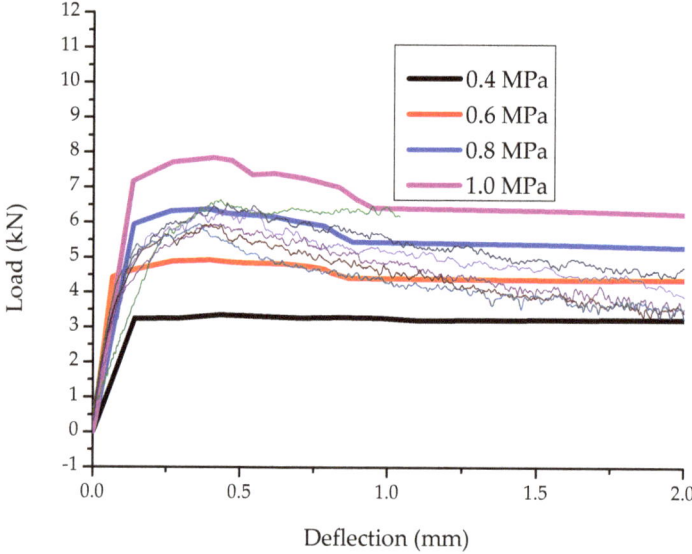

Figure 14. Experimental vs. numerical results of the examined prisms.

The results of Figure 14 show that a value of 0.8 MPa ultimate tensile strength and the constitutive model of Figure 12 can be used to accurately simulate the behavior of SFRRC. This model was used to evaluate the effectiveness of SFRRC as strengthening material.

4.2. Strengthening of Reinforced Concrete Beams Using SFRRC Layer

Reinforced Concrete beams have been examined to evaluate the efficiency of the use of SFRRC as a strengthening material. The examined specimens are based on a previous experimental study [42] where the performance of large-scale beams strengthened with additional concrete layers was studied. The cross section of the initial beam was 150 mm by 250 mm, the thickness of the additional layer was 50 mm and the total length of the examined specimens was equal to 2200 mm while the span length was 2000 mm. The cylinder compressive strength of the concrete of the initial beam was equal to 39.5 MPa while the respective strength value for the additional reinforced concrete layer was 45.4 MPa. B500 steel reinforcement was used in both the initial beam and the additional layer, and the reinforcement details are illustrated in Figure 15.

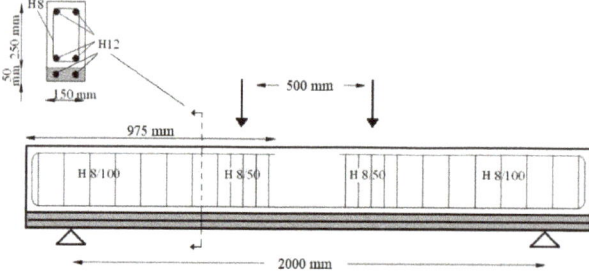

Figure 15. Geometry and reinforcement details of the strengthened beams. Adapted from [42].

The experimental setup and the failure pattern of one of the strengthened beams with conventional reinforced concrete layer are presented in Figure 16a,b.

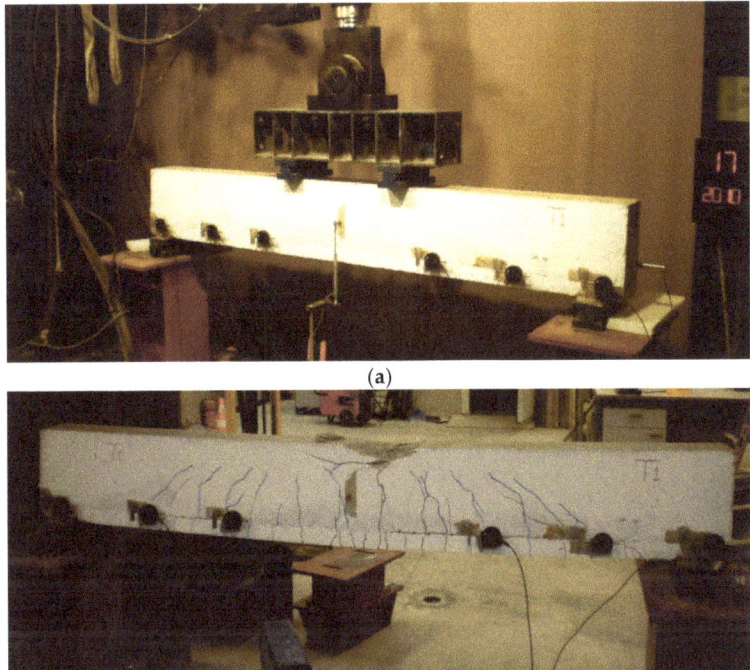

Figure 16. (a) Test setup and (b) failure mode of the beam strengthened with conventional reinforced concrete layer.

Numerical models were also developed for the initial and strengthened beams and the reliability of the models when conventional concrete was used was validated using the experimental results [9].

The same assumptions were used in this study for the simulation of the initial beams. For the strengthening layer, the assumptions presented in Section 4.1 were used with a tensile strength of 0.8 MPa, which was found to be the most appropriate value. Regarding the connection between the strengthening layer and the initial beam, contact elements with coefficient of friction 1.5 and cohesion 1.9 MPa were used, while a shrinkage strain equal to 565 microstrains was also applied to the new concrete layer [9].

The strain and crack distribution at the SFRRC strengthened beam at the maximum load stage are presented in Figure 17.

The load mid-span deflection of the SFRRC strengthened beam (ST_0.8 MPa) is compared with the respective experimental results of the beam strengthened with conventional reinforced concrete layer (ST_RC) and the initial beam prior to strengthening (IB) (Figure 18).

The results of Figure 18 indicate that the SFRRC layer can be effectively used to enhance the structural performance of reinforced concrete structural elements since the application of SFRRC led to an 83% increment of the ultimate load as compared to the initial prior-to-strengthening beam. This is quite close to the performance of the strengthened element with conventional reinforced concrete layer (ST_RC), which is attributed to the fact that the main flexural enhancement is due to the presence of steel bars in the strengthening layer. The examined application has a great potential since the use of additional high-damping SFRRC could be effective for the improvement of energy dissipation in addition to the enhancement of ultimate load capacity of existing structures.

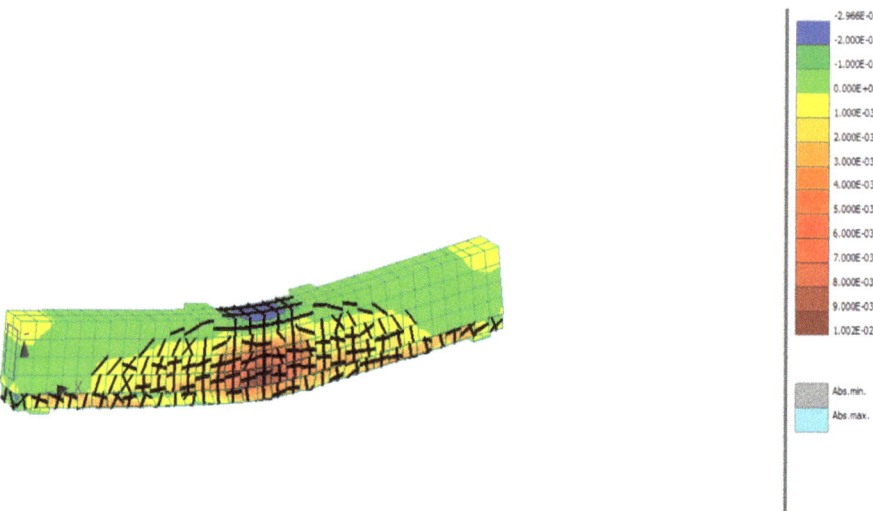

Figure 17. Crack and strain distribution of the SFRRC strengthened beam.

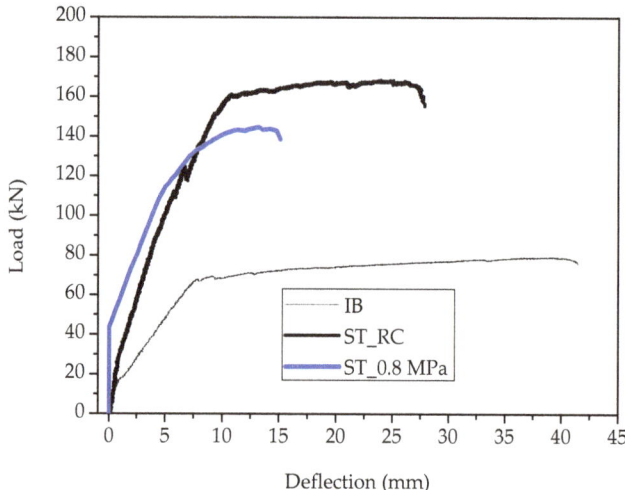

Figure 18. Load-mid span deflection results.

5. Conclusions

SFRRC that includes a high content of rubber and steel fibers obtained from End-of-life tires is a promising material for sustaining cyclic loading and attains an exceptional strain capacity. Not only does it provide such unique response to loading, but SFRRC that incorporates such high amounts of recycled steel fibers and tire particles obtained from End-of-life tires has a significant environmental value and supports the green deal initiative by promoting a circular economy.

Despite its low compressive strength, SFRRC attains the capacity to withstand long-lasting cyclic loads with significant energy absorption, as indicated by the area under each cycle. The significant energy absorbing capability of the material is also evident from the high damping value computed using the cyclic frequency ω and the mass of the specimen. It was shown that a 60% by volume aggregate replacement by rubber particles in combination with steel fiber reinforcement by recycled steel fibers from End-of-life tires led to improving the damping ratio of the material by 75.4%. Higher damping can lead to the absorption of large force energies and thus constitutes this material appropriate for sustaining impact and earthquake loading. Moreover, the SFRRC mix tested in this study has shown to be durable in freeze–thaw cycles and corrosive environments.

Most importantly, SFRRC is not only an environmentally friendly material that can absorb large volumes of waste (and specifically End-of-life tire products) but has also been proven capable of enhancing the structural performance of reinforced concrete structural elements, as indicated by the results obtained through numerical assessment and comparison to conventional methods. In addition to increasing the load capacity of existing RC structural elements, the high-damping capabilities of SFRRC can enhance the energy dissipation of existing structures. Future studies in this field should be focused on the performance of full structures retrofitted with additional SFRRC elements while the application of SFRRC seismic isolation foundation systems could also be explored. While currently there is a lack of previous research in evaluating SFRRC as a strengthening material for existing RC members, additional research in this area could lead to enabling the use of such novel materials in structural engineering applications.

Author Contributions: Conceptualization, T.P., K.N., N.K., A.L., O.T. and K.P.; methodology, T.P., N.K., A.L., O.T. and K.N.; formal analysis, T.P., A.L., O.T. and N.K.; investigation, T.P., A.L., O.T., K.N., N.K. and R.V.; data curation, T.P., A.L., O.T. and N.K.; writing—original draft preparation, T.P.; writing—review and editing, T.P., N.K., A.L., O.T. and K.N.; supervision, K.P. and D.G.H.; project administration, K.N., K.P. and D.G.H.; funding acquisition, N.K. All authors have read and agreed to the published version of the manuscript.

Funding: This research was funded by the Startup Grant for Nicholas Kyriakides, through the Cyprus University of Technology Internal Fund 3/319 and the European Union's Horizon 2020 research and innovation program under the Marie Skłodowska-Curie Individual Fellowship (MSCA-IF) grant agreement No. 748600.

Institutional Review Board Statement: Not applicable.

Data Availability Statement: Data presented in this study are available upon request.

Acknowledgments: The authors would like to thank the University of Sheffield student Abdulaziz Alsaif for his contribution to the experimental work and express their gratitude to the ERATOSTHENES Centre of Excellence.

Conflicts of Interest: The authors declare no conflict of interest. The funders had no role in the design of the study; in the collection, analyses, or interpretation of data; in the writing of the manuscript, or in the decision to publish the results.

References

1. United Nations Conference on Trade and Development (UNCTAD). *Circular Economy: The New Normal?* UNCTAD: Geneva, Switzerland, 2018.
2. Ismail, M.K.; Hassan, A.A.A. Impact resistance and mechanical properties of self-consolidating rubberized concrete reinforced with steel fibers. *J. Mater. Civ. Eng.* **2017**, *29*, 04016193. [CrossRef]
3. Ozbay, E.; Lachemi, M.; Sevim, U.K. Compressive strength, abrasion resistance and energy absorption capacity of rubberized concretes with and without slag. *Mater. Struct./Mater. Constr.* **2011**, *44*, 1297–1307. [CrossRef]
4. Wang, Z.; Hu, H.; Hajirasouliha, I.; Guadagnini, M.; Pilakoutas, K. Tensile stress-strain characteristics of rubberised concrete from flexural tests. *Constr. Build. Mater.* **2020**, *236*, 117591. [CrossRef]
5. Pajak, M.; Krystek, M.; Zakrzewski, M.; Domski, J. Laboratory investigation and numerical modelling of concrete reinforced with recycled steel fibers. *Materials* **2021**, *14*, 1828. [CrossRef]
6. Polydorou, T.; Constantinides, G.; Neocleous, K.; Kyriakides, N.; Koutsokeras, L.; Chrysostomou, C.; Hadjimitsis, D. Effects of pre-treatment using waste quarry dust on the adherence of recycled tyre rubber particles to cementitious paste in rubberised concrete. *Constr. Build. Mater.* **2020**, *254*, 119325. [CrossRef]

7. Brühwiler, E.; Denarié, E. Rehabilitation of concrete structures using Ultra-High Performance Fibre Reinforced Concrete. In Proceedings of the Second International Symposium on Ultra High Performance Concrete, Kessel, Germany, 5–7 March 2008; pp. 1–8.
8. Al-Osta, M.A.; Isa, M.N.; Baluch, M.H.; Rahman, M.K. Flexural behavior of reinforced concrete beams strengthened with ultra-high performance fiber reinforced concrete. *Constr. Build. Mater.* **2017**, *134*, 279–296. [CrossRef]
9. Lampropoulos, A.P.; Paschalis, S.A.; Tsioulou, O.T.; Dritsos, S.E. Strengthening of reinforced concrete beams using ultra high performance fibre reinforced concrete (UHPFRC). *Eng. Struct.* **2016**, *106*, 370–384. [CrossRef]
10. Paschalis, S.A.; Lampropoulos, A.P.; Tsioulou, O. Experimental and numerical study of the performance of Ultra High Performance Fiber Reinforced Concrete-Reinforced Concrete for the flexural strengthening of full scale members. *Constr. Build. Mater.* **2018**, *186*, 351–366. [CrossRef]
11. Ana, B.; Dubravka, B.; Marijan, S. Hybrid Fiber–Reinforced Concrete with Unsorted Recycled-Tire Steel Fibers. *J. Mater. Civ. Eng.* **2017**, *29*, 6017005.
12. Hu, H.; Papastergiou, P.; Angelakopoulos, H.; Guadagnini, M.; Pilakoutas, K. Mechanical properties of SFRC using blended Recycled Tyre Steel Cords (RTSC) and Recycled Tyre Steel Fibres (RTSF). *Constr. Build. Mater.* **2018**, *187*, 553–564. [CrossRef]
13. Lavagna, L.; Nisticò, R.; Sarasso, M.; Pavese, M. An analytical mini-review on the compression strength of rubberized concrete as a function of the amount of recycled tires crumb rubber. *Materials* **2020**, *13*, 1234. [CrossRef] [PubMed]
14. Bignozzi, M.C.; Sandrolini, F. Tyre rubber waste recycling in self-compacting concrete. *Cem. Concr. Res.* **2006**, *36*, 735–739. [CrossRef]
15. Hisham, B.; Bakar, A.; Noaman, A.T.; Akil, H.M. Cumulative Effect of Crumb Rubber and Steel Fiber on the Flexural Toughness of Concrete. *Eng. Technol. Appl. Sci. Res.* **2017**, *7*, 1345–1352.
16. Najim, K.B.; Hall, M.R. Mechanical and dynamic properties of self-compacting crumb rubber modified concrete. *Constr. Build. Mater.* **2012**, *27*, 521–530. [CrossRef]
17. Graeff, A.G.; Pilakoutas, K.; Neocleous, K.; Peres, M.V.N.N. Fatigue resistance and cracking mechanism of concrete pavements reinforced with recycled steel fibres recovered from post-consumer tyres. *Eng. Struct.* **2012**, *45*, 385–395. [CrossRef]
18. Alsaif, A.; Koutas, L.; Bernal, S.A.; Guadagnini, M.; Pilakoutas, K. Mechanical performance of steel fibre reinforced rubberised concrete for flexible concrete pavements. *Constr. Build. Mater.* **2018**, *172*, 533–543. [CrossRef]
19. AbdelAleem, B.H.; Hassan, A.A.A. Use of rubberized engineered cementitious composite in strengthening flexural concrete beams. *Eng. Struct.* **2022**, *262*, 114304. [CrossRef]
20. Raffoul, S.; Garcia, R.; Pilakoutas, K.; Guadagnini, M.; Medina, N.F. Optimisation of rubberised concrete with high rubber content: An experimental investigation. *Constr. Build. Mater.* **2016**, *124*, 391–404. [CrossRef]
21. EN 197-1:2011; Cement—Part 1: Composition, Specifications and Conformity Criteria for Common Cements. European Committee for Standardization: Brussels, Belgium, 2007.
22. EN 450-1:2012; Fly Ash for Concrete Part 1: Definition, Specifications and Conformity Criteria. European Committee for Standardization: Brussels, Belgium, 2012.
23. EN 13263-1; Silica Fume for Concrete. CEN (European Commitee for Standardization): Brussels, Belgium, 2005.
24. EN 934-2:2009; Admixtures for Concrete, Mortar and Grout. CEN (European Commitee for Standardization): Brussels, Belgium, 2009.
25. EN 1097-6:2013; Tests for Mechanical and Physical Properties of Aggregates—Part 6: Determination of Particle Density and Water Absorption. CEN (European Commitee for Standardization): Brussels, Belgium, 2013.
26. Twincon Ltd. *Technical Data Sheet Product: Reused Tyre Steel Fibre (RTSF)*; Twincon Ltd.: Sheffield, UK, 2018.
27. EN 206:2013+A1:2016; Concrete—Specification, Performance, Production and Conformity. CEN (European Commitee for Standardization): Brussels, Belgium, 2016.
28. EN 12390-3:2009/AC:2011; Testing Hardened Concrete—Part 3: Compressive Strength of Test Specimens. CEN (European Commitee for Standardization): Brussels, Belgium, 2011.
29. BS1881-121:1983; Testing Concrete—Part 121: Method for Determination of Static Modulus of Elasticity in Compression. British Standards Institution (BSI): London, UK, 1983.
30. JSCE-SF4; Method of Tests for Flexural Strength and Flexural Toughness of Steel Fiber Reinforced Concrete. JSI: Tokyo, Japan, 2016.
31. Vandewalle, L.; Nemegeer, D.; Balázs, G.L.; Barr, B.; Barros, J.A.O.; Bartos, P.; Banthia, N.; Criswell, M.; Denarie, E.; di Prisco, M.; et al. RILEM RILEM TC 162-TDF: "Test and design methods for steel fibre reinforced concrete" σ-ε-design method (final recommendation). *Mater. Struct./Mater. Constr.* **2003**, *36*, 560–567.
32. EN 14651; Test Method for Metallic Fibered Concrete—Measuring the Flexural Tensile Strength (Limit of Proportionality (LOP), Residual). CEN (European Commitee for Standardization): Brussels, Belgium, 2007.
33. BSI PD CEN/TS 12390-9:2016; Testing Hardened Concrete—Freeze-Thaw Resistance with De-Icing Salts—Scaling. CEN (European Commitee for Standardization): Brussels, Belgium, 2016; p. 36.
34. Nisticò, R.; Lavagna, L.; Boot, E.A.; Ivanchenko, P.; Lorusso, M.; Bosia, F.; Pugno, N.M.; D'Angelo, D.; Pavese, M. Improving rubber concrete strength and toughness by plasma-induced end-of-life tire rubber surface modification. *Plasma Processes Polym.* **2021**, *18*, 2100081. [CrossRef]
35. American Wood Council. *Beam Design Formulas With Shear and Moment Diagrams. Design AID No. 6*; American Wood Council: Washington, DC, USA, 2005; Volume 20.

36. fib Special Activity Group 5. *fib Bulletin 55: Model Code 2010*; fib: Lausanne, Switzerland, 2010.
37. fib Special Activity Group 5. *fib Bulletin 56: Model Code 2010*; fib: Lausanne, Switzerland, 2010.
38. Chopra, A.K. *Dynamics of Structures: Theory and Applications to Earthquake Engineering*; Pearson/Prentice Hall: Upper Saddle River, NJ, USA, 2007; ISBN 013156174X.
39. *EN 1998-1:2004*; Eurocode 8: Design of Structures for Earthquake Resistance—Part 1: General Rules, Seismic Actions and Rules for Buildings. European Committee for Standardization: Brussels, Belgium, 2004.
40. *SS 137244:2019*; Betongprovning—Hårdnad betong—Avflagning vid frysning Concrete testing—Hardened concrete—Scaling at freezing. SIS: Stockholm, Sweden, 2005.
41. Pryl, D.; Červenka, J. *ATENA Program Documentation Part 1*; Cervenka Consulting: Prague, Czech, 2013.
42. Tsioulou, O.T.; Lampropoulos, A.P.; Dritsos, S.E. Experimental investigation of interface behaviour of RC beams strengthened with concrete layers. *Constr. Build. Mater.* **2013**, *40*, 50–59. [CrossRef]

Article

Influence of Cross-Section Shape and FRP Reinforcement Layout on Shear Capacity of Strengthened RC Beams

Muhammad Ahmed, Piero Colajanni and Salvatore Pagnotta *

Dipartimento di Ingegneria, Università di Palermo, 90128 Palermo, Italy; muhammad.ahmed@unipa.it (M.A.); piero.colajanni@unipa.it (P.C.)
* Correspondence: salvatore.pagnotta@unipa.it

Abstract: The evaluation of the shear capacity of an FRP-strengthened reinforced-concrete beam is challenging due to the complex interaction between different contributions provided by the concrete, steel stirrup and FRP reinforcement. The shape of the beam and the FRP inclination can have paramount importance that is not often recognized by the models that are suggested by codes. The interaction among different resisting mechanisms has a significant effect on the shear capacity of beams, since it can cause a reduction in the efficiency of some resisting mechanisms. A comparative study of the performance in the shear resistance assessment provided by three models with six different effectiveness factors (R) is performed, considering different cross-section shapes, FRP wrapping schemes, inclination and anchorage systems. The results revealed that the cross-section shape, the FRP inclination and the efficiency of the FRP anchorages have a significant effect on the shear strength of beams. The analysis results show that the three models are able to provide an accurate average estimation of shear strength (but with a coefficient of variation up to 0.35) when FRP reinforcement orthogonal to the beam axis is considered, while a significant underestimation (up to 19%) affected the results for inclined FRP reinforcement. Moreover, all the models underestimated the resistance of beams with a T section.

Keywords: FRP; shear strengthening; average shear strength; effectiveness factor; inclination

---

Citation: Ahmed, M.; Colajanni, P.; Pagnotta, S. Influence of Cross-Section Shape and FRP Reinforcement Layout on Shear Capacity of Strengthened RC Beams. *Materials* 2022, 15, 4545. https://doi.org/10.3390/ma15134545

Academic Editor: Andreas Lampropoulos

Received: 9 May 2022
Accepted: 22 June 2022
Published: 28 June 2022

Copyright: © 2022 by the authors. Licensee MDPI, Basel, Switzerland. This article is an open access article distributed under the terms and conditions of the Creative Commons Attribution (CC BY) license (https://creativecommons.org/licenses/by/4.0/).

1. Introduction

Shear failure in RC members is one of the most critical and undesired failure phenomena. Beams and columns of existing RC Moment Resisting Frames (MRFs) usually do not satisfy the current code requirements regarding shear strength; thus, often it becomes necessary to strengthen the existing RC structural member in order to protect it from unwanted shear failure [1,2].

Over the last two decades, innovative strengthening techniques such as the use of externally bonded (EB) or near-surface-mounted (NSM) fiber-reinforced polymer (FRP) and externally bonded fiber-reinforced cementitious matrix (FRCM) have been widely used in the axial and shear strengthening of RC members [3,4].

Yet, it is quite difficult to accurately design a strengthening intervention of RC members by means of externally bonded FRP because member strength evaluation is still a topic of debate [5,6]. Many experiments confirm that the shear failure of FRP-strengthened beams usually occurs due to debonding of the FRP [7,8], but different failure modes can occur in a strengthened RC member due to the very presence of the FRP, which is affected by brittle failure [9].

The shear strength evaluation of FRP-reinforced RC beams is also quite complex due to the presence and interaction of the three main contributors, i.e., concrete, externally bonded FRP and transverse steel reinforcement. Experimental results have proved that the presence of FRP modifies the shear contributions provided by concrete and transverse steel reinforcement [10].

In this connection, the brittle failure of FRP reinforcement, which can occur before the yielding of steel stirrups, can have a negative effect on the shear strength provided by the transverse steel reinforcement [11,12]. Some researchers have proved that, when members having a significant amount of steel stirrups have to be strengthened, the transverse reinforcement provides a greater contribution than the FRP because the bond between the steel and concrete is stronger than between the FRP and the concrete surface [13]. However, experiments have also revealed that sometimes the efficiency of transverse reinforcement decreases with the presence of FRP due to brittle failure of the latter, which hinders the yielding of all the steel stirrups intersected by the shear-critical crack, as well as limits the strain achieved by the stirrups at failure [12,13].

The shear contribution provided by the FRP depends on the strengthening scheme. It can be done in different ways: (1) complete wrapping (C) of the member; (2) partial wrapping (U-shape); (3) side wrapping [14]. U-shape and side wrapping are more prone to debonding failure, while there are negligible chances of debonding failure for completely shaped wrapping. The side-wrapping scheme is not considered here because it does not provide a significant increment in the shear capacity of FRP-strengthened RC beams. To avoid debonding failure, a proper anchorage length or proper mechanical connector that prevents debonding between the FRP and the concrete surface should be provided [15].

In the case of the U-shaped scheme, several researchers have introduced different types of anchorages for FRP which have proved to be effective in increasing the shear contribution of the FRP [12,16]. However, their effectiveness is significantly variable due to different arrangements and other technological issues.

Two different approaches are pursued by codes for the shear resistance evaluation of strengthened members. According to a first group of international codes (CSA 2006 [17], ACI 440.2R/17 [18]), the strength of the RC beam reinforced by FRP is evaluated by an additive method. The overall shear resistance of the RC beam is considered as the sum of V_c (shear resistance of concrete), V_s (shear resistance of steel stirrups) and V_f (shear resistance provided by FRP). Regarding the last two contributions, each of them is evaluated by separately taking into account the orientation of each reinforcement, namely (β) for the FRP and (α) for the transverse reinforcement.

By contrast, according to the European approach, the contributions of all the components are considered using the truss mechanism with variable inclination of the concrete strut. Thus, the inclination angles of the FRP (β) and transverse reinforcement (α) are parameters of paramount importance, since they affect the shear strength and contribution of all three of the components discussed above.

Colajanni et al. [19] analyzed a large database comparing the experimental shear and the analytically calculated shear strength of different models. It was found that the angle of inclination of the FRP has a significant effect on the shear strength of RC beams. Moreover, by changing the inclination angle of the FRP with respect to the beam axis, there is a significant change in the interaction between the V_f and V_s.

Oller et al. [20] found that there is some difference between the sum of V_f, V_s, and V_c and the total shear force. The experimental results show that there is a significant contribution of the flange to the shear strength in the case of T-cross-sectional members. In some cases, it was found to be up to 45% of the total shear strength. None of the code models recognizes the effect of the flange in a T cross section as relevant in modifying member shear resistance.

However, despite the presence of such a complex framework, it is unanimously recognized that among the factors affecting the estimation of shear strength, the effective strain of the FRP, expressed through the reduction factor R, plays a predominant role.

In the light of the foregoing discussion, three main models for an RC shear critical beam strengthened with FRP are considered in this research: the model of Colajanni et al. [19], ACI 440.2R-17 [18] and CNR-DT-200/R1 [21]. Each model is analyzed with six different formulations that incorporate the effective FRP strain (R factor). Two of the approaches, namely (1: Khalifa and Nanni (2000, 2002), Pellegrino Modena (2006), and 2: Chen and

Teng (2003)) [7,22–25] are adopted in the Colajanni et al. model, which also considers the reduction factor for steel stirrups [19].

These models are well suited for rectangular RC beams, but the effectiveness in calculating the shear strength of T sections is still a matter of discussion, since the effect of the contribution of flanges is not incorporated in all of them.

To fully understand the effect of a cross-section shape and the influence of the inclination angle of FRP on the shear resistance of the strengthened beam, experimental results on strengthened beam specimens with different section shapes or FRP inclination angles and equal values of the other geometrical and mechanical parameters should be available. Failing these, the influence can be indirectly detected by the variation in the ability of the different analytical models to predict the experimental results.

To this aim, a large database was collected, considering rectangular and T sections, with vertical and inclined FRP, also including specimens characterized by mechanical anchorages between FRP and the RC beam. The experimental values were compared with the shear-strength values obtained from the three above-mentioned models [18,19,21]. The results are discussed focusing on the influence of the cross-section shape and the FRP angle of inclination on both the shear strength of the specimen and the reliability of the shear models, also considering the effect of the various effectiveness-factor models for FRP.

Two different analyzing approaches were adopted. The first approach was to cover the influence of the cross-section shape on the shear strength of the beam externally bonded by FRP. Then a comparison was made between the assessment of the shear strength of R and the T-cross-sectional members with the FRP and steel web reinforcement having the same inclination ($\alpha = \beta$) by means of a fixed model. To cover the influence of FRP inclination, a comparison was made between the same shape but with different inclinations ($\alpha \neq \beta$), again by means of a fixed model.

2. Shear Models

Three different models—ACI 440.2R-17 [18], CNR-DT-200/R1 [21] and Colajanni et al. [19]—are reported and discussed below. The code models are reported hereinafter as presented in their original form. It has to be stressed that in the shear prediction discussed in Section 6, neither the safety factors for the FRP reinforcement, steel stirrups and concrete were considered, nor the strength-reduction factor φ.

2.1. ACI 440.2R-17

ACI 440.2R-17 [18] gives guidelines for the design and evaluation of the shear strength for an RC beam strengthened with externally bonded FRP based on an additive approach.

ACI 318-14 (ACI 2014) [26] is used for the evaluation of concrete (V_c), steel stirrup (V_s) and FRP (V_f) contributions. According to the symbol notation in Figure 1 and Abbreviations, V_c (nominal shear strength provided by shear reinforcement) is calculated as:

$$V_c = 0.167 \, f'^{0.5}_c \, b_w \, d \qquad (1)$$

while V_s (nominal shear strength provided by shear reinforcement) is calculated as:

$$V_s = (A_v \, f_{yt} \, d)/s \qquad (2)$$

and the shear strength provided by the FRP is calculated as:

$$V_f = (A_{fv} \, d_{fv} \, f_{fe} \, (\sin \beta + \cos \beta))/s_f \qquad (3)$$

In the V_f equation, $A_{fv} = 2nt_fw_f$, $f_{fe} = \varepsilon_{fe} \, E_f$ and different safety factors are used for different wrapping schemes. The shear strength of the retrofitted RC beam is equal to:

$$\varphi V = \varphi \, (V_c + V_s + \psi_f \, V_f) \qquad (4)$$

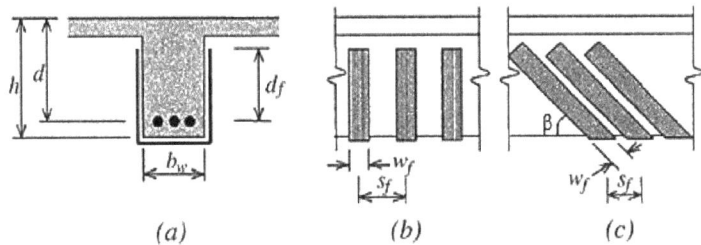

Figure 1. Variables used in the ACI 440.2R model for shear-strengthening calculations (**a**) cross-sectional parameters (**b**) spacing and width of FRP (**c**) inclination angle between FRP strip and beam axis.

For the completely wrapped scheme, $\psi_f = 0.95$ while for other schemes $\psi_f = 0.85$, while φ is a strength-reduction factor. The effective strain of the FRP (ε_{fe}) is calculated based on the different configurations. It should not be more than 0.75 of the ultimate strain ε_{fu}, while for the design it should be limited to 4×10^{-3}. The FRP effective depth is considered as the distance between the centroid of tensile reinforcement and the top free edge of the FRP. It must be stressed that the ACI model takes into account the actual height of the FRP reinforcement by the parameter d_{fv} (Figure 1).

To obtain the effective strain ε_{fe} of partly wrapped sections, the ultimate strain of the FRP is multiplied by a bond-reduction factor k_v, as $\varepsilon_{fe} = k_v \, \varepsilon_{fu} \leq 4 \times 10^{-3}$. k_v can be calculated as $k_v = k_1 k_2 L_e / (11{,}900 \varepsilon_{fu}) \leq 0.75$, where the modification factors k_1 and k_2 can be calculated by using $k_1 = (f'_c/27)^{2/3}$ and $k_2 = (d_{fv} - \gamma L_e)/d_{fv}$ ($\gamma = 1$ for the U-wrapped scheme and $\gamma = 2$ when both sides are wrapped), where the effective length is $L_e = 23{,}300/(E_f \, t_f)^{0.58}$.

2.2. CNR Model

The CNR-DT 200 R1/2013 [21] is the model established by the Italian Research Council (CNR) and it deals with two types of wrapping: U-shaped and full. The equations given are the extension of the equations provided in EN1992-1-1 [27] to evaluate the shear strength of reinforced-concrete beams. The model is derived according to the truss mechanism with variable inclination of the concrete strut, in which the shear capacity of the FRP is calculated using:

$$V_{Rd,f} = (1/\gamma_{Rd}) \, 0.9 \, d \, f_{fed} \, 2 \, t_f \, (\cot \theta + \cot \beta) \, (b_f / p_f) \sin^2 \beta \tag{5}$$

Equation (5) is reported and used consistently with the equation reported in the new version of the CNR-DT code for strengthening by fiber/fabric-reinforced cementitious matrix/mortar [28].

In this equation, $p_f = \bar{p}_f \sin \beta$ represents the spacing of the FRP measured perpendicular to the direction of fiber. The shear capacity of the stirrup and concrete strut is given as:

$$V_{Rd,s} = 0.9 \, d \, (A_{sw}/s) \, f_{ywd} \, (\cot \theta + \cot \alpha) \sin \alpha \tag{6}$$

$$V_{Rd,c} = 0.9 \, d \, b \, \alpha_c \, 0.5 \, f_{cd} \, (\cot \alpha + \cot \theta)/(1 + \cot^2 \theta) \tag{7}$$

In Equation (7), $\alpha_c = 1$ has to be retained for the beam, and the angle ψ, yet to be determined, can be introduced by replacing the angle α listed in the code, in order to stress that the evaluation of the shear strength of the compressed concrete strut is not a trivial issue, as will be shown below. The strengthened member shear resistance is computed as:

$$V_{Rd} = \min (V_{Rd,s} + V_{Rd,f}, V_{Rd,c}) \tag{8}$$

In evaluating the shear strength in a beam with FRP reinforcement inclined with an angle $\beta \neq \alpha$, the CNR-DT 200 R1/2013 model assumes $\psi = \beta$, taking into account in the truss scheme the inclination of the FRP reinforcement (β) only. Thus, in evaluating the strength of the compressed concrete strut, the model neglects the presence of two orders of web reinforcements and the amount of their contributions.

In this regard in [19], it is shown that in a beam which is to be shear-strengthened in which the existing shear reinforcement provides a significant contribution to the shear strength as the FRP, the angle of inclination of the concrete strut should be evaluated assuming ψ as a weighted value between α and β, where $V_{Rd,s}$ and $V_{Rd,f}$ are the weights.

2.3. Colajanni et al. Model

Colajanni et al. [19] proposed a model with variable inclination of the compressed concrete action based on stress field theory. It was derived on the basis of a previous model for a concrete beam reinforced by stirrups with two different inclinations [29]. In the same paper, it was validated against experimental results on ordinary RC beams and hybrid steel-trussed concrete beams (HSTCBs) [30].

This model is able to correctly represent the shear strength of FRP-strengthened RC beams with shear reinforcement arranged in any direction. In order to evaluate the shear strength of the member, three different segments of beams are selected based on the stress field direction. They are obtained by sections parallel to the stress field of the FRP, concrete strut and steel stirrups which are demonstrated in Figure 2. The shear strength is calculated using three different equations by evaluating the vertical equilibrium of each beam segment:

$$V = (b\,0.9\,d\,0.5\,f_c)\left\{\left[R\tilde{\sigma}_f\,f_{fu}\left(A_{fv}/s_f\right)(\cot\theta + \cot\beta)\sin\beta\right] + \left[r\,\tilde{\sigma}_s\,f_{yt}\,(A_v/s)(\cot\theta + \cot\alpha)\sin\alpha\right]\right\} \quad (9)$$

$$V = (b_w\,0.9\,d\,0.5\,f_c)\left\{\left[\tilde{\sigma}_c\,(\cot\theta + \cot\alpha)\sin^2\theta\right] + \left[R\,\tilde{\sigma}_f\,f_{fu}\left(A_{fv}/s_f\right)(\cot\beta - \cot\alpha)\sin\beta\right]\right\} \quad (10)$$

$$V = (b_w\,0.9\,d\,0.5\,f_c)\left\{\left[\tilde{\sigma}_c\,(\cot\theta + \cot\beta)\sin^2\theta\right] + \left[r\,\tilde{\sigma}_s\,f_{yt}\,(A_v/s)(\cot\alpha - \cot\beta)\sin^2\alpha\right]\right\} \quad (11)$$

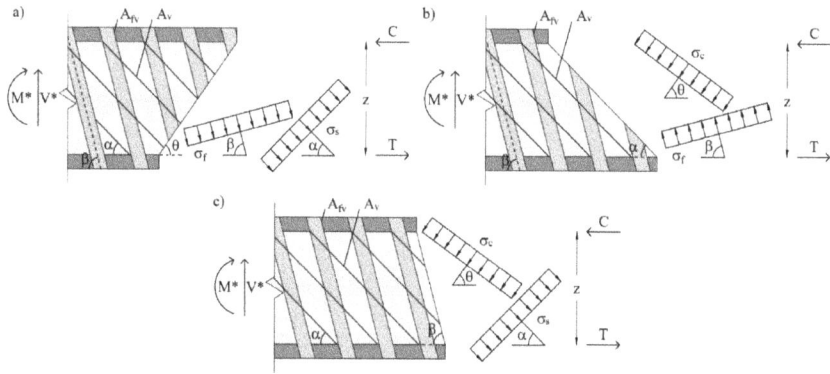

Figure 2. Beam segments identified via three sections parallel to stress field directions of (**a**) concrete strut; (**b**) steel stirrups; (**c**) FRP reinforcement. M* and V* represent the moment and shear acting on the considered section.

In the equations above, $\tilde{\sigma}_f = \sigma_f/f_{fu}$, and $\tilde{\sigma}_s = \sigma_s/f_{yt}$ are the non-dimensional stresses of the FRP reinforcement and steel stirrups, respectively. R is the coefficient for effective strain and stress for the FRP at failure, where the effective stress is $f_{fe} = f_{fu}\,R = E_f\,\varepsilon_{fe}$ and the effective strain is $\varepsilon_{fe} = \varepsilon_{fu}\,R$; f_{fu} is the ultimate stress of the fiber; r is the efficiency coefficient for the steel stirrups, which considers the efficiency of the steel stirrups involved by the shear-critical crack; β represents the angle of the FRP and α represents the angle of the shear reinforcement with the beam axis.

The static theorem of plasticity is used to evaluate the shear strength of RC beam. It means that the shear strength is the maximum value among solutions, and it should satisfy all the equations including the plastic admissibility equations given below:

$$0 \leq \tilde{\sigma}_c, \tilde{\sigma}_f \leq 1, -1 \leq \tilde{\sigma}_c \leq 1 \tag{12}$$

By combining (9), (10) and (12), Equation (13) is obtained. It shows the relation between the FRP, the transverse steel reinforcement and the stress field of the concrete strut.

$$0 \leq \tilde{\sigma}_c = \left(R\tilde{\sigma}_f A_{fu} f_{fu}/\left(b_w s_f 0.5 f_c\right) \sin \beta + r \tilde{\sigma}_s A_v f_{yt}/(b_w s\ 0.5 f_c) \sin \alpha\right)\left(1 + \cot^2 \theta\right) \leq 1 \tag{13}$$

According to the code's suggestion, the lower limit of $\cot \theta = 1.0$, and the upper limit of $\cot \theta = 2.5$ hold. Based on this limitation of inclination of the concrete strut, the shear strength can be evaluated in two steps: (1) Initially, it is assumed that at the failure phase all the three stress fields could reach their stress limit simultaneously. Hence, using the inequality given in Equation (13), the inclination of the concrete strut can be evaluated as:

$$\cot \theta = ((R\ A_{fv} f_{fu}/(b_w s_f 0.5 f_c) \sin \beta + r\ A_v f_{yt}/(b_w s\ 0.5 f_c) \sin \alpha)^{-1} - 1)^{1/2} \tag{14}$$

Step 2: three different cases can occur, depending on the amount of steel or FRP shear reinforcement mechanical ratio ω_{sw} and ω_{fw}, respectively.

Case 1 (small ω_{sw} and ω_{fw} values):

$\cot \theta > 2.5$: It implies that $\cot \theta = 2.5$ must be assumed, and the concrete strut does not fail due to the presence of a small amount of shear reinforcement. So, the shear strength is calculated using Equation (9), in which $\tilde{\sigma}_f = \tilde{\sigma}_s = 1$, while the stresses on the concrete strut can be obtained by using Equation (13).

Case 2 (intermediate ω_{sw} and ω_{fw} values):

$1 \leq \cot \theta \leq 2.5$: It implies that all the stress fields simultaneously achieve their stress limit. Equation (9) is thus used to find the shear strength considering $\tilde{\sigma}_f = \tilde{\sigma}_s = \tilde{\sigma}_c = 1$.

Case 3 (very large ω_{sw} and ω_{fw} values):

$\cot \theta < 1$: It implies that $\cot \theta = 1.0$ must be assumed, and the failure is due to reaching the stress limit in the concrete strut and in one of the shear reinforcements. If it is assumed that $\beta < \alpha$, then the maximum shear strength of the beam is given by the minimum value obtained via Equation (10) assuming that the FRP reinforcement attains the maximum effective strain in tension $\tilde{\sigma}_f = 1$ or by Equation (11), considering that the yielding in the steel stirrups is attained, having $\tilde{\sigma}_s = 1$. In Appendix B, the above three different cases are elucidated by calculation examples.

3. Reduction Factors for Steel Stirrups "r"

Different research has revealed that the simultaneous presence of FRP and steel stirrups decreases the contribution to the shear strength provided by shear reinforcement. It was also found that the increase in the axial rigidity ratio between the steel and FRP causes a reduction in the shear contribution provided by the externally bonded FRP [12,13,31–33].

Due to this interaction, some models were developed; the one proposed by Modifi and Chaallal [34] considers the interaction between the two reinforcement systems and their rigidities, while the model developed by Pellegrino and Modena [32] assumes a fixed reduction coefficient.

In order to model the interaction between the FRP and steel stirrup, Colajanni et al. [19] also included a similar factor in their model, which is able to take into account the possible different inclinations of the FRP and the pre-existing steel web reinforcement. "r" is defined as a bilinear expression that links the reduction in the contribution to the shear strength provided by transverse reinforcement to the ratio between the FRP effective strain in the direction of the steel reinforcement $\varepsilon_{fe,s} = \varepsilon_{fe} \cos(\alpha - \beta)$ and the yield strain of the steel stirrup (ε_{syw}). If $\varepsilon_{fe,s}/\varepsilon_{syw} \leq 1.33$, then $r = 0.75\ \varepsilon_{fe,s}/\varepsilon_{syw}$, otherwise it is considered as $r = 1$.

4. Effectiveness Factor "R"

The failure of a shear-strengthened RC beam with externally bonded FRP is due to several factors, including peeling of the concrete cover, failure of the FRP, debonding of the FRP from the concrete surface, the loss of aggregate interlock, etc. Most of these phenomena precede attainment of the ultimate strain in the FRP. Thus, in order to limit the contribution of the FRP reinforcement, the effectiveness factor "R" is applied to the ultimate strain of the FRP fiber, which reduces the ultimate FRP tensile stresses.

To evaluate shear strength, six different effectiveness factors "R" were used. All of these have different approaches to deal with the strengthening of FRP. The first two R factors were used in [19]; the first was derived according to Khalifa and Nanni and Pellegrino and Modena [22–24]. The effectiveness factor is taken as the minimum among the four coefficients (R_1, R_2, R_3, R_4), which represent different modes of failure. R_1 considers the tensile failure of the FRP, while R_2 and R_3 represent the debonding phenomenon and failure of the FRP due to shear crack width, respectively. Lastly, R_4 considers failure due to peeling of the concrete cover.

$$R_1 = 0.56(\rho_f E_f)^2 - 1.22(\rho_f E_f) + 0.78$$

$$R_2 = [(f_{ck})^{2/3}(d_{fv} - \eta L_e)\,[738.93 - 4.06(E_f t_f)]]/\varepsilon_{fu} d_{fv} 10^6$$

$$R_3 = 6 \times 10^{-3}/\varepsilon_{fu}$$

$$R_4 = (2 f_{ct} A_c \cos^2 \beta b_{c,v})/(n_f t_f L_f E_f\,[(h_f - L_e)/(h_f)]b_f\,\varepsilon_{fu})$$

The second R factor was proposed by Chen and Teng [7]. It is the minimum between two factors (R_5 and R_6). One represents the tensile rupture of the FRP across the crack, and the other represents the debonding failure of the FRP due to insufficient bond length.

$$R_5 = (1 + (d - d_{fv})/z)/2$$

For $\lambda < 1$ ($\lambda = L_{max}/L_e$)

$$R_6 = (\sigma_{f,max}/E_f \varepsilon_{fu}) \times (2/\pi \lambda \times ((1-\cos \pi \lambda/2)/(\sin \pi \lambda /2)))$$

For $\lambda \geq 1$

$$R_6 = (\sigma_{f,max}/E_f \varepsilon_{fu}) \times (1 - (\pi-2)/\pi \lambda)$$

The definition of the other four effectiveness factors can be found in ACI [18], CNR [21] fib [35] and Mofidi and Challal (M&C) [36].

5. Description of Data Sets and Analysis Steps

To cover the main aspects of the research (influence of cross-section shape and angle of FRP inclination), two data sets and two comparison approaches were adopted which are explained in Figure 3. The beams included in the database have effective depths of the cross section in the range of 155 mm and 831 mm, while the shear span ranges between 2.3 m and 3.8 m. The transverse internal steel reinforcement is constituted by vertical steel stirrups whose maximum geometrical ratio is 0.48%.

The FRP reinforcement geometrical ratio varies between 0.04% and 3.00%. The ultimate FRP tensile strength ranges between 106 and 4361 MPa, while the Young's modulus varies between 8 and 640 GPa. In order to stress the influence of the FRP inclination (β), two data sets were analyzed, namely Data Set 1 (DS1) with $\alpha = \beta$ and Data Set 2 (DS2) with $\alpha \neq \beta$, where α is the inclination of the steel reinforcement with the beam axis while β represents the angle of the FRP with the beam axis.

Both data sets contain results regarding R and T members with different wrapping schemes. On the basis of the wrapping scheme, rectangular RC beams are divided into three subsets, namely RF, RU and RU*. The first subset F represents full/complete wrapping, the U represents U-jacketing without anchorages, and U* represents U-jacketing with partially efficient anchorages. Analogously, T beams are divided into TU, TU* and TU/F, the latter representing U-jacketing with fully efficient anchorages.

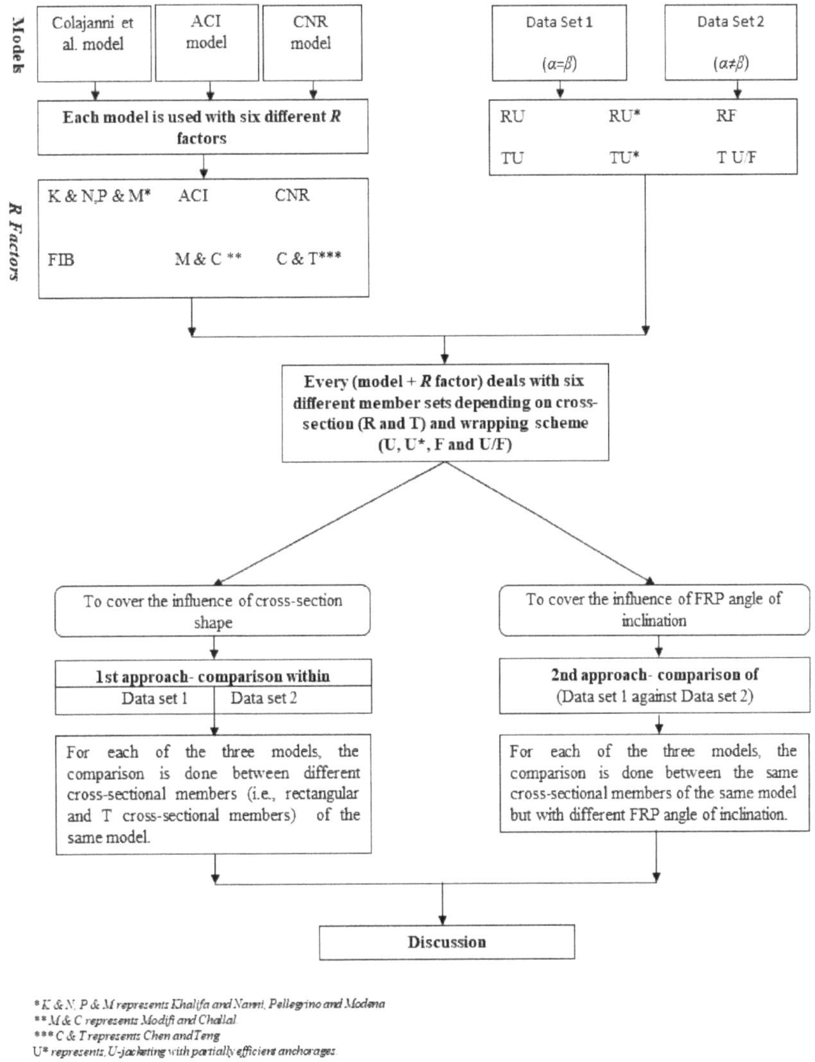

Figure 3. Flow chart representing data-analysis methodology.

* K & N, P & M represents Khalifa and Nanni, Pellegrino and Modena
** M & C represents Modifi and Challal
*** C & T represents Chen and Teng
U* represents U-jacketing with partially efficient anchorages

It is pointed out that, for the U*-wrapping scheme, the shear strength was assessed as if the beam was strengthened by ordinary U-jacketing, since the increase in shear capacity provided by partially efficient anchors cannot be assessed. Moreover, in the case of U/F, the beams were considered as strengthened by complete wrapping (as done in [20]).

In (DS1), where $\alpha = \beta$, there are 40 rectangular beams with U-shaped wrapping, 7 reinforced-concrete beams with partially efficient anchorages (U*), and 10 beams with the complete/full wrapping scheme. Similarly, there are 52 T reinforced beams with U-wrapping, 18 T beams with U*-wrapping, and 11 TU/F beams with fully efficient anchorage.

In (DS2), where $\alpha \neq \beta$, there are 10 rectangular beams with U-wrapping, 7 rectangular reinforced-concrete beams RU* with partially efficient anchorages and 1 rectangular beam with complete wrapping. In DS2, for T beams there are two beams with U-wrapping, while for TU* and T U/F no experimental shear values are available.

During the analysis, v_{exp} of the data sets was used. v_{the} is the analytical assessment of dimensionless shear, which for each model was calculated using all six different R-factor models, while v_{exp} is the experimental value which can be expressed as $v_{exp} = V_{exp}/(b_w\, 0.9d\, 0.5 f_c)$. τ_{avg} is the average value of the ratio $\tau_{avg} = V_{exp}/V_{the}$.

The following steps are performed in the analysis:

1. Colajanni et al. Model [19], ACI model [18] and CNR model [21] are used for the analysis with six different R factors.
2. Every model + R factor deals with six different member sets, which differ in the type of cross section (R and T) and wrapping scheme (U, U*, F and U/F).
3. For each data set, the results provided by the Colajanni et al. model with the six different formulations of the R effectiveness factor are discussed.
4. For each of the three models: in the first approach, to cover the influence of the cross-section shape, a comparison is made between the R and T sections (i.e., between RU and TU, between RU* and TU*, between RF and TU/F within Data Set 1 and within Data Set 2).
5. For each of the three models: in the second approach, to recognize the influence of the FRP inclination angle, a comparison is made of Data Set 1 against Data Set 2 for the effectiveness of each model in the strength assessment of members having the same cross-section shape but with different inclination angles (i.e., RU of DS1 and RU of DS2, RU*of DS1 and RU* of DS2, RF of DS1 and RF of DS2).

6. Results and Discussion

Different effectiveness factors (R) are used in the model proposed by Colajanni et al. [19] and their efficiency is compared in order to determine the influence of the effectiveness factor on the shear capacity assessment. To evaluate the reliability and efficiency of the model, the average ratio $\tau_{Avg} = V_{exp}/V_{the}$ and its CoV are analyzed. In the reliability assessment, R as proposed by ACI [18], CNR [21], fib [35] and Mofidi and Challal [36] (M&C) is used in its original form, considering the steel-stirrup reduction factor equal to r = 1, as well as two R-factor models proposed in Colajanni et al. [19] including the steel-stirrup reduction factor.

In Table 1, the results for the whole database reported in Table A1 of Appendix A are summarized, proving that the R factor has a major effect on evaluating the shear strength of strengthened beams. By using the effectiveness factor proposed by CNR or ACI, the best average values (mean efficiency ratio τ_{avg} = 0.97) were obtained, while the worst results were obtained in the case of using the R factor of fib.

Table 1. Evaluation of combined results of Data Set 1 and Data Set 2 using the Colajanni et al. model with different R factors.

	R (K&N, P&M)	R (ACI)	R (CNR)	R fib	R M&C	R C&T
τ_{avg}	0.94	0.97	0.97	0.82	1.12	0.95
CoV	0.27	0.32	0.25	0.25	0.29	0.20

The M&C model yielded a 12% underestimation of the average shear strength. The effectiveness factors of Khalifa and Nanni + Pellegrino and Modena, and Chen and Teng yielded better results as compared to fib but with a slight overestimation of the shear strength. A coefficient of variation parameter was also used to analyze the results. From the results, it is seen that the highest scattering of data was observed in the case of R as given by the ACI model with (CoV = 0.32) while it is concluded that the highest accuracy and reliability was obtained in the case of the Chen and Teng R-factor model, since it provides the smallest CoV (0.20) with a τ_{avg} = 0.95, very close to the best ones.

A more effective analysis can be performed if the whole database is split into two subsets, according to the values of the inclination of the steel (α) and FRP (β) web reinforcement. From Figure 4, it can be seen that, for $\alpha = \beta$, similarly as for the whole database, a very accurate estimation of the shear strength was achieved by using the R factor of ACI and CNR, while the worst results were obtained in the case of using the effectiveness factor of fib and M&C. The R-factor model proposed by Chen and Teng slightly (4%) overestimated the shear strength, but less scattering was observed as (CoV = 0.20).

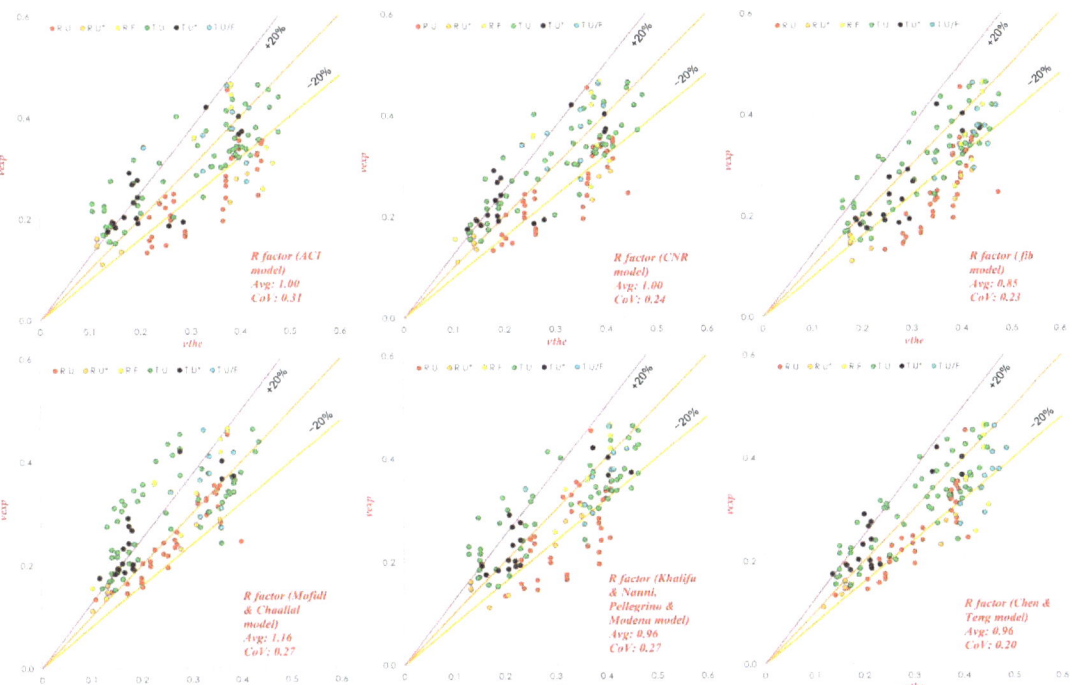

Figure 4. Experimental vs. theoretical shear strength for the Colajanni et al. model with six different R factors. (Data Set 1, $\alpha = \beta$). RU* and TU* represents Rectangular and T beam having U-jacketing with partially efficient anchorages.

By contrast, from Figure 5, where $\alpha \neq \beta$, it can be seen that a large over estimation of shear strength was found for all the models; the largest one ($\tau_{avg} = 0.61$) as well as the largest scattering of data was observed when the R factor proposed by the fib model was used. Similar results can be observed in the case of $\alpha = \beta$.

The highest accuracy among the others with less overestimation and scattering ($\tau_{avg} = 0.87$, CoV = 0.14) was observed for the effectiveness factor proposed by Chen and Teng. This unfavorable large overestimation was less marked in the models with the effectiveness factors of Khalifa and Nanni + Pellegrino and Modena, or Chen and Teng because of the introduction of the "r" factor. In Table 2, the results are further subdivided into the two databases described in Sections 6.1–6.3.

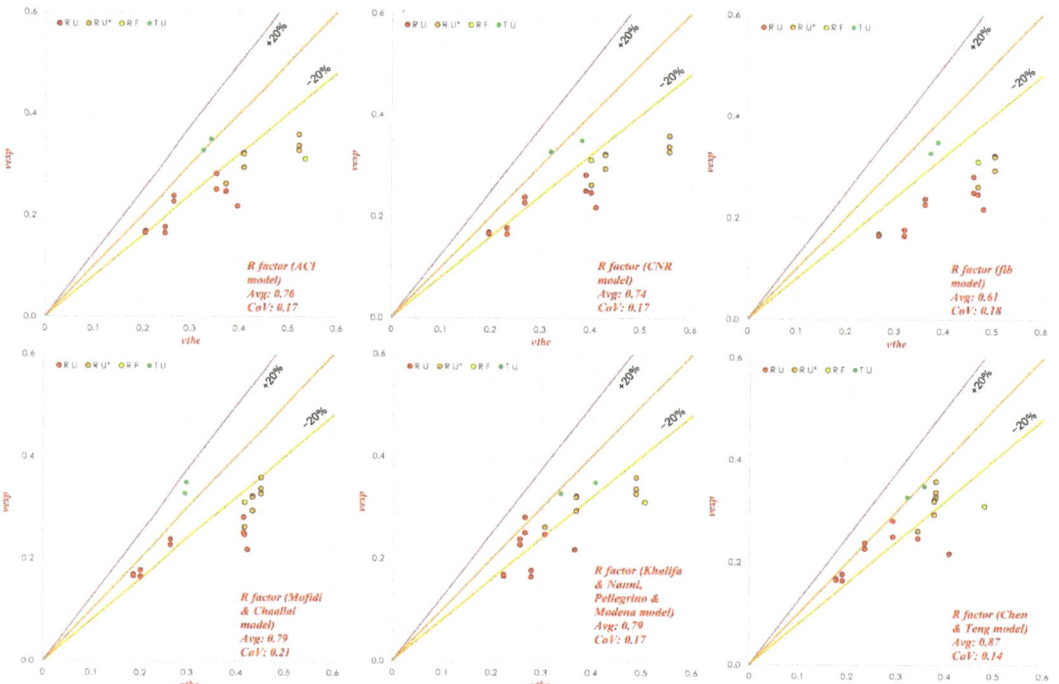

Figure 5. Shear strength calculation (experimental vs. theoretical) for the Colajanni et al. model with six different R factors. (For Data Set 2, $\alpha \neq \beta$). RU* represents Rectangular while TU* represents T beam having U-jacketing with partially efficient anchorages.

6.1. Conclusion Based on 1st Approach for DS1 ($\alpha = \beta$)

The results were compared by taking the average value provided by the six different R-factor models, divided on the basis of section shape and efficiency of the anchorage. In the Colajanni et al. model, the results in the case of RU revealed that there was an overestimation of the shear strength on average of about 19%, while in the TU scheme, the shear strength on average was underestimated by about 9%. The first lack of accuracy, namely the overestimation of the rectangular sections, can be attributed to the inefficiency of the model in taking into account the reduced length of the FRP fiber, which in many specimens did not reach the top end of the beam. For T beams, the overestimation was mitigated by the presence of the shear contribution provided by the flange; this contribution produced a significant increment in the strength of the tested specimens that was not predicted by the model.

In the case of RU* and TU*, there was the same trend as for RU and TU. There was almost a 13% overestimation on average in the case of RU* and 11% in the case of TU*. There was less of a dispersion of the data observed for TU* with an average CoV of 0.18.

In the case of fully wrapped rectangular or equivalent fully wrapped T sections, namely RF and TU/F, the average values were almost the same, and τ_{avg} was close to unity. This is due to the fact that for RF and TU/F, some approaches underestimated while others overestimated the shear strength. It was observed that overall, more accurate estimation of the shear strength was obtained for RF and TU/F as compared to the previous comparisons, with less dispersion. This is due to the fact that the effective fiber strength of the completely wrapped sections was more accurately estimated than that of the partially wrapped ones, and possibly the partial ineffectiveness of the anchorage in the TU/F section was compensated by the flange contribution.

Table 2. Results of calculation on the basis of the 1st and 2nd approaches from Database 1 and Database 2.

		R Factors	RU		RU*		RF				R Factors	TU		TU*		TU/F	
			τ_{avg}	CoV	τ_{avg}	CoV	τ_{avg}	CoV				τ_{avg}	CoV	τ_{avg}	CoV	τ_{avg}	CoV
For $\alpha = \beta$	Colajanni et al. Model	K&N, P&M [1]	0.78	0.25	0.87	0.25	1.01	0.10		Colajanni et al. Model	K&N, P&M	1.06	0.27	1.04	0.16	1.01	0.22
		ACI	0.79	0.24	0.92	0.33	1.01	0.24			ACI	1.10	0.31	1.16	0.22	1.09	0.29
		CNR	0.82	0.18	0.89	0.23	1.13	0.24			CNR	1.05	0.21	1.17	0.21	1.13	0.18
		FIB	0.70	0.21	0.71	0.17	0.89	0.14			FIB	0.97	0.20	0.91	0.15	0.85	0.13
		M&C [2]	0.95	0.13	1.00	0.15	1.25	0.20			M&C	1.30	0.29	1.27	0.17	1.12	0.18
		C&T [3]	0.82	0.13	0.84	0.15	0.88	0.17			C&T	1.04	0.19	1.10	0.15	0.81	0.11
		Average	0.81	0.19	0.87	0.21	1.03	0.18			Average	1.09	0.24	1.11	0.18	1.00	0.19
	ACI Model	K&N, P&M	0.99	0.21	0.90	0.11	1.10	0.08		ACI Model	K&N, P&M	0.95	0.19	1.03	0.20	0.91	0.30
		ACI	0.95	0.15	0.86	0.09	1.06	0.26			ACI	0.96	0.19	1.07	0.24	0.96	0.35
		CNR	0.99	0.15	0.86	0.08	1.16	0.17			CNR	0.96	0.18	1.07	0.23	1.00	0.32
		FIB	0.91	0.17	0.79	0.07	0.94	0.11			FIB	0.94	0.19	1.00	0.23	0.83	0.34
		M&C	1.10	0.18	0.95	0.07	1.26	0.11			M&C	1.04	0.20	1.11	0.23	1.02	0.38
		C&T	1.01	0.14	0.88	0.06	0.91	0.15			C&T	0.98	0.19	1.07	0.22	0.77	0.29
		Average	0.99	0.17	0.87	0.08	1.07	0.15			Average	0.97	0.19	1.06	0.23	0.91	0.33
	CNR Model	K&N, P&M	0.78	0.25	0.87	0.25	1.01	0.10		CNR Model	K&N, P&M	1.06	0.27	1.04	0.16	1.01	0.22
		ACI	0.79	0.24	0.92	0.33	1.01	0.24			ACI	1.10	0.31	1.16	0.22	1.09	0.29
		CNR	0.82	0.18	0.89	0.23	1.13	0.24			CNR	1.05	0.21	1.17	0.21	1.13	0.18
		FIB	0.70	0.21	0.71	0.17	0.89	0.14			FIB	0.97	0.20	0.91	0.15	0.85	0.13
		M&C	0.95	0.13	1.00	0.15	1.25	0.20			M&C	1.30	0.29	1.27	0.17	1.12	0.18
		C&T	0.82	0.13	0.84	0.15	0.88	0.17			C&T	1.04	0.19	1.10	0.15	0.81	0.11
		Average	0.81	0.19	0.87	0.21	1.03	0.18			Average	1.09	0.24	1.11	0.18	1.00	0.19
For $\alpha \neq \beta$	Colajanni et al. Model	K&N, P&M	0.79	0.20	0.79	0.11	0.62	xxxxx		Colajanni et al. Model	K&N, P&M	0.91	0.09	xxxxx	xxxxx	xxxxx	xxxxx
		ACI	0.75	0.14	0.71	0.09	0.58	xxxxx			ACI	1.01	0.01	xxxxx	xxxxx	xxxxx	xxxxx
		CNR	0.74	0.16	0.67	0.10	0.78	xxxxx			CNR	0.97	0.08	xxxxx	xxxxx	xxxxx	xxxxx
		FIB	0.57	0.11	0.58	0.07	0.66	xxxxx			FIB	0.89	0.02	xxxxx	xxxxx	xxxxx	xxxxx
		M&C	0.77	0.20	0.73	0.08	0.75	xxxxx			M&C	1.14	0.04	xxxxx	xxxxx	xxxxx	xxxxx
		C&T	0.88	0.17	0.85	0.07	0.65	xxxxx			C&T	1.00	0.03	xxxxx	xxxxx	xxxxx	xxxxx
		Average	0.75	0.16	0.72	0.09	0.67	xxxxx			Average	0.99	0.04	xxxxx	xxxxx	xxxxx	xxxxx
	ACI Model	K&N, P&M	0.84	0.12	0.90	0.06	0.60	xxxxx		ACI Model	K&N, P&M	1.04	0.01	xxxxx	xxxxx	xxxxx	xxxxx
		ACI	0.83	0.13	0.83	0.07	0.56	xxxxx			ACI	1.09	0.05	xxxxx	xxxxx	xxxxx	xxxxx
		CNR	0.82	0.14	0.79	0.08	0.78	xxxxx			CNR	1.07	0.02	xxxxx	xxxxx	xxxxx	xxxxx
		FIB	0.71	0.13	0.71	0.07	0.65	xxxxx			FIB	1.04	0.06	xxxxx	xxxxx	xxxxx	xxxxx
		M&C	0.82	0.16	0.85	0.08	0.75	xxxxx			M&C	1.13	0.06	xxxxx	xxxxx	xxxxx	xxxxx
		C&T	0.88	0.13	0.95	0.09	0.64	xxxxx			C&T	1.09	0.04	xxxxx	xxxxx	xxxxx	xxxxx
		Average	0.82	0.14	0.84	0.07	0.66	xxxxx			Average	1.08	0.04	xxxxx	xxxxx	xxxxx	xxxxx
	CNR Model	K&N, P&M	0.79	0.20	0.78	0.13	0.57	xxxxx		CNR Model	K&N, P&M	0.89	0.12	xxxxx	xxxxx	xxxxx	xxxxx
		ACI	0.75	0.15	0.68	0.11	0.54	xxxxx			ACI	1.01	0.01	xxxxx	xxxxx	xxxxx	xxxxx
		CNR	0.74	0.17	0.63	0.10	0.76	xxxxx			CNR	0.97	0.08	xxxxx	xxxxx	xxxxx	xxxxx
		FIB	0.56	0.13	0.54	0.07	0.61	xxxxx			FIB	0.89	0.01	xxxxx	xxxxx	xxxxx	xxxxx
		M&C	0.76	0.21	0.68	0.06	0.72	xxxxx			M&C	1.14	0.04	xxxxx	xxxxx	xxxxx	xxxxx
		C&T	0.87	0.17	0.85	0.07	0.61	xxxxx			C&T	1.00	0.03	xxxxx	xxxxx	xxxxx	xxxxx
		Average	0.75	0.17	0.69	0.09	0.64	xxxxx			Average	0.98	0.05	xxxxx	xxxxx	xxxxx	xxxxx

K&N, P&M [1] represents Khalifa and Nanni, Pellegrino and Modena, MC [2] represents Modifi and Challal. C&T [3] represents Chen and Teng. U* represents beam having U-jacketing with partially efficient anchorages.

Overall, the Colajanni et al. model slightly overestimated the shear strength for R-cross-sectional members while it underestimated it for T-cross-sectional members, but generally speaking, the underestimation in the case of the T members was less as compared to the overestimation of the R members.

The ACI model predicted the results very accurately in the case of both RU- and TU-cross-sectional members. Just a 1% overestimation was observed in the case of the RU cross sections while a 3% overestimation was observed in the case of the TU members, which reveals that this model is less sensitive to the contribution of flanges. More accurate results were obtained by the C&T R-factor model with CoV = 0.14 for R members and CoV = 0.19 for T members.

In the case of RU* there was less of a dispersion of the data (Avg value of the CoV = 0.08) while for TU* there was more of a dispersion of the data (CoV = 0.23). A 13% overestimation was observed in the case of RU* and a 6% underestimation for TU*.

In general, less accurate results were observed in the case of TU/F (Avg CoV= 0.33) due to the flange effect. Additionally, it was observed that the ACI model yielded better results in the case of the R member than the T-cross-sectional ones. When $\alpha = \beta$, the CNR model yielded the same results as the Colajanni et al. model when the same value of R was used, and it also performed well enough in the prediction of the average shear strength of RF and TU/F.

6.2. Conclusion Based on First Approach for DS 2 ($\alpha \neq \beta$)

The Colajanni et al. model, similarly to the CNR model, in the case of RU overestimated the shear strength by about 25%, but it yielded a better estimation of the average shear strength in the case of TU. Actually, the results referring to the T database for ($\alpha \neq \beta$) were very few, with only two specimens for the TU series, so the values of CoV were almost meaningless. Additionally, due to the lack of data on TU*, a comparison cannot be made.

In general, the ACI model was the performed best in the case of the rectangular beams, because it only overestimated the shear strength by about 18%, but it underestimated it by about 8% in the case of T members.

6.3. Conclusion Based on the 2nd Approach

A comparison of the effectiveness of each model in the strength assessment of specimens having the same cross-section shape but with different inclination angles stresses that both the Colajanni et al. and CNR models for RU and RU* overestimated more τ_{avg} in DS2 than in DS1, but with less of a dispersion of the data, while the average overestimation for RU* was larger in the CNR model than in the Colajanni one. This is due to the ability of the Colajanni et al. model to take into account the difference of steel and FRP reinforcement orientation, and to properly evaluate their contributions in determining the inclination of the concrete strut.

For TU, excellent results were obtained for both models in the case of DS2, with the best prediction of the average shear strength being only a 1% overestimation with less of a dispersion of the data (CoV 0.04) for both the Colajanni and CNR models. In the case of the ACI model, better results were obtained in the case of DS1 for RU and RU*, but in the case of TU, less accurate prediction but less of a dispersion was found in DS2.

The CNR model provided the same results in the case of RU for both DS1 and DS2. In the case of RU* in DS1, it provided better values of τ_{avg}, but more scattering of the data as compared to DS2. In the case of TU, excellent results were achieved in DS2.

7. Conclusions

The large values of the CoV (up to 0.35) that affect the assessment of the shear strength in a large database that includes specimens with different cross-section shapes and FRP reinforcement inclinations suggest that a deep analysis of the results can provide insights into the merits and demerits of the analyzed models, as well as into the effect of different characteristics of the strengthened specimens. The analysis shows that most of the models are ineffective, with the exception of the ACI model (just 1% overestimation in the case of RU cross sections while a 3% overestimation in the case of TU), because of their inability to take into account a reduced height of the fiber with respect to the total effective depth of the section, due to the presence of the top flange of the section represented by the slab. Similarly, none of the analyzed models can adequately take into account the effect of the presence of flanges in the T section, which experimental results have proved to be effective in increasing the strength of the FRP-strengthened beam.

Analyzing a database containing specimens with mechanical anchorages of the FRP, another source of uncertainty derives from the efficiency of the anchorages, which in many cases is not able to ensure that failure only occurs when the ultimate strain of the FRP is reached, i.e., by preventing FRP debonding failure.

Regarding the effect of FRP fiber inclination, the Colajanni et al. model is the only one among those based on the variable inclination of the action of the compressed concrete that is able to consistently take into account any different inclination of the FRP (β) and steel reinforcement (α). This characteristic makes it possible to mitigate the overestimation of the resistance (15% for R section and 1% for T section) that affects all the analysis models for $\alpha \neq \beta$. This circumstance is favored by the presence of the effectiveness factor of the steel reinforcement in the model, which takes into account the different orientation of the two reinforcements, and the consistent evaluation of the strength of the compressed concrete.

Author Contributions: Conceptualization, P.C.; Data curation, M.A.; Formal analysis, M.A. and P.C.; Investigation, S.P.; Methodology, P.C. and S.P.; Supervision, P.C.; Visualization, M.A.; Writing—original draft, M.A., P.C. and S.P.; Writing—review & editing, M.A., P.C. and S.P. All authors have read and agreed to the published version of the manuscript.

Funding: This research received no external funding.

Institutional Review Board Statement: Not applicable.

Informed Consent Statement: Not applicable.

Data Availability Statement: The data presented in this study are available on request from the corresponding author.

Conflicts of Interest: The authors declare no conflict of interest.

Abbreviations

Following Notations are used in this paper:

A_{fv}, A_{sw}	Area of steel stirrups
R	Reduction coefficient (ratio of effective average stress/strain in FRP sheet to its ultimate strength)
R	Rectangular beam
T	T beam
V_c	Shear resistance of concrete
V_s	Shear resistance of steel stirrups
V_f	Shear resistance provided by FRP
$V_{Rd,f}$	FRP contribution to the shear capacity
$V_{Rd,s}$	Steel contribution to the shear capacity
$V_{Rd,c}$	Steel contribution to the shear capacity
τ_{avg}	Average shear strength ($\tau_{avg} = V_{exp}/V_{the}$)
α, β	Angle of steel and FRP transverse reinforcement
a	Shear span
b_f, b_w	Web widths of FRP and concrete
d, d_{fv}	Effective depth of beam and FRP
f_c, f'_c	Characteristic compressive strength of concrete
E_f, E_{sw}	FRP and steel elastic modulus
f_{bd}	Design resistance of the adhesion between FRP and concrete
f_{ywd}	Design steel stirrup strength
f_{fed}	Effective design strength of the FRP shear reinforcement
f_{yt}	Characteristic yield strength of transverse reinforcement
f_{fe}	Effective stress in the FRP; stress level attained at section failure
f_{fu}	Design ultimate tensile strength of FRP
f_{ck}, f_{ctm}	Characteristic cylinder compressive and mean concrete tensile strength of concrete
f_{sy}, f_{yt}	Yielding stresses of longitudinal steel reinforcement and steel stirrups
h_w	Beam cross-section height
k_v, k_1, k_2	Bond-reduction coefficient and modification factors
L_{max}, L_e	Maximum and effective length
r	Reduction factor for steel stirrups
w_f	Spacing, thickness, and width of the FRP strip

s_f, t_f	Spacing and thickness of FRP strip
$\bar{s}_f$	Spacing of FRP strips measured perpendicular to FRP strip axis
s	Spacing of the steel stirrups
V, V_n	External, and nominal shear forces
v_{exp}, v_{the}	Experimental and theoretical nondimensional shear strengths, where $v_{exp} = (V_{exp}/(b_w\, 0.9d\, 0.5 f_c))$
z	Inner lever arm
ε_{syw}	Yield strain of steel stirrup
ε_{fe}	Effective FRP strain
ε_{fu}	Nominal FRP strain
$\varepsilon_{fe,s}$	Effective strain in the direction of transverse steel reinforcement
θ	Angle between member axis and concrete stress
λ	Maximum bond length (normalized)
$\tilde{\sigma}_c$	Stress of the web concrete (non-dimensional)
$\tilde{\sigma}_f$	Tensile stress of transverse FRP (non-dimensional)
$\tilde{\sigma}_s$	Stress in transverse reinforcement (non-dimensional)
φ	Angle between the FRP reinforcement direction and steel stirrups
ρ_f, ρ_s	Transverse geometrical ratio of fiber and steel reinforcement
$\sigma_{f,max}$	Maximum stress along the bond length
ψ	Fictitious angle of reinforcement incorporating FRP and transverse steel reinforcement
ψ_f	Reduction factor equal to 0.95 in case of wrapping scheme, 0.85 for the other schemes
ω_{fw}, ω_{sw}	Mechanical ratio of transverse FRP and stirrups reinforcement
ω_{fw}	$(2 b_f t_f f_{fu})/(b_w\, s_f \sin\beta\, f_c)$
ω_{sw}	$(A_v f_{yt})/(b_w\, s \sin\alpha\, f_c)$

Appendix A. Specimen Details and Experimental Results

Table A1. Details of specimens along with the results obtained after experimental tests.

	Specimen no.	f_c (MPa)	b_w (mm)	d (mm)	a/d	ρ_s (%)	f_{yt} (MPa)	E_{stw} (GPa)	t_f (mm)	β (°)	ρ_f (%)	f_{fu} (MPa)	E_f (GPa)	wrap U,C	v_{exp} (-)
Sato et al. (1997) [37]	No.2	35.7	150	240	2.5	0.42	387	183	0.11	90	0.15	3480	230	T, U	0.39
	No.3	35.3	150	240	2.5	0.42	387	183	0.11	90	0.15	3480	230	T, U/C	0.46
Deniaud & Cheng (2001) [38]	T6S4-C90	44.1	140	528	2.8	0.10	520	260	0.11	90	0.08	3400	230	T, U	0.19
	T6S4-G90	44.1	140	528	2.8	0.10	520	260	1.80	90	2.57	106	18	T, U	0.20
	T6S2-C90	44.1	140	528	2.8	0.20	520	260	0.11	90	0.08	3400	230	T, U	0.21
Deniaud & Cheng (2003) [39]	T4S4-G90	30.0	140	362	3.0	0.10	520	200	1.80	90	2.57	106	18	T, U	0.30
	T4S2-G90	30.3	140	362	3.0	0.20	520	200	1.80	90	2.57	106	18	T, U	0.33
	T4S2-C45	29.4	140	362	3.0	0.20	520	200	0.70	45	0.50	442	45	T, U	0.33
	T4S2-Tri	30.4	140	362	3.0	0.20	520	200	2.10	60	3.00	124	8	T, U	0.35
Bousselham & Chaallal (2006) [40]	SB-S1-0.5L	25.0	152	356	3.0	0.38	650	215	0.06	90	0.08	3100	243	T,U	0.46
	SB-S1-1L	25.0	152	356	3.0	0.38	650	215	0.11	90	0.14	3100	243	T, U	0.42
	SB-S1-2L	25.0	152	356	3.0	0.38	650	215	0.21	90	0.28	3100	243	T, U	0.44
Pellegrino & Modena (2006) [24]	A-U1-C-17	41.4	150	250	3.0	0.39	534	210	0.17	90	0.22	3450	230	R, U	0.34
	A-U1-C-20	41.4	150	250	3.0	0.34	534	210	0.17	90	0.22	3450	230	R, U	0.32
	A-U1-S-17	41.4	150	250	3.0	0.39	534	210	0.17	90	0.22	3450	230	R, U	0.35
	A-U1-S-20	41.4	150	250	3.0	0.34	534	210	0.17	90	0.22	3450	230	R, U	0.34
	A-U2-C-17	41.4	150	250	3.0	0.39	534	210	0.33	90	0.44	3450	230	R, U	0.35
	A-U2-C-20	41.4	150	250	3.0	0.34	534	210	0.33	90	0.44	3450	230	R, U	0.33
	A-U2-S-17	41.4	150	250	3.0	0.39	534	210	0.33	90	0.44	3450	230	R, U	0.31
	A-U2-S-20	41.4	150	250	3.0	0.34	534	210	0.33	90	0.44	3450	230	R, U	0.30
Leung et al. (2007) [41]	SB-U1	27.4	75	155	2.9	0.28	550	210	0.11	90	0.10	4200	235	R, U	0.45
	SB-F1	27.4	75	155	2.9	0.28	550	210	0.11	90	0.10	4200	235	R, C	0.46
	SB-F2	27.4	75	155	2.9	0.28	550	210	0.11	90	0.10	4200	235	R, C	0.46
	MB-U1	27.4	150	305	3.0	0.28	550	210	0.22	90	0.10	4200	235	R, U	0.27
	MB-U2	27.4	150	305	3.0	0.28	550	210	0.22	90	0.10	4200	235	R, U	0.28
	MB-F1	27.4	150	305	3.0	0.28	550	210	0.22	90	0.10	4200	235	R, C	0.34
	MB-F2	27.4	150	305	3.0	0.28	550	210	0.22	90	0.10	4200	235	R, C	0.44
	LB-U1	27.4	300	660	2.7	0.14	550	210	0.44	90	0.10	4200	235	R, U	0.23
	LB-U2	27.4	300	660	2.7	0.14	550	210	0.44	90	0.10	4200	235	R, U	0.23
	LB-F1	27.4	300	660	2.7	0.14	550	210	0.44	90	0.10	4200	235	R, C	0.36
	LB-F2	27.4	300	660	2.7	0.14	550	210	0.44	90	0.10	4200	235	R, C	0.36
Monti & Liotta (2007) [12]	UF90	11.0	250	410	3.5	0.10	500	210	0.22	90	0.18	2600	390	R, U	0.25
	US60	11.0	250	410	3.5	0.10	500	210	0.22	60	0.08	2600	390	R, U	0.22
	US45+	11.0	250	410	3.5	0.10	500	210	0.22	45	0.06	2600	390	R, U	0.25
	US45++	11.0	250	410	3.5	0.10	500	210	0.22	45	0.06	2600	390	R, U*	0.26
	UF45+ A	11.0	250	410	3.5	0.10	500	210	0.22	45	0.12	2600	390	R, U*	0.33
	UF45++ B	11.0	250	410	3.5	0.10	500	210	0.22	45	0.12	2600	390	R, U*	0.34
	UF45++ C	11.0	250	410	3.5	0.10	500	210	0.22	45	0.12	2600	390	R, U*	0.36
	US45+ D	11.0	250	410	3.5	0.10	500	210	0.22	45	0.09	2600	390	R, U*	0.32
	US45++ E	11.0	250	410	3.5	0.10	500	210	0.22	45	0.09	2600	390	R, U*	0.32
	US45++ F	11.0	250	410	3.5	0.10	500	210	0.22	45	0.09	2600	390	R, U*	0.29
	WS45+	11.0	250	410	3.5	0.10	500	210	0.22	45	0.06	2600	390	R, C	0.31
	USVA	10.6	250	400	3.5	0.10	500	200	0.22	45	0.09	3000	390	R, U	0.25
	USVA+	10.6	250	400	3.5	0.10	500	200	0.22	45	0.09	3000	390	R, U	0.28

Table A1. Cont.

	Specimen no.	f_c (MPa)	b_w (mm)	d (mm)	a/d	ρ_s (%)	f_{yt} (MPa)	E_{sw} (GPa)	t_f (mm)	β (°)	ρ_f (%)	f_{fu} (MPa)	E_f (GPa)	wrap U,C	v_{exp} (-)
Pellegrino & Modena (2008) [32]	B-U1-C-14	46.2	150	240	3.0	0.48	534	210	0.17	90	0.22	3450	230	R, U	0.34
	B-U2-C-14	46.2	150	240	3.0	0.48	534	210	0.33	90	0.44	3450	230	R, U	0.35
	B-U1-C-17	46.2	150	240	3.0	0.39	534	210	0.17	90	0.22	3450	230	R, U	0.32
	B-U2-C-17	46.2	150	240	3.0	0.39	534	210	0.33	90	0.44	3450	230	R, U	0.33
Grande et al. (2009) [33]	RS4Wa	21.0	250	411	3.4	0.10	476	210	0.19	90	0.15	2600	392	R, C	0.26
	RS3Wa	21.0	250	411	3.4	0.13	476	210	0.19	90	0.15	2600	392	R, C	0.34
	RS2Wa	21.0	250	411	3.4	0.20	476	210	0.19	90	0.15	2600	392	R, C	0.31
	RS4Ub	21.0	250	411	3.4	0.10	476	210	0.19	90	0.15	2600	392	R, U*	0.23
	RS3Ua	21.0	250	411	3.4	0.13	476	210	0.19	90	0.15	2600	392	R, U*	0.28
	RS2Ua	21.0	250	411	3.4	0.20	476	210	0.19	90	0.15	2600	392	R, U*	0.29
Belarbi et al. (2012) [16]	RC-8-S90-NA	20.7	457	831	3.3	0.15	276	200	0.22	90	0.06	3792	228	T, U	0.24
	RC-8-S90-DMA	23.8	457	831	3.3	0.15	276	200	0.22	90	0.06	3792	228	T, U*	0.15
	RC-12-S90-NA	28.9	457	831	3.3	0.10	276	200	0.22	90	0.06	3792	228	T, U	0.15
	RC-12-S90-DMA	30.5	457	831	3.3	0.10	276	200	0.22	90	0.06	3792	228	T, U*	0.18
	RC-12-S90-PC	19.2	457	831	3.3	0.10	276	200	0.22	90	0.06	3792	228	T, U*	0.29
	RC-12-S90-HS-PC	18.3	457	831	3.3	0.10	276	200	0.22	90	0.06	3792	228	T, U*	0.27
Panda et al.(2013) [42]	S300-1L-SZ-U-90	40.4	100	230	3.2	0.19	252	200	0.36	90	0.72	160	13	T, U	0.22
	S300-1L-SZ-UA-90	40.4	100	230	3.2	0.19	252	200	0.36	90	0.72	160	13	T, U	0.23
	S200-1L-SZ-U-90	42.1	100	230	3.2	0.28	252	200	0.36	90	0.72	160	13	T, U	0.22
	S200-1L-SZ-UA-90	42.1	100	230	3.2	0.28	252	200	0.36	90	0.72	160	13	T, U	0.23
Baggio et al. (2014) [43]	6-G-N	50.1	150	310	2.9	0.21	384	200	0.51	90	0.34	575	26	R, U	0.16
	7-PD-G-N	50.1	150	310	2.9	0.21	384	200	0.51	90	0.34	575	26	R, U	0.15
	8-PD-G-CA	50.1	150	310	2.9	0.21	384	200	0.51	90	0.34	575	26	R, U*	0.15
	9-PD-G-GA	50.1	150	310	2.9	0.21	384	200	0.51	90	0.34	575	26	R, U*	0.16
Colalillo & Sheikh (2014) [44]	S5-US	47.6	400	545	3.1	0.07	501	195	1.00	90	0.25	961	95	R, U*	0.11
	S5-UA	47.6	400	545	3.1	0.07	501	195	1.00	90	0.50	961	95	R, U	0.13
	S5-CS	47.6	400	545	3.1	0.07	501	195	1.00	90	0.25	961	95	R, C	0.16
	S2-US	47.5	400	545	3.1	0.14	501	195	1.00	90	0.25	961	95	R, U*	0.13
	S2-UA	47.5	400	545	3.1	0.14	501	195	1.00	90	0.50	961	95	R, U	0.15
Ozden et al. (2014) [45]	FBwoA-CFRP	12.4	120	339	3.8	0.14	249	200	0.13	90	0.05	4300	238	T, U	0.27
	FBwA-CFRP	12.4	120	339	3.8	0.14	249	200	0.13	90	0.05	4300	238	T, U/C	0.36
	PBwA-CFRP	12.4	120	339	3.8	0.14	249	200	0.13	90	0.05	4300	238	U, U/C	0.29
	FBwoA-GFRP	12.4	120	339	3.8	0.14	249	200	0.16	90	0.06	3400	73	T, U	0.27
	FBwA-GFRP	12.4	120	339	3.8	0.14	249	200	0.16	90	0.06	3400	73	T, U/C	0.34
	PBwA-GFRP	12.4	120	339	3.8	0.14	249	200	0.16	90	0.06	3400	73	T, U/C	0.34
	FBwoA-Hi-CFRP	12.4	120	339	3.8	0.14	249	200	0.14	90	0.05	2600	640	T, U	0.24
	FBwA-Hi-CFRP	12.4	120	339	3.8	0.14	249	200	0.14	90	0.05	2600	640	T, U/C	0.27
	PBw-Hi-C	12.4	120	339	3.8	0.14	249	200	0.14	90	0.05	2600	640	T, U/C	0.31
Mofidi & Chaallal (2014) [36]	WT-ST-50	31.0	152	350	3.0	0.38	540	206	0.11	90	0.07	3450	230	T, U	0.33
	WT-ST-70	31.0	152	350	3.0	0.38	540	206	0.11	90	0.10	3450	230	T, U	0.34
	WT-SH-100	31.0	152	350	3.0	0.38	540	206	0.11	90	0.14	3450	230	T, U	0.34
Mofidi et al. (2014) [46]	S1-LS-NE	33.7	152	350	3.0	0.38	650	205	2.00	90	0.60	1350	90	T, U	0.34
	S1-LS-PE	33.7	152	350	3.0	0.38	650	205	2.00	90	0.60	1350	90	T, U*	0.37
	S1-EB-NA	33.7	152	350	3.0	0.38	650	205	0.11	90	0.14	3450	230	T, U	0.36
El-Saikaly et al. (2015) [47]	S1-EB	28.0	152	350	3.0	0.25	580	200	0.38	90	0.50	894	65	T, U	0.32
	S1-LS	28.0	152	350	3.0	0.25	580	200	1.40	90	0.21	2250	120	T, U	0.30
	S1-LS-Rope	28.0	152	350	3.0	0.25	580	200	1.40	90	0.21	2250	120	T, U/C	0.38
	S3-EB	28.0	152	350	3.0	0.38	580	200	0.38	90	0.50	894	65	T, U	0.38
	S3-LS	28.0	152	350	3.0	0.38	580	200	1.40	90	0.21	2250	120	T, U	0.36
	S3-LS-Rope	28.0	152	350	3.0	0.38	580	200	1.40	90	0.21	2250	120	T, U/C	0.42
Qin et al. (2015) [48]	S00	29.6	125	295	3.1	0.29	542	210	1.00	90	1.60	986	96	T, U	0.37
Chen et al. (2016) [49]	S8-U	46.1	200	320	3.0	0.25	416	200	0.17	90	0.08	4361	226	T, U	0.23
	S8-UFA1	46.1	200	320	3.0	0.25	416	200	0.17	90	0.08	4361	226	T, U*	0.24
	S8-UFA2	46.1	200	320	3.0	0.25	416	200	0.17	90	0.08	4361	226	T, U*	0.28
Frederick et al. (2017) [50]	TB2	27.2	130	235	3.2	0.17	415	200	0.15	90	0.23	1400	119	T, U	0.37
	TB4	27.2	130	235	3.2	0.17	415	200	0.15	90	0.23	1400	119	T, U*	0.42
El-Saikaly et al. (2017) [51]	EBS-BL	28.0	152	350	3.0	0.25	580	200	0.38	90	0.50	894	65	T, U	0.37
	EBS-ER	28.0	152	350	3.0	0.25	580	200	0.38	90	0.50	894	65	T, U*	0.40
	EBL-RF	28.0	152	350	3.0	0.25	580	200	2.00	90	0.30	1350	90	T, U/C	0.4
	EBS-NA	28	152	350	3.0	0.25	580	200	0.38	90	0.5	894	65	T, U	0.32
	EBL-NA	28	152	350	3.0	0.25	580	200	2.00	90	0.3	1350	90	T, U	0.30
	EBL-RW	28	152	350	3.0	0.25	580	200	2.00	90	0.3	1350	90	T, U/C	0.38

Table A1. Cont.

	Specimen no.	f_c (MPa)	b_w (mm)	d (mm)	a/d	ρ_s (%)	f_{yt} (MPa)	E_{stw} (GPa)	t_f (mm)	β (°)	ρ_f (%)	f_{fu} (MPa)	E_f (GPa)	wrap U,C	v_{exp} (-)
Nguyen-Minh et al. (2018) [52]	P-A1-2.3-C	30.6	120	406	2.3	0.16	342	205	1.00	90	0.83	986	96	T, U	0.41
	P-A1-2.3-G	30.6	120	406	2.3	0.16	342	205	1.30	90	1.08	575	26	T, U	0.40
	P-A1-2.3-G-Cont.	30.6	120	406	2.3	0.16	342	205	1.30	90	2.17	575	26	T, U	0.43
	P-A1-2.3-C-Cont.	30.6	120	406	2.3	0.16	342	205	1.00	90	1.67	986	96	T, U	0.45
	P-A2-2.3-C	30.6	120	406	2.3	0.16	342	205	2.00	90	1.67	986	96	T, U	0.43
	P-B1-2.3-C	44.4	120	406	2.3	0.16	342	205	1.00	90	0.83	986	96	T, U	0.32
	P-B1-2.3-G	44.4	120	406	2.3	0.16	342	205	1.30	90	1.08	575	26	T, U	0.32
	P-B1-2.3-G-Cont.	44.4	120	406	2.3	0.16	342	205	1.30	90	2.17	575	26	T, U	0.34
	P-B1-2.3-C-Cont.	44.4	120	406	2.3	0.16	342	205	1.00	90	1.67	986	96	T, U	0.36
	P-B2-2.3-C	44.4	120	406	2.3	0.16	342	205	2.00	90	1.67	986	96	T, U	0.34
	P-C1-2.3-C	58.7	120	406	2.3	0.16	342	205	1.00	90	0.83	986	96	T, U	0.29
	P-C1-2.3-G	58.7	120	406	2.3	0.16	342	205	1.30	90	1.08	575	26	T, U	0.27
	P-C1-2.3-G-Cont.	58.7	120	406	2.3	0.16	342	205	1.30	90	2.17	575	26	T, U	0.31
	P-C1-2.3-C-Cont.	58.7	120	406	2.3	0.16	342	205	1.00	90	1.67	986	96	T, U	0.32
	P-C2-2.3-C	58.7	120	406	2.3	0.16	342	205	2.00	90	1.67	986	96	T, U	0.30
Oller et al. (2019) [20]	M1-a	42.8	200	493	3.0	0.12	646	200	0.17	90	0.04	3400	230	T, U	0.15
	M1-b	42.8	200	493	3.0	0.12	646	200	0.17	90	0.04	3400	230	T, U	0.15
	M1A	39.0	200	493	3.0	0.12	646	200	0.17	90	0.04	3400	230	T, U*	0.16
	M1B	38.5	200	493	3.0	0.12	646	200	0.17	90	0.04	3400	230	T, U*	0.17
	M2A	39.0	200	493	3.0	0.12	646	200	0.17	90	0.07	3400	230	T, U*	0.21
	M2B	38.5	200	493	3.0	0.12	646	200	0.17	90	0.07	3400	230	T, U*	0.21
	H1-a	44.4	200	493	3.0	0.12	646	200	0.17	90	0.04	3400	230	T, U	0.15
	H2-a	44.4	200	493	3.0	0.12	646	200	0.17	90	0.07	3400	230	T, U	0.19
	H2-b	49.7	200	493	3.0	0.12	646	200	0.17	90	0.07	3400	230	T, U	0.17
	H2A	44.7	200	493	3.0	0.12	646	200	0.17	90	0.07	3400	230	T, U*	0.19
	H2B	49.6	200	493	3.0	0.12	646	200	0.17	90	0.07	3400	230	T, U*	0.17
	H3A	44.7	200	493	3.0	0.12	646	200	0.17	90	0.17	3400	230	T, U*	0.23
	H3B	49.6	200	493	3.0	0.12	646	200	0.17	90	0.17	3400	230	T, U*	0.21
Alzate et al. (2013) [53]	U90S5-a(L)	37.0	250	420	3.5	0.11	500	200	0.29	90	0.14	4000	240	R, U	0.16
	U90S5-a(S)	37.0	250	420	3.5	0.11	500	200	0.29	90	0.14	4000	240	R, U	0.14
	U90S5-b(L)	28.0	250	420	3.5	0.11	500	200	0.29	90	0.14	4000	240	R, U	0.21
	U90S5-b(S)	28.0	250	420	3.5	0.11	500	200	0.29	90	0.14	4000	240	R, U	0.20
	U90C5-a(L)	24.5	250	420	3.5	0.11	500	200	0.29	90	0.23	4000	240	R, U	0.22
	U90C5-a(S)	24.5	250	420	3.5	0.11	500	200	0.29	90	0.23	4000	240	R, U	0.20
	U90C5-b(L)	22.6	250	420	3.5	0.11	500	200	0.29	90	0.23	4000	240	R, U	0.26
	U90C5-b(S)	22.6	250	420	3.5	0.11	500	200	0.29	90	0.23	4000	240	R, U	0.24
	U90S3-a(L)	20.5	250	420	3.5	0.11	500	200	0.17	90	0.08	3800	240	R, U	0.25
	U90S3-b(L)	20.5	250	420	3.5	0.11	500	200	0.17	90	0.08	3800	240	R, U	0.23
	U90S3-b(L)	22.6	250	420	3.5	0.11	500	200	0.17	90	0.08	3800	240	R, U	0.22
	U90S3-b(S)	22.6	250	420	3.5	0.11	500	200	0.17	90	0.08	3800	240	R, U	0.24
	U90S3-c(L)	28.0	250	420	3.5	0.11	500	200	0.17	90	0.08	3800	240	R, U	0.20
	U90S3-c(S)	28.0	250	420	3.5	0.11	500	200	0.17	90	0.08	3800	240	R, U	0.16
	U90C3-a(L)	30.2	250	420	3.5	0.11	500	200	0.17	90	0.13	3800	240	R, U	0.17
	U90C3-a(S)	30.2	250	420	3.5	0.11	500	200	0.17	90	0.13	3800	240	R, U	0.18
	U90C3-b(L)	30.2	250	420	3.5	0.11	500	200	0.17	90	0.13	3800	240	R, U	0.16
	U90C3-b(S)	30.2	250	420	3.5	0.11	500	200	0.17	90	0.13	3800	240	R, U	0.17
	U45S5(L)	30.7	250	420	3.5	0.11	500	200	0.29	45	0.14	4000	240	R, U	0.17
	U45S5(S)	30.7	250	420	3.5	0.11	500	200	0.29	45	0.14	4000	240	R, U	0.18

In the table, U* represents beam having U-jacketing with partially efficient anchorages.

Appendix B. Calculation Examples of the Colajanni et al. Model

With regard to the shear model proposed by Colajanni et al. [19], in this section three calculation examples will be carried out using a step-by-step procedure based on the equations reported in Section 2.3. For each of the three possible cases (namely $\cot\theta > 2.5$, $1 \leq \cot\theta \leq 2.5$, $\cot\theta < 1$) a calculation example is given below. With the exception of the third case (which is quite difficult to find in a real application since it would require an amount of fiber and/or stirrups not compatible with engineering applications), these examples are developed starting from one of the specimens listed in Appendix A, using the R factor based on the equations given in [7].

Case 1:

One of the specimens tested by Alzate et al. [53] is used. Beam U90S5-a(L) has a rectangular cross section with dimensions equal to 250×420 mm, and a length of 4300 mm. It has stirrups with a diameter of 8 mm arranged at a spacing of 380 mm, while it is retrofitted with a U-shaped scheme made with CFRP strips having a thickness of 0.29 mm, a width of 300 mm and a spacing of 500 mm, arranged at right angles with respect to the beam axis. Using the data reported in Appendix A, first of all the R factor is calculated according to the equations reported in Section 4 (i.e., R_5 and R_6, equal to 0.65 and 0.18, respectively). Then, starting from the R-factor value, the r factor is computed based on the procedure described in Section 3, equal to 0.92. After that, the inclination of the concrete strut is calculated via Equation (14), which provides a value of $\cot\theta > 2.5$; thus, due to the limitations of $\cot\theta$ values, $\cot\theta = 2.5$ is assumed. Consequently, assuming $\widetilde{\sigma}_f = \widetilde{\sigma}_s = 1$, the shear capacity can be calculated using Equation (9), which provides a value of 310 kN.

Case 2:

One of the specimens tested by Pellegrino and Modena [24] is used. Beam A-U1-C-17 has a rectangular cross section with dimensions equal to 150×300 mm, and a length of 4800 mm. It has stirrups with a diameter of 8 mm arranged at a spacing of 170 mm, while it is retrofitted with a U-shaped scheme made with continuous CFRP sheets having a thickness of 0.17 mm, arranged at right angles with respect to the beam axis. Using the data reported in Appendix A, first of all the R factor is calculated according to the equations reported in Section 4 (i.e., R_5 and R_6, equal to 0.50 and 0.23, respectively). Then, starting from the R-factor value, the r factor is computed based on the procedure described in Section 3, equal to 1. After that, the inclination of the concrete strut is calculated via

Equation (14), which provides a value of cot θ = 2.09. Therefore, assuming $\tilde{\sigma}_c = \tilde{\sigma}_f = \tilde{\sigma}_s = 1$, the shear capacity can be calculated using one of Equations (9)–(11), which provide a value of 272 kN.

Case 3:

As already stated before, none of the specimens analyzed in the table reported in Appendix A provides a cot θ value lower than 1. Therefore, to carry out the comparison, the specimen with the lowest cot θ is selected and then the spacing of FRP is properly modified to obtain a cot θ less than 1. To this aim, one of the specimens tested by El-Saikaly et al. [47] is used. Beam S3-LS-Rope has a T cross section with dimensions equal to 152 × 406 mm, with a flange width and thickness of 508 mm and 102 mm, respectively. It has stirrups with a diameter of 8 mm arranged at a spacing of 175 mm, while it is retrofitted with a U-shaped scheme made with CFRP strips having a thickness of 1.4 mm, a width of 20 mm and a spacing of 175 mm, arranged at right angles with respect to the beam axis. Anchorages made with carbon-fiber ropes are added to each strip to prevent debonding failure. In fact, the experimental test showed an FRP tensile failure, equivalent to a complete wrapping scheme. Using the data reported in Appendix A, first of all the R factor is calculated according to the equations reported in Section 4 (i.e., only R_5 because the section is considered fully wrapped thanks to the presence of the anchorages, and it is equal to 0.66). Then, starting from the R-factor value, the r factor is computed based on the procedure described in Section 3. After that, the inclination of the concrete strut is calculated via Equation (14), which provides a value of cot θ = 1.28. Therefore, the spacing of the CFRP is reduced to 110 mm, with which, again using Equation (14), a cot θ = 0.97 is obtained. Thus, cot θ = 1 is assumed, so Equation (10) is used considering $\tilde{\sigma}_f = 1$ and Equation (11) is used with $\tilde{\sigma}_s = -1$. The shear strength of the beam is the minimum one obtained through the above two equations, and it is equal to 335 kN.

References

1. Colajanni, P.; Pagnotta, S.; Recupero, A.; Spinella, N. Shear resistance analytical evaluation for RC beams with transverse re-inforcement with two different inclinations. *Mater. Struct.* **2020**, *53*, 18. [CrossRef]
2. Colajanni, P.; Recupero, A.; Spinella, N. Shear strength degradation due to flexural ductility demand in circular RC columns. *Bull. Earthq. Eng.* **2015**, *13*, 1795–1807. [CrossRef]
3. Koutas, L.N.; Tetta, Z.; Bournas, D.A.; Triantafillou, T.C. Strengthening of concrete structures with textile reinforced mortars: State-of-the-art review. *J. Compos. Constr.* **2019**, *23*, 03118001. [CrossRef]
4. Sudhakar, R.; Partheeban, P. Strengthening of RCC Column Using Glass Fibre Reinforced Polymer (GFRP). *Int. J. Appl. Eng. Res.* **2017**, *12*, 4478–4483.
5. Sas, G.; Täljsten, B.; Barros, J.; Lima, J.; Carolin, A. Are available models reliable for predicting the FRP contribution to the shear resistance of RC beams? *J. Compos. Construct.* **2009**, *13*, 514–534. [CrossRef]
6. Campione, G.; Colajanni, P.; La Mendola, L.; Spinella, N. Ductility of R.C. Members Externally Wrapped With Frp Sheets. *J. Compos. Constr.* **2007**, *11*, 279–290. [CrossRef]
7. Chen, J.F.; Teng, J.G. Shear capacity of FRP-strengthened RC beams: FRP debonding. *Constr. Build. Mater.* **2003**, *17*, 27–41. [CrossRef]
8. Kotynia, R.; Oller, E.; Marí, A.; Kaszubska, M. Efficiency of shear strengthening of RC beams with externally bonded FRP materials—State-of-the-art in the experimental tests. *Compos. Struct.* **2021**, *267*, 113891. [CrossRef]
9. Spinella, N.; Colajanni, P.; Recupero, A.; Tondolo, F. Ultimate shear of RC beams with corroded stirrups and strengthened with FRP. *Buildings* **2007**, *9*, 34. [CrossRef]
10. Ferreira, D.; Oller, E.; Marí, A.; Bairán, J. Numerical Analysis of Shear Critical RC Beams Strengthened in Shear with FRP Sheets. *J. Compos. Constr.* **2013**, *17*, 04013016. [CrossRef]
11. Colajanni, P.; Pagnotta, S. Influence of the effectiveness factors in assessing the shear capacity of RC beams strengthened with FRP. In Proceedings of the COMPDYN 2021, Athens, Greece, 27–30 June 2021.
12. Monti, G.; Liotta, M. Tests and design equations for FRP-strengthening in shear. *Construct. Build. Mater.* **2007**, *21*, 799–809. [CrossRef]
13. Bousselham, A.; Chaallal, O. Mechanisms of shear resistance of concrete beams strengthened in shear with externally bonded FRP. *J. Compos. Constr.* **2008**, *12*, 499–512. [CrossRef]
14. Cao, S.Y.; Chen, J.F.; Teng, J.G.; Hao, Z.; Chen, J. Debonding in RC Beams Shear Strengthened with Complete FRP Wraps. *J. Compos. Constr.* **2005**, *9*, 417–428. [CrossRef]
15. Oller, E.; Kotynia, R.; Marí, A. Assessment of the existing models to evaluate the shear strength contribution of externally bonded frp shear reinforcements. *Compos. Struct.* **2021**, *266*, 113641. [CrossRef]
16. Belarbi, A.; Bae, S.W.; Brancaccio, A. Behavior of full-scale RC T-beams strengthened in shear with externally bonded FRP sheets. *Constr. Build. Mater.* **2012**, *32*, 27–40. [CrossRef]
17. CAN/CSA (Canadian Standards Association). *Canadian Highway Bridge Design Code. S6-06*; CAN/CSA: Mississagua, ON, Canada, 2006.
18. ACI Committee 440. *ACI 440.2R-17; Guide for the Design and Construction of Externally Bonded FRP Systems for Strengthening Concrete Structures*. ACI Committee 440: Farmington Hills, MI, USA, 2017.
19. Colajanni, P.; Guarino, V.; Pagnotta, S. Shear capacity model with variable orientation of concrete stress field for RC beams strengthened by FRP with different inclinations. *J. Compos. Constr.* **2021**, *25*, 04021037. [CrossRef]
20. Oller, E.; Pujol, M.; Marí, A. Contribution of externally bonded FRP shear reinforcement to the shear strength of RC beams. *Compos. Part B Eng.* **2019**, *164*, 235–248. [CrossRef]

21. CNR-DT-200/R1; Istruzioni per la Progettazione, l'Esecuzione ed il Controllo di Interventi di Consolidamento Statico Mediante l'Utilizzo di Compositi Fibrorinforzati. CNR (Consiglio Nazionale delle Ricerche—National Research Council): Rome, Italy, 2013. (In Italian)
22. Khalifa, A.; Nanni, A. Improving shear capacity of existing RC T-section beams using CFRP composites. *Cem. Concr. Compos.* **2000**, *22*, 165–174. [CrossRef]
23. Khalifa, A.; Nanni, A. Rehabilitation of rectangular simply supported RC beams with shear deficiencies using CFRP composites. *Constr. Build. Mater.* **2002**, *16*, 135–146. [CrossRef]
24. Pellegrino, C.; Modena, C. Fiber-reinforced polymer shear strengthening of reinforced concrete beams: Experimental study and analytical modeling. *ACI Struct. J.* **2006**, *103*, 720–728. [CrossRef]
25. Chen, J.F.; Teng, J.G. Shear capacity of fiber-reinforced polymer-strengthened reinforced concrete beams: Fiber reinforced polymer rupture. *J. Struct. Eng.* **2003**, *129*, 615–625. [CrossRef]
26. ACI (American Concrete Institute). *ACI 318-14*; Building Code Requirements for Structural Concrete and Commentary. ACI: Farmington Hills, MI, USA, 2014.
27. CEN (European Committee for Standardization). *EN1992-1-1*; Design of Concrete Structures, Part 1.1: General Rules and Rules for Buildings. CEN: Brussels, Belgium, 2004.
28. *CNR-DT 215/2018*; Guide for the Design and Construction of Externally Bonded Fibre Reinforced Inorganic Matrix Systems for Strengthening Existing Structures. CNR (Advisory Committee on Technical Recommendations for Construction—National Research Council): Rome, Italy, 2020.
29. Colajanni, P.; La Mendola, L.; Mancini, G.; Recupero, A.; Spinella, N. Shear capacity in concrete beams reinforced by stirrups with two different inclinations. *Eng. Struct.* **2014**, *81*, 444–453. [CrossRef]
30. Colajanni, P.; La Mendola, L.; Monaco, A.; Pagnotta, S. Seismic Performance of Earthquake-Resilient RC Frames Made with HSTC Beams and Friction Damper Devices. *J. Earthq. Eng.* 2021; 1–27, in press. [CrossRef]
31. Ali, M.S.M.; Oehlers, D.J.; Seracino, R. Vertical shear interaction model between external FRP transverse plates and internal steel stirrups. *Eng. Struct.* **2006**, *28*, 381–389. [CrossRef]
32. Pellegrino, C.; Modena, C. An experimentally based analytical model for the shear capacity of FRP-strengthened reinforced concrete beams. *Mech. Compos. Mater.* **2008**, *44*, 231–244. [CrossRef]
33. Grande, E.; Imbimbo, M.; Rasulo, A. Effect of transverse steel on the response of RC beams strengthened in shear by FRP: Experimental study. *J. Compos. Constr.* **2009**, *13*, 405–414. [CrossRef]
34. Mofidi, A.; Chaallal, O. Shear strengthening of RC beams with EB FRP: Influencing factors and conceptual debonding model. *J. Compos. Constr.* **2011**, *15*, 62–74. [CrossRef]
35. FIB (Fédération Internationale du Béton—International Federation for Structural Concrete). *Externally Applied FRP Reinforcement for Concrete Structures*; Fib Bulletin 90; FIB: Lausanne, Switzerland, 2019.
36. Mofidi, A.; Chaallal, O. Tests and design provisions for reinforced-concrete beams strengthened in shear using FRP sheets and strips. *J. Concrete Struct. Mater.* **2014**, *8*, 117–128. [CrossRef]
37. Sato, Y.; Ueda, T.; Kakuta, Y.; Ono, S. Ultimate shear capacity of reinforced concrete beams with carbon fiber sheet. In Proceedings of the Third International Symposium of Non-Metallic(FRP)Reinforcement for Concrete Structures, Sapporo, Japan, 14–16 October 1997; Japan Concrete Institute: Tokyo, Japan, 1997; pp. 499–506.
38. Deniaud, C.; Cheng, J.J.R. Shear behavior of reinforced concrete T-beams with externally bonded fiber-reinforced polymer sheets. *ACI Struct. J.* **2001**, *98*, 386–394. [CrossRef]
39. Deniaud, C.; Cheng, J.J.R. Reinforced concrete T-beams strengthened in shear with fiber reinforced polymer sheets. *J. Compos. Constr.* **2003**, *7*, 302–310. [CrossRef]
40. Bousselham, A.; Chaallal, O. Behavior of reinforced concrete T-beams strengthened in shear with carbon fiber-reinforced polymer—An experimental study. *ACI Struct. J.* **2006**, *103*, 339–347. [CrossRef]
41. Leung, C.K.Y.; Chen, Z.; Lee, S.; Ng, M.; Xu, M.; Tang, J. Effect of size on the failure of geometrically similar concrete beams strengthened in shear with FRP strips. *J. Compos. Constr.* **2007**, *11*, 487–496. [CrossRef]
42. Panda, K.C.; Bhattacharyya, S.K.; Barai, S.V. Effect of transverse steel on the performance of RC T-beams strengthened in shear zone with GFRP sheet. *Constr. Build. Mater.* **2013**, *41*, 79–90. [CrossRef]
43. Baggio, D.; Soudki, K.; Noël, M. Strengthening of shear critical RC beams with various FRP systems. *Constr. Build. Mater.* **2014**, *66*, 634–644. [CrossRef]
44. Colalillo, M.A.; Sheikh, S.A. Behavior of shear-critical RC beams strengthened with FRP—experimentation. *ACI Struct. J.* **2014**, *111*, 1373–1384. [CrossRef]
45. Ozden, S.; Atalay, H.M.; Akpinar, E.; Erdogan, H.; Vulaş, Y.Z. Shear strengthening of reinforced concrete T-beams with fully or partially bonded fibre-reinforced polymer composites. *Struct. Concr.* **2014**, *15*, 229–239. [CrossRef]
46. Mofidi, A.; Thivierge, S.; Chaallal, O.; Shao, Y. Behavior of reinforced concrete beams strengthened in shear using L-shaped CFRP plates: Experimental investigation. *J. Compos. Constr.* **2014**, *18*, 04013033. [CrossRef]
47. El-Saikaly, G.; Godat, A.; Chaallal, O. New anchorage technique for FRP shear-strengthened RC T-beams using CFRP rope. *J. Compos. Constr.* **2015**, *19*, 04014064. [CrossRef]
48. Qin, S.; Dirar, S.; Yang, J.; Chan, A.H.C.; Lshafie, M. CFRP shear strengthening of reinforced-concrete T-beams with corroded shear links. *J. Compos. Constr.* **2015**, *19*, 04014081. [CrossRef]

49. Chen, G.M.; Zhang, Z.; Li, Y.L.; Li, X.Q.; Zhou, C.Y. T-section RC beams shear-strengthened with anchored CFRP U-strips. *Compos. Struct.* **2016**, *144*, 57–79. [CrossRef]
50. Frederick, F.F.R.; Sharma, U.K.; Gupta, V.K. Influence of end anchorage on shear strengthening of reinforced concrete beams using CFRP composites. *Curr. Sci.* **2017**, *112*, 973–981. [CrossRef]
51. El-Saikaly, G.; Chaallal, O.; Benmokrane, B. Comparison of anchorage systems for RC T-beams strengthened in shear with EB-CFRP. In Proceedings of the 6th Asia-Pacific Conference on FRP in Structures, Singapore, 19–21 July 2017; International Institute for FRP in Construction: Singapore, 2017; pp. 1–5.
52. Nguyen-Minh, L.; Vo-Le, D.; Tran-Thanh, D.; Pham, T.M.; Ho-Huu, C.; Rovnák, M. Shear capacity of unbonded post-tensioned concrete T-beams strengthened with CFRP and GFRP U-wraps. *Compos. Struct.* **2018**, *184*, 1011–1029. [CrossRef]
53. Alzate, A.; Arteaga, A.; De Diego, A.; Cisneros, D.; Perera, R. Shear strengthening of reinforced concrete members with CFRP sheets. *Mater. Constr.* **2013**, *63*, 251–265. [CrossRef]

Article

Seismic Assessment and Retrofitting of Existing Road Bridges: State of the Art Review

Dominik Skokandić *, Anđelko Vlašić, Marija Kušter Marić, Mladen Srbić and Ana Mandić Ivanković

Department for Structures, Faculty of Civil Engineering, University of Zagreb, 10 000 Zagreb, Croatia; andjelko.vlasic@grad.unizg.hr (A.V.); marija.kuster.maric@grad.unizg.hr (M.K.M.); mladen.srbic@grad.unizg.hr (M.S.); ana.mandic.ivankovic@grad.unizg.hr (A.M.I.)
* Correspondence: dominik.skokandic@grad.unizg.hr

Abstract: The load-carrying capacity assessment of existing road bridges, is a growing challenge for civil engineers worldwide due to the age and condition of these critical parts of the infrastructure network. The critical loading event for road bridges is the live load; however, in earthquake-prone areas bridges generally require an additional seismic evaluation and often retrofitting in order to meet more stringent design codes. This paper provides a review of state-of-the-art methods for the seismic assessment and retrofitting of existing road bridges which are not covered by current design codes (Eurocode). The implementation of these methods is presented through two case studies in Croatia. The first case study is an example of how seismic assessment and retrofitting proposals should be conducted during a regular inspection. On the other hand, the second case study bridge is an example of an urgent assessment and temporary retrofit after a catastrophic earthquake. Both bridges were built in the 1960s and are located on state highways; the first one is a reinforced concrete bridge constructed monolithically on V-shaped piers, while the second is an older composite girder bridge located in Sisak-Moslavina County. The bridge was severely damaged during recent earthquakes in the county, requiring urgent assessment and subsequent strengthening of the substructure to prevent its collapse.

Keywords: seismic assessment; retrofitting; existing road bridges; bridge assessment; Croatia earthquake; urgent strengthening

Citation: Skokandić, D.; Vlašić, A.; Kušter Marić, M.; Srbić, M.; Mandić Ivanković, A. Seismic Assessment and Retrofitting of Existing Road Bridges: State of the Art Review. *Materials* **2022**, *15*, 2523. https://doi.org/10.3390/ma15072523

Academic Editor: Andreas Lampropoulos

Received: 21 February 2022
Accepted: 25 March 2022
Published: 30 March 2022

Copyright: © 2022 by the authors. Licensee MDPI, Basel, Switzerland. This article is an open access article distributed under the terms and conditions of the Creative Commons Attribution (CC BY) license (https://creativecommons.org/licenses/by/4.0/).

1. Introduction

The civil infrastructure network in the United States and Western Europe was built primarily in the post-World War II era, most notably in the 1960s and 1970s, and has therefore reached the end its designed service life. In addition to the ageing process, the deterioration of materials along with increased traffic volume and weight contribute to the alarming condition of critical infrastructure. Due to their complexity and exposure, bridges and viaducts are often defined as critical parts of transport networks, and many were built according to old codes that did not impose such strict durability requirements as today [1]. Although the evaluation and maintenance of bridges has been the subject of numerous research projects and papers over the past two decades, public awareness of the safety of existing bridges has been heightened by the two recent catastrophic collapses. The first was the Morandi Bridge in Genoa [2], which collapsed during a storm, and the second was a pedestrian overpass in Miami [3] that collapsed during construction, both in 2018. In both cases, the collapse resulted in multiple fatalities and enormous property damage. These types of bridge failures do not occur often; however, because of the serious consequences they attract widespread attention and are often used as lessons for the future. One of the causes which can trigger the partial or complete collapse of deteriorated existing bridges is an extreme natural event, such as flooding, landslides, or earthquakes, as presented in [4].

Bridges in seismically active areas are sensitive to the effects of earthquakes, most notably displacements and vibrations due to ground acceleration. Therefore, they must be appropriately designed and constructed with a certain degree of ductility.

The current design standard for new bridges in the EU, the Eurocodes [5], provides an approach based on modal analysis and behavioral factors. This means that it is uneconomical to design bridges that provide a full elastic response to seismic actions. As a solution, the design method takes into account the reduced seismic loads (with a behavioral factor) while focusing on structural detailing to ensure ductile behavior of the bridge system [6]. While such an approach is justified for the design of new bridges, it is not cost-effective when used in the assessment of existing ones which were designed according to old codes. These codes did not include guidelines for ductile behavior or structural robustness, and older codes did not consider seismic actions at all. As the Eurocode does not cover the assessment of existing bridges, the purpose of this paper is to provide the state of the art with respect to the methods used by researchers and engineers worldwide.

In many cases, design codes and strategies for bridge maintenance have been modified and improved after a major earthquake. In California, for example, the Department of Transportation initiated a bridge retrofit program after the 1971 San Fernando earthquake, and after the 1989 Loma Prieta earthquake this program was made mandatory for all public bridges in the state of California [7]. In Japan, the catastrophic Great Hanshin Earthquake in 1995 resulted in the complete collapse of the 18-span highway bridge in Kobe. Due to the extreme cases of structural failure during the Kobe earthquake, the lessons learned have influenced design codes both in Japan and worldwide [8].

In Croatia, most of the medium and large span road bridges were built between the 1960s and 1980s using the design codes that took into account a proportion of the seismic load [9]. Therefore, they need to be re-evaluated and retrofitted due to their age, deterioration, the aggressive marine environment, increased traffic load, improper maintenance, etc. On the other hand, there are a number of short-span bridges on some local and less frequented state roads that were built before seismic loads were part of the design specifications. These bridges are in many cases supported by masonry abutments and/or piers, while the superstructure consists of either steel girders or reinforced concrete slabs. Several similar bridges are located in Sisak-Moslavina County (SMC), one of the largest counties in Croatia, located about 50 km southeast of Zagreb. Both the city of Zagreb and SMC were hit by strong earthquakes in 2020. In March 2020, a strong earthquake hit Zagreb and its surroundings, followed by numerous aftershocks. The epicenter was located about 7 km from the city center, the magnitude was M_L = 5.5, and the intensity according to the EMS−98 scale was VII [10]. The second earthquake occurred on 29 December, with an epicenter about 3 km from the town of Petrinja in Sisak-Moslavina County. The magnitude of the earthquake was M_L = 6.2, and it had an intensity between VIII and IX according to the EMS−98 scale [11]. Both earthquakes resulted in massive property damage and, unfortunately, fatalities. The total financial damage as estimated by the World Bank was around EUR 16.5 billion [12].

This paper is conceived as a state-of-the-art literature review focused on the methods for seismic assessment, analysis, and retrofitting of existing road bridges. Additionally, two Case Study bridges are presented as examples of the application of the described methods. Both bridges are located in Croatia, and they represent different approaches to seismic assessment, namely, as a part of regular maintenance in the first case and urgent repair following a significant earthquake in the second.

The first bridge is a reinforced concrete hinged strut frame bridge over four spans [13] located in the coastal part of Croatia. The second bridge is a two-span composite girder bridge located in Sisak Moslavina County, and was damaged during the 2020 earthquakes. Both bridges were built according to the old codes without taking into consideration the active seismic zones where they are located. The second bridge is one of eight bridges that were inspected after the December 2020 earthquake [14].

2. Seismic Assessment of Existing Road Bridges

2.1. Analysis Methods

The current EU seismic design standards, EN 1998, consist of six parts, the first of which addresses new buildings [15], the second of which addresses the seismic design of new bridges [5], and the third of which covers procedures and guidelines for the assessment and retrofit of existing buildings [16]. Parts four through six deal with special engineering structures such as silos, pipelines, foundations, towers, etc. The assessment of existing bridges is generally not covered by the current generation of Eurocodes, and EU Member States (and other countries around the world) use different approaches, as presented in [17]. During the assessment procedure for existing bridges, the greatest uncertainties are associated with the modelling of traffic loads, as these represent a dominant live load on the bridge structure [9,18]. However, seismic actions have a significant effect on bridges in seismic active areas, and can cause a sudden partial or complete failure if a capacity design methodology is not taken into account in the initial design [19].

Seismic evaluation is generally performed in a manner similar to the design of new bridges. It uses the numerical model of the existing bridge and site-specific seismic and soil data to define both the demand and the capacity of the bridge. This requires the use of one of the available seismic analysis methods listed in Table 1. Regardless of the choice of analysis method, the majority of existing bridges designed without consideration of seismic loads or with moderate seismic loads can be described as structures with inadequate lateral stiffness and ductility in their substructure elements, i.e., the piers and abutments.

In order to develop a numerical model of a bridge that does not deviate from realistic behavior, the first step in the assessment method is data collection. This includes the original design projects (if available in the archives), any previous general and special bridge inspection reports, and any records of previous strengthening or other work performed on the bridge [20]. For existing bridges, site-specific traffic load models can be extrapolated using Bridge Weigh-in-Motion methods along with additional structural information such as influence lines, dynamic properties, etc. [1]. Traffic measurements can provide modal properties (modal shapes, frequencies, etc.) of the bridge, as described in [21]. In addition to traffic and structural data, there are numerous other nondestructive (NDT) methods of collecting information on the properties of the built-in materials, the location and quantity of reinforcement, and numerous other bridge features [22]. These and similar methods are well-covered by a number of authors (e.g., [23–26]) and will not be described in detail, as they exceed the scope of this paper.

Table 1. List of the most commonly used methods for seismic analysis of existing road bridges.

Analysis Method	Type	Source
Response spectrum method		
Fundamental mode method	Linear analysis	[5,13,27]
Time series analysis		
Time history analysis	Non–linear analysis	[5,28]
Pushover analysis		[5,6,13,27,29–31]
Probabilistic and sampling methods	Non–linear analysis	[32–35]

The selection of an analysis method is based on a number of parameters, mainly the bridge type, as different structural systems are subjected to different failure modes. For example, masonry arch bridges and integral bridges are usually short-spanned, very massive, and have relatively squat piers. On the other hand, cable-stayed bridges, tall and long viaducts, and slender concrete arch bridges are much more sensitive to lateral forces and dynamic effects. In addition to the method of analysis, the performance of similar bridges under seismic loading depends on the number of piers, cross-section, type and amount of reinforcement, etc.

The disadvantage of linear modal analysis is that it does not consider the redistribution of forces that occurs after the development of the plastic joint(s). Therefore, it does not take into account the new corresponding failure modes, and consequently risks not identifying all the critical structural elements of the selected bridge [30]. On the other hand, a nonlinear analysis takes into account the formation of plastic joints and the corresponding seismic force which they dissipate through deformations. Such analyses are based on the rotational capacity of the structural elements, the M–φ curve, which exploits the nonlinear behavior of the material, leading to "hidden" reserves in the load-bearing capacity of the structural element and the corresponding bridge. On the other hand, this requires a high degree of knowledge about the behavior of the elements in terms of their rotational capacity [36]. For existing bridges, it is not always possible to provide sufficient information about the material properties and built-in reinforcement to develop an accurate M–φ curve. While NDT tools can be used to determine the material properties of the concrete [22], the amount and yield stress of the built-in reinforcement is often estimated from available documentation. The most common nonlinear analysis used in the assessment procedures for existing bridges is the static nonlinear analysis (pushover) method, as the dynamic method (cyclic loading-time history analysis) is time-consuming and requires data sets from realistic earthquakes.

Pushover analysis is performed by subjecting the numerical bridge model to a gradually increasing lateral force until the displacement reaches a predetermined threshold. It is convenient for existing structures where the predominant failure mode has been identified, because it does not consider higher modal shapes. Therefore, it can be used for most bridges for which the M–φ curve of structural elements is available or can be approximated.

Considering the large number of bridges in transportation networks, a number of authors have developed fragility curves for bridge stocks, which provide valuable data for bridge management and decision making in terms of prioritizing repairs and planning [37].

Fragility curves have proven to be a very convenient tool for evaluating typical bridges that make up the majority of the existing bridge stock in the EU and the US. For more complex bridge types, such as concrete arch bridges, cable-stayed bridges, and extremely long-span bridges in general, seismic evaluation again requires a more detailed analysis, as is always the case for landmark bridges. The most common bridge types on roads and motorways in the EU are reinforced (RC) or prestressed (PSC) concrete girder bridges with spans of up to 30 m, either simply supported or continuous over several spans, according to a survey conducted as part of the SERON research project [38]. Choi et al. [39] present an inventory of bridges in Central and Middle America with similar data. The most common superstructure material is reinforced concrete, followed by steel girders with concrete decking and PSC girders.

The seismic behavior of these bridges generally depends on the fragility of bearings, piers, abutments, and foundations, which should be analyzed using a nonlinear approach. The superstructure can generally be modeled with a linear elastic analysis, as its stiffness does not have a significant effect on the behavior of the bridge. In this analysis, it is assumed that the superstructure remains in the linear elastic region of the stress–strain curve during longitudinal seismic actions and transmits only the shear force to the substructure [39]. On the other hand, the piers and abutments should provide a significant measure of ductility and transversal stiffness in order to withstand seismic actions. Current design codes for new bridges [5] require high ductility levels for RC piers, achieved by confining the reinforcement. The ductile behavior in the compression zone of the cross-section should be ensured within the potential plastic hinge regions; at the same, time buckling of longitudinal reinforcement must be avoided, even after multiple cycles of seismic effects.

For existing bridges, which do not meet the listed detailing criteria, the seismic resilience (fragility curves) of RC piers should be developed based on the moment–curvature relationship, the M–φ curve. The M–φ curve for each cross-section can be developed analytically or numerically, if sufficient data is available, and subsequently used as an input in the nonlinear analysis. The procedure for the development of M–φ curves for existing RC piers with insufficient ductility is described in [6,36]. M–φ curves can be developed empirically using various scale models in laboratory experiments, as presented in [6,40].

2.2. Fragility Curves

Seismic fragility curves in general represent the ability of a structural system (e.g., a bridge) to withstand a seismic event. They provide a relationship between the capacity of the system (C) and demand (D), depending on the peak ground acceleration (PGA), in terms of the conditional probability of failure. Failure is defined as an event where the demand (D) exceeds the structural capacity (C) [41]. The fragility of the bridge can be expressed as

$$\text{Fragility} = P[D > C \mid PGA]. \tag{1}$$

Fragility curves can be developed empirically based on historical data on past earthquakes or analytically using one of the methods listed in Table 1. The latter approach is more common, because there is insufficient data to empirically produce reliable curves [37]. Fragility curves are, in general, developed for one of the bridge components, mainly for the substructure parts, e.g., piers, abutments, and bearings. For each element, demand and capacity parameters are defined as a part of the analytical process in terms of displacements, lateral forces, rotation, etc., along with defined thresholds. A number of authors have dealt with this issue in the last two decades, as presented in detail in [37]. For example, Choi et al. [39] consider four limit states for bridge piers in terms of the damage index (d_{pl}), which is defined as a ratio between the current horizontal displacement (e.g., of the top of the pier) and a displacement at the same point in a significant cross-section (e.g., the bottom of the pier). Theoretical fragility curves based on [39] are presented in Figure 1, each for one of the defined damage states.

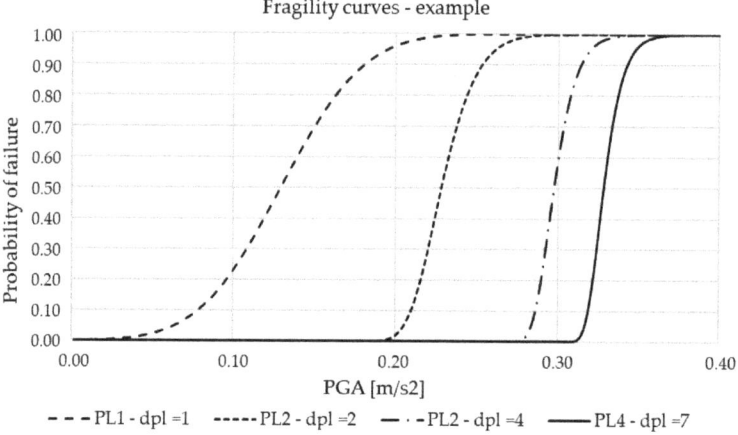

Figure 1. Example of fragility curves for bridge piers without adequate seismic capacity.

2.3. Literature Review: Seismic Assessment

A number of authors have studied the seismic behavior of bridges as part of infrastructure networks, several of them using fragility curves. Choi et al. [39] analyzed a large bridge inventory in the U.S.; based on this, they defined the four most common bridge types and developed a fragility curve for each of them. Zanini et al. [42] developed a fragility curve for the typical bridge at risk of corrosion in the northeastern Italian transportation network. In the second part of their study, they used the curves to determine the vulnerability of the whole network, taking into account the change in traffic flow when necessary. Stefanidou and Kappos [37] applied generic fragility curves to the larger landmark bridges and concluded that generic curves are applicable to simply supported bridges, while they tend to over- or underestimate the seismic resilience of more complex bridge structures. They provide the detailed state of the current state of practice in fragility curve development based on the selected element, limit state, analysis method, and seismic input [37].

While they do not use fragility curves, Sheikh and Légeron [43] have proposed a method for seismic assessment of bridges designed according to the Canadian Highway Bridge Design Code where the design method is force-based and strength is the measure of seismic performance. They developed a PGA-to-displacement ratio to replace the design code force-to-displacement ratio in order to take into account the seismic characteristic of the transportation network. In Italy, new "Guidelines for risk classification, safety assessment and structural health monitoring of existing structures" were recently adopted, including seismic assessment, and are presented on a Case Study PSC bridge by Cosenza and Losanno [44].

Di Sarno et al. [45] conducted a study on the seismic behavior of RC and Masonry bridges after the 2016 series of earthquakes in the Italian Apennine region. At that time, three large earthquakes with magnitudes $M_L = 6.1$, $M_L = 5.9$, and $M_L = 6.5$, each with several aftershocks, occurred over the period of August to October 2016. They concluded that the masonry bridges proved to be robust enough to withstand the seismic effects without global damage to the structure. The only bridges that were closed to traffic were those where the wing walls of the abutments were damaged, which was noted after the 2020 earthquakes in Croatia as well. Unreinforced masonry arch bridges are widely used in the European road and rail network and their assessment has been the subject of numerous studies, with several methods for their assessment developed over the last 25 years [31]. Due to their short spans and large overall mass and section dimensions, these bridges are generally very susceptible to the out-of-plane mechanism that can lead to complete bridge failure, as in the case of the Rio Claro Bridge in Chile [45]. Most commonly used analytical approaches to the assessment of masonry arch bridges are provided and compared in a study by Pelà, Aprile, and Benedetti [28]. Implementation of experimental testing for calibration of the numerical model of masonry arch bridges is presented in [26].

In addition to the displacement-based seismic capacity for existing RC piers, Huang and Huang [46] investigated other failure modes in their study of aging bridges in seismically active areas. They defined a practical framework for developing fragility curves for multiple failure modes (e.g., shear, bending shear, bending failure, etc.) based on the FE model, which was verified by six cyclic loading experiments. To simulate the deterioration mechanism (chloride infusion degradation), they used a probabilistic approach to reduce the cross-sectional area and yield strength of the corroded reinforcement. Alipour and Shafei [47] addressed the seismic resistance of deteriorated highway transportation networks. They used a nonlinear pushover analysis to estimate the decrease in the base shear capacity of the RC over a fifty-year lifespan due to corrosion of the reinforcement. The average decrease is presented in ten-year intervals, with the total decrease after fifty years being about 51%. Additional findings on the seismic resilience of concrete structures under corrosion can be found in [42,48]. Fragility curves are used to present the efficiency of the retrofitting strategy in a study by Padgett and DesRoches [41], in which they calculated the increase of the bridge's resilience due to the selected retrofitting measures and limit state.

In the Croatian coastal region, there are a number of long-span RC arch bridges built between the 1960s and 2000s that are exposed to a very aggressive environment due to the strong winds and chlorides from the sea. As the older bridges (e.g., the Pag Bridge and the Šibenik Bridge) were constructed with only moderate seismic loads, there was a need for a practical and efficient seismic load assessment method. A multi-step method was developed in [27], with the first step being linear multimodal analysis, followed by the limit state checks. If the bridge does not meet the requirements, the evaluation is performed again at the second stage, where nonlinear pushover analysis is used. As the spandrel piers of certain RC arch bridges have non-typical cross-sections, they were assessed using both numerical and experimental approaches with scale models subjected to vertical and horizontal forces in the laboratory [6].

Zelaschi, Monteiro, and Pinho [49] used a different approach to assess the seismic fragility of transport networks at the macro level. They do not use the classical approach of a generic fragility curve, rather applying a statistical tool to develop a large-scale seismic assessment of RC bridges in Italy. Using the known geometric and material properties of the bridges in the network, they use parametric characterization and Latin hyper-cube sampling to define a two-variable formula for the preliminary estimation of the period of vibration and seismic stress for selected bridges. While this approach presents an interesting method for large-scale assessment, the results are biased based on the uncertainty modelling, as is often the case with probabilistic methods.

3. Seismic Retrofitting of Existing Bridges

After successfully assessing and evaluating the existing bridge at risk of seismic activity, the optimal retrofit strategy must be selected, a process that differs significantly from the design process for a new bridge. Therefore, the selection of a strategy must be made with the aim of minimizing construction work and traffic disruption on the bridge, keeping the overall cost in reasonable proportion to the value of the bridge, and providing the retrofitted bridge with adequate seismic resilience [50]. The cost–benefit analysis of bridge assessment, maintenance, and repair is a topic that has been studied by a number of authors, and will not be further developed in this paper as it is beyond our scope. For more information on estimating the value of bridges as part of the transportation network, see, e.g., [51]; for the cost optimization algorithm for bridge repairs, see, e.g., [1].

After an extensive literature review of the topic of retrofitting methods was conducted, the majority of the papers focus either on a certain retrofitting method, e.g., jacketing [52], or a certain bridge type, e.g., arch bridges [50]. Therefore, state-of-the-art review presented here includes all available seismic retrofitting methods for road bridges. The list, along with corresponding literature sources, is summarized in Table 2 based on the structural element the method refers to. The most vulnerable structural elements of the bridge and corresponding retrofitting methods are described in detail in Sections 3.1–3.3, with practical examples where available.

Table 2. List of common methods for seismic retrofitting of existing road bridges.

Retrofit Method	Bridge Element	Bridge Type	Source
Steel jacketing	RC piers	Any bridge with RC piers	
Concrete/mortar jacketing			[52–55]
CFRP jacketing			[52,54–57]
ECC jacketing			[58,59]
AFRP jacketing			[52,60]
FRCM jacketing			[57,61]
GFRP jacketing			[52]
UHPFRC jacketing/repair			[62–64]
Bracing or infill walls between piers in the transverse direction			[65]
External prestressing with unbonded tendons	Superstructure	girder bridge, cable stayed bridge, slab bridge, box-girder bridge,	[13,66,67]
Span restrainers		any bridge with sliding bearings	[53,65]
Reinforced concrete jacking	Cap beams/ RC joints	Any bridge with RC piers/cap beams	[53,65,68–70]
Transverse external prestressing			[66,67]
Seat extenders	Cap beams/Abutments		[53,65]
Seismic isolation	Bearings	All bridges	[71–79]
Foundation cap confinement			[65]
Restrainers			[65,80]
Bumper blocks			[4,53]
Dampers			[13,75,81–83]
Replacement			[76,84–86]

Table 2. Cont.

Retrofit Method	Bridge Element	Bridge Type	Source
Spandrel wall strengthening	Spandrel walls	Masonry bridges	[50,87]
Abutment wing walls stabilization	Abutment	Any bridge with massive abutments	[53]

3.1. Bridge Columns/Piers

The most common technique for seismically strengthening RC columns is to physically increase their cross-section and increase their ductility by adding encasement jackets made of new reinforced concrete, steel, or high-performance materials. Frequently used materials are Carbon Fiber Reinforced Polymer (CFRP), Aramid Fiber Reinforced Polymer (AFRP), Glass Fiber Reinforced Polymer (GFRP), Fiber Reinforced Cementitious Matrix (FCRM), Engineered Cementitious Composite (ECC), Ultra-High-Performance Fiber-Reinforced Concrete (UHPRC), etc.; a detailed review of jacketing techniques and the materials used can be found in [52].

The principle of this technique is quite simple; the objective is to increase the ductility and shear capacity of the existing column either by adding new reinforcement and a new layer of concrete or by using prefabricated jackets connected to the existing column by grouting. However, it is important to consider the increase in mass and stiffness of the retrofitted section, as their increase leads to shorter periods and increased base shear, thus increasing the seismic demands on the selected column and structural system. Traditionally, concrete jacketing is the oldest and most common method. A new layer of reinforcement is placed around the existing section over part or all of the length of the RC column and the concrete is poured using traditional formwork. The connection between the old and new sections is achieved by anchoring the new reinforcement in the existing concrete or by using high-strength bolts. An example of concrete encasement can be found in Figure 2.

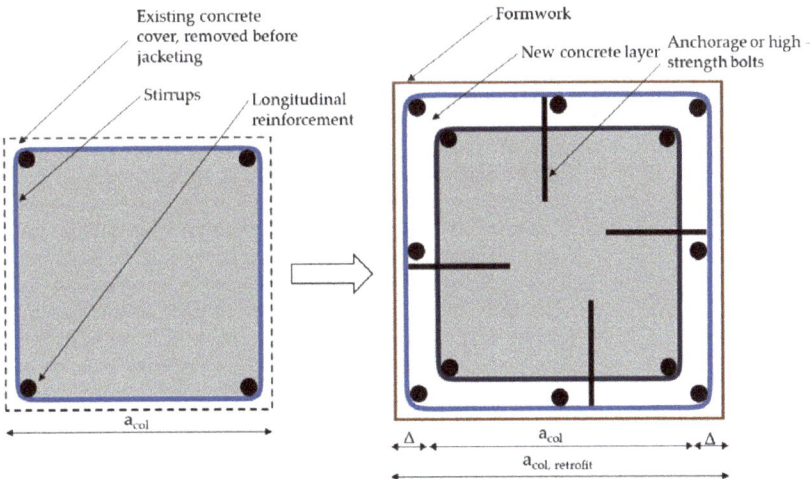

Figure 2. An example of reinforced concrete jacketing of existing columns.

The obvious disadvantages of RC jacketing are its cost and time inefficiency due to formwork installation, especially for taller piers. Moreover, the increase in ductility depends only on the additional longitudinal and transverse reinforcement (stirrups), as concrete is a brittle material. Another disadvantage of this technique is the relatively large increase in cross-section compared to other methods, which can be crucial in existing bridges where the passage under the bridge for crossing roads, railways, etc. is restricted. Taking into account the typical cross-sections of bridge piers and the diameter of the reinforcement used, the thickness of the cross-section with the RC sheathing increases by about 0.5 m. On the other hand, steel or high-performance material sheathing can increase the ductility and resilience of the cross-section by only a few centimeters [54]. Steel jacketing is most effective on circular columns where the sheathing is prefabricated in two halves that are positioned and welded around the column. The gap between the steel plate and the existing concrete is filled with grout to ensure the bond. In this way, a new composite cross-section is created that features increased shear strength, ductility, and buckling resistance [65]. The disadvantage of steel jacketing is its lack of cost-effectiveness due to more complex procedures and corrosion protection. Moreover, similar to RC jacketing, it increases the stiffness of the cross-section, which changes the seismic demands on the pier/column [52]. There are some examples of the combined use of RC and steel jacketing, where the steel plates are used as formwork. This method was applied to the columns of the Pag Bridge in Croatia, as shown in Figure 3. The Pag Bridge, built in the late 1960s, is a concrete arch bridge with very slender columns in a very aggressive maritime environment. During major reconstruction in the late 1990s, these columns were strengthened by adding a new layer of reinforcement and concrete encased in a steel plate with a thickness of 12 mm [88].

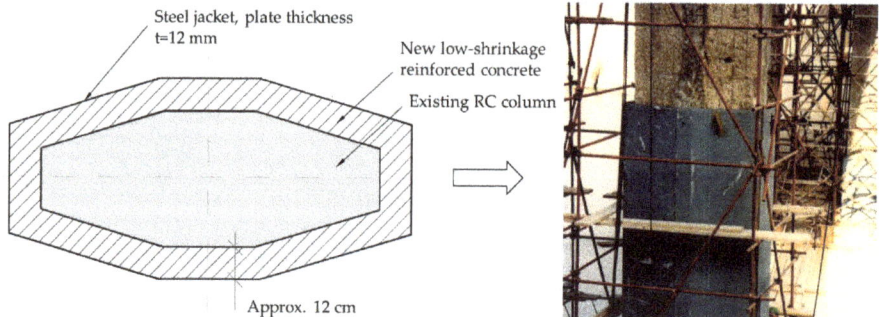

Figure 3. Seismic retrofitting of RC columns on Pag Bridge in Croatia showing the combination of RC and steel jacketing.

In addition to traditional RC and steel jacketing, high-performance materials based on fiber-reinforced polymers (FRP) have become an increasingly popular method for retrofitting existing RC columns over the past three decades. The obvious advantages of these materials are speed and simplicity of installation, a high strength-to-weight ratio, and minimal increase in the cross-section of the structural element. Therefore, columns retrofitted with FRP materials do not have a significant increase in stiffness or mass, and the aesthetics of the bridge remain visually unchanged. Furthermore, these materials are environmentally friendly when compared to traditional reinforced concrete or steel. On the other hand, the efficiency of these materials is lower due to premature bonding; their utilization is only 30–35% [52]. Bonding is achieved externally using epoxy resins.

As shown in Table 2, there are a number of different FRP composites used for retrofitting RC columns, each with certain advantages and disadvantages. In addition to those listed, several composites are in various research phases and have not yet been used in practice. A comparison of different jacketing methods can be found in the detailed review by Raza et al. [52], which describes all common methods along with their effects on the structural element, cost, and other characteristics. Recapitulation of the three jacketing methods is provided in Table 3, based on [52].

Table 3. Comparison of the main characteristics of the described jacketing methods.

Method	Effect on the Structural Element			Cost
	Strength	Ductility	Stiffness	
RC jacketing	Increase	Increase	Increase	Very high
Steel jacketing	Significant increase	Significant increase	Increase	High
FRP jacketing	Increase	Significant increase	No effect	Moderate

3.2. Cap Beams/Concrete Joints

Many road bridges are supported on RC bents consisting of two or more columns connected with a transverse cap beam, as shown in Figure 4. The concrete joint between the top of the column and cap beam is subjected to both shear and flexural stresses during cyclic seismic loading. In older bridges, this joint is often designed without sufficient anchorage length in the longitudinal direction and sufficient reinforcement in the transverse direction, and can therefore be designated as a critical point because of the possibility of brittle failure. As described in Chapter 2, the approach to bridge design is that the superstructure should remain in the elastic region to force plastic hinges in the columns. This is achieved by increasing the stiffness and strength of the cap beam and the concrete joint between the beam and the column [65]. The retrofitting technique is similar to RC jacketing of columns, as the new concrete layer is poured around the joint along with sufficient reinforcement to be anchored in both the beam and a column; an example is shown in Figure 4.

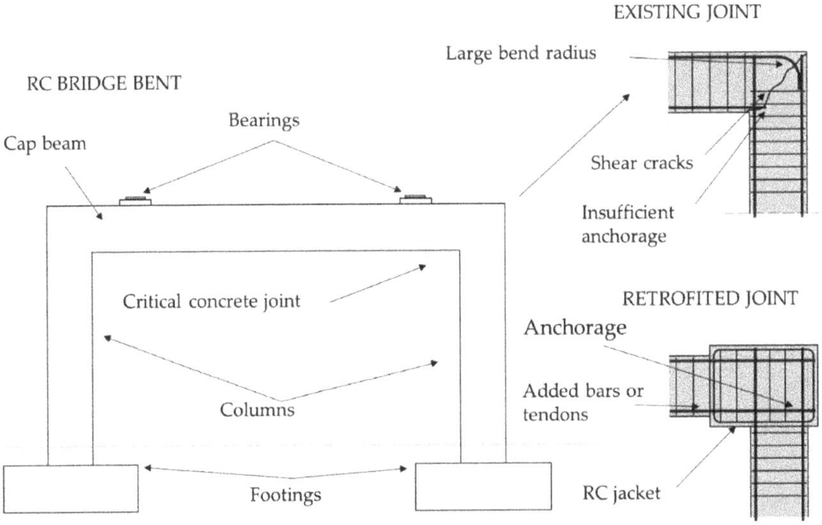

Figure 4. Seismic retrofitting of cap beams and concrete joints in RC bridge bents.

RC bents for new bridges are not generally designed as shown in Figure 4 because the cap beam is usually extended beyond the columns, which provides more space for the reinforcement of the critical joint section. Therefore, while on some older bridges the cap beams are retrofitted similar to Figure 4, the new beam section is extended beyond the columns, creating a cantilever. This method is called a seat extender [53].

The lateral stiffness of the cap beam can be increased by external prestressing, a strength enhancement technique normally used for flexural strengthening of the superstructure [89]. An alternative approach to increasing the lateral stiffness of RC bents is to provide X bracing between the columns or to construct infill walls connected to both the columns and the cap beams. These retrofit methods are rarely used today because they are costly and affect both mass and stiffness; however, examples can be found among older bridges in the USA [53,65].

3.3. Seismic Isolation/Damping

Seismic isolation, often referred to as the most effective and successful structural earthquake protection measure, is now commonly used for all types of structures in seismically active areas. The principle is very simple, aiming to "decouple" the ground motion and vibration of the structure and thereby reduce the lateral forces on the structure. The concept of seismic isolation is several centuries old, and is not described in this paper; it can be found in a detailed report by Makris [72].

In the design and the assessment of road bridges, seismic isolator bearings (SIB) are placed between the superstructure and the substructure, and can be positioned under the foundations of the piers as well, though this is less common. The principle is to reduce the seismic response of the bridge by "decoupling" the superstructure and substructure, thus reducing the displacement of the superstructure and consequently the lateral forces in the piers and abutments. Both experimental and analytical studies have proven that there is a substantial decrease in the seismic response of the isolated and non-isolated bridge [76,79]. The disadvantage of isolation is that while it reduces the stiffness and corresponding lateral forces, it increases the natural period and total displacements of the structure. To account for this, all SIBs are equipped with built-in or external energy dissipators called dampers that reduce the Eigenperiod by about 30% [78].

The most common bridges where seismic isolation has been used for retrofitting are those with massively heavy superstructures supported on a relatively slender substructure (piers and abutments). Prior to the introduction of design codes with integrated seismic loads, most bridges were supported on unreinforced elastomeric bearings used primarily to transfer extensive vertical gravitational and traffic loads and horizontal loads due to thermal expansion and creep. Therefore, in the event of an earthquake there was a risk of brittle failure of piers due to extensive lateral forces, especially for irregular continuous bridges where the piers have a significant height difference. Application of seismic isolation in the USA was initiated after the 1971 San Fernando earthquake, and was widely accepted in the practice with new Caltrans guidelines in the aftermath of the Loma Prieta earthquake in 1989 [71].

SIB can be divided into two main groups, those based on elastomers (rubber-based) and those based on friction (sliding-based), depending on the principle of energy dissipation. The elastomer-based bearings are more commonly used for simply supported bridges, while plain bearings (also known as pot bearings) are used for longer continuous bridges spanning multiple spans. A schematic of each type is shown in Figure 5.

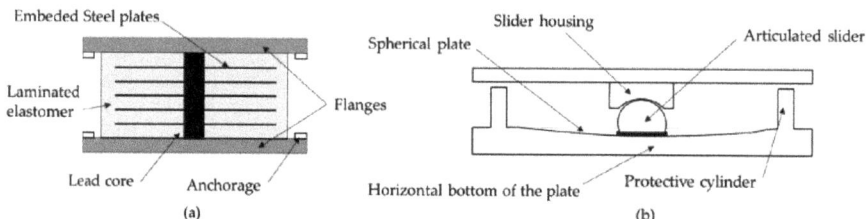

Figure 5. Isolation bearing schematics (**a**) elastomer (rubber)-based; (**b**) friction (sliding)-based.

The selection of bearings for seismic isolation in bridge retrofitting depends on numerous bridge characteristics, such as the structural system, number, height of piers, etc. The dimensions of the bearings depend on the required vertical load-carrying capacity (bearing dimensions) and the allowable total displacement (bearing height). A number of authors have performed both numerical and experimental analyses of commonly used

SIBs, e.g., [75,76,78,79,90]. Compared to the previously-listed retrofit techniques, seismic isolation is the most effective in terms of cost–benefit analysis, although replacement of the bearings requires complete closure of the bridge. In addition, many older bridges do not have columns designed for the installation of a hydraulic jack to elevate the superstructure during replacement. Therefore, the piers or cap beams on these bridges must be reinforced prior to seismic isolation.

An alternative measure to mitigate lateral forces on longer bridges is the installation of dampers, which are referred to as "passive energy dissipation devices". However, dampers are mainly used in the construction of new landmark bridges or to control the vibration of railway bridges carrying high-speed trains. They are rarely used for retrofitting typical road bridges (see e.g., [13,81–83]). The first Case Study is an example of a seismic retrofit using dampers for restriction of longitudinal displacement of the superstructure.

Seismic restrainers and bumper blocks can be considered as a retrofit technique for RC bridges, where severe displacement of the superstructure can cause the girders to slip off the bearings. Bumper blocks are installed in most new bridges in seismic areas as a protective measure and to increase structural robustness [4]. Restraint systems are a protective measure that prevents the superstructure from slipping off its supports, and are one of the most cost-effective retrofit techniques. Cable restraints were used extensively on California bridges after the San Fernando earthquake [65]. Restrainers are suitable for bridges with multiple simply supported spans because they "couple" the spans together in both the longitudinal and transverse directions without affecting the structural system.

There are a number of other methods that are used today for existing bridges, more commonly for special bridges (masonry, cable-stayed bridges, listed bridges, etc.) which are beyond the scope of this paper. Several of them are listed in Table 2, together with relevant sources offering a more detailed review.

4. First Case Study Bridge

4.1. General Bridge Description

The first Case Study [13] bridge is a reinforced concrete bridge constructed monolithically in the 1960s. The bridge is comprised of five spans with dimensions of 19.0 + 4.0 + 27.0 + 4.0 + 19.0 m, and follows a complex road axis that features horizontal and vertical curves. The superstructure is comprised of a voided RC slab supported on V-shaped piers, visible in Figure 6; a cross-section of the slab is presented in Figure 7.

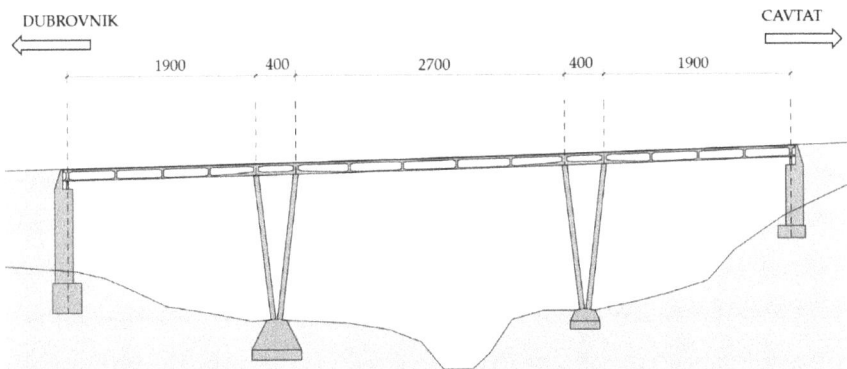

Figure 6. Longitudinal section of the first Case Study bridge (units in cm).

The connection between the superstructure and the abutments is considered hinged, as the slab is supported on "Pendl" bearings. However, the connection of the slab and V-shaped piers (four in total) is hinged only in the longitudinal direction; in the transversal direction it is rigid and the two piers and deck form a rigid frame. The resulting statical system of the bridge is called a hinged strut frame.

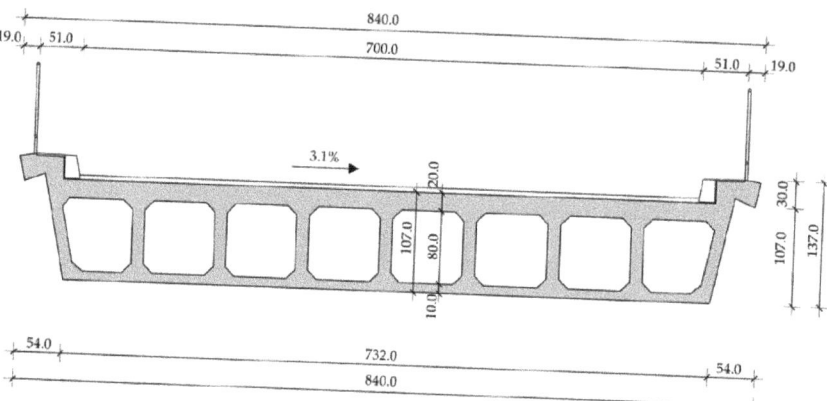

Figure 7. Cross-section (in the middle of the span) of the first Case Study bridge (units in cm).

The bridge was designed according to 1960s codes prior to the Skopje earthquake in 1963, after which the design codes were updated; thus, seismic loads were not taken into account. A review of the available documentation shows that only dead loads, traffic loads, temperature variations, and concrete shrinkage were considered.

4.2. Assessment Procedure

The height of the voided slab is 110 cm, the concrete class is C20/25, and the reinforcement is smooth with a yield strength of 220/360 N/mm². In the middle of the span, the bottom flange has a thickness of 10 cm and is reinforced with three layers of Ø20 mm plain bars; above the supports, the flange is 20 cm thick and is reinforced with both Ø10 and Ø20 mm bars. The top flange has a constant thickness of 20 cm and is reinforced with both Ø12 and Ø16 mm bars in the middle of the span and Ø20 mm bars above the supports. The webs have a constant thickness of 14 cm and are reinforced with Ø12 bars.

The cross-section of V-shaped piers has a constant depth of 50 cm with variable width, ranging from 100 cm at the bottom to 200 cm at the top of the piers. The longitudinal reinforcement is comprised of a total of 30 Ø20 bars; the stirrups are positioned with variable spacing ranging from 7 to 30 cm. Additional stirrups are placed at the same spacing for the confinement of the longitudinal bars. All stirrups in both superstructure and piers have a diameter of 8 mm.

The bridge was assessed for both modern-day traffic loads and for seismic effects, as it is located in an earthquake-prone area in the southern coastal part of Croatia where the peak ground acceleration is 0.290 g for a return period of 475 years and the PGA is 0.146 g for a 95-year return period. As the traffic load assessment exceeds the scope of this paper, only seismic assessment and retrofitting proposals are presented. The numerical model of the bridge was developed using software for structural analysis and beam elements for both superstructure and piers. All bearings were modelled as springs with variable longitudinal and transverse stiffness to simulate the realistic behavior of the bridge. In accordance with EN 1998-3 [16], the cracked state of concrete was taken into account by a reduction in the structural stiffness. To simulate this reduction, the concrete modulus of elasticity in the numerical model was reduced to 50% of its value. The numerical model is presented in Figure 8.

As the connection between the piers and superstructure is hinged in the longitudinal direction, the longitudinal resulting force is transferred to abutments. The seismic analysis was conducted both using linear modal analysis and nonlinear static pushover analysis, using the multi-level method developed by Franteović et al. [27] in accordance with current design codes. In the first step of the analysis, the target displacement for both the longitudinal and transverse direction was obtained using linear modal analysis. A total of fifty modal shapes were calculated; the results show that the first dominant mode is shaped as a translation in the longitudinal direction, while the second (with a smaller period) is a translation in the transverse direction. The first modal shape is presented in Figure 8, while the resulting modal parameters (periods, target displacements, and response spectrum values for obtained periods) for the first two modal shapes are provided in Table 4.

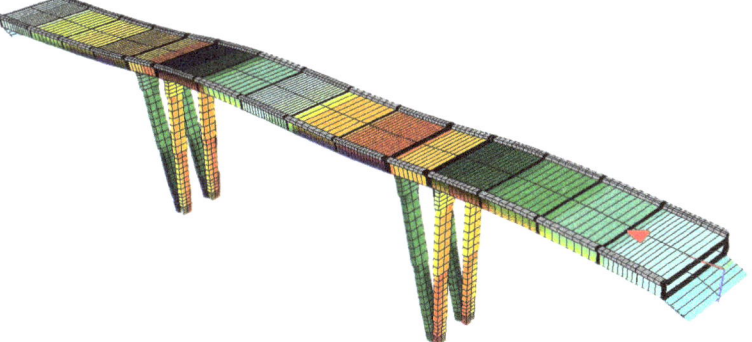

Figure 8. Numerical model of the Case Study bridge, first modal shape (longitudinal).

Table 4. Modal parameters for first Case Study bridge, obtained using linear analysis.

Modal Shape	Direction	Period	Spectral Response	Target Displacement
1	Longitudinal (x)	$T_x = 0.94$ s	$S(T_x) = 0.305$ g	$d_{E,x} = 76$ mm
2	Transverse (y)	$T_y = 0.36$ s	$S(T_y) = 0.706$ g	$d_{E,y} = 35$ mm

The target displacement for the first modal shape (Table 1) is 76 mm; while the existing expansion joint was designed to allow a maximum longitudinal displacement of 70 mm, when exceeded this would cause a collision between the superstructure and abutment wall. Therefore, the Case Study bridge was reassessed using the static nonlinear pushover analysis. A total of four cases were calculated with two types of load distributions for each direction, as presented in Figure 9 for first modal shape.

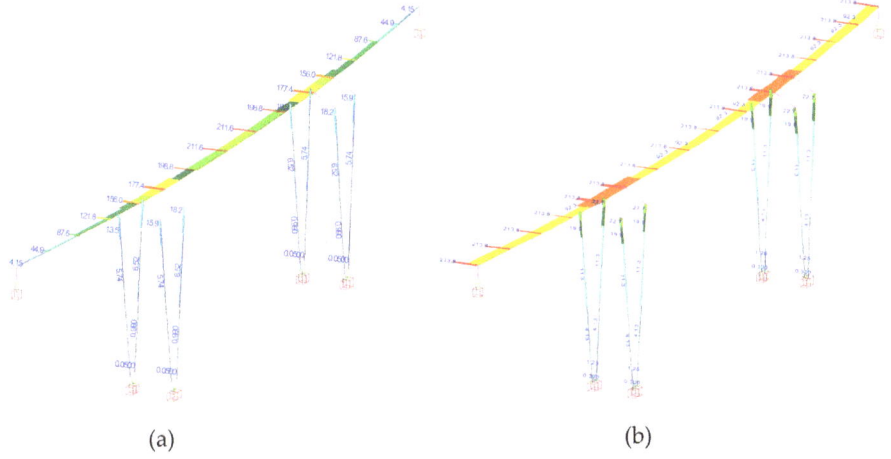

(a) (b)

Figure 9. (a) horizontal load distribution along the superstructure; (b) horizontal load in proportion with dominant modal shape.

Pushover analysis results are presented in the form of load–displacement curves for each direction in Figure 10; the less favorable of two load distribution cases are presented. Two spectral responses are provided for each modal shape; the first ($S_{E,dx}$ and $S_{E,dy}$) shows the seismic load in which the target displacements (d_{Ex} and d_{Ey}) are reached, while the second ($S_{E,T1,x}$ and $S_{E,T1,y}$) shows the maximum seismic load based on linear analysis along with the corresponding maximum displacements ($d_{x,T1}$ and $d_{y,T1}$).

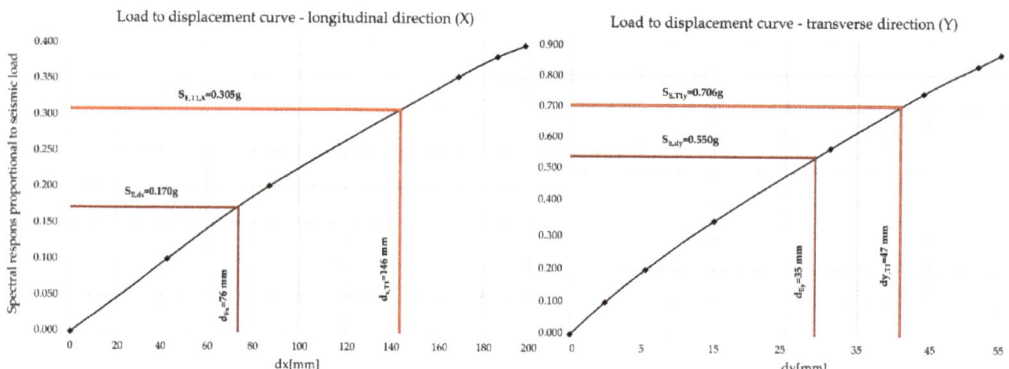

Figure 10. Load–displacement curves for the First Case Study bridge, longitudinal (**left**) and transverse (**right**) direction.

The curves in Figure 10 show that the Case Study bridge does not have sufficient seismic resilience, as displacements corresponding to the two dominant modal shapes (Tx and Ty) exceed the target displacements ($d_{x,T1} > d_{Ex}$ and $d_{y,T1} > d_{Ey}$).

In addition to the displacements, the load-carrying capacity of the piers, abutments, and bearings was assessed. The pier shear failure occurs when the maximum seismic load is applied, while the "Pendl" bearings in the abutments cannot withhold the resulting forces.

4.3. Retrofitting Proposal

In accordance with the assessment results (Figure 10), a proposal for seismic retrofitting is presented [13], focusing on the limitation of the displacements and the shear failure of the piers.

In order to restrict the longitudinal displacement of the superstructure, a total of three dampers with a maximum +/−50 mm should be placed on the abutment which transfers the longitudinal reaction, as shown in Figure 10. Each of the dampers has a capacity for transfer of longitudinal force of 1250 kN; however, analysis of the existing abutments showed that such loads exceed its load-bearing capacity. Therefore, the abutment should be strengthened; this can be conducted using RC jacketing with additional reinforcement or with the addition of geotechnical anchors which would transfer the load directly to the embankment. Due to the limited space beneath the bridge, the solution with geotechnical anchors was chosen, as can be seen in Figure 11.

As the supports on the abutment are not able to withstand the transverse seismic forces, a transversal displacement restraint to be placed on both abutments is proposed. These are present in Figure 12 as additional steel profiles (HE550M) which have sufficient shear resistance to withstand imposed seismic forces. The shear strengthening of the piers should be conducted using FRP jacketing, as RC (due to limited space) and steel (due to variable cross-section) are not optimal solutions and would not be cost-effective.

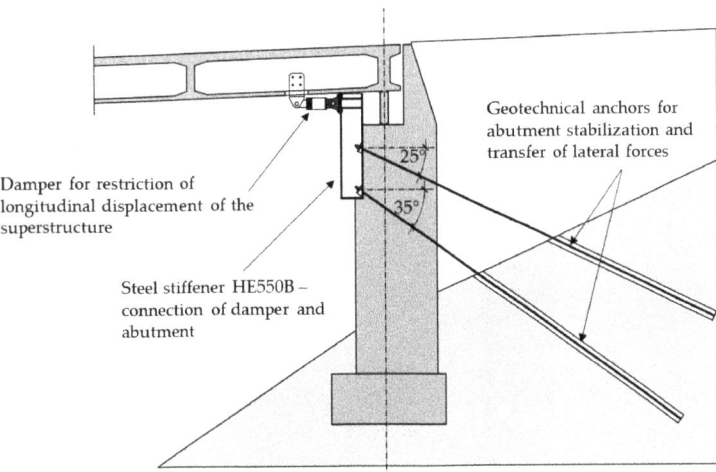

Figure 11. The proposed damper installation for the restriction of longitudinal displacement with geotechnical anchors, first Case Study bridge.

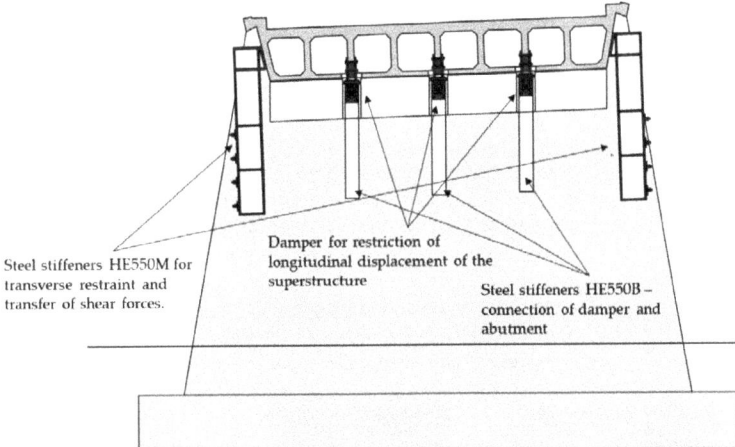

Figure 12. Transverse view of the abutment with both longitudinal and transverse stiffeners and displacement restraints, proposal for seismic retrofitting of the first Case Study bridge.

5. Second Case Study Bridge

5.1. Rapid Visual Assessment after the December 2020 Earthquake

The bridge over the Maja River at the southern entrance to the village of Majske Poljane was a three-span continuous girder bridge built in the 1950s, although this is based on conjecture as no data were available. The bridge superstructure consists of three cast-iron girders and a monolithic reinforced concrete deck resting directly on two RC piers and abutments. The abutment towards the town of Glina is masonry, the other is reinforced concrete. The superstructure and the abutment are presented in Figure 13.

Nine months after the Zagreb earthquake, Sisak Moslavina County was hit by two strong earthquakes, the first, in the early morning of 28 December, had a magnitude of $M_L = 5.4$, and the next day an even stronger earthquake struck, with a magnitude of $M_L = 6.2$ and an intensity between VIII and IX according to the EMS−98 scale. The resulting damage was devastating and claimed seven lives. The village of Majske Poljane suffered the most damage; almost all houses were destroyed. Immediately after the first earthquake, teams of structural engineers were sent out to conduct a preliminary rapid assessment of the structural damage using the procedures introduced after the Zagreb earthquake, as described in [91]. The authors were tasked with the preliminary rapid assessment of all bridges in the area [14], as heavy construction equipment, trucks with humanitarian aid, etc., were needed.

(a) (b)

Figure 13. (a) Masonry abutment and (b) composite superstructure of the bridge to Majske Poljane.

During the visual inspection of the bridge to Majske Poljane, it was found that the masonry abutment had structural damage affecting its stability, with immediate action needed in order to allow heavy traffic over the bridge. The damage to the abutment is shown in Figure 14. The stone blocks from both the abutment wall and the wing walls were damaged and started to fall off. As the abutment was constructed as shallow with land infill that to transfer the vertical loads to the foundations, there was a danger that the collapse of the wing walls would result in the collapse of the bridge.

(a) (b) (c)

Figure 14. (a,b) Damage to the masonry walls of the abutment; (c) escarpment of infill material from the abutment.

5.2. Urgent Retrofitting and Decision for Further Actions on the Bridge

Both the wing and abutment walls required stabilization, and the only possible solution due to urgency and the available resources in the area at the time was to impose an additional dead load on the abutment and surrounding embankment. This was conducted by adding stone blocks that would act as stabilizing weights to allow traffic on the bridge, which was critical as the bridge is one of the only two entrances to the village. This urgent and temporary retrofitting is shown in Figure 15.

Figure 15. (**a**) Stone material for urgent stabilization; (**b**) placing of the additional weight on the abutment and embankment.

In addition to the damage caused by the earthquake, it was clear that the bridge was in an overall poor condition due to irregular maintenance and scour damage. Therefore, several months after the earthquake, when the surrounding area stabilized, a detailed inspection and assessment were conducted. As the bridge is currently under construction, the details of the project cannot be disclosed.

6. Conclusions

This paper reviews the state of the art in seismic assessment and retrofitting methods for existing road bridges, an important challenge for engineers which is not covered by the Eurocode.

There are a number of methods available for analysis of the structural behavior and response of road bridges during a seismic event; a list is provided in Table 1, with no general conclusion as to the optimal choice.

The multistep procedure for the seismic assessment of typical existing road bridges is defined and presented, and is shown in the flowchart in Figure 16. The procedure is based on a step-by-step analysis in cases both where a bridge is inspected as part of regular maintenance (as in the first case study) and where inspection is urgent due to a recent severe seismic event (as in the second case study). In both instances, the first step in the analysis is a visual inspection, followed by a decision as to whether further evaluation is required. If it is, the bridge is assessed at the first level by developing a numerical model based on documentation and on-site measurements and using linear modal analysis. For bridges damaged by recent seismic activity, there is an additional step where decisions as to urgent stabilization measures precede the first assessment level.

If a bridge meets the required ULS checks (codified-EN 1998-2 [5]), it is considered reliable and can be used without restrictions. However, if it does not meet the ULS checks, it is re-evaluated at the second level, where the numerical model is calibrated using material properties obtained with in situ NDT tools and Bayesian updating. The analysis at this level is based on a nonlinear pushover analysis, and additional checks are performed. Similar to the first level, the bridge is considered safe for use if it meets the verifications. If not, the seismic retrofit strategy is selected and applied based on the identification of the critical structural elements. First, the local checks (at the element level) are performed, and if they are satisfied, evaluation at the system level (global model) is performed. If at any of the steps the selected measure is not considered effective, the procedure is repeated with a revised retrofit strategy.

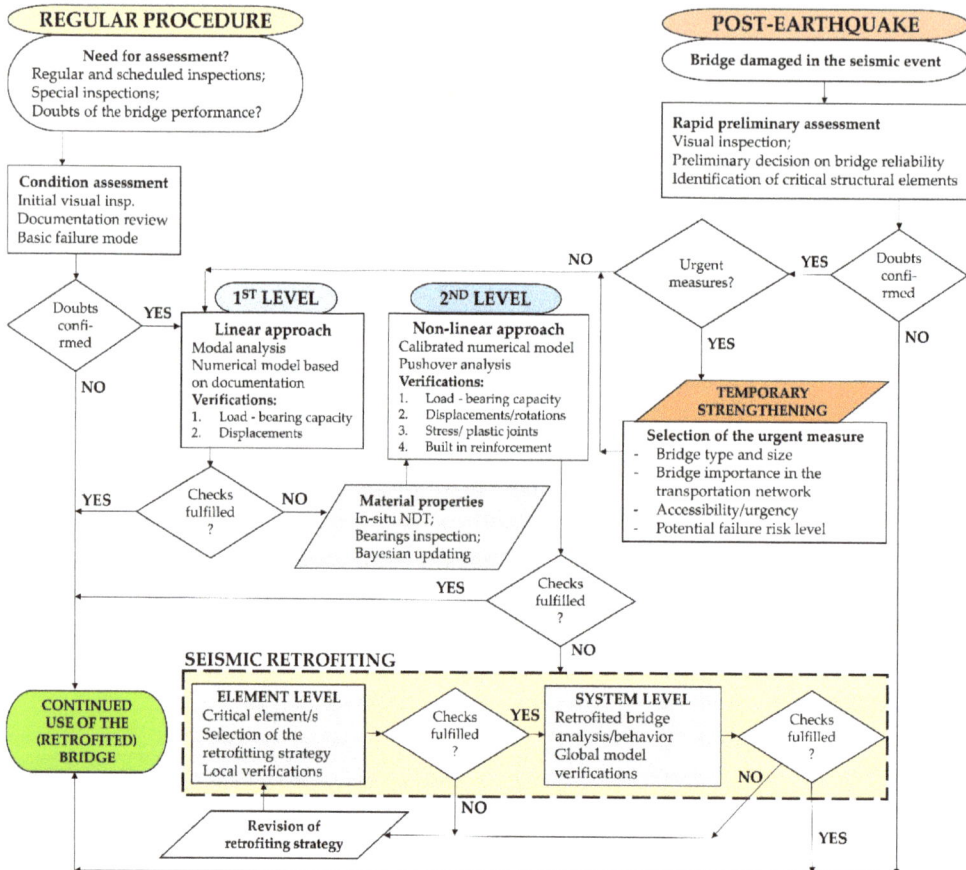

Figure 16. Proposed multi-level assessment method for typical existing road bridges.

The retrofitting methods for existing road bridges in earthquake-prone areas are summarized in Table 2, based on the corresponding structural element. The most vulnerable parts of the common bridge types are piers and bearings, as they are sensitive to lateral loads caused by seismic events.

As a continuation of this research, fragility curves will be developed for the bridges in Sisak-Moslavina County to be used for priority ranking. However, the detailed assessment and retrofitting for those bridges which are deemed as vulnerable requires a more site-specific approach, similar to the second Case Study bridge.

The new generation of Eurocodes currently in development are expected to be expanded, and will have more focus on pre-existing structures. Existing bridges, both road and railway, should be covered in detail in terms of both analysis and retrofitting methods. It is recommended that modern materials based on polymers are included, along with ultimate and serviceability limit state verification, as this would expand their practical application to real bridges.

Author Contributions: Conceptualization, D.S., A.V., M.K.M., M.S. and A.M.I.; methodology, D.S.; software, D.S., A.V. and M.K.M.; validation, D.S., A.V., M.K.M., M.S. and A.M.I.; formal analysis, D.S.; investigation, D.S., A.V., M.K.M., M.S. and A.M.I.; resources, D.S., M.K.M. and A.M.I.; data curation, D.S.; writing—original draft preparation, D.S.; writing—review and editing, D.S., M.K.M. and A.M.I.; visualization, D.S.; supervision, D.S., A.V., M.K.M., M.S. and A.M.I.; project administration, D.S. and M.K.M. All authors have read and agreed to the published version of the manuscript.

Funding: This research was founded in the framework of the project "Key Performance Indicators for Existing Road Bridges". The project team consists of project leader Ana Mandić Ivanković and team members, which include all remaining authors.

Institutional Review Board Statement: Not applicable.

Informed Consent Statement: Not applicable.

Data Availability Statement: Not applicable.

Conflicts of Interest: The authors declare no conflict of interest.

References

1. Skokandić, D.; Ivanković, A.M. Value of additional traffic data in the context of bridge service-life management. *Struct. Infrastruct. Eng.* **2020**, *18*, 456–475. [CrossRef]
2. Calvi, G.M.; Moratti, M.; O'Reilly, G.J.; Scattarreggia, N.; Monteiro, R.; Malomo, D.; Calvi, P.M.; Pinho, R. Once upon a Time in Italy: The Tale of the Morandi Bridge. *Struct. Eng. Int.* **2019**, *29*, 198–217. [CrossRef]
3. Ayub, M. *Pedestrian Bridge Collapse at University in Miami*; US Department of Labor: Washington, DC, USA, 2019.
4. Milić, I.; Ivanković, A.M.; Syrkov, A.; Skokandić, D. Bridge failures, forensic structural engineering and recommendations for design of robust structures. *J. Croat. Assoc. Civ. Eng.* **2021**, *73*, 717–737. [CrossRef]
5. CEN: EN 1998-2; Design of Structures for Earthquake Resistance—Part 2: Bridges. European Committee for Standardization (CEN): Brussels, Belgium, 2005.
6. Srbić, M.; Ivanković, A.M.; Vlašić, A.; Kovačević, G.H. Plastic Joints in Bridge Columns of Atypical Cross-Sections with Smooth Reinforcement without Seismic Details. *Appl. Sci.* **2021**, *11*, 2658. [CrossRef]
7. Mitchell, D.; Bruneau, M.; Saatcioglu, M.; Williams, M.; Anderson, D.; Sexsmith, R. Performance of bridges in the 1994 Northridge earthquake. *Can. J. Civ. Eng.* **1995**, *22*, 415–427. [CrossRef]
8. Sakellariadis, L.; Anastasopoulos, I.; Gazetas, G. Fukae bridge collapse (Kobe 1995) revisited: New insights. *Soils Found.* **2020**, *60*, 1450–1467. [CrossRef]
9. Skokandić, D.; Ivanković, A.M.; Žnidarič, A.; Srbić, M. Modelling of traffic load effects in the assessment of existing road bridges. *J. Croat. Assoc. Civ. Eng.* **2020**, *71*, 1153–1165. [CrossRef]
10. Stepinac, M.; Lourenço, P.B.; Atalić, J.; Kišiček, T.; Uroš, M.; Baniček, M.; Novak, M.Š. Damage classification of residential buildings in historical downtown after the ML5.5 earthquake in Zagreb, Croatia in 2020. *Int. J. Disaster Risk Reduct.* **2021**, *56*, 102140. [CrossRef]
11. Lulić, L.; Ožić, K.; Kišiček, T.; Hafner, I.; Stepinac, M. Post-Earthquake Damage Assessment—Case Study of the Educational Building after the Zagreb Earthquake. *Sustainability* **2021**, *13*, 6353. [CrossRef]
12. Government of Croatia. *World Bank Report: Croatia Earthquake Rapid Damage and Needs Assessment*; Government of Croatia: Zagreb, Croatian, 2020.
13. Vlašić, A.; Srbić, M.; Skokandić, D.; Ivanković, A.M. Post-Earthquake Rapid Damage Assessment of Road Bridges in Glina County. *Buildings* **2022**, *12*, 42. [CrossRef]
14. Marin, F.; Darko, M.; Zlatko, Š. Seismic Assessment of Obod Bridge in Croatia//Durability of Concrete Structures. In Proceedings of the 8th Central European Congresses on Concrete Engineering/Radić, Jure, PLitvice Lakes, Croatia, 4–6 October 2012; Marija, K., Ed.; pp. 65–7214.
15. CEN: EN 1998-1; Design of Structures for Earthquake Resistance—Part 1: General Rules, Seismic Actions and Rules for Buildings. European Committee for Standardization (CEN): Brussels, Belgium, 2004.
16. CEN: EN 1998-3; Design of Structures for Earthquake Resistance—Part 3: Assessment and Retrofitting of Buildings. European Committee for Standardization (CEN): Brussels, Belgium, 2005.
17. Wiśniewski, D.F.; Casas, J.R.; Ghosn, M. Codes for Safety Assessment of Existing Bridges—Current State and Further Development. *Struct. Eng. Int.* **2012**, *22*, 552–561. [CrossRef]
18. Ivanković, A.M.; Skokandić, D.; Žnidarič, A.; Kreslin, M. Bridge performance indicators based on traffic load monitoring. *Struct. Infrastruct. Eng.* **2019**, *15*, 899–911. [CrossRef]
19. Prendergast, L.J.; Limongelli, M.P.; Ademovic, N.; Anžlin, A.; Gavin, K.; Zanini, M. Structural Health Monitoring for Performance Assessment of Bridges under Flooding and Seismic Actions. *Struct. Eng. Int.* **2018**, *28*, 296–307. [CrossRef]
20. Stepinac, M.; Kisicek, T.; Renić, T.; Hafner, I.; Bedon, C. Methods for the Assessment of Critical Properties in Existing Masonry Structures under Seismic Loads—The ARES Project. *Appl. Sci.* **2020**, *10*, 1576. [CrossRef]

21. Kong, X.; Cai, C.S.; Deng, L.; Zhang, W. Using Dynamic Responses of Moving Vehicles to Extract Bridge Modal Properties of a Field Bridge. *J. Bridg. Eng.* **2017**, *22*, 04017018. [CrossRef]
22. Marić, M.K.; Ivanković, A.M.; Vlašić, A.; Bleiziffer, J.; Srbić, M.; Skokandić, D. Assessment of reinforcement corrosion and concrete damage on bridges using non-destructive testing. *J. Croat. Assoc. Civ. Eng.* **2019**, *71*, 843–862. [CrossRef]
23. Lee, S.; Kalos, N.; Shin, D.H. Non-destructive testing methods in the U.S. for bridge inspection and maintenance. *KSCE J. Civ. Eng.* **2014**, *18*, 1322–1331. [CrossRef]
24. Liu, L.; Guo, T. Seismic non-destructive testing on a reinforced concrete bridge column using tomographic imaging techniques. *J. Geophys. Eng.* **2005**, *2*, 23–31. [CrossRef]
25. Lantsoght, E.; van der Veen, C.; de Boer, A.; Hordijk, D. State-of-the-art on load testing of concrete bridges. *Eng. Struct.* **2017**, *150*, 231–241. [CrossRef]
26. Conde, B.; Ramos, L.F.; Oliveira, D.V.; Riveiro, B.; Solla, M. Structural assessment of masonry arch bridges by combination of non-destructive testing techniques and three-dimensional numerical modelling: Application to Vilanova bridge. *Eng. Struct.* **2017**, *148*, 621–638. [CrossRef]
27. Seismic assessment of existing reinforced concrete arch bridges. *J. Croat. Assoc. Civ. Eng.* **2014**, *66*, 691–703. [CrossRef]
28. Pelà, L.; Aprile, A.; Benedetti, A. Comparison of seismic assessment procedures for masonry arch bridges. *Constr. Build. Mater.* **2013**, *38*, 381–394. [CrossRef]
29. Mwafy, A.; Kwon, O.-S.; Elnashai, A. Seismic assessment of an existing non-seismically designed major bridge-abutment–foundation system. *Eng. Struct.* **2010**, *32*, 2192–2209. [CrossRef]
30. Paraskeva, T.S.; Kappos, A.; Sextos, A. Extension of modal pushover analysis to seismic assessment of bridges. *Earthq. Eng. Struct. Dyn.* **2006**, *35*, 1269–1293. [CrossRef]
31. Pelà, L.; Aprile, A.; Benedetti, A. Seismic assessment of masonry arch bridges. *Eng. Struct.* **2009**, *31*, 1777–1788. [CrossRef]
32. Homaei, F.; Yazdani, M. The probabilistic seismic assessment of aged concrete arch bridges: The role of soil-structure interaction. *Structures* **2020**, *28*, 894–904. [CrossRef]
33. Xie, Y.; Zheng, Q.; Yang, C.-S.W.; Zhang, W.; DesRoches, R.; Padgett, J.E.; Taciroglu, E. Probabilistic models of abutment backfills for regional seismic assessment of highway bridges in California. *Eng. Struct.* **2019**, *180*, 452–467. [CrossRef]
34. Monteiro, R.; Delgado, R.; Pinho, R. Probabilistic Seismic Assessment of RC Bridges: Part I—Uncertainty Models. *Structures* **2016**, *5*, 258–273. [CrossRef]
35. MonteiroR Sampling based numerical seismic assessment of continuous span RC bridges. *Eng. Struct.* **2016**, *118*, 407–420. [CrossRef]
36. Bending moment curvature relationship as an indicator of seismic resistance of older bridge piers. *J. Croat. Assoc. Civ. Eng.* **2019**, *71*, 481–488. [CrossRef]
37. Stefanidou, S.P.; Kappos, A.J. Bridge-specific fragility analysis: When is it really necessary? *Bull. Earthq. Eng.* **2018**, *17*, 2245–2280. [CrossRef]
38. Kaundinya, I.; Heimbecher, F. Identification and Classification of European Bridge and Tunnel Types. In *Proceedings of the Taller, Longer, Lighter—Meeting Growing Demand with Limited Resources, Report of the IABSE IASS 2011 Symposium, IABSE, London, UK, 20–23 September 2011*; International Association for Bridge and Structural Engineering (IABSE)/IASS: Zurich, Switzerland, 2011; ISBN 978-0-7079-7122-3.
39. Choi, E.; DesRoches, R.; Nielson, B. Seismic fragility of typical bridges in moderate seismic zones. *Eng. Struct.* **2004**, *26*, 187–199. [CrossRef]
40. Saiidi, M. Managing seismic performance of highway bridges—Evolution in experimental research. *Struct. Infrastruct. Eng.* **2011**, *7*, 569–586. [CrossRef]
41. Padgett, J.E.; DesRoches, R. Methodology for the development of analytical fragility curves for retrofitted bridges. *Earthq. Eng. Struct. Dyn.* **2008**, *37*, 1157–1174. [CrossRef]
42. Zanini, M.A.; Pellegrino, C.; Morbin, R.; Modena, C. Seismic vulnerability of bridges in transport networks subjected to environmental deterioration. *Bull. Earthq. Eng.* **2013**, *11*, 561–579. [CrossRef]
43. Sheikh, M.N.; Légeron, F. Performance based seismic assessment of bridges designed according to Canadian Highway Bridge Design Code. *Can. J. Civ. Eng.* **2014**, *41*, 777–787. [CrossRef]
44. Cosenza, E.; Losanno, D. Assessment of existing reinforced-concrete bridges under road-traffic loads according to the new Italian guidelines. *Struct. Concr.* **2021**, *22*, 2868–2881. [CrossRef]
45. Di Sarno, L.; da Porto, F.; Guerrini, G.; Calvi, P.M.; Camata, G.; Prota, A. Seismic performance of bridges during the 2016 Central Italy earthquakes. *Bull. Earthq. Eng.* **2019**, *17*, 5729–5761. [CrossRef]
46. Huang, C.; Huang, S. Seismic resilience assessment of aging bridges with different failure modes. *Structures* **2021**, *33*, 3682–3690. [CrossRef]
47. Alipour, A.; Shafei, B. Seismic Resilience of Transportation Networks with Deteriorating Components. *J. Struct. Eng.* **2016**, *142*, 1–12. [CrossRef]
48. Biondini, F.; Camnasio, E.; Titi, A. Seismic resilience of concrete structures under corrosion. *Earthq. Eng. Struct. Dyn.* **2015**, *44*, 2445–2466. [CrossRef]
49. Zelaschi, C.; Monteiro, R.; Pinho, R.J.S.M. Parametric Characterization of RC Bridges for Seismic Assessment Purposes. *Structures* **2016**, *7*, 14–24. [CrossRef]

50. Modena, C.; Tecchio, G.; Pellegrino, C.; da Porto, F.; Donà, M.; Zampieri, P.; Zanini, M.A. Reinforced concrete and masonry arch bridges in seismic areas: Typical deficiencies and retrofitting strategies. *Struct. Infrastruct. Eng.* **2014**, *11*, 415–442. [CrossRef]
51. Ivanković, A.M.; Skokandić, D.; Marić, M.K.; Srbić, M. Performance-Based Ranking of Existing Road Bridges. *Appl. Sci.* **2021**, *11*, 4398. [CrossRef]
52. Raza, S.; Khan, M.K.I.; Menegon, S.J.; Tsang, H.-H.; Wilson, J.L. Strengthening and Repair of Reinforced Concrete Columns by Jacketing: State-of-the-Art Review. *Sustainability* **2019**, *11*, 3208. [CrossRef]
53. Wright, T.; DesRoches, R.; Padgett, J.E. Bridge Seismic Retrofitting Practices in the Central and Southeastern United States. *J. Bridg. Eng.* **2011**, *16*, 82–92. [CrossRef]
54. Ogata, T.; Osada, K. Seismic retrofitting of expressway bridges in Japan. *Cem. Concr. Compos.* **2000**, *22*, 17–27. [CrossRef]
55. Billah, A.M.; Alam, M.S. Seismic performance evaluation of multi-column bridge bents retrofitted with different alternatives using incremental dynamic analysis. *Eng. Struct.* **2014**, *62–63*, 105–117. [CrossRef]
56. Abdessemed, M.; Kenai, S.; Bali, A.; Kibboua, A. Dynamic analysis of a bridge repaired by CFRP: Experimental and numerical modelling. *Constr. Build. Mater.* **2011**, *25*, 1270–1276. [CrossRef]
57. Del Zoppo, M.; Di Ludovico, M.; Balsamo, A.; Prota, A. Comparative Analysis of Existing RC Columns Jacketed with CFRP or FRCC. *Polymers* **2018**, *10*, 361. [CrossRef]
58. Li, X.; Chen, K.; Hu, P.; He, W.; Xiao, L.; Zhang, R. Effect of ECC jackets for enhancing the lateral cyclic behavior of RC bridge columns. *Eng. Struct.* **2020**, *219*, 110714. [CrossRef]
59. Hung, C.-C.; Chen, Y.-S. Innovative ECC jacketing for retrofitting shear-deficient RC members. *Constr. Build. Mater.* **2016**, *111*, 408–418. [CrossRef]
60. Seyhan, E.C.; Goksu, C.; Uzunhasanoglu, A.; Ilki, A. Seismic Behavior of Substandard RC Columns Retrofitted with Embedded Aramid Fiber Reinforced Polymer (AFRP) Reinforcement. *Polymers* **2015**, *7*, 2535–2557. [CrossRef]
61. Zanini, M.A.; Toska, K.; Faleschini, F.; Pellegrino, C. Seismic reliability of reinforced concrete bridges subject to environmental deterioration and strengthened with FRCM composites. *Soil Dyn. Earthq. Eng.* **2020**, *136*, 106224. [CrossRef]
62. Dagenais, M.-A.; Massicotte, B.; Boucher-Proulx, G. Seismic Retrofitting of Rectangular Bridge Piers with Deficient Lap Splices Using Ultrahigh-Performance Fiber-Reinforced Concrete. *J. Bridg. Eng.* **2018**, *23*, 04017129. [CrossRef]
63. Tong, T.; Yuan, S.; Zhuo, W.; He, Z.; Liu, Z. Seismic retrofitting of rectangular bridge piers using ultra-high performance fiber reinforced concrete jackets. *Compos. Struct.* **2019**, *228*, 111367. [CrossRef]
64. Reggia, A.; Morbi, A.; Plizzari, G.A. Experimental study of a reinforced concrete bridge pier strengthened with HPFRC jacketing. *Eng. Struct.* **2020**, *210*, 110355. [CrossRef]
65. Mitchell, D.; Sexsmith, R.; Tinawi, R. Seismic retrofitting techniques for bridges—A state-of-the-art report. *Can. J. Civ. Eng.* **1994**, *21*, 823–835. [CrossRef]
66. Markogiannaki, O.; Tegos, I.; Papadrakakis, M. Seismic Retrofitting of R/C Bridges with the Use of Unbonded Tendons. In Proceedings of the 5th International Conference on Computational Methods in Structural Dynamics and Earthquake Engineering (COMPDYN 2015), Crete, Greece, 25–27 May 2015; pp. 1799–1811.
67. Recupero, A.; Spinella, N.; Colajanni, P.; Scilipoti, C.D. Increasing the Capacity of Existing Bridges by Using Unbonded Prestressing Technology: A Case Study. *Adv. Civ. Eng.* **2014**, *2014*, 1–10. [CrossRef]
68. Billah, A.M.; Alam, M.S. Performance-based prioritisation for seismic retrofitting of reinforced concrete bridge bent. *Struct. Infrastruct. Eng.* **2013**, *10*, 929–949. [CrossRef]
69. Moustafa, M.A.; Mosalam, K.M. Seismic response of bent caps in as-built and retrofitted reinforced concrete box-girder bridges. *Eng. Struct.* **2015**, *98*, 59–73. [CrossRef]
70. Billah, A.H.M.M.; Alam, M.S.; Bhuiyan, M.A.R. Fragility Analysis of Retrofitted Multicolumn Bridge Bent Subjected to Near-Fault and Far-Field Ground Motion. *J. Bridg. Eng.* **2013**, *18*, 992–1004. [CrossRef]
71. ImbsenRA Use of Isolation for Seismic Retrofitting Bridges. *J. Bridg. Eng.* **2001**, *6*, 425–438. [CrossRef]
72. MakrisN Seismic isolation: Early history. *Earthq. Eng. Struct. Dyn.* **2019**, *48*, 269–283. [CrossRef]
73. Javanmardi, A.; Ibrahim, Z.; Ghaedi, K.; Khan, N.B.; Ghadim, H.B. Seismic isolation retrofitting solution for an existing steel cable-stayed bridge. *PLoS ONE* **2018**, *13*, e0200482. [CrossRef] [PubMed]
74. Dicleli, M.; Mansour, M.Y.; Constantinou, M.C. Efficiency of Seismic Isolation for Seismic Retrofitting of Heavy Substructured Bridges. *J. Bridg. Eng.* **2005**, *10*, 429–441. [CrossRef]
75. Xie, Y.; Zhang, J. Design and Optimization of Seismic Isolation and Damping Devices for Highway Bridges Based on Probabilistic Repair Cost Ratio. *J. Struct. Eng.* **2018**, *144*, 04018125. [CrossRef]
76. Castaldo, P.; Priore, R.L. Seismic performance assessment of isolated bridges for different limit states. *J. Civ. Struct. Health Monit.* **2018**, *8*, 17–32. [CrossRef]
77. Matsagar, V.A.; Jangid, R.S. Base Isolation for Seismic Retrofitting of Structures. *Prac. Period. Struct. Des. Constr.* **2008**, *13*, 175–185. [CrossRef]
78. Nguyen, X.-D.; Guizani, L. Optimal seismic isolation characteristics for bridges in moderate and high seismicity areas. *Can. J. Civ. Eng.* **2021**, *48*, 642–655. [CrossRef]
79. Dolce, M.; Cardone, D.; Palermo, G. Seismic isolation of bridges using isolation systems based on flat sliding bearings. *Bull. Earthq. Eng.* **2007**, *5*, 491–509. [CrossRef]

80. MalekiS Effect of Side Retainers on Seismic Response of Bridges with Elastomeric Bearings. *J. Bridg. Eng.* **2004**, *9*, 95–100. [CrossRef]
81. Moliner, E.; Museros, P.; Martínez-Rodrigo, M.D. Retrofit of existing railway bridges of short to medium spans for high-speed traffic using viscoelastic dampers. *Eng. Struct.* **2012**, *40*, 519–528. [CrossRef]
82. Martínez-Rodrigo, M.D.; Lavado, J.; Museros, P. Dynamic performance of existing high-speed railway bridges under resonant conditions retrofitted with fluid viscous dampers. *Eng. Struct.* **2010**, *32*, 808–828. [CrossRef]
83. Andrawes, B.; Desroches, R. Comparison between Shape Memory Alloy Seismic Restrainers and Other Bridge Retrofit Devices. *J. Bridg. Eng.* **2007**, *12*, 700–709. [CrossRef]
84. Lian, Q.; Yuan, W.; Yu, J.; Dang, X. Traffic efficiency of post-earthquake road network in fault region retrofitted by friction core rubber bearing. *Structures* **2021**, *33*, 54–67. [CrossRef]
85. Khan, A.K.M.T.A.; Bhuiyan, M.A.R.; Ali, S.B. Seismic Responses of a Bridge Pier Isolated by High Damping Rubber Bearing: Effect of Rheology Modeling. *Int. J. Civ. Eng.* **2019**, *17*, 1767–1783. [CrossRef]
86. Chen, X.; Li, C. Seismic performance of tall pier bridges retrofitted with lead rubber bearings and rocking foundation. *Eng. Struct.* **2020**, *212*, 110529. [CrossRef]
87. Bayraktar, A.; Hökelekli, E. Seismic Performances of Different Spandrel Wall Strengthening Techniques in Masonry Arch Bridges. *Int. J. Arch. Herit.* **2020**, *15*, 1722–1740. [CrossRef]
88. Šavor, Z.; Mujkanovic, N.; Hrelja Kovacevic, G.; Bleiziffer, J. Reconstruction of the Pag Bridge. In *Proceedings of the Chinese-Croatian Joint Colloquium-Long Arch Bridges, Brijuni Islands, Croatia, 10–14 July 2008*; Radić, J., Chen, B., Eds.; SECON HDGK: Brijuni Islands, Croatia, 2008; pp. 241–252.
89. Ghallab, A.; Khafaga, M.; Farouk, M.; Essawy, A. Shear behavior of concrete beams externally prestressed with Parafil ropes. *Ain Shams Eng. J.* **2013**, *4*, 1–16. [CrossRef]
90. Sánchez, J.; Masroor, A.; Mosqueda, G.; Ryan, K. Static and Dynamic Stability of Elastomeric Bearings for Seismic Protection of Structures. *J. Struct. Eng.* **2013**, *139*, 1149–1159. [CrossRef]
91. Uroš, M.; Šavor Novak, M.; Atalić, J.; Sigmund, Z.; Baniček, M.; Demšić, M.; Hak, S. Post-earthquake damage assessment of buildings-procedure for conducting building inspections. *J. Croat. Assoc. Civ. Eng.* **2021**, *72*, 1089–1115. [CrossRef]

Article

Experimental and Numerical Study of the Ultimate Flexural Capacity of a Full-Size Damaged Prestressed Concrete Box Girder Strengthened with Bonded Steel Plates

Yong Li, Zijie Yu * and Yongqian Liu

The Key Laboratory of Roads and Railway Engineering Safety Control, School of Civil Engineering, Shijiazhuang Tiedao University, Shijiazhuang 050043, China
* Correspondence: yuzijie2240170626@163.com; Tel.: +86-18832407186

Abstract: Using steel plates attached with epoxy resin adhesive to strengthen prestressed reinforced concrete bridges has become a common method to increase bearing capacity in engineering because of the simple technology, low cost and good strengthening effects. The strengthening method of steel plates has been gradually applied to repair damaged bridges in practical engineering. After a cross-line box girder bridge was struck by a vehicle, the steel bars and concrete of a damaged girder were repaired and strengthened by steel plates, and then the ultimate bending bearing capacity was studied through a destructive test. The results of the destructive test were compared with those of an undamaged girder to verify the effect of the repair and strengthening of the damaged girder. The results showed that the actual flexural bearing capacity of the repaired girder strengthened by steel plates was 1.63 times the theoretical bearing capacity, 36.7% more than that of the damaged girder and 95.3% of that of an undamaged girder. The flexural cracking moment of the repaired girder strengthened by steel plates reached 66.3% of that of the undamaged girder. The maximum crack width decreased by 24.6%, and the maximum deflection increased by 2.7%, compared with the undamaged girder when the repaired girder strengthened by steel plates finally failed. Moreover, this method of attaching steel plates can increase the ductility of bridges and reduce the degree of cracking. Additionally, the actual safety factor of the repaired girder was greater than three, and it had a large safety reserve.

Keywords: prestressed concrete box girder; concrete damage; steel bars and strand fracture; ultimate flexural bearing capacity; destructive test; repair and strengthening

1. Introduction

With the accelerated pace of urban construction and the continuous rise in the number of cars, the cross-line bridge has become an essential structure to improve city traffic, relieve the pressure of urban traffic and improve the efficiency of road transport [1]. However, cross-line bridges are often struck by vehicles to varying degrees [2], which will cause cracks in the girder body and even damage to the steel bars, concrete and post-tensioned strands [3–5]; as a result, their flexural capacity does not meet the design load specified in the current design specification [6]. To reduce the effects of traffic on in-service bridges after they undergo collisions, and prevent large areas of traffic paralysis, there is a need for new methods to repair and strengthen or demolish and reconstruct bridges [7]. Demolition and reconstruction require very large financial input and cause severe environmental pollution. Compared with demolition and reconstruction, repairing and strengthening in-service bridges have the advantages of economy, environmental protection and less impact on road traffic. Therefore, repairing and strengthening in-service bridges have become common methods to improve the bearing capacity and restore damaged in-service bridges [8,9].

At present, reinforcement by steel plates is widely used because of its advantages, such as fast construction, low cost and remarkable strengthening effect, for prestressed bridges [10]. The steel plates are generally bonded to the girder by a structural adhesive or epoxy resin, and anchored on the tensile edge or the weak surface of the bridge to form a common force with the bridge as a whole, and then play the role of steel bars which increase the strength [11]. Compared with strengthening by bonding fibers, strengthening by bonding steel plates can make full use of the mechanical properties of steel plates [12]. Steel plates are easy to obtain and relatively inexpensive [13], and they have the material characteristics of uniform stress and good plasticity [14]. Strengthening by steel plates has been proven to effectively improve the stiffness, reduce the deformation under live loads [15], enhance the crack resistance [16], and more importantly, effectively improve the bending [17] and shear performance of the main girder [18], and it has no significant impact on the appearance of the structure and headroom [19]. Most of the previous conclusions on the strengthening effect of bonded steel plates were obtained for undamaged girders, but there were few relevant studies on the damaged and cracked prestressed concrete box girders that have been repaired [20,21]. Even if there are, the bearing capacity is analyzed through refined finite element analysis, and no full-bridge destructive test is carried out [22,23]. Therefore, this paper evaluates the bearing capacity of the damaged girder after repair through the full-bridge destructive test.

We take a prestressed concrete box girder bridge as the research object, specifically the third and fourth spans of the cross-line that were severely damaged when struck by a truck. The mechanical properties of the damaged and undamaged fourth span box girders were compared previously [24]. The technical status of the bridge was assessed, and the bearing capacity of the third span in the damaged girder was checked according to the specification [6]. The decision was made to use steel with the same strength as the steel bars to repair the damaged main steel bars, and stirrup and concrete with the same strength as the original concrete was used to repair the concrete of the bottom plate, and at the same time, the cracks were closed and strengthened with steel plates [25]. To study the ultimate bearing capacity of the damaged girder after repair and strengthening, destructive tests were conducted on an undamaged girder and the repaired girder strengthened by steel plates [26–30], and the destructive process was simulated through detailed analysis [31,32]. The actual bearing capacity of the repaired girder strengthened by steel plates was evaluated by contrasting the data of the destructive tests and a thorough study of the two girders. The results can serve as a guide for future evaluations of the bearing capacity of similarly damaged bridges.

2. Engineering Situations

2.1. Bridge Information

The basic information of the bridge is detailed in a reference [24], and Figure 1 shows the whole layout of the bridge in detail. The expressway is crossed by the third and fourth spans of the bridge, and the traffic flow under the bridge is large. Due to a collision by an overhigh vehicle, a girder was severely damaged in the third span and another in the fourth span, as shown in Figure 2. The damaged #4-2 girder and the undamaged #4-1 girder of the fourth span were compared and studied [24]. In this paper, the bearing capacity of the damaged #3-2 girder of the third span, which was repaired and strengthened with steel plates, was studied. The undamaged #3-1 girder of the third span, which was strengthened with carbon fiber reinforced plastic (CFRP) plates, will be researched in a future study.

(a)

(b)

(c)

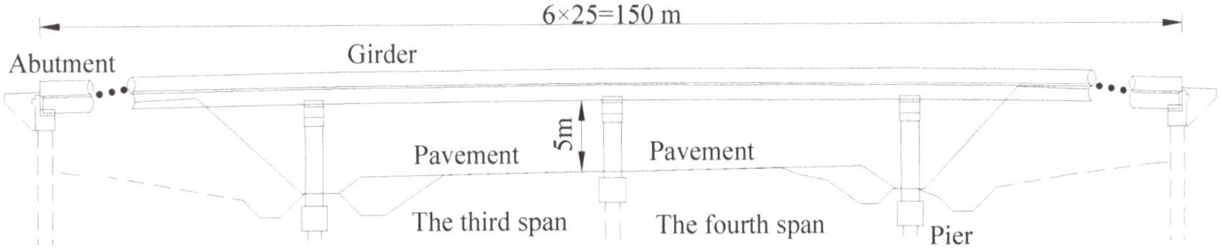

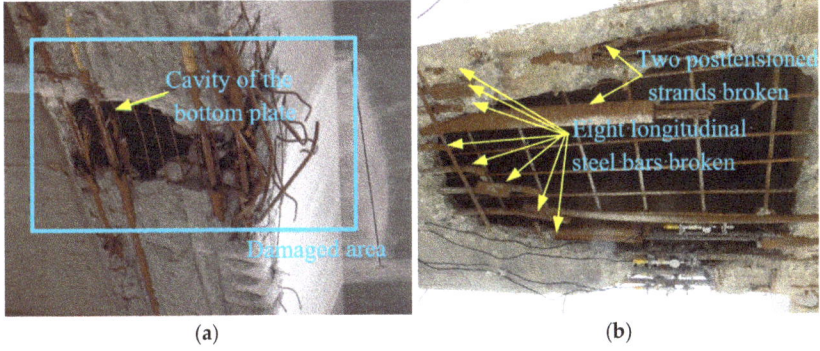

Figure 1. Details of the entire bridge arrangement. (**a**) Elevation arrangement (Unit: m); (**b**) Plane layout; (**c**) Cross-sectional layout (Unit: cm).

(**a**)

(**b**)

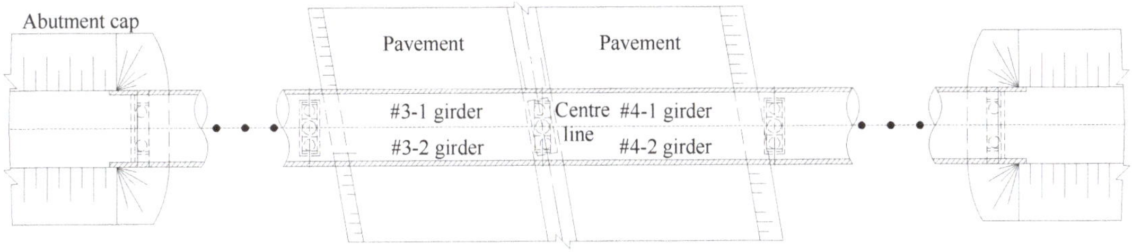

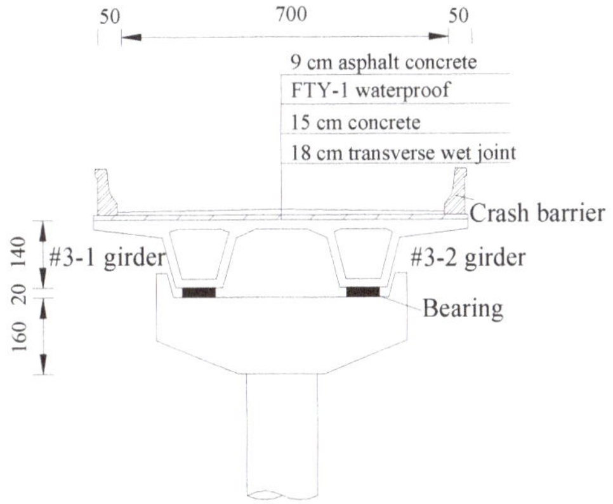

Figure 2. *Cont.*

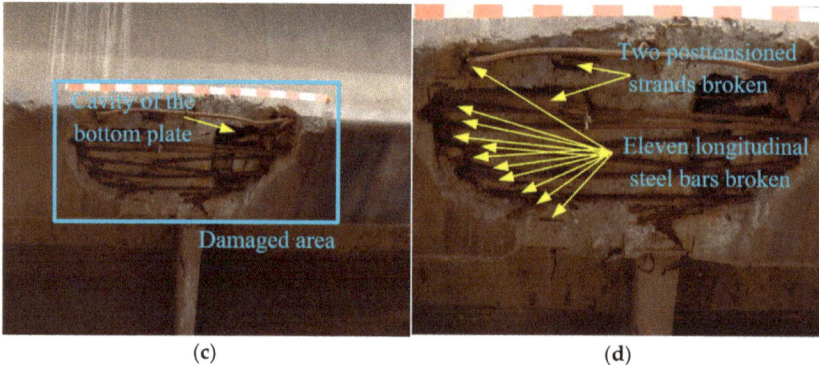

Figure 2. Modes of damage at the bottom of box girders. (**a**) Damaged area of the #4-2 girder; (**b**) Steel bars and pretensioned strands broken in the #4-2 girder; (**c**) Damaged area of the #3-2 girder (**d**) Steel bars and pretensioned strands broken in the #3-2 girder.

2.2. Appearance of Damage

Due to the collision with the overhigh vehicle, the prestressed concrete box #4-2 and #3-2 girders were severely damaged. The significant modes of damage of the #4-2 girder are shown in Figure 2a,b [17], and the main modes of damage of the #3-2 girder are shown in Figure 2c,d. A total of 1.82 m² of concrete fell off, and a cavity appeared in the bottom plate of the #4-2 girder. Additionally, 2 m² of concrete were damaged 2~4.5 m from the middle span of the #3-2 girder, but a cavity of only 0.4 m² formed. In the bottom plate of the #4-2 girder, eight longitudinal steel bars and two pretensioned strands broke. In addition to the fracture of these steel bars and pretensioned strands, three more longitudinal steel bars broke in the #3-2 girder than in the #4-2 girder. Moreover, the unbroken steel bars and pretensioned strands in the damaged areas of the two girders were exposed to air due to the loss of concrete.

To facilitate the transportation of the prestressed concrete box girders and the destructive tests, and considering that the overstretched flange plates of box girders had little influence on bearing capacity, the flange plates on both sides of the box girders were partially removed. The cut girders were supported simply at the girder end, as indicated in Figure 3. Figure 4a,b show the arrangement of the steel bars and pretensioned strands of the B-B section in the box girder. Figure 4c,d show the steel bars and pretensioned strands that broke in the C-C section of the #4-2 and #3-2 girders, respectively, whereas Figure 4e shows the longitudinal arrangement of the pretensioned strands.

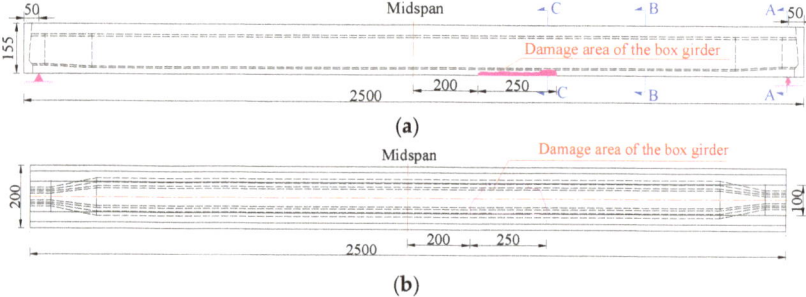

Figure 3. *Cont.*

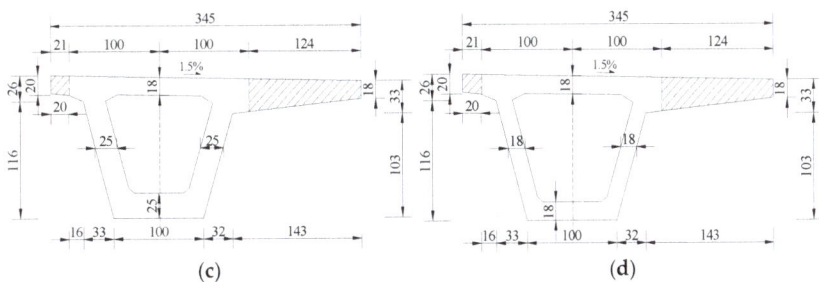

Figure 3. Cross-sectional diagrams of the box girders at the third and fourth spans (Unit: cm). (**a**) Front view; (**b**) Bottom view; (**c**) A-A section; (**d**) B-B section.

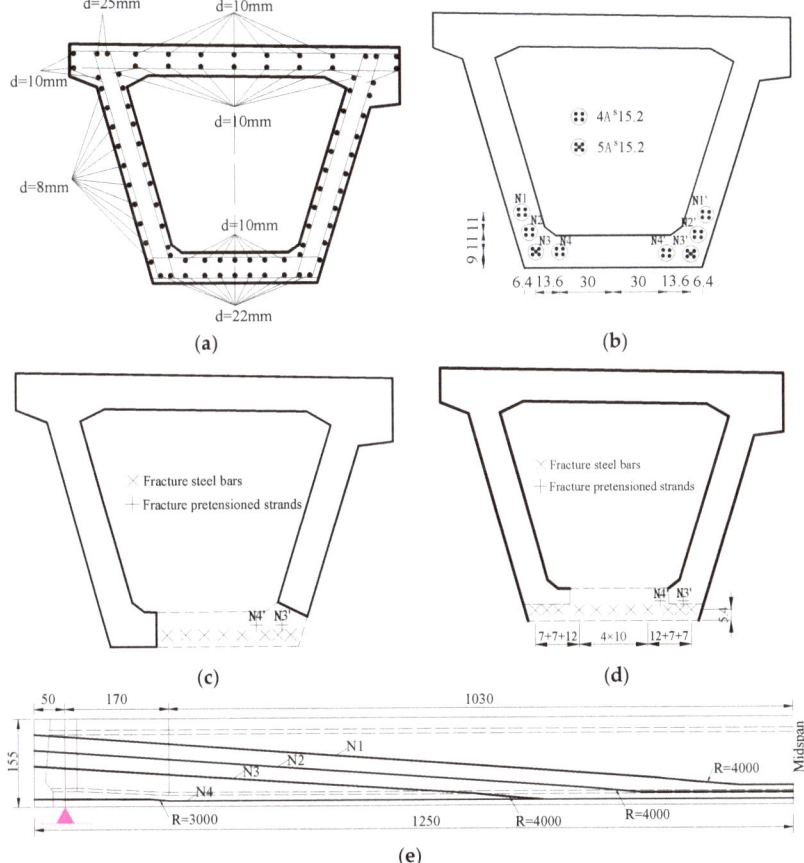

Figure 4. Arrangement of steel bars and strands in the girder (Unit: cm). (**a**) Steel bars in the undamaged girder (B-B); (**b**) Pretensioned strands in the undamaged girder (B-B); (**c**) Steel bars in the damaged region of the #4-2 girder (C-C); (**d**) Steel bars in the damaged region of the #3-2 girder (C-C); (**e**) Strands arrangement.

3. Theoretical Analysis

3.1. Flexural Capacity

The bearing capacity of the #3-2 girder in undamaged, damaged and repaired states was calculated to determine whether it needed to be strengthened. First, the calculated section was determined before the ultimate bending capacity was calculated. Due to the adoption of a simply supported system at the girder end, the computed section of the undamaged #3-2 girder was selected as the midspan section. The calculated section was determined to be the section with the most severe damage for the damaged #3-2 girder and the repaired #3-2 girder since the damaged area was close to the midspan. In the repaired girder, the same strengths of steel bars and concrete were used to repair the broken longitudinal steel bars, stirrups and concrete of the bottom plate, and the cracks were closed. The box section was equivalent to an I-shape when the flexural bearing capacity was computed, according to the specification [6]. Figure 5a depicts the equivalent sections of the undamaged and repaired box girders, and Figure 5 shows the equivalent section of the damaged girder (b).

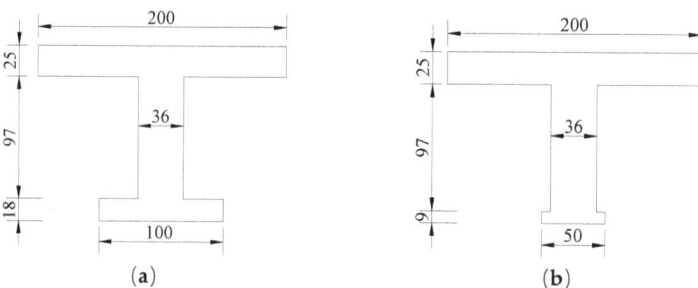

Figure 5. Equivalent sections of the #3-2 girder in undamaged, damaged and repaired states (Unit: cm). (**a**) Undamaged girder and repaired girder; (**b**) Damaged girder.

According to the design drawings, specifications and results of the appearance inspection, the calculated parameters of the #3-2 girder in undamaged, damaged and repaired states were obtained, as shown in Table 1. According to the specification [6], Equation (1) was satisfied by the calculated sections of the damaged, undamaged and repaired girders, and Equation (2) was employed to calculate the flexural bearing capacity Mu. The undamaged girder's calculated section had a flexural bearing capacity Mu of 8442 kN/m and a cracking bending moment of 6984 kN/m. The flexural bearing capacities of the damaged and repaired girders were 5127 kN·m and 6432 kN·m, respectively, which were 39.3% and 23.8% less than that of the undamaged girder. However, the result for the repaired girder was 25.5% larger than that of the damaged girder.

$$f_{sd}A_s + f_{pd}A_p \leq f_{cd}b'_f h'_f + f'_{sd}A'_s \tag{1}$$

$$M_u = f_{cd}b'_f x(h_0 - \frac{x}{2}) + f'_{sd}A'_s(h_0 - a'_s) \tag{2}$$

where f_{sd} and f'_{sd} are design tensile strength and compressive strength of longitudinal steel bars. f_{pd} is design tensile strength of strands. f_{cd} is design compressive strength of concrete. A_s and A'_s are areas of longitudinal reinforcement. A_p is section area stands. h'_f and b'_f are thickness and width of flanges for the equaled I-shape. h_0 is the effective height. x is the height of the compressive zone. a'_s is the distance between the area centroid of compression reinforcement and concrete edge.

Table 1. Parameters for calculating the bearing capacity of the #3-2 girder in its undamaged, damaged and repaired states.

#3-2 Girder	f_{cd}/MPa	f_{sd}/MPa	f'_{sd}/MPa	f_{pd}/MPa	A_s/mm²	A'_s/mm²	A_p/mm²	h_0/mm	a'_s/mm	x/mm
Undamaged	22.4	280	280	1260	4181	2670	4726	1253	45	142.4
Damaged	22.4	195	280	1260	707	2670	3475	1178	45	84.1
Repaired	22.4	280	280	1260	4181	2670	3475	1211	45	107.2

3.2. Load Effect under the Designed Load

The value of the design load effect of the prestressed concrete box girder was calculated in order to figure out if the flexural bearing capacity of the damaged #3-2 girder and repaired #3-2 girder satisfied the requirements of the current bridge design specification [6]. According to the design drawings of the girder, the design load was determined to be road-II level, and the calculated bridge load effect combination was determined to be 1.2 times the dead load plus 1.4 times the road-II level, according to the specification. The effect of the prestressed concrete box girder's design load was 6390 kN·m. The comparison between the theoretical values of the flexural bearing capacities of the three box girders and the design load effect value is shown in Figure 6.

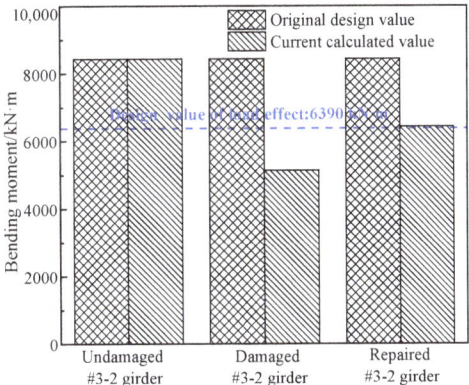

Figure 6. Comparison between the calculated flexural capacities and the design load effect value.

Figure 6 shows that the theoretical value of the damaged #3-2 girder's flexural bearing capacity was 19.8% less than the value of the design load effect of the original #3-2 girder, which indicated that the damaged #3-2 girder was a hidden danger and a great risk to safety, and the vehicle impact had a massive effect on the bridge's bearing capacity. The bearing capacity of the repaired #3-2 girder was calculated to be 6432 kN·m, which was similar to the design load effect of the original #3-2 girder of 6390 kN·m. This indicated that the safety reserve of the repaired #3-2 girder was insufficient, which limited its capacity. It was necessary to strengthen the repaired #3-2 girder by steel plates.

3.3. Strengthening Scheme of the Repaired Girder

Table 2 shows the performance parameters of steel plate Q345 that was selected for strengthening. Considering the width of the bottom of the prestressed concrete girder, two steel plates were selected for strengthening the girder. The width of a single steel plate was 30 cm, and the effective length was 2200 cm, as shown in Figure 7. According to the specification [33], the elastic modulus was 210 GPa, the design value of the axial tensile strength was 275 MPa and the measured tensile strength of the strengthening steel plates was 385 MPa. According to the strengthened specification [34], the strengthened effect of the thickness of the 6-10 mm thick steel plate on the repaired box girder was calculated using Equations (3) and (5), and Figure 8 shows the results. Considering the field test conditions and the results of the strength analysis, an 8 mm thick steel plate was finally selected to reinforce the repaired #3-2 girder.

$$f_{cd}bx + f'_{sd}A'_s = f_{sd}A_s + f_{pd}A_p + \psi_{sp}f_{sp}A_{sp} \tag{3}$$

$$2a'_s \leq x \leq \xi_b h_0 \tag{4}$$

$$M'''_u = f_{cd}bx\left(h_0 - \frac{x}{2}\right) + f'_{sd}A'_s(h_0 - a'_s) + \psi_{sp}f_{sp}A_{sp} \tag{5}$$

where f_{sp} is the design tensile strength of the steel plates. A_{sp} is the area of the steel plates. A_p is the section area of the strands. Ψ_{sp} is the influence coefficient of the bearing capacity of the bonded steel plate when second-stage stress and cracks in the strengthened girder are considered; the value ranged from 0.85 to 0.95, according to the maximum width of the surface cracks before strengthening. Finally, ξ_b is the balanced relative depth of the compressive area.

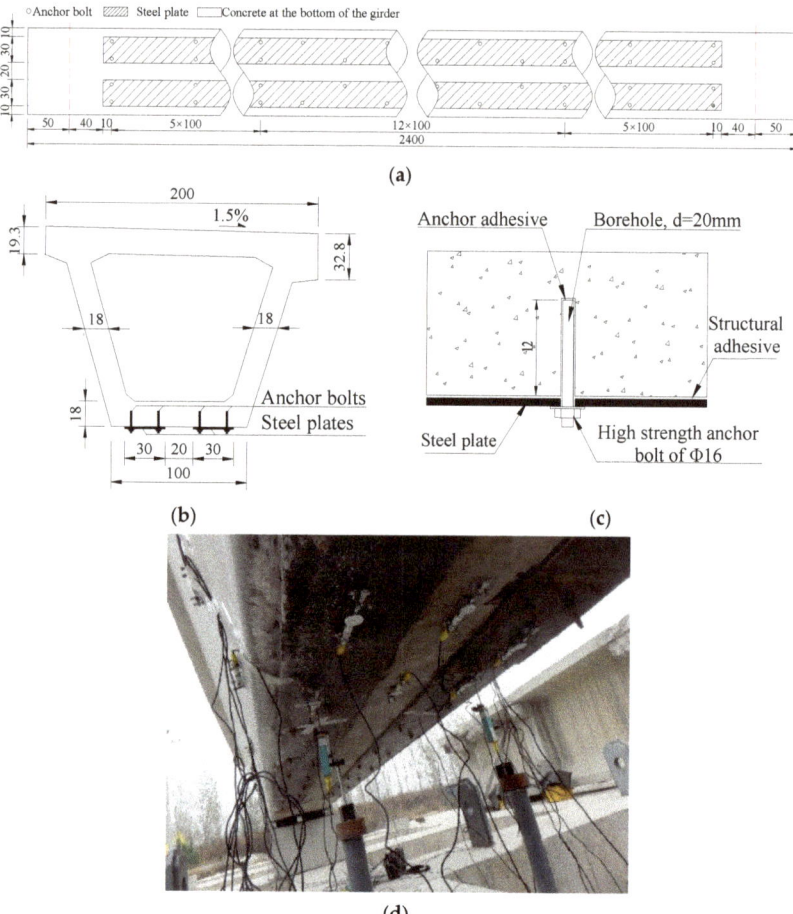

Figure 7. Layout of steel plates. (**a**) Layout at the bottom of the girder (Unit: cm); (**b**) Cross-sectional view of the steel plates layout (Unit: cm); (**c**) Details of the layout of steel plates; (**d**) Actual steel plates at the bottom of the girder.

Table 2. Performance characteristics of the strengthening materials.

Materials	Design Strength/MPa		Measured Strength/MPa	Elastic Modulus/N·mm^{-2}	Density/N·mm^{-3}
	Tensile	Shear	Tensile		
Q345 steel plates	275	160	385	2.1×10^5	7.7×10^{-5}
Q345 anchor	160	–	–	2.1×10^5	7.7×10^{-5}

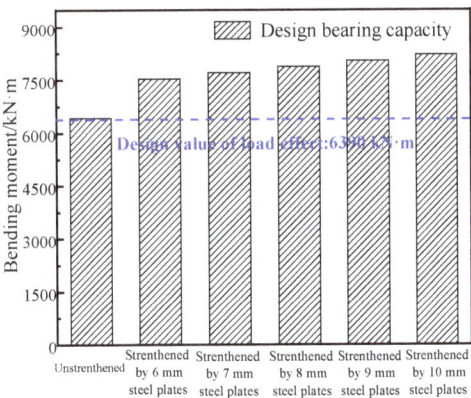

Figure 8. Comparison of the bearing capacities of the #3-2 girder strengthened by various steel plates.

Considering the strengthened analysis results and field test conditions, an 8 mm thick steel plate was finally selected to reinforce the repaired #3-2 girder. The main construction steps of attaching the steel plates were as follows: First, the bottom concrete of the girder was grinded and cleaned to make the surface smooth and flat. Then, the positions of two steel plates were determined at the bottom of the girder, and they were anchored with Q345 high-strength anchor. Glue was spread between the steel plates as well as the concrete using the pressure glue injection method to firmly bond the two surfaces. Finally, the steel plates were brushed with protective paint to prevent corrosion.

4. The Refined Finite Element Analysis
4.1. Establishment of the Model of the Undamaged #3-2 Girder

To predict the change in the ultimate bearing capacity and mechanical properties of the repaired #3-2 girder strengthened by steel plates during the destructive process, and to compare the difference between the undamaged #3-2 girder and the damaged #3-2 girder, a model of the #3-2 girder in undamaged, damaged and repaired states was established by ABAQUS for a detailed analysis [35]. The modeling process of the undamaged #3-2 girder complied with the modeling process of the undamaged #4-1 girder. The modeling details of the undamaged #4-1 girder were described in Section 3.3 of another paper [24], and Figure 9 shows the stress-strain curves of the concrete, pretensioned strands and steel bars. Linear truss elements and eight-node hexahedral reduction integral elements were applied to establish the steel bars, pretensioned strands model and the concrete model, respectively. Considering the accuracy, convergence and computational efficiency of the numerical simulation, the mesh sizes of the concrete, steel bars and pretensioned strands were determined to be 120, 240 and 240 mm, respectively, and the number of divided grids were 20,533, 18,065 and 810, respectively. The interface constraint of the steel bars, pretensioned strands and concrete are simulated by the embedding function. The effective stress of the pretensioned strands considering the prestress loss is shown in Table 3. For the box girder parts that were built, the pretensioned strands were subjected to a drop-temperature load to cause them to shrink, and the adjacent boundary was naturally strained to resist contraction, which realized the application of a prestressed load. The boundary condition applied to the girder end was a simply supported system; one end was a fixed hinge support, and the other end was a unidirectional sliding hinge support. Static-General in ABAQUS was used as a computational solver in analysis. The detailed finite element model of the undamaged #3-2 girder is shown in Figure 10.

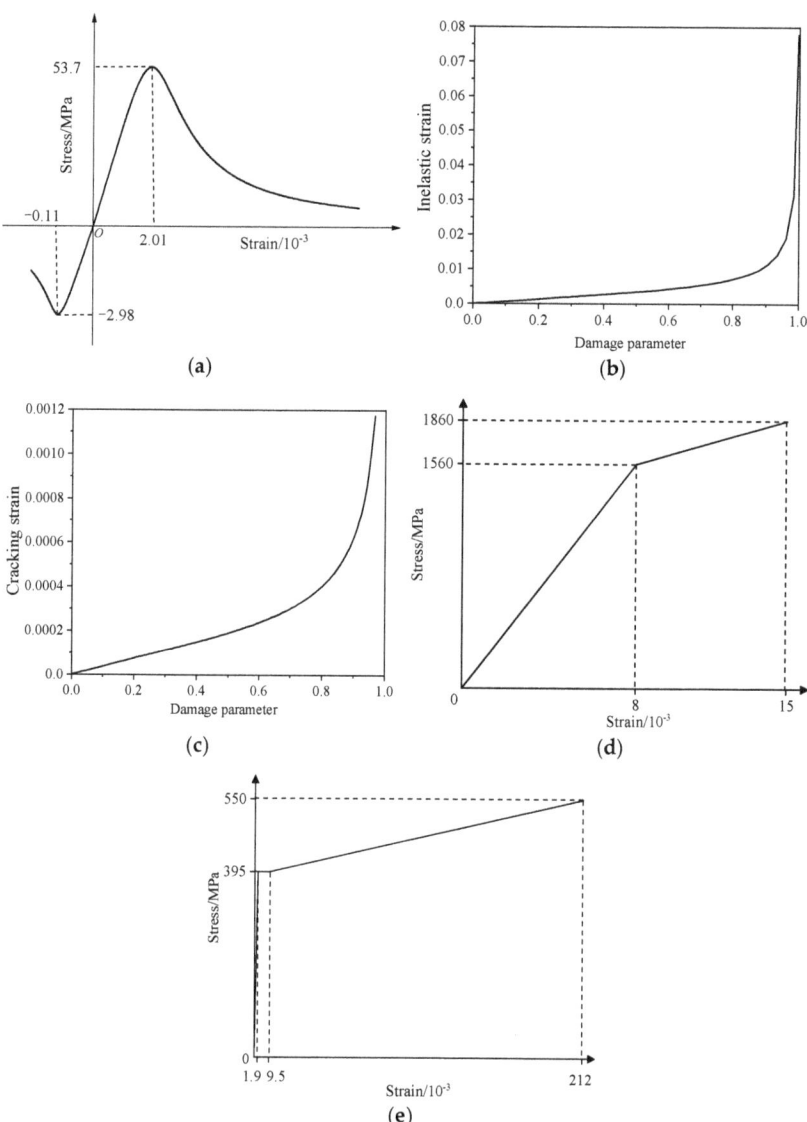

Figure 9. Constitutive curves of the finite element model. (**a**) Stress vs. strain of concrete; (**b**) Compression damage of concrete; (**c**) Tension damage of concrete; (**d**) Stress vs. strain of the pretensioned strands; (**e**) Stress vs. strain of steel bars.

Table 3. Prestress loss.

Number of Strands	σ_{con} /MPa	σ_{l1} /MPa	σ_{l2} /MPa	σ_{l4} /MPa	σ_{l5} /MPa	σ_{l6} /MPa	σ_{pe} /MPa
N1	1395	56.71	120.18	14.33	24.92	65.36	1113.49
N2	1395	56.71	120.18	14.33	24.92	65.36	1113.49
N3	1395	56.71	120.18	14.33	24.92	65.36	1113.49
N4	1395	33.63	92.67	14.94	30.71	68.38	1154.67

where σ_{con} is the strands' tension controlling stress, σ_{l1} is the frictional losses, σ_{l2} is the anchorage losses, σ_{l4} is the prestress loss caused by the concrete's elastic compression, σ_{l5} is the prestress loss caused by the prestressed strands relaxing, σ_{l6} is the time-dependent loss caused by the concrete's creep and shrinkage and σ_{pe} is the strands' actual stress upon anchoring.

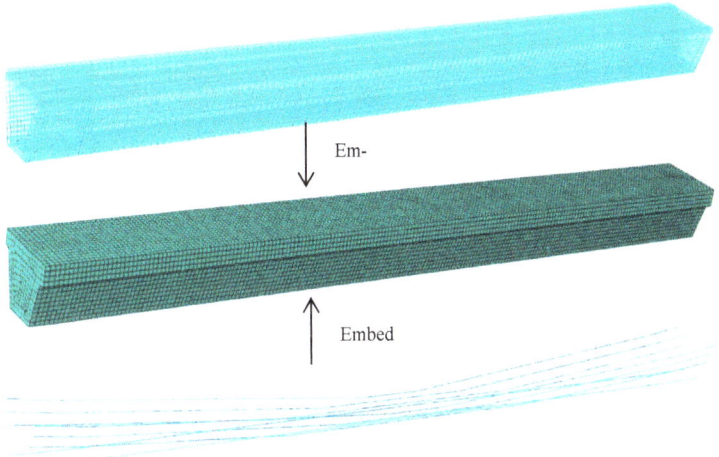

Figure 10. Refined model of the undamaged #3-2 girder.

4.2. Establishment of the Refined Model of the Repaired #3-2 Girder Strengthened by Steel Plates

The process of modeling the repaired #3-2 girder strengthened by steel plates was basically the same as that of modeling the undamaged girder. The difference was in the simulation of broken pretensioned strands and strengthened steel plates. The broken pretensioned strands in the strengthened box girder were divided into two parts. Some of the strands were well bonded to the concrete, and they were regarded as the longitudinal steel bars. The temperature field was not applied, and the embedding function was used to constrain the strands with the concrete. The other strands separated from the concrete, and they were modeled using the life and death unit method. That is, the Model Change function in the software was used to passivate the prestress as it was applied. A four-node curved shell element was used to establish the strengthened steel plate model, and the stress–strain curves of the steel plates are shown in Figure 11a. According to relevant strengthening specifications [34], before the strengthened girder reached the ultimate bearing capacity, bond stripping failure between the strengthened steel plates and concrete was not allowed to occur, and the bond strength and shear strength of the adhesive were greater than the tensile strength and shear strength of the concrete. Therefore, the tie constraint was directly used to connect the reinforced steel plates and concrete, and the failure of the concrete was used to simulate the bond stripping failure between the strengthened steel plates and concrete. Figure 11 shows the finite element model of the repaired #3-2 girder strengthened by steel plates.

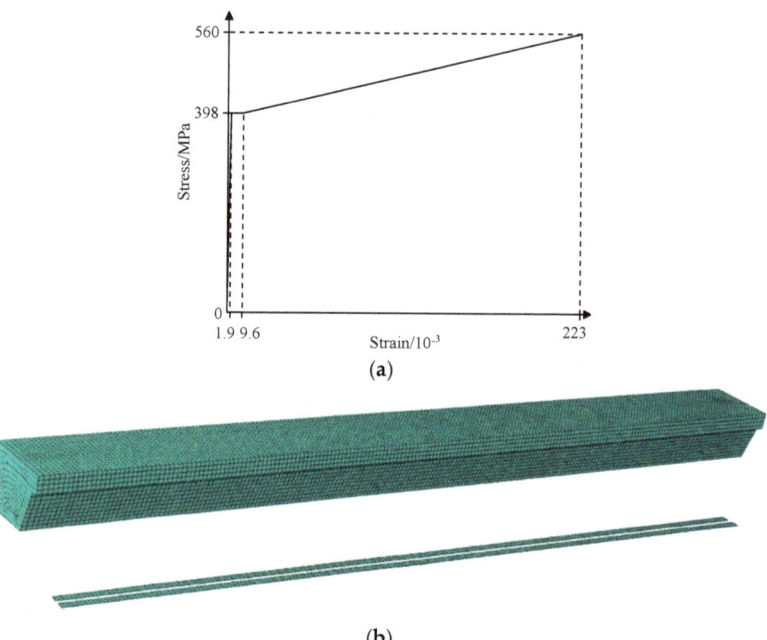

Figure 11. Refined model of the repaired #3-2 girder strengthened by steel plates. (**a**) Stress vs. strain curve of steel plates; (**b**) Refined model.

4.3. Establishment of the Model of the Damaged #3-2 Girder

The modeling of the damaged #3-2 girder was based on the model of the repaired #3-2 girder strengthened by steel plates. The main difference between them was that the simulation of the damaged #3-2 girder did not include reinforced steel plates, but the simulation included the spalling of damaged concrete and broken longitudinal steel bars. According to the actual condition of the damaged 3-2# girder, the damaged concrete spalled and the broken ordinary steel bars were cut, and the Model Change function of ABAQUS was used to model them. The model of the damaged #3-2 girder is shown in Figure 12.

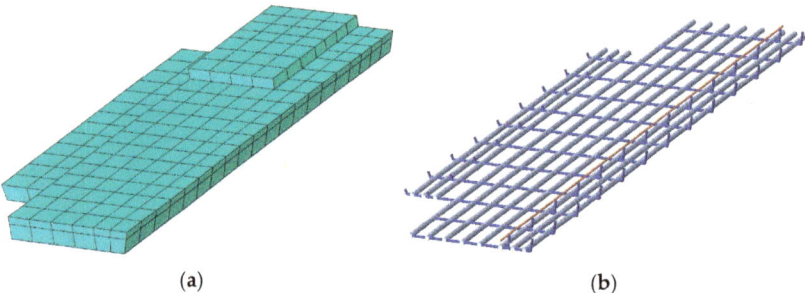

Figure 12. *Cont.*

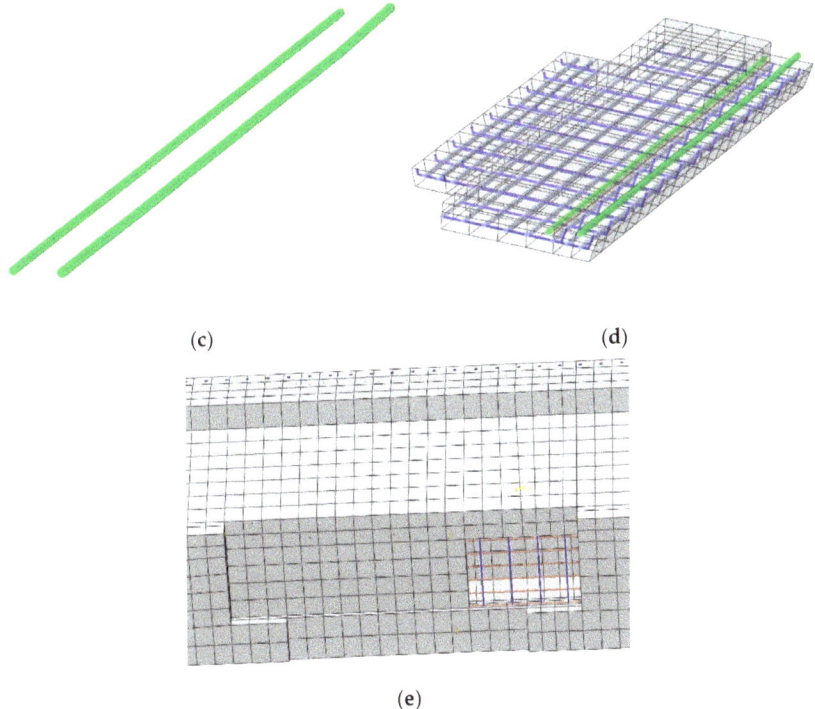

Figure 12. Details in the damaged district of the #3-2 girder. (**a**) The damaged part of concrete; (**b**) The damaged part of the steel bars; (**c**) The damaged part of the pretensioned strands; (**d**) The damaged part; (**e**) The damaged area of the girder.

4.4. The Results of Refined Finite Element Analysis

Figure 13 shows the load vs. deformation curves of the repaired #3-2 girder strengthened by steel plates, the undamaged #3-2 girder and the damaged #3-2 girder were obtained in accordance with the nonlinear analysis. It shows that the destructive processes of the repaired #3-2 girder strengthened by steel plates and the damaged #3-2 girder were the same as that of the undamaged #3-2 girder, which included the elastic stage, the working stage with cracks and the destructive stage.

The deformation of the undamaged #3-2 girder was approximately 2.6 mm in each stage loaded with 50 kN, the deformation of the repaired #3-2 girder strengthened by steel plates was approximately 2.2 mm in each stage loaded with 40 kN and the deformation of the damaged #3-2 girder was approximately 3.3 mm in each stage loaded with 40 kN in the elastic stage. This indicated that the stiffness of the undamaged #3-2 girder in the elastic stage after being struck by a vehicle decreased by as much as 36%, but after being strengthened by steel plates, the girder that was injured had a 50% improvement in stiffness. The undamaged #3-2 girder's cracking load was 361 kN, while the damaged #3-2 girder's cracking load was 200 kN, or 44.6% less than the undamaged #3-2 girder's cracking load. The cracking load of this repaired #3-2 girder strengthened by steel plates was 265 kN, which was 32.5% more than that of the damaged #3-2 girder, but there was still a large gap between the undamaged #3-2 girder, indicating that strengthened steel improved the cracking performance of the structure, but the improvement range was limited. When the load was increased to 875 kN, the concrete at the top of the undamaged #3-2 girder was crushed. The damaged #3-2 girder's failure load was 552 kN, which was 36.9% less than that of the undamaged #3-2 girder. The failure load of the repaired #3-2 girder strengthened by steel plates was 840 kN, which was 52.2% more than that of the damaged #3-2 girder but 4% less than that of the undamaged #3-2 girder.

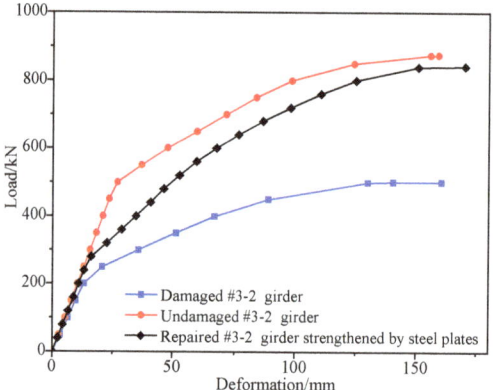

Figure 13. Deformation vs. load from the refined finite element analysis of the #3-2 girder in undamaged, damaged and repaired states.

According to the refined analysis, Figure 14 shows the concrete stress nephograms of the three states of the #3-2 girder when loaded to failure, and Figure 15 shows the stress nephograms of longitudinal steel bars, pretensioned strands and steel plates. Figure 14 shows that the maximum concrete compressive stress of the undamaged #3-2 girder reached 40.08 MPa, while that of the damaged #3-2 girder was 32.83 MPa, which was 18.08% less, and that of the repaired #3-2 girder strengthened by steel plates was 39.32 Mpa, which was 19.77% more than that of the damaged #3-2 girder. When the failure load was reached, the maximum tensile stress of the longitudinal steel bars of the three girders was 395 Mpa, which proved that the steel bars yielded. However, the pretensioned strands did not yield at this time. The tensile stress of the pretensioned strands of the damaged #3-2 girder was 1688.72 Mpa, while that of the repaired #3-2 girder strengthened by steel plates was 1586.09 MPa and that of the undamaged #3-2 girder was 1541.49 MPa. The results showed that the strengthening methods had a remarkable strengthening impact, effectively reducing the tensile stress of the pretensioned strands.

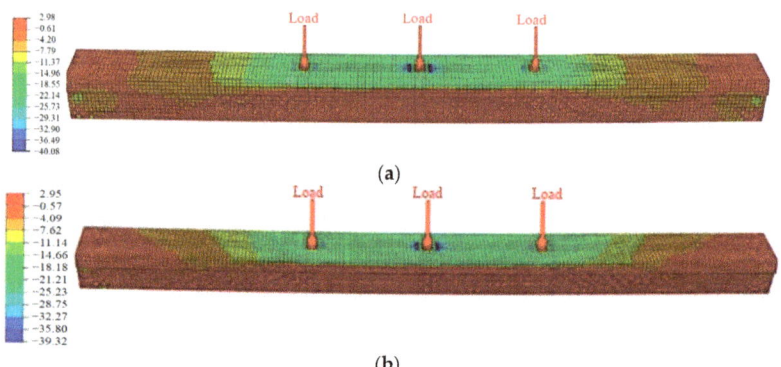

Figure 14. *Cont.*

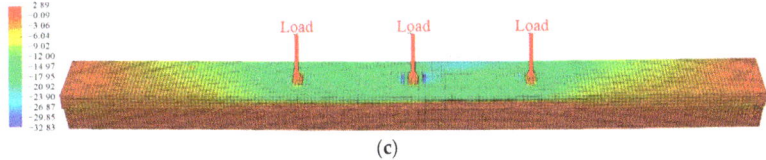

(c)

Figure 14. Stress diagrams of the concrete. (**a**) The undamaged #3-2 girder; (**b**) The repaired #3-2 girder strengthened by steel plates; (**c**) The damaged #3-2 girder.

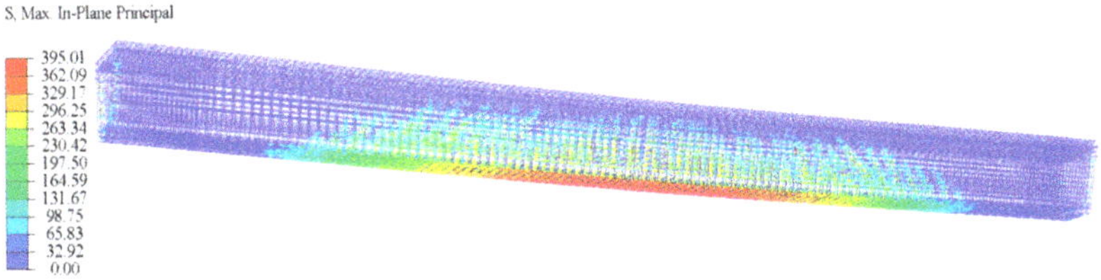

(a)

(b)

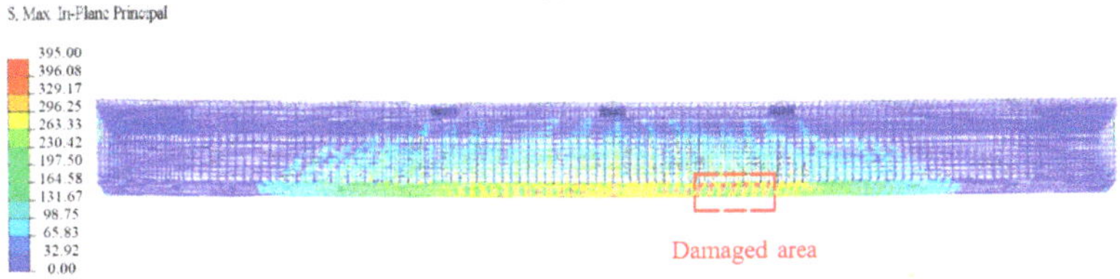

(c)

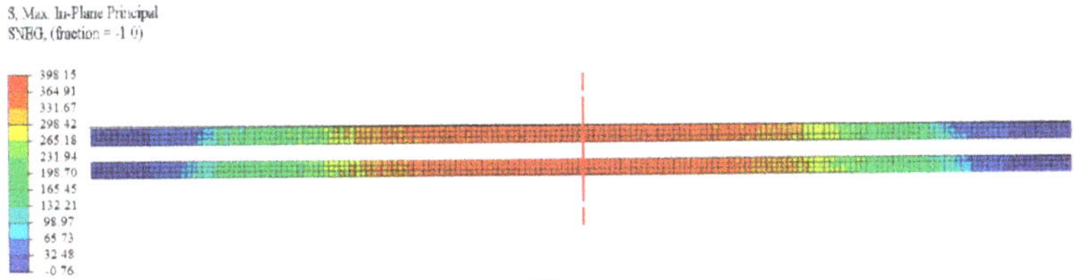

(d)

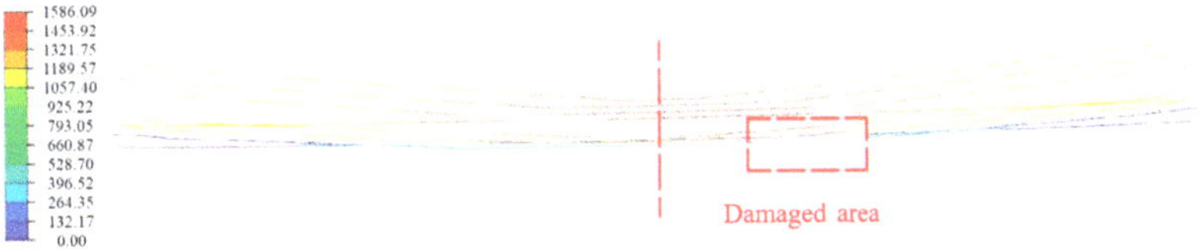

(e)

Figure 15. *Cont.*

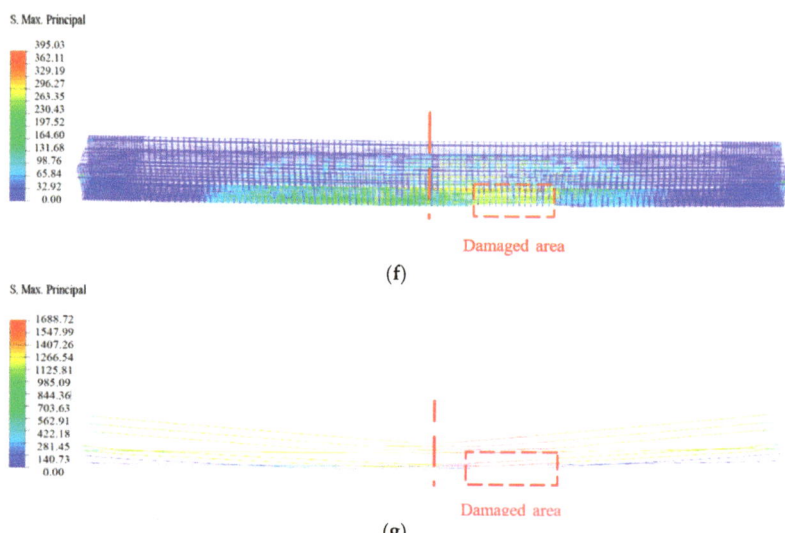

Figure 15. Stress diagrams of the steel bars, steel plates and pretensioned strands. (**a**) The steel bars of the undamaged #3-2 girder; (**b**) The pretensioned strands of the undamaged #3-2 girder; (**c**) The steel bars of the repaired #3-2 girder strengthened by steel plates; (**d**) The steel plates of the repaired #3-2 girder strengthened by steel plates; (**e**) The pretensioned strands of the repaired #3-2 girder strengthened by steel plates; (**f**) The steel bars of the damaged #3-2 girder; (**g**) The pretensioned strands of the damaged #3-2 girder.

4.5. Ultimate Flexural Capacity

Table 4 shows the predicted values of the ultimate flexural capacity and cracking moment of the repaired #3-2 girder strengthened by steel plates, the undamaged #3-2 girder and the damaged #3-2 girder based on the refined analysis. It shows that the bearing capacity and the cracking moment decreased by 30.9% and 32.8%, respectively, for the damaged #3-2 girder compared with the undamaged #3-2 girder due to the collision by the overhigh vehicle. This indicates that structural damage such as concrete shedding, the fracture of steel bars and the fracture of pretensioned strands greatly reduced the flexural bearing capacity and crack resistance of the girder.

Table 4. The predicted values of the ultimate bending capacity and cracking moment of the #3-2 girder.

Bending Moment	Undamaged #3-2 Girder Value/kN·m	Damaged #3-2 Girder Value/kN·m	Refined Model of the Repaired #3-2 Girder Strengthened by Steel Plates Value/kN·m
Cracking moment	7272	4887	5748
Ultimate bending moment	13,831	9551	13,367

5. Destructive Test

5.1. Loading System

To explore the recovery degree of the flexural capacity and the failure process of the repaired girder strengthened by steel plates, the repaired #3-2 girder strengthened by steel plates and the undamaged #4-1 girder were selected for a destructive test. Limited by the test site and conditions, three loading points were set for the destructive test of the repaired

#3-2 girder strengthened by steel plates and the undamaged #4-1 girder, which were loaded by three jacks with a peak output force of 2000 kN. The distance between adjacent loading points was 4.0 m, and Figure 16a shows the loading position of the test box girder. The loading frame was assembled with slot steel and anchored on the concrete foundation, and its design bearing capacity was greater than the design load in the test, which ensured the safety and accuracy of the test. To prevent damage to the concrete near the loading position prior to the test from stress concentration, a steel plate with the following measurements was placed between the jacks and the girder: 60 cm × 60 cm × 3 cm. A pressure sensor with a range of 0~2000 kN was installed between the jack and the girder on the reaction frame. The test load and the test loading rate were controlled through measurements made by the pressure sensor during the test. The loading system is shown in Figure 16.

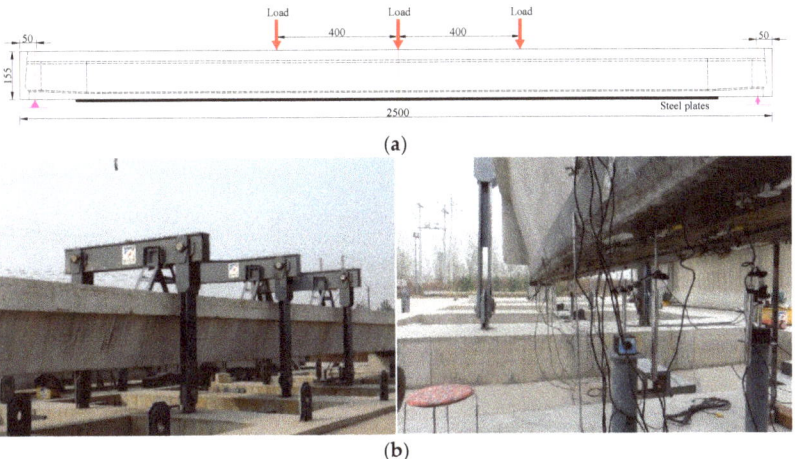

Figure 16. Loading layout of the destructive test. (**a**) Loading location (Unit: cm); (**b**) Field loading system.

5.2. Loading Scheme

The loading procedure used the hierarchical loading method and was divided into three loading stages. First, the girder was loaded to 70% of the predicted value of the cracking load of the undamaged #4-1 girder and then unloaded. The second load was applied to 85% of the predicted value of the ultimate load and then unloaded, and the third load was loaded until the girder was damaged. During the loading of the undamaged #4-1 girder, 50 kN was applied at each stage. Considering that the repaired #3-2 girder strengthened by steel plates was repaired and strengthened based on the damaged girder, the load step was adjusted from 50 kN to 40 kN to ensure the safe and smooth progress of the test. When the load of each jack exceeded 450 kN, displacement control was employed to deliver the load until the box girder was damaged.

The strain and deformation measurement points were arranged at the key section of the repaired #3-2 girder strengthened by steel plates and the undamaged #4-1 girder. To facilitate comparison and analysis with the load test results of the undamaged #4-1 girder, the web strain measurement points, bottom displacement measurement points and bottom steel bar strain measurement points of the repaired #3-2 girder strengthened by steel plates were consistent with those of the undamaged #4-1 girder. The concrete strain measurement points on the bottom plate were slightly adjusted. Figure 17 shows the precise locations and labels of the strain and deformation measurement points. The protective concrete layer at the installation area was removed before installing the steel bar strain sensors, and the steel bars' surfaces were then sanded until they were smooth. Throughout the destructive test, all kinds of sensor data were monitored and collected.

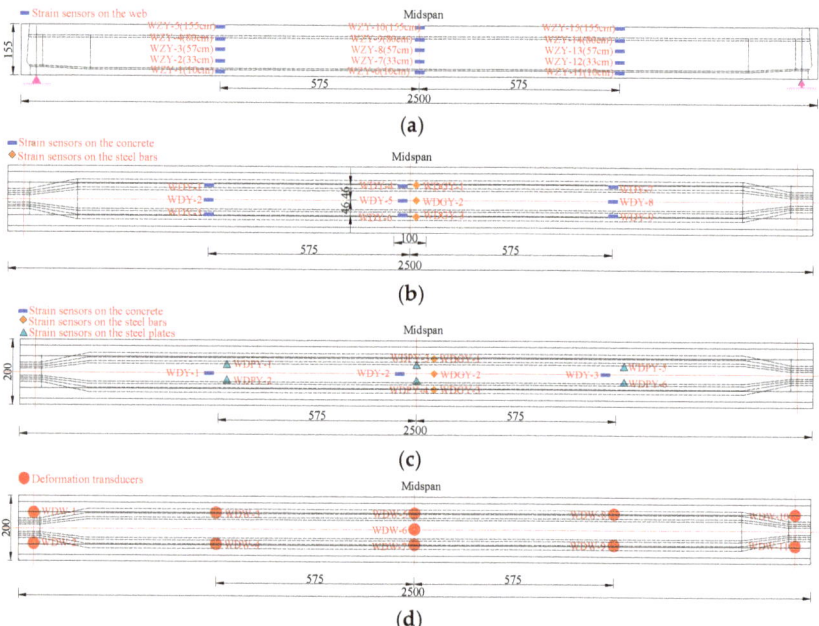

Figure 17. Sensor arrangements for the destructive test (Unit: cm). (**a**) Strain sensor arrangement at the concrete web; (**b**) Strain sensor arrangement at the concrete bottom and steel bars of the undamaged #4-1 girder; (**c**) Strain sensor arrangement at the concrete bottom and steel bars of the repaired #3-2 girder strengthened by steel plates; (**d**) Deformation sensor arrangement at the bottom of the girder.

5.3. Test Results

Section 3,4 of a previous paper provided details of the destructive tests of the undamaged #4-1 girder [24]. Figure 18 shows the load vs. deformation curves of the midspan section of the repaired #3-2 girder strengthened by steel plates under each cyclic loading, which were obtained in accordance with the destructive test and compared with the results of the refined analysis, and Figure 19 shows the deformation curves along the girder length at each load stage. Figure 18 indicates that the destructive process of the repaired #4-1 girder strengthened by steel plates, which included the elastic stage, the working stage with cracks and the destructive stage when the steel bars succumbed, was exactly the same as the undamaged #3-2 girder. Moreover, the finite element analysis results for the girders after improvement were similar to the destructive test results. When each stage was loaded with 50 kN, the undamaged #4-1 girder deformed by around 2.9 mm in the elastic stage. When loaded with 40 kN, the repaired #3-2 girder that had been strengthened by steel plates deformed by approximately 2.5 mm in each stage. Cracks appeared at the bottom of the undamaged #4-1 girder when the load was 362 kN, and the cracking load of the repaired #3-2 girder strengthened by steel plates was 240 kN, which was 33.7% less than that of the undamaged #4-1 girder. The steel bars of the undamaged #4-1 girder began to yield when the load was 801 kN, and the steel bars' yield load in the repaired #3-2 girder strengthened by steel plates was 740 kN, which was 7.6% less than that of the undamaged #4-1 girder. When the load reached 850 kN, the top concrete of the undamaged #4-1 girder was scrunched, and the failure load of the repaired #3-2 girder strengthened by steel plates was 802 kN, which was 5.6% less than that of the undamaged #4-1 girder. When the repaired #3-2 girder strengthened by steel plates was damaged, the maximum crack width was 1.53 mm, which was 24.6% less than that of the undamaged #4-1 girder, and the maximum deformation was 145.8 mm, which was 2.7% more than that of the undamaged

#4-1 girder. Obviously, the girder's ductility has improved, and the cracking degree was reduced by strengthening using steel plates. Figure 14 indicates that under symmetric load conditions, the deformation of the two girders maintained good symmetry.

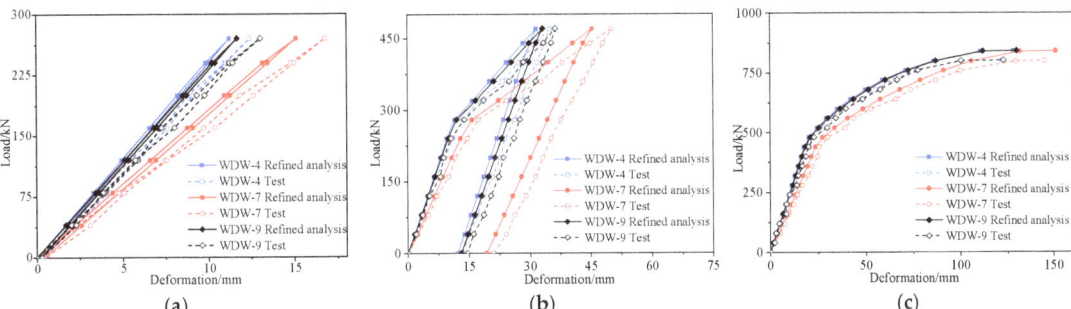

Figure 18. Deformation vs. load of the repaired #3-2 girder strengthened by steel plates. (**a**) Elastic stage; (**b**) Working stage with cracks; (**c**) Failure stage.

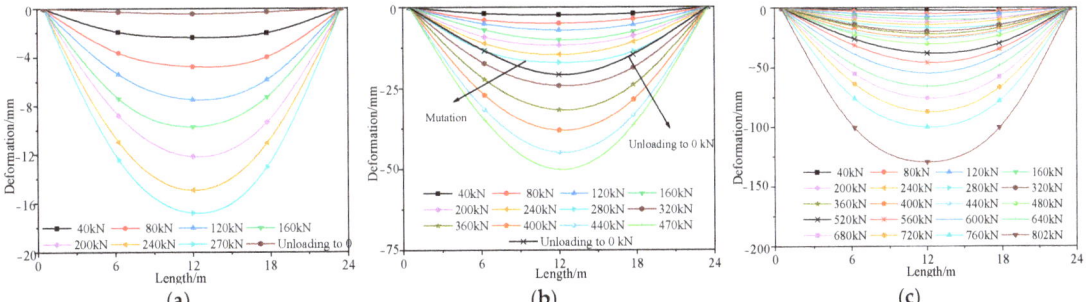

Figure 19. Deformation curve of the bottom plate of the repaired #3-2 girder strengthened by steel plates. (**a**) Elastic stage; (**b**) Working stage with cracks; (**c**) Failure stage.

Figure 18 also shows how the deformation of the undamaged #4-1 girder during the elastic stage in the refined model was approximately 2.6 mm in each stage loaded with 50 kN, which was 10.3% less than the destructive test result. The deformation of the repaired #3-2 girder strengthened by steel plates was approximately 2.2 mm in each stage loaded with 40 kN, which was 12% less than the result of the destructive test. The cracking load of the undamaged #4-1 girder was 361 kN, and that of the repaired #3-2 girder strengthened by steel plates was 260 kN. The failure loads of the undamaged #4-1 girder and the repaired #3-2 girder strengthened by steel plates were, respectively, 875 and 840 kN, which were 2.9% and 4.7% more than the destructive test results.

Figures 20 and 21 show, respectively, the strain vs. load curves for the web and bottom plate of the midspan section of the repaired #3-2 girder strengthened by steel plates that were obtained during the destructive test. Figure 20 indicates that the concrete strains of the right web under the same load were somewhat larger for the undamaged #4-1 girder and the repaired #3-2 girder strengthened by steel plates because the left web is taller than the right web. Figures 20 and 21 show that the strains of the two girders exhibited an obvious three-stage development process. The strain vs. load curves of the undamaged #4-1 girder and the repaired #3-2 girder strengthened by steel plates, respectively, each showed an inflection point because of the appearance of cracks when the loads reached 362 and 240 kN. Before this inflection point, both the steel bars and concrete were in the elastic stage, and the strains and load states were basically linear. The reason that the crack load of the repaired #3-2 girder strengthened by steel plates was 33.7% less than that of the undamaged #4-1 girder was because of the fracture of the pretensioned strands. The steel bars of the undamaged #4-1 girder began to yield when the load reached 801 kN during the destructive test, and the steel bars' strain was 2008 µε. The yield load of the steel bars of the repaired #3-2 girder strengthened by steel plates was reduced by 7.6%, and the strain of the steel bars was 2009 µε. In the serviceability state, the midspan strain of the repaired #3-2 girder strengthened by steel plates was 984.3 µε, which was 29.6% more than that of the undamaged #4-1 girder.

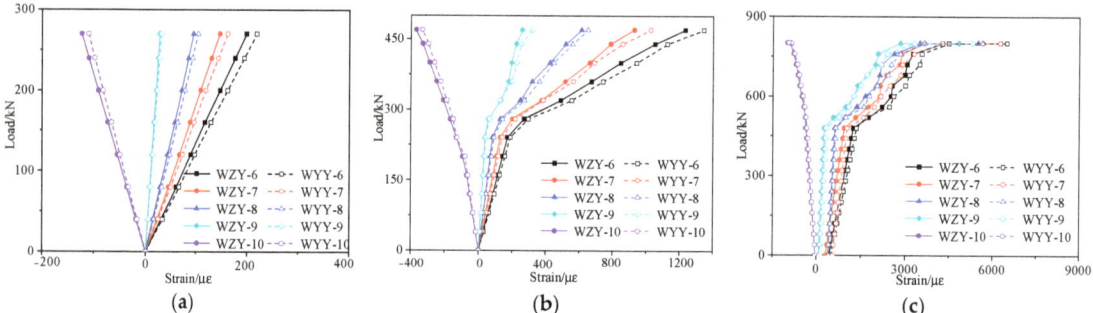

Figure 20. Strain vs. load curves of the concrete web at midspan of the repaired #3-2 girder strengthened by steel plates. (**a**) Elastic stage; (**b**) Working stage with cracks; (**c**) The failure stage.

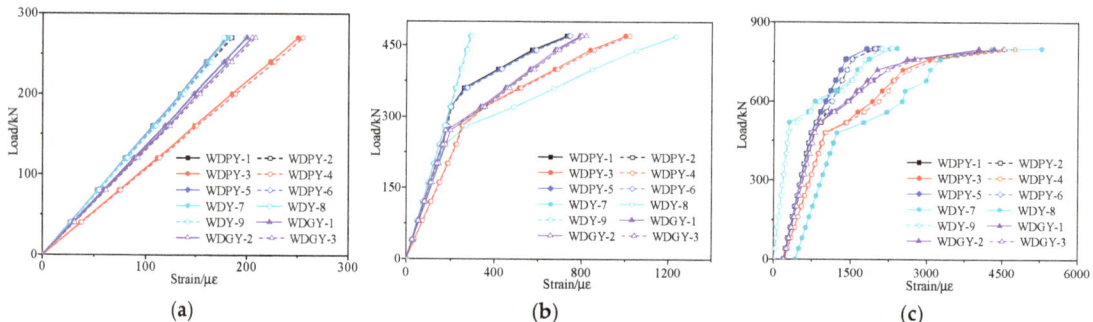

Figure 21. Strain vs. load curves of the repaired #3-2 girder strengthened by steel plates. (**a**) Elastic stage; (**b**) Working stage with cracks; (**c**) The failure stage.

5.4. Ultimate Flexural Capacity

The cracking moment and the actual ultimate flexural capacity of the box girders was calculated from the sum of bending moments by the combination of cracking load, actual failure load and dead load, and then compared with the theoretical value of the material's design strength and the predicted value derived from the finite element analysis, as shown in Table 5. It shows that the values of the cracking moment and ultimate moment predicted by the refined analysis were similar to the destructive test results. The errors of the cracking moment and ultimate moment of the undamaged #4-1 girder were 3.4% and 2.5%, respectively, and the errors of the repaired #3-2 girder strengthened by steel plates were 5.4% and 3.9%, respectively. These results showed that the refined analysis accurately predicted the actual failure process and ultimate flexural capacity of the undamaged #4-1 girder and the repaired #3-2 girder strengthened by steel plates. The actual bearing capacities of the undamaged #4-1 girder and the repaired #3-2 girder strengthened by steel plates were approximately 1.60 and 1.63 times the theoretical values calculated according to the design strength, respectively, which indicated that each of the two girders had a large safety reserve. Compared with the undamaged #4-1 girder, the ultimate flexural capacity of the repaired #3-2 girder strengthened by steel plates reached 95.3%, the maximum deformation was 103% when the girder was damaged and the cracking load was only 66.3%, which indicated that the structural repair and strengthening measures basically restored the ultimate bearing capacity of the damaged girder to the level of the undamaged girder and improved the structural ductility. However, the cracking load of the damaged box girder did not improve to the desired level. The maximum bending moment of the undamaged single girder of the bridge under the combination of the six-axle vehicle load and the dead load was 6 084 kN·m, which was less than the design value of the bearing capacity of the component and met the specification. Additionally, the bearing capacities of the undamaged #4-1 girder and the repaired #3-2 girder strengthened by steel plates from the destructive test were approximately 2.2 and 2.1 times the maximum load effect, respectively.

Table 5. Comparison of the predicted and experimental results.

Girder	Bending Moment	Theoretical Value/kN·m	Refined Analysis Value/kN·m	Refined Analysis Error/%	Destructive Test Value/kN·m	Destructive Test Error/%
Undamaged #4-1 girder	Cracking moment	6984	7272	4.1	7034	0.7
	Ultimate bending moment	8442	13,831	63.8	13,500	60.0
Repaired #3-2 girder strengthened by steel plates	Cracking moment	5124	5682	10.9	5417	5.7
	Ultimate bending moment	7913	13,367	63	12,864	53.6

6. Comparative Analysis

To further research the effect of the strengthening measures by steel plates for the improvement of the bearing capacity of the damaged girder and the degree of the recovery of the bearing capacity to the level of the undamaged #4-1 girder, the girders were compared and analyzed in terms of their flexural stiffness, flexural deformation and stress and crack width in accordance with the load combination given in the specification. Because the destructive test was performed after the girder had been strengthened with steel plates, the destructive test of the damaged #3-2 girder was lacking. However, according to research on the damaged #4-2 girder [24], which found that the refined analysis could simulate the actual damage process and bearing capacity of the damaged girder accurately, the results of the refined analysis of the damaged #3-2 girder were used instead of the results of the destructive test. The load included the dead load, secondary load, design live load and loads of three-axle, five-axle and six-axle vehicles given in the specification, which were denoted DL, SL, DLL, Type 1, Type 2 and Type 3, respectively. Schematic diagrams and lateral layouts of the vehicles are shown in Figure 22.

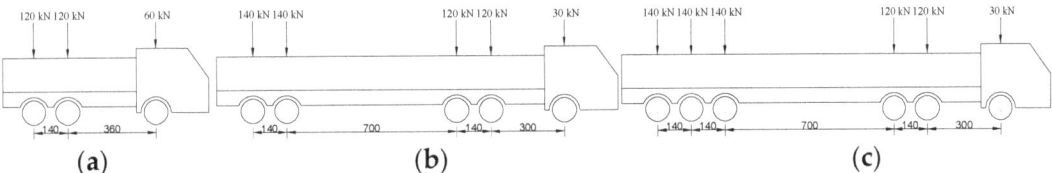

Figure 22. Cont.

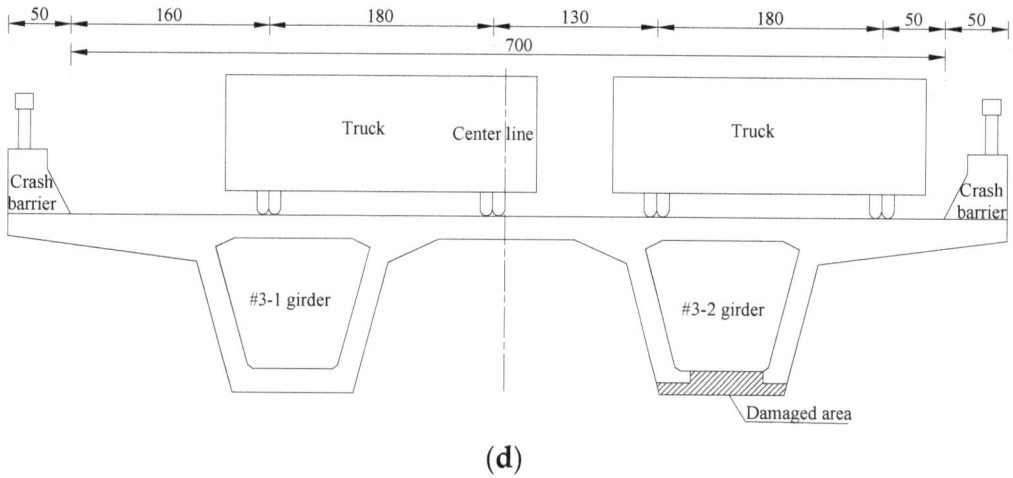

Figure 22. Different types of vehicles and horizontal arrangement. (**a**) Type 1; (**b**) Type 2; (**c**) Type 3; (**d**) Horizontal arrangement.

6.1. Stiffness

The flexural stiffness of bridges is an essential index used to assess the ability of bridges to resist deformation. Bridge damage is often accompanied by stiffness degradation, which can indirectly reflect the damage in the process of the bridge destructive test. According to the destructive test of the repaired #3-2 girder strengthened by steel plates, the undamaged #4-1 girder and the refined analysis of the damaged #3-2 girder, Figure 23 shows the load vs. stiffness curves for the three girders. It indicates that in the elastic stage, the flexural stiffness of the girders was the same as the initial flexural stiffness. Because of the repair of the broken steel bars and damaged concrete and strengthening by steel plates, the stiffness of the repaired #3-2 girder strengthened by steel plates increased by 6.6% compared with the damaged #3-2 girder, and reached 94.3% of that of the undamaged #4-1 girder. When the undamaged #4-1 girder, the repaired #3-2 girder strengthened by steel plates and the damaged #3-2 girder were loaded to 362, 240 and 200 kN, respectively, the flexural stiffness of the box girders decreased significantly, indicating that new cracks formed in the box girders, and the girders arrived at the working stage with cracks. The stiffness of the three girders showed a decreasing trend due to the continuous development of concrete cracking in the whole working stage with cracks. However, unlike the slow decline of the undamaged #4-1 girder and the damaged #3-2 girder, the stiffness dropped sharply due to the yielding of the steel plates when the repaired #3-2 girder strengthened by steel plates was loaded to 540 kN. When the undamaged #4-1 girder, the repaired #3-2 girder strengthened by steel plates and the damaged #3-2 girder were respectively loaded to 801, 740 and 410 kN, the flexural stiffness decreased sharply due to the yielding of the steel bars. Under the combination of different live loads, secondary load and dead load, the stiffness of the repaired #3-2 girder strengthened by steel plates increased by 27.1%, 17.5% and 18.3% of the stiffness of the damaged #3-2 girder and reached 72.7%, 51.9% and 56.4% of the stiffness of the undamaged #4-1 girder.

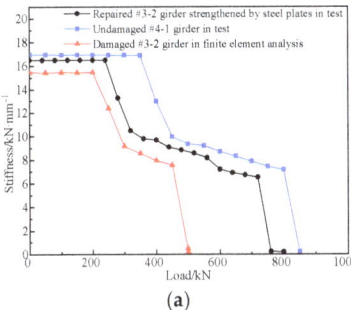
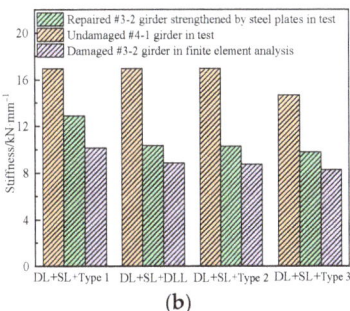

Figure 23. Stiffness comparison. (**a**) Load vs. stiffness curve; (**b**) Stiffness degradation under different live loads.

6.2. Flexural Deformation

Basis on the refined analysis of the damaged #3-2 girder and the destructive test of the undamaged #4-1 girder and the repaired #3-2 girder strengthened by steel plates, Figure 24 shows the load vs. midspan deformation curves of the girders. For 0.7 times the 5-axle vehicle load, the midspan deformations of the undamaged #4-1 girder, the repaired #3-2 girder strengthened by steel plates and the damaged #3-2 girder were 7.6, 8.5 and 9.9 mm, and the flexure to span ratios were 1/3158, 1/2824 and 1/2424, respectively. The deformation of the repaired #3-2 girder strengthened by steel plates was 14.1% less compared with that of the damaged #3-2 girder and 11.8% more than the undamaged #4-1 girder, but both met the 1/600 requirement of the specification. For the 0.7 times six-axle vehicle load, the midspan deformations of the three girders were 10.8, 11.2 and 12.8 mm, and the flexure-to-span ratios were 1/2222, 1/2143 and 1/1875, respectively. The deformation of the repaired #3-2 girder strengthened by steel plates was 12.5% less compared with that of the damaged #3-2 girder and 3.7% more compared with that of the undamaged #4-1 girder. Thus, the deformations of the repaired #3-2 girder strengthened by steel plates for different live loads were significantly less than those of the damaged #3-2 girder, but slightly more than those of the undamaged #4-1 girder. The former difference was because the strengthening measures by the steel plates effectively increased the stiffness of the damaged #3-2 girder. The latter difference was because the broken pretensioned strands of the repaired #3-2 girder strengthened by steel plates were not repaired, which decreased the stiffness. However, the safety reserve of the repaired #3-2 girder strengthened by steel plates was still large.

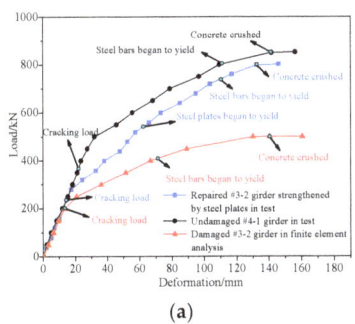
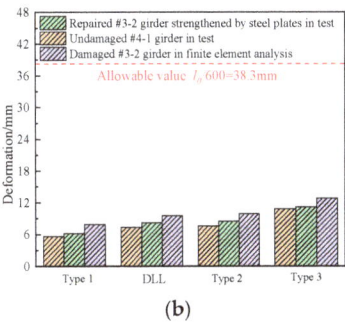

Figure 24. Midspan deformation comparison. (**a**) Load vs. deformation curve; (**b**) Midspan deformation under different live loads.

6.3. Stress

Basis on the refined analysis of the damaged #3-2 girder and the destructive test of the undamaged #4-1 girder and the repaired #3-2 girder, Figure 25 shows the stress diagrams of the girders. It indicates that for the combination of 0.7 times the six-axle vehicle load, secondary load and dead load, the maximum concrete stress of the repaired #3-2 girder strengthened by steel plates was approximately 8.4 MPa, which was 31.9% less than that of the damaged #3-2 girder and 11.1% more than that of the undamaged #4-1 girder, however, all were far less than the allowable stress of 16.2 MPa in the design specification. The maximum steel bar stress of the repaired #3-2 girder strengthened by steel plates was approximately 69.8 MPa, which was 60.3% less than that of the damaged #3-2 girder and 13.8% more than that of the undamaged #4-1 girder. Thus, the steel plates were an effective strengthening measure. The reason that the stresses of the repaired #3-2 girder strengthened by steel plates were reduced compared to those of the damaged #3-2 girder was that the areas of the concrete and steel bars increased, which caused the stiffness of the whole section to increase, and the repaired concrete and steel bars partially bore the stress. The reason that the stresses of the repaired #3-2 girder strengthened by steel plates increased compared to those of the undamaged #4-1 girder was that the area of the strands decreased, which led to the reduction of the stiffness of whole section, and the remaining steel bars, strands and concrete bore greater stress.

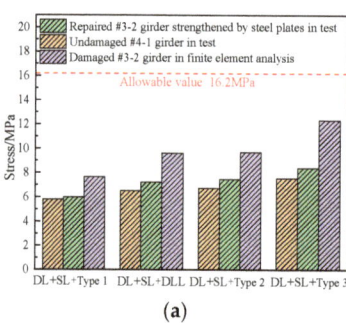
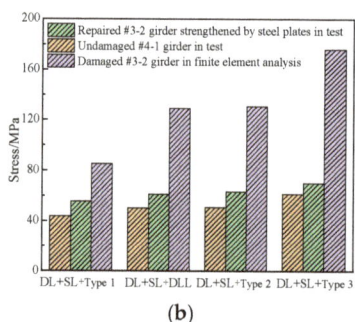

Figure 25. Stress comparison (**a**) Stress of concrete; (**b**) Stress of longitudinal steel bars.

6.4. Crack Width

Basis on the refined analysis of the damaged #3-2 girder and the destructive test of the undamaged #4-1 girder and the repaired #3-2 girder, Figure 26 shows the load vs. crack width diagrams of the girders. The maximum crack width of the repaired #3-2 girder strengthened by steel plates were approximately 0.05 mm and 0.08 mm, which were 87.3% and 82.2% less than that of the damaged #3-2 girder and 28.6% and 20% less than that of the undamaged #4-1 girder. This indicated that the crack resistance of the repaired #3-2 girder strengthened by steel plates not only recovered to the level of the undamaged #4-1 girder, but also improved to some extent. This was because the steel plates attached to the girder effectively protected the concrete and limited the development of cracks.

6.5. Traffic Load Capacity

To verify the recovery degree of the bearing capacity of the repaired #3-2 girder strengthened by steel plates, three standard vehicle loads were considered, i.e., three-axle, five-axle and six-axle vehicles. Then, the maximum bending moment effect under the combined action of the vehicle load, secondary load and dead load was obtained and compared with the flexural capacity of the repaired #3-2 girder strengthened by steel plates, the damaged #3-2 girder and the undamaged #4-1 girder to calculate the safety factor of the vehicles, as shown in Table 6.

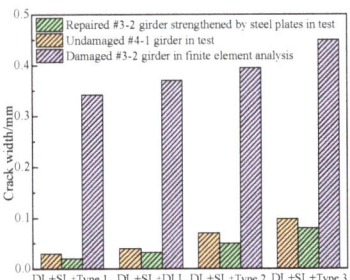

Figure 26. Crack width comparison.

Table 6. Safety coefficient for different types of vehicles.

	Bending Moment		Safety Coefficient					
			Ultimate Bearing Capacity of the Undamaged #4-1 Girder		Ultimate Bearing Capacity of the Repaired #3-2 Girder Strengthened by Steel Plates		Ultimate Bearing Capacity of the Damaged #3-2 Girder	
Type of Truck	The Dead Load	The Different Types of Vehicles	Theoretical Prediction	Destructive Test	Theoretical Prediction	Destructive Test	Theoretical Prediction	Refined Model
	/kN·m	/kN·m	(8442 kN·m)	(13,500 kN·m)	(7913 kN·m)	(12,864 kN·m)	(5127 kN·m)	(9551 kN·m)
1	2 053	1 720	3.71	6.66	3.41	6.29	1.79	4.36
2	2 053	2 301	2.78	4.97	2.55	4.70	1.34	3.26
3	2 053	2 947	2.17	3.88	1.99	3.67	1.04	2.54

Table 6 shows that the flexural capacity of the repaired #3-2 girder strengthened by steel plates through the destructive test was 62.6% more than that obtained according to the specification [32] and 36.7% more than that of the damaged #3-2 girder, reaching 95.3% of that of the undamaged #4-1 girder. This was because the strengthening measures by steel plates made the concrete, steel bars and strengthened steel plates bear the same force. The stiffness of the girder improved. The theoretical flexural bearing capacity of the repaired #3-2 girder strengthened by steel plates reached 93.7% of the level of the undamaged #4-1 girder, and the safety coefficient of the six-axle vehicle was less than 2.0. However, the actual safety coefficients of the three-axle, five-axle and six-axle vehicles were greater than 3.0 according to the destructive test, which indicated that the girder had a good capacity for vehicles.

7. Conclusions

The damaged #3-2 girder of an in-service prestressed concrete bridge with continuous girders was obtained by demolishing, repairing and strengthening. A destructive test of the strengthened #3-2 girder and the undamaged #4-1 girder was performed. Refined f analysis was conducted on the repaired #3-2 girder strengthened by steel plates, the undamaged #4-1 girder and the damaged #3-2 girder. The ultimate bending capacities of the three girders were compared and analyzed, and the following conclusions were drawn.

(1) The ultimate flexural bending capacities of the repaired #3-2 girder strengthened by steel plates and the undamaged #4-1 girder were obtained from a destructive test, and that of the damaged #3-2 girder was obtained by a refined analysis. The actual flexural bearing capacity of the repaired #3-2 girder strengthened by steel plates was 1.63 times the theoretical bearing capacity, which was calculated according to the specification [25], and 36.7% more than that of the damaged #3-2 girder, reaching 95.3% of that of the undamaged #4-1 girder. Although the theoretical flexural capacity of the repaired #3-2 girder strengthened by steel plates failed to meet the traffic requirements for six-axle vehicles, the actual safety factor was greater than three, which indicated a good capacity for vehicle traffic.

(2) The analysis showed that compared with the damaged #3-2 girder, the cracking moment of the repaired #3-2 girder strengthened by steel plates was 12.5% less, the stiffness in the elastic stage was 6.6% more and the maximum deformation was 9.2% less. Compared with the undamaged #4-1 girder, the cracking moment of the repaired #3-2 girder strengthened by steel plates reached 66.3%, the maximum crack width decreased by 24.6% and the stiffness in the elastic stage reached 94.3%. When the repaired #3-2 girder strengthened by steel plates finally failed, the maximum deflection was 2.7% more than the undamaged #4-1 girder. This showed that repaired and strengthening the damaged #3-2 girder improved the ductility, reduced the degree of cracking and increased the durability. Additionally, these results showed that first repairing the damaged #3-2 girder and then strengthening it with steel plates was an effective maintenance and strengthening measure.

(3) The finite element numerical analysis method based on the material testing results of the bridge reflected the mechanical properties and destruction process of the repaired #3-2 girder strengthened by steel plates and the undamaged #4-1 girder. The load vs. deformation curves obtained by this method were basically consistent with those of the failure test. The bearing capacities of the repaired #3-2 girder strengthened by steel plates and the undamaged #4-1 girder were accurately determined, and the errors with respect to the actual bearing capacities were 2.5% and 3.9%, respectively.

(4) Through destructive tests and a refined analysis, some conclusions about the repaired #3-2 girder strengthened by steel plates, damaged #4-2 girder and undamaged #4-1 girder were drawn. However, the improvement of the bearing capacity of an undamaged girder after strengthening has not been explored. Therefore, the destructive test and comparative analysis of the undamaged girder strengthened by CFRP will be carried out in the future.

Author Contributions: Conceptualization, Y.L. (Yong Li) and Y.L. (Yongqian Liu); methodology, Y.L. (Yong Li), Y.L. (Yongqian Liu) and Z.Y.; software, Z.Y.; formal analysis, Y.L. (Yong Li), Y.L. (Yongqian Liu) and Z.Y.; experiment, Y.L. (Yong Li), Y.L. (Yongqian Liu) and Z.Y.; data curation, Y.L. (Yong Li), Y.L. (Yongqian Liu) and Z.Y.; writing—original draft preparation, Y.L. (Yong Li), Y.L. (Yongqian Liu) and Z.Y.; writing—review and editing, Y.L. (Yong Li), Y.L. (Yongqian Liu) and Z.Y.; supervision, Y.L. (Yong Li); project administration, Y.L. (Yong Li); funding acquisition, Y.L. (Yong Li) and Z.Y. All authors have read and agreed to the published version of the manuscript.

Funding: This study is supported by the S & T program of Hebei Province in China (Grant Nos: 21375403D, 20375410D), the Natural Science Foundation of Hebei Province in China (Grant No: E2020210017), the Excellent Youth Project of Shijiazhuang Tiedao University (2018), and the Graduate Innovation Grant Projects of Hebei Province (Grant No: CXZZSS2022110).

Institutional Review Board Statement: Not applicable.

Informed Consent Statement: Not applicable.

Data Availability Statement: No new data were created or analyzed in this study. Data sharing is not applicable to this article.

Conflicts of Interest: The authors declare no conflict of interest.

References

1. Shubbar, A.; Al-khafaji, Z.; Nasr, M.; Falah, M. Using non-destructive tests for evaluating flyover footbridge: Case study. *Knowl. -Based Eng. Sci.* **2020**, *1*, 23–39. [CrossRef]
2. Russo, F.M.; Wipf, T.J.; Klaiber, F.W. Diagnostic Load Tests of a Prestressed Concrete Bridge Damaged by Overheight Vehicle Impact. *Transp. Res. Rec.* **2000**, *1696*, 103–110.
3. Tho, N.C.; Van Thom, D.; Cong, P.H.; Zenkour, A.M.; Doan, D.H.; Van Minh, P. Finite element modeling of the bending and vibration behavior of three-layer composite plates with a crack in the core layer. *Compos. Struct.* **2023**, *305*, 116529.
4. Thai, D.N.; Van Minh, P.; Hoang, C.P.; Duc, T.T.; Cam, N.N.T. Bending of symmetric sandwich FGM beams with shear connectors. *Math. Probl. Eng.* **2021**, *2021*, 1–15. [CrossRef]
5. Dung, N.T.; Van Minh, P.; Hung, H.M.; Tien, D.M. The third-order shear deformation theory for modeling the static bending and dynamic responses of piezoelectric bidirectional functionally graded plates. *Adv. Mater. Sci. Eng.* **2021**, *2021*, 1–15. [CrossRef]

6. *JTG 3362-2018*; Specifications for Design of Highway Reinforced Concrete and Prestressed Concrete Bridges and Culverts. China Communication Press: Beijing, China, 2018.
7. Abbas, H.Q.; Al-Zuhairi, A.H. Flexural Strengthening of Prestressed Girders with Partially Damaged Strands Using Enhancement of Carbon Fiber Laminates by End Sheet Anchorages. *Eng. Technol. Appl. Sci. Res.* **2022**, *12*, 8884–8890. [CrossRef]
8. Shukla, A.K.; Goswami, P.; Maiti, P.R.; Rai, G. Damage Analysis of Simple RC Bridge Girder and its Strengthening Measures. *J. Fail. Anal. Prev.* **2021**, *21*, 1374–1386. [CrossRef]
9. Hou, W.; Huang, F.; Zhang, K. Flexural Behavior on Damaged Steel Beams Strengthening with CFRP Sheets Subjected to Overloading. *Polymers* **2022**, *14*, 1419. [CrossRef]
10. Lei, D.; Chen, G.; Chen, Y.; Ren, Q. Experimental research and numerical simulation of RC beams strengthening with bonded steel plates. *Sci. China Technol. Sci.* **2012**, *55*, 3270–3277. [CrossRef]
11. Al-Hassani, H.M.; Al-Ta'an, S.A.; Mohammed, A.A. Behavior of damaged reinforced concrete beams strengthening with externally bonded steel plate. *Tikrit J. Eng. Sci.* **2013**, *20*, 48–59. [CrossRef]
12. Daouadji, T.H. Analytical and numerical modeling of interfacial stresses in beams bonded with a thin plate. *Adv. Comput. Des.* **2017**, *2*, 57–69. [CrossRef]
13. Altin, S.; Anil, Z.; Kara, M.E. Improving shear capacity of existing RC beams using external bonding of steel plates. *Eng. Struct.* **2005**, *27*, 781–791. [CrossRef]
14. Aykac, S.; Kalkan, L.; Uysal, A. Strengthening of reinforced concrete beams with expoxy-bonded perforated steel plates. *Struct. Eng. Mech.* **2012**, *44*, 735–751. [CrossRef]
15. Alam, M.A.; Sami, A.; Mustapha, K.N. Embedded connectors to eliminate debonding of steel plate for optimal shear strengthening of RC beam. *Arab. J. Sci. Eng.* **2017**, *42*, 4053–4068. [CrossRef]
16. Esfahani, M.; Kianoush, M.; Tajari, A. Flexural behaviour of reinforced concrete beams strengthening by CFRP sheets. *Eng. Struct.* **2007**, *29*, 2428–2444. [CrossRef]
17. Rasheed, H.A.; Abdalla, J.A.; Hawileh, R.A.; Al-Tamimi, A.K. Flexural behavior of reinforced concrete beams strengthening with externally bonded Aluminum Alloy plates. *Eng. Struct.* **2017**, *147*, 473–485. [CrossRef]
18. Abdalla, J.A.; Abu-Obeidah, A.S.; Hawileh, R.A.; Rasheed, H.A. Shear strengthening of reinforced concrete beams using externally-bonded aluminum alloy plates: An experimental study. *Constr. Build. Mater.* **2016**, *128*, 24–37. [CrossRef]
19. Rakgate, S.M.; Dundu, M. Strength and ductility of simple supported R/C beams retrofitted with steel plates of different width-to-thickness ratios. *Eng. Struct.* **2018**, *40*, 192–202. [CrossRef]
20. Duc, D.H.; Van Minh, P.; Tung, N.S. Finite element modelling for free vibration response of cracked stiffened FGM plates. *Vietnam. J. Sci. Technol.* **2020**, *58*, 119–129.
21. Van Phung, M.; Nguyen, D.T.; Doan, L.T.; Van Duong, T. Numerical investigation on static bending and free vibration responses of two-layer variable thickness plates with shear connectors. *Iran. J. Sci. Technol. Trans. Mech. Eng.* **2022**, *46*, 1047–1065. [CrossRef]
22. Kralovanec, J.; Bahleda, F.; Moravcik, M. State of prestressing analysis of 62-year-old bridge. *Materials* **2022**, *15*, 3583. [CrossRef]
23. Mateckova, P.; Bilek, V.; Sucharda, O. Comparative study of high-performance concrete characteristics and loading test of pretensioned experimental beams. *Crystals* **2021**, *11*, 427. [CrossRef]
24. Li, Y.; Yu, Z.; Wu, Q.; Liu, Y.; Wang, S. Experimental-Numerical Study on the Flexural Ultimate Capacity of Prestressed Concrete Box Girders Subjected to Collision. *Materials* **2022**, *15*, 3949. [CrossRef] [PubMed]
25. Nie, J.; Wang, Y.; Cai, C.S. Experimental research on fatigue behavior of RC beams strengthening with steel plate-concrete composite technique. *J. Struct. Eng.* **2011**, *137*, 772–781. [CrossRef]
26. Li, Y.; Zhang, J.; Chen, J.; Wu, L. Ultimate flexural capacity of a severely damaged reinforced concrete T-girder bridge. *J. Bridge Eng.* **2017**, *22*, 05017003. [CrossRef]
27. Bagge, N.; Popescu, C.; Elfgren, L. Failure tests on concrete bridges: Have we learnt the lessons. *Struct. Infrastruct. Eng.* **2018**, *14*, 292–319. [CrossRef]
28. Zhang, J.; Ren, H.; Yu, B. Failure Testing of a Full-Scale Reinforced Concrete T-Girder Bridge. *Adv. Mater. Res.* **2011**, *1269*, 1767–1773.
29. Miller, R.A.; Aktan, A.E.; Shahrooz, B.M. Destructive Testing of Decommissioned Concrete Slab Bridge. *J. Struct. Eng.* **1994**, *120*, 2176–2198. [CrossRef]
30. Zhang, J.; Peng, H.; Cai, C.S. Destructive Testing of a Decommissioned Reinforced Concrete Bridge. *J. Bridge Eng.* **2013**, *18*, 564–569. [CrossRef]
31. Murray, C.; Arancibia, M.D.; Okumus, P.; Floyd, R.W. Destructive testing and computer modeling of a scale prestressed concrete I-girder bridge. *Eng. Struct.* **2019**, *183*, 195–205. [CrossRef]
32. Martina, Š.; David, L.; Jiří, D.; Drahomír, N. Modeling of degradation processes in concrete: Probabilistic lifetime and load-bearing capacity assessment of existing reinforced concrete bridges. *Eng. Struct.* **2016**, *119*, 49–60.
33. *GB 50017-2017*; Standard for Design of Steel Structures. Ministry of housing and Urban-Rural Development of China: Beijing, China, 2017.

34. *JTGT522-2008*; Specifications for Strengthening Design of Highway Bridges. Ministry of Transport of the People's Republic of China: Beijing, China, 2008.
35. Abaqus, Inc. *ABAQUS Analysis User's Manual*; Abaqus, Inc.: Providence, RI, USA, 2004.

Disclaimer/Publisher's Note: The statements, opinions and data contained in all publications are solely those of the individual author(s) and contributor(s) and not of MDPI and/or the editor(s). MDPI and/or the editor(s) disclaim responsibility for any injury to people or property resulting from any ideas, methods, instructions or products referred to in the content.

Article

Numerical Analysis on Transverse Splicing Structure for the Widening of a Long Multi-Span Highway Concrete Continuous Box Girder Bridge

Wenqing Wu [1,*], Hui Zhang [2], Zheng Liu [1] and Yunpeng Wang [1]

1 School of Transportation, Southeast University, Nanjing 210096, China
2 Quality and Safety Technology Department, Jiangsu Provincial Transportation Engineering Construction Bureau, Nanjing 210004, China
* Correspondence: wuwenqing@seu.edu.cn

Abstract: For the bridge widening of long multi-span highway concrete continuous box girder with a conventional splicing structure, due to the large longitudinal difference deformation by concrete shrinkage and creep between the existing and new ones, the widened structure will have an overlarge bending deformation after widening, especially an obvious transverse deformation at the end of girder, which will lead to structural damage to the newly widened structure. To effectively absorb the difference deformation mentioned above, this study proposes a novel transverse splicing structure based on the folding effect of a corrugated steel plate (CSP) (hereinafter referred to as "the CSP splicing structure"). Then, a finite element structural analysis was performed on the mechanical properties of the widened structure with the CSP splicing structure, and compared to those of a widened structure adopting the conventional concrete splicing mode, to clarify the transverse load transferring mechanism of the structure. Finally, by conducting a sensitivity analysis on CSP thickness, corrugation length, splicing stitch width, and other structural parameters, a sound parameter combination scheme was put forward. According to the research results, to ensure effective utilization of the CSP folding effect, the corrugation pattern direction of CSP should be set as horizontal, and the wave angle as the degree of 90°. In addition, it mitigated the transverse tensile stress to effectively avoid concrete cracking feasibility on the top flange of the box girder at the end of the girder. This study offers a feasible way of avoiding the structural damage produced by an excess transverse deformation at the end of the girder after bridge widening of a long multi-span concrete continuous box girder.

Keywords: bridge widening; transverse stress; corrugated sheet plate (CSP); shrinkage and creep; settlement difference; parameter optimization

1. Introduction

Economic development and population growth have put tremendous pressure on transportation services. Therefore, many existing highway bridges cannot meet the current increased traffic demands. Widening existing highway bridges not only requires less time to improve their capacity but also saves investment. A common way to do this is to build a new bridge side by side with the existing one and structurally connect the two bridge decks with an in situ concrete stitching slab. At present, the huge demand for bridge widening mainly comes from highway bridges, not railway bridges.

Currently, integral widening, in which the new section of the structure is made continuous with the existing one, is attractive because it avoids the complication of a movement joint that would often be below a running carriageway [1]. Under this scheme, the substructures of new and existing bridges are separated from each other, while their superstructures are transversely spliced together through structural connections and collectively subjected to overall stress, as shown in Figure 1.

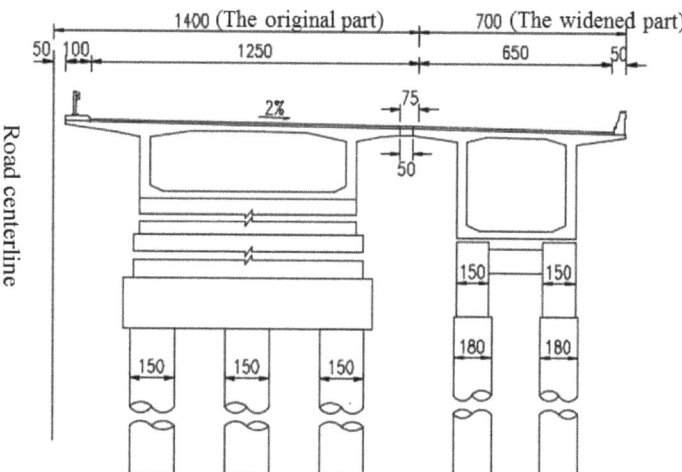

Figure 1. Transverse widening mode of connection superstructures but disconnection substructures.

The interactions between the new and existing bridges are incurred by their different completion times, producing restraint stresses in the new and existing structures. For instance, the significant differences in their different completion times between the new and existing bridges would result in a series of deformation differences between them, including the longitudinal shrinkage and creep deformation of concrete, vertical foundation settlement, and so forth [1]. To be specific, the longitudinal concrete shrinkage and creep deformation of the new bridge are restrained by the existing bridge, which not only produces secondary structural internal forces but also causes overall transverse bending deformation in the widened bridge [2–6], as shown in Figure 2.

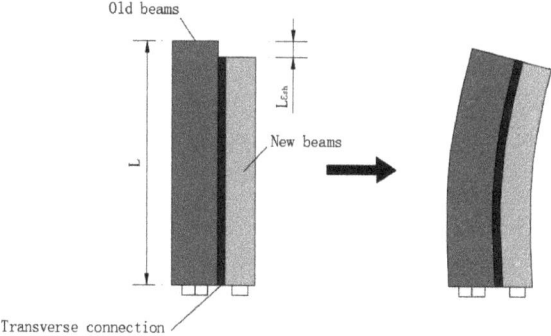

(a) Shrink before splicing   (b) Shrinkage after splicing

Figure 2. Transverse deformation of widening structure induced by concrete shrinkage.

A practical engineering case described in the literature [3] involves a prestressed concrete continuous box girder bridge with a multi-span length of 38.5 + 65 + 38.5 m, a total length of 142 m, and a net single-deck width of 12.5 m. In the present widening mode, a prestressed concrete continuous box girder with the same span layout was placed on one side of the existing bridge. After a deck width increase of 8.25 m, the single-deck bridge had a full deck width of 20.75 m. Within one year after splicing widening, the structure showed an obvious transverse bending deformation, characterized by outward transverse deformation at both ends of the continuous girder. Due to the huge deformation (as large as 45 mm), the end of the box girder seriously squeezed the external anti-knock block, cracking the anti-knock block and causing serious deformation and partial failure of supports, as illustrated in Figure 3. With the increasing length of the continuous girder bridge, this structural deformation increased significantly. This poses a huge challenge to the transverse splicing of long-unit concrete continuous girder bridges [7]. Therefore, analyzing the transverse effect of the structure induced by concrete shrinkage and creep after multi-span bridge widening is very valuable.

Figure 3. Anti-knock block failure because of transverse deformation of the girders.

To mitigate the adverse effect because of the different completion times of the new and existing bridges, the method of delayed splicing of the new and existing bridges for a period of six months is usually adopted worldwide to reduce the adverse effects of differences in deformation between the new and existing bridges on the widened structure [8–10]. However, for the widened structures of multi-span long-unit bridges, a delay of six months alone does not fundamentally solve the problem [11]. Shi et al. [7] analyzed the effects of long-term loads on the mechanical performance of three widened long-unit prestressed concrete girder bridges. Their results indicated that, under four long-term loads (i.e., concrete shrinkage and creep, temperature gradient, and overall temperature rise and decline), long-unit bridges would produce huge transverse deformation, and the maximum deformations occurred on the abutments or the transition piers of one-unit continuous girder bridges, as shown in Figure 2. Moreover, the transverse displacement caused by concrete shrinkage accounted for 69% of the total transverse displacement produced by the four long-term loads, indicating that concrete shrinkage is the dominant factor affecting the transverse deformation of a widened long-unit bridge. Wen [6] analyzed an example of a prestressed concrete box-girder bridge widening. According to the calculated results, longitudinal tensile stress at the flange of the new girder can reach 2.84 MPa due to shrinkage and creep effects of the new box girder after 10 years, which may lead to concrete cracks. Chen et al. [12] proposed an "indirect splicing" method for the splicing widening of new and existing bridges, in which a longitudinal spliced segment is set by intervals along the widened bridge to reduce the integrity of the widened bridge after splicing; however, in that case, it would be difficult to coordinate the deformation of the new and existing structures in the unspliced segments. Thus, in-depth studies are still needed to understand the stress mechanism of transverse internal force transfer.

On the whole, as far as the widening structures of long concrete continuous girder bridges are concerned, the study on the interaction mechanism between the new and existing bridges and the countermeasures is still in the start-up stage, and the lack of related research results has severely hindered the further development of bridge-widening technologies and should be urgently addressed.

2. Research Background

In this study, a bridge of Beijing-Shanghai Expressway was taken as the engineering background. The bridge is a prestressed concrete continuous box girder bridge with a variable cross-section and a total length of 214 m, including a four-span of 40 + 2 × 67 + 40 m.

The bridge deck is divided into 4-lane double-way, and the single-way box girder uses a single-box single-cell section, as shown in Figure 4. The existing bridge was widened by constructing a new box girder by one side of the existing bridge, and then the superstructure of the new and existing box girder was connected by a concrete stitching slab, while substructures are disconnected. Only the flange plates of the new and existing bridge girders are transversely spliced but not the transverse diaphragms. Figure 5 shows a traditional transverse splicing structure. According to the relevant literature [3,13], if the transverse splicing structure shown in Figure 5 is adopted because of the longitudinal deformation differences between the new and existing bridges in concrete shrinkage and creep, the widened structure will produce an obvious transverse deformation, and the maximum transverse deformation at both ends of the continuous girder will reach 68 mm, severely impairing its structural safety.

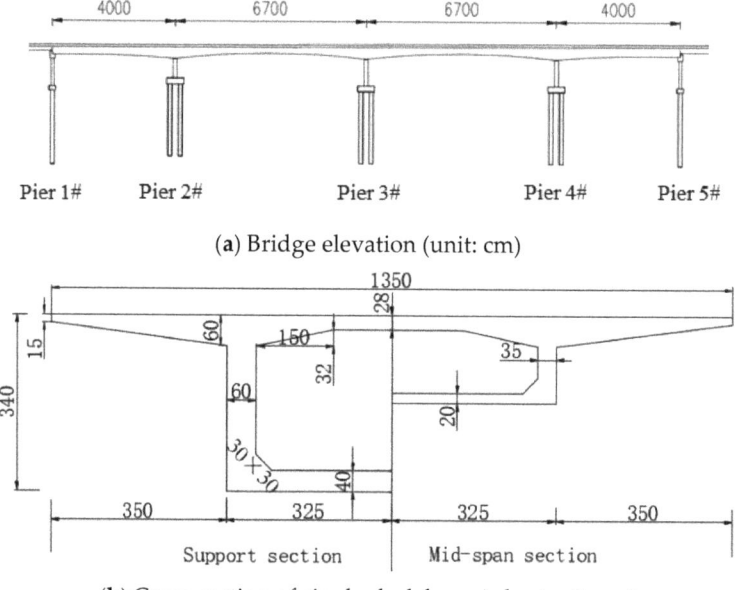

(**a**) Bridge elevation (unit: cm)

(**b**) Cross-section of single-deck box girder (unit: cm)

Figure 4. General layout of a bridge.

To effectively deal with the differences between the new and existing bridges in the longitudinal shrinkage and creep deformation of concrete after widening, this study proposes a novel transverse splicing structure based on the folding effect of a corrugated steel plate (CSP). By investigating its technical feasibility and force transferring mechanism and exploring its reasonable structural parameters, this study aims to offer a new solution to the splicing widening of long multi-span concrete continuous bridges.

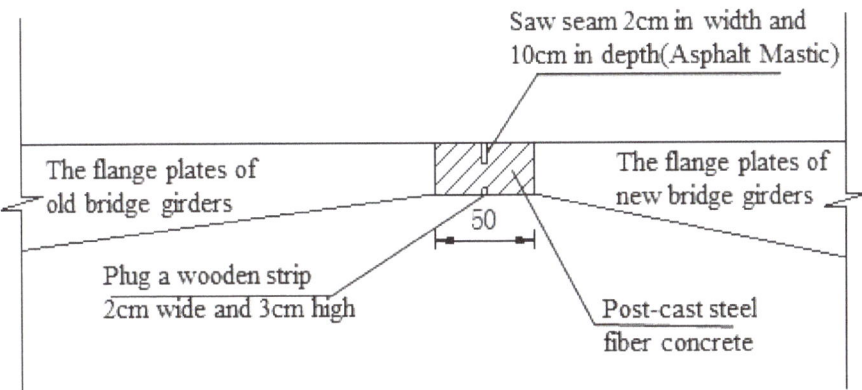

Figure 5. Schematic diagram of the traditional transverse splicing structure.

3. Effectiveness of CSP Splicing Structure

CSP is widely used in continuous box girder bridges with corrugated steel webs, mainly by virtue of its sound longitudinal folding effect. On the basis of such a longitudinal folding effect, two possible splicing structures were put forward, as shown in Figure 6. To comparatively study different structural layout schemes, two transverse connection schemes are proposed according to the different corrugation pattern directions of CSP: (i) vertical corrugation pattern, under which CSP is vertically corrugated, and (ii) horizontal one, where CSP is horizontally corrugated. Depending on the specific wave angle of the corrugated steel plate, each type is further divided into two forms, i.e., right angle and acute angle according to the size of the wave angle (see Figure 7).

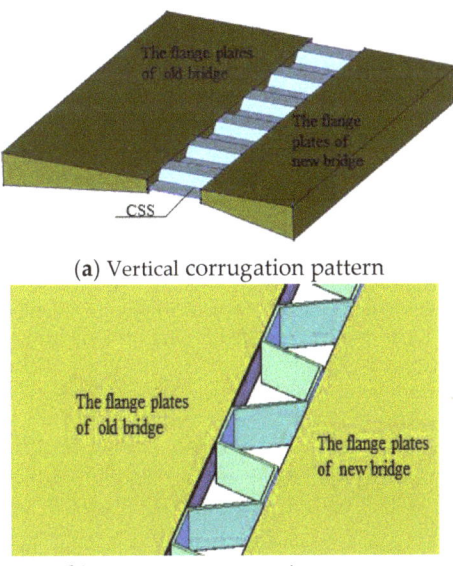

(**a**) Vertical corrugation pattern

(**b**) Horizontal corrugation pattern

Figure 6. Schematic diagram of a comparison of CSP connection schemes.

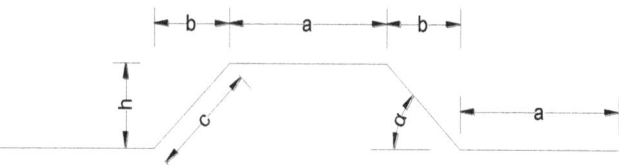

Figure 7. Schematic diagram of the wave angle of the corrugated steel plate. (a: length of straight segment, b: projection length of slant segment, c: length of slant segment, h: wave height, α: wave Angle).

Before building the FE model, the mechanical properties of steel and concrete are listed in Table 1. All standard values of the cube compressive strength f_{cu}, the axial compressive strength f_{ck}, and tensile strength f_{tk}, the elastic modulus E_c of C50 concrete can be determined according to the reference [13,14]. All standard values of the yielding strength f_y, the ultimate strength f_u and the elastic modulus E_s of Q345 steel plate concrete can be determined according to the reference [14].

Table 1. Specifcation value of material mechanical strength/MPa.

Material	Category	Specifcation Value
C50 concrete	f_{ck}	32.4
	f_{cu}	57.5
	f_{tk}	2.64
	E_c	345,000
Q345 Steel plate	f_y	345
	f_u	510
	E_s	200,000

The stress-strain relation of concrete damage constitutive in this paper is the concrete stress-strain relation suggested in Appendix C of the Code for Design of concrete Structures of China [14] (as shown in Figures 8 and 9).

Shrinkage and creep coefficients were adopted according to the CEB-FIP model (CEB, 1990), and the relative humidity of the ambient environment was set at 60% [15].

On this basis, two overall 3D finite element models were built for bridge widening. The overall dimensions of the two models followed the design drawings of the bridge under investigation in this study. Considering the different stress characteristics and thickness of structural members, two types of elements (3D solid elements and plate shell elements) were used for modeling: new and existing concrete box girders are both made of C50 concrete and simulated using eight-node hexahedral elements; CSP uses Q345 steel that is 10 mm in thickness and was simulated using four-node plane elements. Figure 10 shows a schematic diagram of the meshing cross-section of a full-bridge finite element model. As for the connection between the steel plate and the flange plate of the existing and new box girder, in the overall model, it is simplified that the bolt acts as a connection. The connection action was modeled using spring elements that had longitudinal capabilities in 1-D, 2-D, or 3-D applications [16]. The longitudinal spring-damper is a uniaxial tension-compression element with up to three degrees of freedom at each node: translations in the nodal x, y, and z directions.

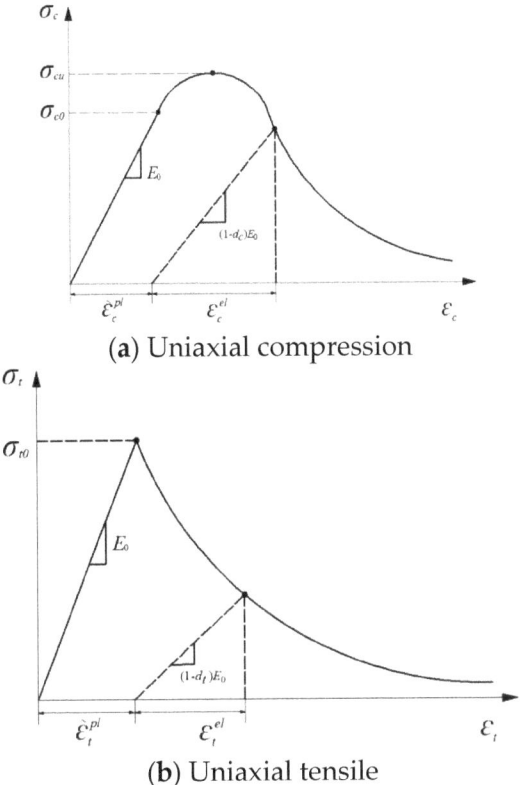

Figure 8. Stress−strain relation of concrete.

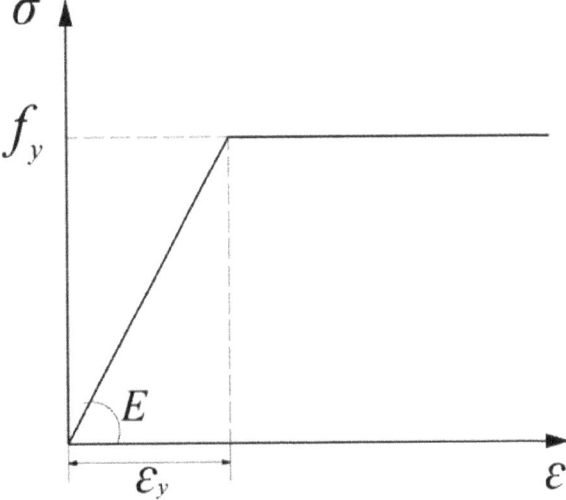

Figure 9. Stress−strain relation of steel.

Figure 11 shows various pivot support restraint conditions, where SX, ZX, HX, and GD denote bilaterally slidable support, longitudinally slidable support, transversely slidable support, and fixed support, respectively.

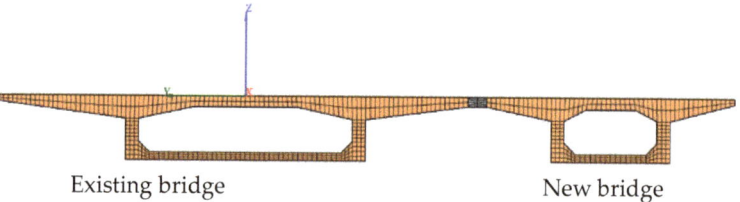

Figure 10. Schematic diagram of meshing cross-section of overall bridge model.

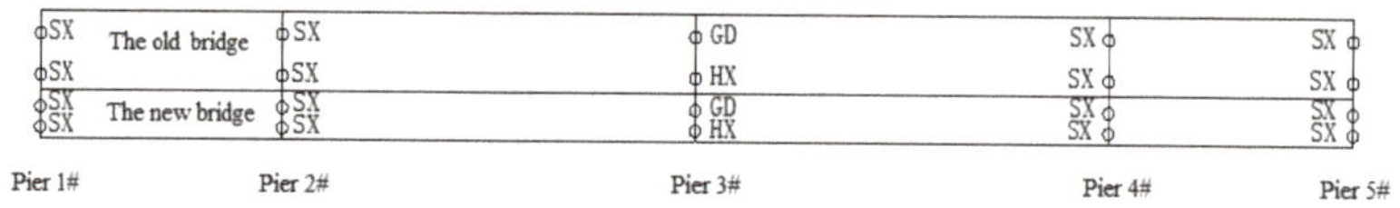

Figure 11. Support layout of the model.

The load case considers only the differences in concrete shrinkage and creep between the new and existing bridges, for which the structure produces transverse deformation, as shown in Figure 2. Through the established finite element analysis model, the paper first assumed that the existing bridge had been working for 10 years and then considered the creep and shrinkage deformation of the new and old bridge for 10 years of joint work after the completion of bridge widening [5,7]. This paper focuses on the longitudinal shrinkage and creep deformation difference between old and new structures, as well as the stress and deformation state of the widened structure under the effect of the deformation difference mentioned above.

Table 2 provides the transverse deformation values at the end of the girder produced by the widened bridge under the action of the differences in deformation between the new and existing bridges in concrete shrinkage and creep under different CSP connection schemes. Recent studies have shown that, for a bridge widened using the traditional transverse splicing structure, the girder-end transverse deformation can reach 68 mm under the same working conditions [3].

Table 2. Transverse deformation at the girder end under different CSP connection schemes (mm).

Connection Scheme	Bending Angle	Existing Bridge	New Bridge
Vertical corrugation pattern	Right angle	−67.77	−67.76
	Acute angle	−67.31	−67.34
Horizontal corrugation pattern	Right angle	−20.54	−20.28
	Acute angle	−66.98	−67.06

The data listed in Table 2 lead to the following facts:

(1) When the "CSP connection with vertical corrugation pattern" shown in Figure 6a was adopted, neither the right-angle steel plate nor the acute-angle steel plate could effectively reduce transverse deformation at the end of the girder under the action of shrinkage and creep deformation difference of the new and existing bridges. In this case, the displacement value was nearly the same as that in cases where traditional splicing structures were used, indicating that the corrugated steel plate hardly exerts a "folding effect." This analysis suggests that under the action of the differences in deformation between the new and existing bridges in the longitudinal shrinkage and creep deformation of concrete, CSP is in a state of longitudinal eccentric compression instead of uniform

compression, indicating that CSP can exert a "folding effect" only through the local non-uniform extrusion deformation of the structure; however, the small width of the splicing stitch seriously restricts the "folding effect" of CSP.

(2) Under the "CSP connection scheme with horizontal corrugation pattern," it was possible to effectively reduce the transverse displacement at the end of the girder of the widened bridge only when a right-angle CSP is adopted. According to the analysis, when there was a longitudinal deformation difference between the new and existing box girder flange plates, the CSP set between them produced longitudinal compression, inevitably resulting in transverse expansion of the CSP structure. However, such transverse expansion was resisted by both the new and existing flange plates on both sides, making it impossible for the CSP to bring its "folding effect" into play. In contrast, CSP bent at a right angle ("right-angle CSP" for short) avoided this problem and adapted well to the differences between the new and existing bridges in the longitudinal shrinkage and creep deformation of concrete. The splicing structure with a horizontal CSP for right-angle bending is discussed later.

Based on the analysis of the above two types of CSP layout schemes, in this study, we proposed a horizontal CSP splicing structure bent at a right angle, which is applicable to the widening of a long-unit concrete continuous box girder bridge (hereinafter referred to as "the CSP splicing structure"), as shown in Figure 12. The invention patent was granted [17] (Chinese patent number: 201710545690.2).

In the constitution of the structure, the first right-angle CSP is arranged continuously along the longitudinal splicing stitch, and the corrugation pattern direction was set as horizontal; the peaks and valleys of CSP were connected to the new and existing box girder flange plates using high-strength bolts, thus creating a transverse splicing structure. Next, a thin sheet steel isolated layer was prepared at the top of the CSP structure to bear the bridge deck pavement. Finally, a new bridge deck was paved on the entire spliced segment to form a flat bridge deck. The isolated layer can be welded together with the top surface of the box girder on one side and set free on the other side. In this way, it will neither restrict the longitudinal deformation of CSP nor cause it to slip out of service.

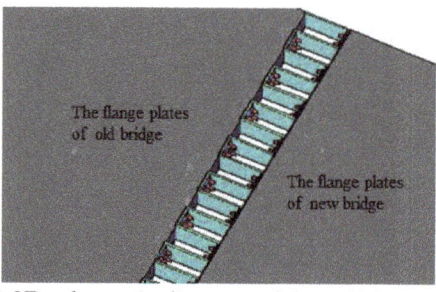

(a) 3D schematic diagram of the splicing part

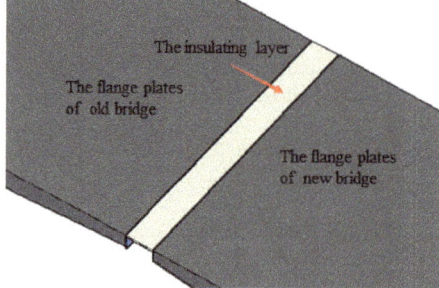

(b) Schematic diagram of the insulating layer for the splicing part

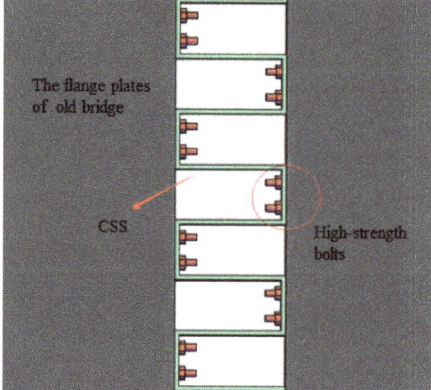

(c) Top view of CSP transverse connection

Figure 12. Schematic diagram of the horizontal CSP splicing structure.

4. Analysis of the Mechanical Behavior of a CSP Transverse Splicing Structure

4.1. Analysis of the Working Mechanism

In order to illustrate the working mechanism of the new splicing jointing structure, an analysis is conducted by taking a local structure with a length of 2 m longitudinally around the end support of the girder and building a refined local FE model, as shown in Figure 13a. The CSP splicing structure uses a right-angle CSP with a thickness of 10 mm, a corrugation height of 0.5 m, and a corrugation length of 0.5 m to analyze the deformation state of the novel splicing structure and its stress change rules under the action of the differences between the new and existing bridges in concrete shrinkage and creep.

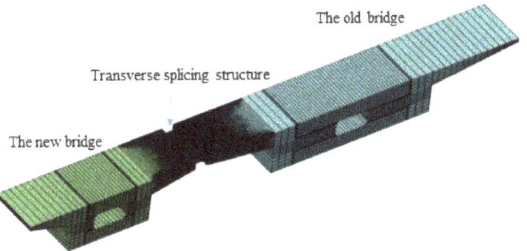

(a) Overall schematic diagram of the local model

(b) Detailed schematic diagram of the splicing part

Figure 13. Schematic diagram of the local finite element model.

In the local refined finite element model, the new and existing box girder concrete, steel plate, and bolts were all simulated using solid elements, and the elements within the splicing part were refined again. The model has a total of 496,455 elements, and its boundary conditions are simulated as a hinging in the four corners at the bottom of the box girder. To be specific, CSP is connected to the new and existing flange plates by using four bolts of 20 mm in diameter and 15 d in anchorage depth, as shown in Figure 13b. The bonding slip between the internal screw and the concrete is not considered in this model, and the sides of the steel plate are connected to the sides of the existing and new concrete flange plate using a spring element with compression only.

To effectively apply the effect of structural deformation difference, all the nodes on the end cross-section within the range of the new bridge are rigidly connected to the centroid node of the cross-section, followed by applying a longitudinal forced displacement at the mass centroid to simulate the longitudinal shrinkage and creep displacement difference between the existing and new bridges. According to the global finite element model shown in Figure 10, the maximum value of the longitudinal forced displacement difference was 13 mm. Figure 14 shows a schematic diagram of the deformation at the splicing part of the new and existing box girder after applying a longitudinal forced displacement to the new bridge.

Figure 14 shows that the deformation of transverse steel plates is dominated by S-shaped bending deformation, which allows the splicing structure to easily absorb the large longitudinal dislocation deformation. This deformation characteristic ensures that the longitudinal deformation difference between the new and existing box girders is almost entirely borne by CSP, resulting in a very small longitudinal restraining effect of the existing bridge on the new bridge. The above deformation characteristics reveal the working mechanism of the splicing structure with a CSP that can adapt well to the longitudinal shrinkage and creep deformation differences between the new and existing bridges.

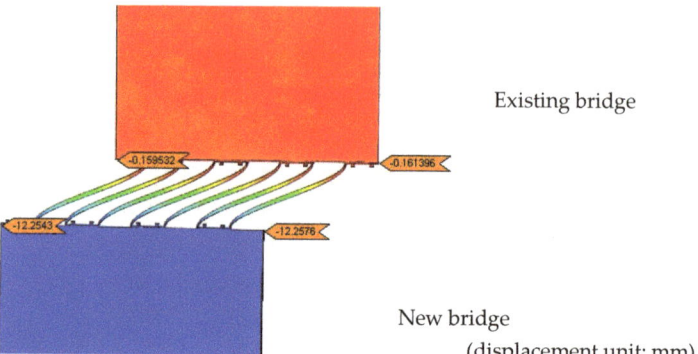

Figure 14. Deformation of CSP after application of longitudinal forced displacement on the new bridge.

4.2. Effects of Different Splicing Modes on the Overall Stress Characteristics of the Widening Structure

To investigate the differences in stress characteristics under the traditional splicing mode and the CSP splicing mode, two finite element models were built to stimulate two types of transverse splicing structures and then to analyze the widening structure. Three key sections were selected, as shown in Figure 15.

Figure 15. Schematic diagram of a longitudinal control section.

The combination value of actions comprises dead weight, prestress, shrinkage creep, foundation post-construction settlement difference, lane load, and so forth. Figure 14 shows the transverse stress variation of the top fiber of the box girder bridge deck under a combination action case on the 1#–3# sections under the action combination. Specifically, the foundation post-construction settlement difference refers to the difference between the new and existing bridges in the foundation settlement deformation after transverse splicing. Generally, it is assumed that the settlement deformation of the existing bridge foundation is 0, and that produced by the new bridge is equal to the post-construction settlement difference. For the bridge under investigation in this study, the foundation post-construction settlement difference was set at 5 mm [13,18,19].

Figure 16 shows that, (1) on 1# and 3# support sections, the transverse tensile stress at the top slab of the existing-bridge box girder increased significantly but remained unchanged basically in the 2# section of the side span. This analysis reveals that the foundation post-construction settlement difference mainly occurs in the support sections but is very small in the mid-span section. This is the reason for the obvious increase in the transverse stress of the support section. (2) Splicing widening has the largest adverse effect on the girder-end (1#) section of the existing bridge. Under traditional hinged splicing, the transverse tensile stress of the inner flange plate of the box girder reaches 4.557 MPa, and the maximum tensile stress occurs at the upper middle edge of the flange plate, far exceeding the standard tensile strength of C50 concrete (2.65 MPa). The inner flange plate of the existing bridge will be cracked in a large area, requiring effective reinforcement measures. In contrast, when CSP splicing is used for splicing widening, the maximum tensile stress on the inner flange plate of the existing bridge box girder is 2.510 MPa, which has a great advantage over the traditional splicing method. Moreover, the maximum principal tensile stress of the corrugated steel plate under comprehensive working conditions was 207.541 MPA, not exceeding the yield strength of the Q345 steel plate.

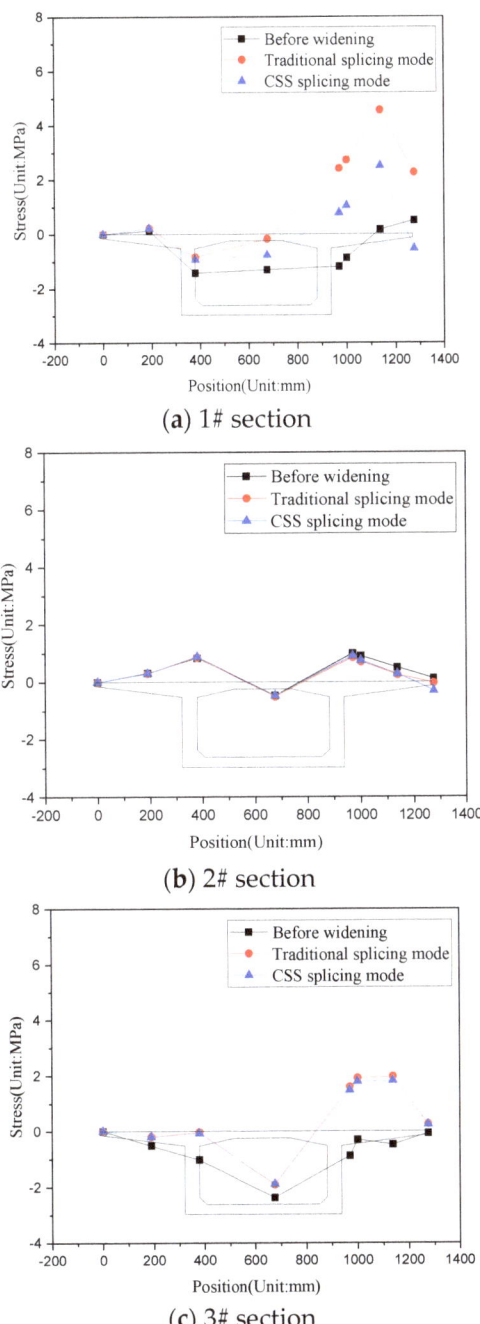

Figure 16. Transverse stress variation of the top fiber of the box girder bridge deck.

5. Parameters and Dimensional Design of CSP Splicing Structures

The selected analysis parameters included CSP thickness, CSP wavelength, and splicing stitch width. Given that the differences between the new and existing bridges in concrete shrinkage and creep and the foundation post-construction settlement are the main factors affecting the stress performance of the splicing widening structure, the parameter sensitivity analysis was performed mainly under the two action cases:

Action case 1: The action of the differences between the new and existing bridges in concrete shrinkage and creep deformation;

Action case 2: The action of the difference between the new and existing bridges in foundation post-construction settlement.

The analysis results of the three parameters are listed in Tables 2–4. The data in the table were analyzed as follows:

Table 3. Sensitivity analysis of CSP thickness.

Action Case	Index	CSP Thickness/mm		
		10	15	20
1	Transverse displacement at the end of girder/mm	20	39	52
	Tensile stress on the existing bridge flange plate at girder end/MPa	0.65	0.85	1.04
	Tensile stress on new-bridge flange plate at the end of the girder/MPa	1.78	1.68	1.57
	Principal tensile stress on the CSP/MPa	197.94	197.15	173.98
2	Deflection difference between the new and existing flange plates at side support/mm	1.289	1.225	1.189
	Maximum tensile stress on existing bridge flange plate at side support/MPa	4.59	4.61	4.60
	Maximum tensile stress on the new-bridge flange plate at the side support/MPa	4.25	4.49	4.67
	Principal tensile stress on CSP/MPa	87.06	66.05	54.33

Table 4. Sensitivity analysis of the splicing stitch width.

Load Case	Index	Splicing Stitch Width/m		
		0.5	0.75	1.0
1	Transverse displacement at girder end/mm	20	9	5
	Tensile stress on existing bridge flange plate at girder end/MPa	0.65	0.52	0.45
	Tensile stress on new-bridge flange plate at girder end/MPa	1.78	1.81	1.81
	Principal tensile stress on CSP/MPa	197.94	137.13	95.49
2	Deflection difference between the new and existing flange plates at side support/mm	1.289	1.834	2.321
	Maximum tensile stress on the existing bridge flange plate at the side support/MPa	4.59	4.06	3.57
	Maximum tensile stress on the new-bridge flange plate at the side support/MPa	4.25	3.71	3.22
	Principal tensile stress on the CSP/MPa	87.06	97.98	88.08

(1) Table 3 shows that with increasing CSP thickness, the transverse displacement at the end of the girder increases significantly, while other structural responses, such as the transverse tensile stress on the concrete flange plate and the tensile stress on the steel sheet, are not sensitive to this change. This shows that when the thickness of the corrugated steel plate increases, its splicing stiffness will increase significantly, significantly decreasing its ability to absorb the longitudinal concrete shrinkage and creep of new and existing bridges. Therefore, the thickness of the steel plate should be relatively small and should be 10 mm or 15 mm.

(2) Table 4 shows that when the width of the splicing stitch increases from 0.5 m to 1.0 m, the transverse displacement value of the girder-end section decreases significantly, while other structural responses are not sensitive to such change. This indicates that increasing splicing stitch width decreases the splicing stiffness of CSP, which further weakens the restraining effect of the existing bridge on the new bridge and significantly decreases transverse displacement. However, the width of the splicing stitch should not be too large. When the width reaches 1.0 m, the deflection difference between the new and existing flange plates, induced by the foundation settlement of the new bridge on the support section, increases significantly and is not conducive to road safety. Based on the relevant literature [19], it is appropriate to control the relative deflection difference of the flange end of the new and old bridges within 2 mm, so the width of the patchwork joints should not be too large. The optimum width of the splicing stitch was set at 0.5 m or 0.75 m.

(3) Table 5 shows that when corrugation length increases from 0.5 m to 1.0 m, the transverse displacement value of the girder-end section decreases. Apparently, an increased corrugation length decreased the splicing stiffness of sheet steel and significantly reduced transverse displacement. However, when the corrugation length reaches 1.0 m, a high degree of concrete stress concentration occurred within the range of the corrugated steel plates and the concrete flange plates of the new and existing box girders connected thereto, and is detrimental to the deformation coordination between the new and existing box girders. Thus, the corrugation length of CSP is still recommended to be 0.5 m.

Table 5. Sensitivity analysis of corrugation length.

Load Case	Index	Corrugation Length/m	
		0.5	1.0
1	Transverse displacement at girder end/mm	20	8.5
	Tensile stress on the existing bridge flange plate at the end of the girder/MPa	0.65	0.49
	Tensile stress on the new-bridge flange plate at the end of the girder/MPa	1.78	1.82
	Principal tensile stress on the CSP/MPa	197.94	161.63
2	Deflection difference between the new and existing flange plates at side support/mm	0.868	1.027
	Maximum tensile stress on the existing bridge flange plate at the side support/MPa	4.63	4.68
	Maximum tensile stress on the new-bridge flange plate at the side pivot/MPa	3.96	7.92
	Principal tensile stress on the CSP/MPa	87.06	126.29

Based on the analysis results of the sensitivity of parameters, in this study, for the sake of achieving a better splicing effect, two parameter value schemes are proposed, i.e., Scheme 1: right-angle CSP with a thickness of 10 mm, a splicing stitch width of 0.5 m, and a corrugation length of 0.5 m; Scheme 2: right-angle CSP with a thickness of 15 mm, a splicing stitch width of 0.75 m, and a corrugation length of 0.5 m. The advantages and disadvantages of the two parameter optimization schemes were further analyzed, as detailed in Table 6.

Table 6. Comparison of the two parameter optimization schemes.

LOAD Case	Index	Scheme 1	Scheme 2
1	Transverse displacement at the end of girder/mm	20	20
	Tensile stress on the existing bridge flange plate at girder end/MPa	0.65	0.63
	Tensile stress on the new-bridge flange plate at the end of girder/MPa	1.78	1.75
	Principal tensile stress on CSP/MPa	197.94	165.32
2	Deflection difference between the new and existing flange plates at the side support/mm	1.289	1.727
	Maximum tensile stress on the existing bridge flange plate at the side support/MPa	4.59	4.14
	Maximum tensile stress on the new-bridge flange plate at the side support/MPa	4.25	3.97
	Principal tensile stress on the CSP/MPa	87.06	74.58

According to Table 6, the maximum transverse displacement values of girder-end section of both schemes are both 20 mm. However, based on the stress states of the new and existing box girder flange plates and CSP, Scheme 2 is considered a better scheme.

6. Conclusions

In conclusion, this study successfully demonstrated the scheme design and working mechanism of a CSP splicing structure based on a case study of a bridge of Beijing-Shanghai Expressway. The sensitivity of the relevant parameters was analyzed, and a better scheme design was put forward for this type of splicing structure. The results of this study are as follows:

(1) When the "CSP connection scheme with a vertical corrugation pattern" is adopted, the corrugated steel plate with a right angle or an acute angle cannot effectively reduce lateral deformation at the end of the widened bridge girder, indicating that the corrugated steel plate almost does not exert a "folding effect." When the "CSP connection scheme with a horizontal corrugation pattern" is adopted, only when the wave of right angle in the steel plate is adopted can the corrugated steel plate effectively exert the "folding effect," and reduce the transverse deformation at the end of girder under the action of concrete shrinkage and creep deformation difference between the new and existing bridge. Based on the analysis results, the CSP splicing structure waved at the right angle is proposed and can be effectively applied to the transverse splicing widening of long multi-span concrete continuous box girder bridges.

(2) When the CSP splicing structure is adopted as the widening stitch structure for the long multi-span concrete continuous box girder bridge, the maximum transverse deformation at the end of girder is about 20 mm, about 1/3 of that under the traditional splicing scheme. This indicates that the CSP splicing structure can adapt well to the action of the differences between the new and existing bridges in concrete shrinkage and creep deformation.

(3) The working mechanism of the right-angle CSP is explained as follows. Due to the longitudinal deformation difference between the new and existing box girders, the deformation of the sheet steel is dominated by S-shaped bending deformation; the splicing

structure has a strong ability to absorb longitudinal displacement, and the longitudinal restraint effect of the existing bridge on the new bridge is very small. As a result, the CSP splicing structure can adapt well to the differences between the new and existing bridges in the longitudinal shrinkage and creep deformation of concrete.

(4) Under the combined action, when traditional splicing is adopted, the transverse tensile stress on the top flange plate at the end of girder of the existing bridge reaches 4.557 MPa, far exceeding the standard tensile strength of C50 concrete, making it necessary to take effective anti-cracking and reinforcement measures. In contrast, when the CSP splicing structure is adopted, both the maximum tensile stress on the top flange plate of the box girder of the existing bridge and the maximum principal tensile stress on the CSP are within the safe range, indicating that the comprehensive stress performance of the CSP splicing structure is obviously superior to that of the traditional splicing structures.

(5) Through investigating the rules of the effects of CSP thickness, corrugation length, splicing stitch width, and other structural parameters on the stress characteristics of the widening structure, a sound parameter combination scheme was successfully established for the CSP splicing structure. The optimum CSP thickness, splicing stitch width, corrugation length, and waving angle were found to be 15 mm, 75 cm, 0.5 m, and right angle, respectively.

Author Contributions: Conceptualization, W.W.; Formal analysis, H.Z.; Data curation, Z.L.; Investigation, Y.W. All authors have read and agreed to the published version of the manuscript.

Funding: This research was funded by National Nature Science Funding of China (Grants number:52278149).

Informed Consent Statement: Not applicable.

Data Availability Statement: Not applicable.

Conflicts of Interest: The authors declare no conflict of interest.

References

1. Hosseini, M.; Jefferson, A.D. Time-dependent behavior of widened reinforced concrete under-bridge. *Mater. Struct.* **1998**, *319*, 714–719. [CrossRef]
2. Sun, Q.X.; Liu, C.; Sha, L.X.; Lu, Y. Experimental study on bending performance of different types of UHPC in bridge stitching joint. *Mater. Struct.* **2021**, *54*, 179. [CrossRef]
3. Wu, W.-Q.; Tang, Z.-X.; Zhang, H.; Zhao, H. Research on Structural Diseases Due to a Joint Widening of Concrete Continuous Box Girder Bridge. *China J. Highw. Transp.* **2018**, *31*, 63–73.
4. Wen, Q.J.; Jing, H.W. Numerical simulation of creep and shrinkage in widened concrete bridges. *Mag. Concr. Res.* **2014**, *66*, 661–673. [CrossRef]
5. Tu, B.; Fang, Z.; Dong, Y.; Frangopol, D.M. Time-variant reliability of widened deteriorating prestressed concrete bridges considering shrinkage and creep. *Eng. Struct.* **2017**, *153*, 1–16. [CrossRef]
6. Wen, Q.J. Long-term effect analysis of prestressed concrete boxgirder bridge widening. *Constr. Build. Mater.* **2011**, *25*, 1580–1586. [CrossRef]
7. Shi, X.; Li, X.; Ruan, X.; Ying, T. Analysis of structural behavior in widened concrete box girder bridges. *Struct. Eng. Int.* **2008**, *18*, 351–355. [CrossRef]
8. Kwan, A.K.H.; Ng, P.L. Reducing damage to concrete stitches in bridge decks. *Bridg. Eng. Proc. Inst. Civ. Eng.* **2006**, *159*, 53–62. [CrossRef]
9. Chai, Y.H.; Hung, H.J. Waiting Period for Closure Pours in Bridge Widening or Staged Construction. *J. Bridge Eng.* **2016**, *21*, 04016006. [CrossRef]
10. ACI Committee 345. *Guide for Widening Highway Bridges (345.2R-13) Farmington Hills*; American Concrete Institute (ACI): Miami, FL, USA, 2013; pp. 1–25.
11. Zhang, L.F.; Yan, J.; Ma, H.Y.; Yu, H.F.; Wang, Y.; Mei, Q.Q. Experimental study on magnesium sulfate cement concrete splices of widened box girder. *KSCE J. Civ. Eng.* **2021**, *25*, 4742–4750. [CrossRef]
12. Chen, K.-M.; Wu, Q.-X.; Chen, B.C.; Zhang, G. Partial Splicing Method for Extended Long-connected Bridge and Experiment on Splicing Structure. *China J. Highw. Transp.* **2016**, *29*, 99–107. (In Chinese)
13. Wu, W.; Shan, H.; Yang, S.; Tang, Z. Key Assumption to Evaluate the Mechanical Performance of Widened Voided-slab Bridge Due to Foundation Settlement. *KSCE J. Civ. Eng.* **2018**, *22*, 1225–1234. [CrossRef]

14. *JTG D60–2004*; General Specifications for Design of Highway Bridges and Culverts. China Communications Press: Beijing, China, 2004. (In Chinese)
15. CEB (Comité Euro-International du Béton). *CEB-FIP Model Code for Concrete Structures*; CEB: Lausanne, Switzerland, 1990.
16. Hawileh, R.A.; Naser, M.Z.; Abdalla, J.A. Finite element simulation of reinforced concrete beams externally strengthened with short-length CFRP plates. *Compos. Part B Eng.* **2013**, *45*, 1722–1730. [CrossRef]
17. Wu, W.Q.; Zhang, H.; Zhao, H.; Yang, Y.; Zhang, X. A Horizontal Splicing Structure and Construction Method of Concrete Continuous Box Girder Bridge. Chinese patent 201710545690.2, 1 April 2019.
18. Zhou, J.; Li, T.; Ye, X.; Shi, X. Safety assessment of widened bridges considering uneven multilane traffic-load modeling: Case study in China. *J. Bridge Eng.* **2020**, *25*, 1–12. [CrossRef]
19. Zhou, J.; Lu, Z.; Zhou, Z. Structural safety assessment and traffic control strategies of widened highway bridges under maintenance works requiring partial lane closure. *KSCE J. Civ. Eng.* **2022**, *26*, 1846–1857. [CrossRef]

Article

Concrete-Filled Prefabricated Cementitious Composite Tube (CFPCCT) under Axial Compression: Effect of Tube Wall Thickness

Bi Kai [1], A. B. M. A. Kaish [2,*] and Norhaiza Nordin [1]

1. Department of Civil Engineering, Infrastructure University Kuala Lumpur, Kajang 43000, Malaysia
2. Department of Civil Engineering, Universiti Kebangsaan Malaysia, Bangi 43600, Malaysia
* Correspondence: amrul.kaish@ukm.edu.my

Abstract: Research on different prefabricated cementitious composites for constructing composite concrete columns is comparatively more limited than that of concrete filled steel tube columns. The main objective of this study was to observe the axial compressive behavior of concrete-filled prefabricated cementitious composite tube (CFPCCT) specimens. In the CFPCCT composite column, the spiral steel bar is arranged as a hoop reinforcement in the cementitious tube before its prefabrication. Following this, the concrete is poured into the prefabricated cementitious composite tube. The tube is able to provide lateral confinement and can carry the axial load, which is attributed to the strength of CFPCCT composite column. The effect of tube wall thickness on the behavior of CFPCCT is studied in this research. A total of eight short-scale CFPCCT composite columns, with three different tube wall thicknesses (25 mm, 30 mm and 35 mm), are tested under axial compressive load. The cementitious composite tube-confined specimens showed a 24.7% increment in load-carrying capacity compared to unconfined specimens. Increasing the wall-thickness had a positive impact on the strength and ductility properties of the composite column. However, poor failure behavior was observed for thicker tube wall. Therefore, concrete-filled cementitious composite tube columns can be considered as an alternative and effective way to construct prefabricated concrete columns.

Keywords: prefabricated construction; concrete-filled column; cementitious composite tube reinforcement; mechanical property; structure behavior

Citation: Kai, B.; Kaish, A.B.M.A.; Nordin, N. Concrete-Filled Prefabricated Cementitious Composite Tube (CFPCCT) under Axial Compression: Effect of Tube Wall Thickness. *Materials* 2022, 15, 8119. https://doi.org/10.3390/ma15228119

Academic Editor: Andreas Lampropoulos

Received: 1 September 2022
Accepted: 3 November 2022
Published: 16 November 2022

Copyright: © 2022 by the authors. Licensee MDPI, Basel, Switzerland. This article is an open access article distributed under the terms and conditions of the Creative Commons Attribution (CC BY) license (https://creativecommons.org/licenses/by/4.0/).

1. Introduction

Concrete has excellent characteristics, such as compressive strength, durability, easy modification, and can be poured in a wide range of shapes. As a load-carrying component of structures, concrete columns play a crucial role in both industrial and civil buildings [1,2]. However, the main shortcomings of concrete materials are their poor deformability and low tensile strength [2]. Therefore, rather than using plain concrete, the cement based composite material would be widely used in future. It is expected that cement-based composites have higher strength, greater elastic modulus and better deformability [3,4]. Due to the birth of superplasticizer, the production and application of high-strength fiber reinforced cementitious composites became possible [5].

The construction component of cylindrical columns today has multiple solutions for structural performance. However, a large number of studies have concentrated on material performance enhancement, adding fiber reinforcement and hoop confinements to achieve better performance [4]. Cementitious composites are one of the most widely used building materials for the construction of infrastructure. Kaish et al. studied cementitious composite ferrocement jacketed circular and square concrete columns under axial compression [6–8]. Later, Kaish et al. proposed L and U shaped prefabricated ferrocement jacket for strengthening square concrete column [9]. Lim et al. studied concrete-filled hollow PC columns and reported the structural performance of initial stiffness, maximum strength and displacement ductility under seismic loading [10]. Hosoya and Asano studied precast concrete shell columns and reported a cracking pattern and failure behavior [11]. Liang-jin et al. studied concrete-filled engineered cementitious composite (ECC) tubular piers and presented the first cracking stress, ultimate loading and ultimate stress [12]. Pan et al. [13] investigated the seismic behavior of plain concrete filled ECC formwork. They have observed the effect of the slenderness ratio of such composite concrete columns. Meng and Khayat [14] investigated RC columns cast in stay-in-place prefabricated ultra-high-performance concrete (UHPC) formwork. They have presented an experimental and numerical study using the ABAQUS finite element package. Xiao and Ma [15] conducted the seismic retrofitting of RC circular columns with inadequate lap length using prefabricated fiber reinforced concrete (FRC) jackets. Shao et al. [16] proposed two different techniques for improving the seismic performance of shear-efficient RC columns, without increasing the cross-section of the column applicable, for mid-rise to high-rise buildings using UHPC jacketing. Hung et al. [17] conducted research on innovative UHPC jacketing solutions for shear-deficient reinforced concrete columns. These solutions included cast-in-place UHPC jackets, as well as prefabricated UHPC panels. Guan et al. [18] recommended for the plastic hinge region of precast columns to be strengthened with locally produced UHPC jackets. Tian et al. [19] reported numerical and experimental studies on the tubular column made of concrete filled ultra-high-performance concrete. Zhu et al. [20] and Zhang et al. [21] investigated the long-term creep and shrinkage behavior of concrete filled UHPC tubular columns under axial compression. They also developed numerical models to determine the creep and shrinkage coefficient. Shan et al. [22] investigated the seismic capacity of concrete filled cement-based tubular composite column systems. Their results confirm that this type of column carries significantly higher lateral load and absorbs higher seismic energy compared to the conventional columns.

All of these studies show the high potential for field application of prefabricated cementitious composite tubes for composite concrete columns. It combines the feasible characteristics of cementitious composites and the versatility of prefabrication, thus creating a high-performance feasible product. However, a limited number of research has been reported in the area of cementitious composites as a prefabricated confining material. Tube thickness is an important parameter that needs to be investigated in detail. Previously reported research did not study this aspect for normal strength cementitious composite tube composite columns. Therefore, this study investigated the tube thickness effect on the behavior of concrete filled prefabricated normal strength cementitious composite tubes. Three different tube thicknesses were investigated in this study under axial compression. The following sections reports the methodologies followed and discusses the obtained results in detail.

2. Experimental Program

For the specimen series, a total of 8 short-scale columns are tested under an axial compression load. All specimens were 300 mm in total height and 210 mm in overall diameter. The main testing parameter was the wall thickness of the composite tube. The spiral reinforcement was positioned closely, at the middle point of the tube. The spiral spacing was selected as constant for all of the specimens, which was 20 mm. The wall thickness varied among 25 mm, 30 mm and 35 mm, respectively, as the main variable in this study. These thicknesses were chosen based on practical considerations. Placing the hoop reinforcement is difficult in the cementitious composite tubes where less than 25 mm thickness was chosen. On the other hand, a thickness above 25 mm is impractical for small size specimens. Sectional details of tested specimens are shown in Figure 1. All other specimen details are shown in Table 1.

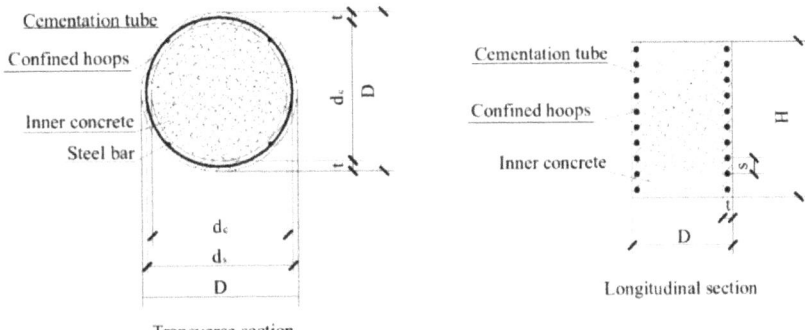

Figure 1. Transverse and longitudinal section of specimens.

Table 1. Tested specimens' details.

Groups	Specimens	Wall Thicknesses (mm)	Hoop Spacing (mm)	Type of Confinements
C-20-25	C-20-25a	25	20	Cementitious composite tube confined
	C-20-25b	25	20	
C-20-30	C-20-30a	30	20	Cementitious composite tube confined
	C-20-30b	30	20	
C-20-35	C-20-35a	35	20	Cementitious composite tube confined
	C-20-35b	35	20	
C-1	C-1a	-	-	Unconfined
	C-1b	-	-	

2.1. Material

2.1.1. Cement

The binding materials used to prepare the concrete and the cementitious composite tube specimens were locally sourced ordinary Portland cement. The strength grade was 42.5 N ordinary Portland cement.

2.1.2. Fine & Coarse Aggregates

The coarse aggregate used in concrete was downgraded crushed gravel with a maximum diameter of 12.5 mm. The locally available fine river sand (percentage passing 600 μm sieve approximate 40%) was used for both the core concrete and tube fabrication.

2.1.3. Confining Hoops

The reinforcement (Figure 2) chosen for the spiral hoops was a 3.5 mm diameter steel bar with 350 MPa yield strength. Hoop spacing is provided in Table 1.

2.1.4. Composite Materials

Ordinary Portland cement and silica fume (specific surface area of 22,500 m^2/kg) were used as a binder in the composite mix. Quartz sand, with a maximum size 0.4 mm, was mixed with the binder, with a ratio of 1:1. Corrugated steel-fibers, with a 0.5 mm nominal diameter and an average length of 35 mm, were used as fiber in the composite mix.

2.2. Mix Design and Cubic Conprssive Strength

The concrete mix design was prepared for obtaining a concrete target strength of grade 35 (35 MPa) with single mix design. The mix in the laboratory was selected specifically as 281.4 kg/m^3 cement, 667.44 kg/m^3 sand, 1001.16 kg/m^3 gravel. The water cement ratio was 0.43. For the cementitious composite tube, the mix ratio was followed as Cement: river sand: quartz sand: water = 1.0:1.0:1.0:0.45. Cement: silica fume: steel fiber: superplasticizer ratio was 1.0:0.1:0.1:0.000992. The compressive strength determined from a 100 mm cube of core concrete and a cementitious composite tube achieved 23.42 MPa and 31.32 MPa at 7 days of curing, respectively.

Figure 2. Confining hoop reinforcements.

2.3. Specimen Fabricaiton

The spiral hoop was placed inside the steel mold, and then the cementitious mixture was cast in three equal layers. The mold was placed on a vibrating table to achieve proper compaction. All specimens were covered by plastic bags after casting to confirm the intense hydration reaction during the first 4–6 h. The molds were removed after 24 h and then the tubes were cured in a water tank at approximately 26($\pm$1) °C or for 3 days. The specimen was taken out of the water tank after 3 days of curing and dried for 2 h to achieve surface dry condition.

Next, the freshly mixed concrete was poured into the cementitious tube in three equal layers and compacted with a 25 mm diameter rod, 30 times each layer. An additional concrete layer was cast over the top of every specimen to achieve a finished surface, and then sealed back into plastic bags for 24 h. All specimens were cured under water to finish the rest of the 28 days of hardening before being tested. Thus, the whole process period lasted 32 days. Fabrication process is depicted in Figure 3.

2.4. Hardened Property Test

All specimens were tested under axial compression with a universal testing machine. The specimens' axial deformation was measured using two linear variable displacement transducers (LVDTs), marked as AS-1 and AS-2, of 100 mm gauge length attached to the middle region of the column. The axial strain for each specimen was calculated based on the average deformations recorded by the LVDTs. In addition, two 100 mm gauge length LVDTs were horizontally installed at mid-height to record the lateral deformations and subsequent strains. Both gauges were placed on two opposite sides of the surface. To record the axial load, a 100-ton load cell was placed at the top of specimen.

All of the LVDTs and load cell sensors were connected to a data logger before starting the test to collect the data. All columns were tested under a loading rate of approximately 1 kN/s, until they reached failure. Figure 4 shows the test setup and instrumentation followed in this experiment.

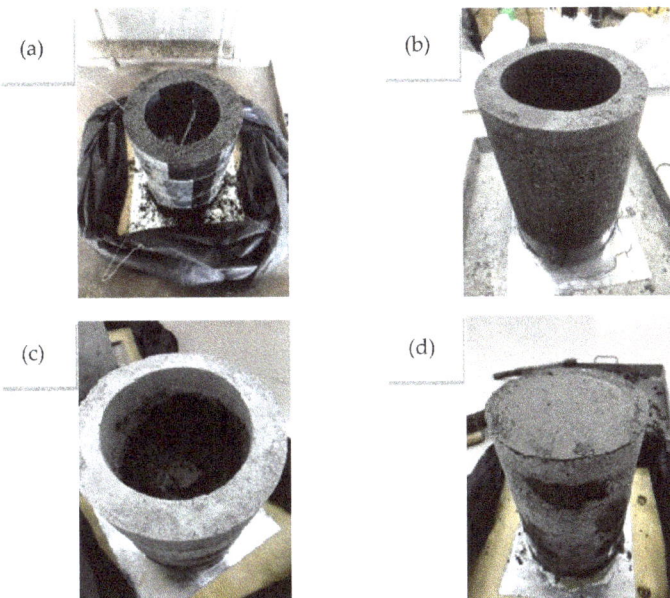

Figure 3. Tube fabrication procedure. (**a**) Tube pouring; (**b**) Surface drying; (**c**) Core concrete pouring; (**d**) Core capping.

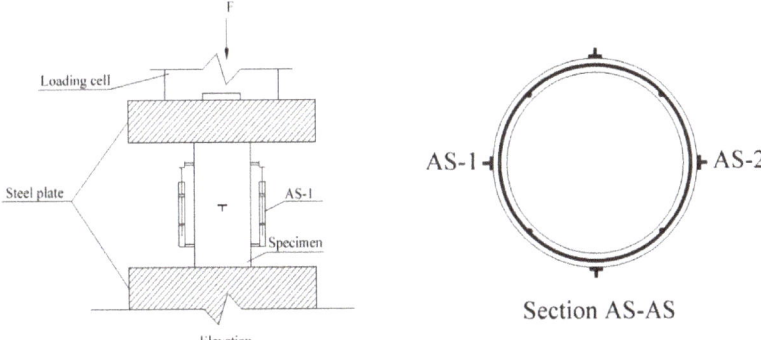

Figure 4. Test setup and instrumentation.

3. Results and Discussions

The test results were collected with data recording software, installed in a laptop that was connected to the data logger. The following sub-sections present and discuss the recorded results.

3.1. Load and Deflection Capacities

The ultimate load carrying capacity of unconfined specimens and cementitious composite tube-confined specimens is tabulated in Table 2. It can be seen that the load carrying capacity of tube-reinforced specimens was progressively improved. The enhanced loading capacity of tube-confined reinforcement specimens is a result of the external tube.

Table 2. Tested specimens' load carrying capacity.

Specimen	Yield Load (kN)		Ultimate Load (kN)		Ultimate Axial Stress (MPa)	Increment in Load to C-1 (%)
	Measured	Average	Measured	Average		
C-1	324.92	329.40 327.16	387.47	389.65 388.56	21.99	-
C-20-35	382.67	375.29 378.98	466.22	481.25 473.74	30.79	40.0
C-20-30	395.86	392.36 394.11	488.50	480.16 484.33	27.42	24.7
C-20-25	391.33	403.84 397.59	483.64	488.72 486.18	24.19	10.0

It directly provides the benefit of a higher load carrying capacity from cementitious composite tube confinement. As the independent variable of tube reinforcement, it can be said that the tube contributes significant lateral confinement to improve the load carrying performance. The increment in the ultimate load carrying capacity of CFPCCT specimens reached up to 40% higher than the unconfined specimens. Similar results were also presented for ferrocement-confined cylindrical concrete columns, where the same size specimens were tested [6]. It should be noted that the outer prefabricated tube carried both the axial load and lateral confining pressure, and thus contributed to an enhanced load carrying capacity.

The greater thickness of the prefabricated tube contributed to a higher load carrying capacity in this experiment. Moreover, it should be noted that the increment in load carrying is a nearly linear increase in this case. However, a tube with 35 mm thickness is already higher, and increases almost 50% of cross-sectional size compared to the diameter of the core concrete. Therefore, it is not recommended to provide higher thickness beyond this.

The axial deflection is related to the stiffness of the member. Therefore, axial deflection is reported in this study even though the column is very short. The ultimate axial deflection capacities of the test specimens are reported in Table 3. It is observed that the axial deflections, at both ultimate load and at failure, of group C-20-35 are higher than rest of the specimen groups. With the wall-thickness increasing, the axial deflections are 1.61, 1.58 and 1.29 times at the ultimate load deflection relative to group C-1, respectively. The higher axial deflection capacity considers that the ductile nature of cementitious composite tubes permits more lateral deformation from the composite materials and the hoop reinforcement.

Table 3. Axial deflections of all specimens.

Specimen	Axial Deflection at Ultimate Load (mm)			Axial Deflection at Failure (mm)			Relative to C-1 in at Ultimate Load
	Measured		Average	Measured		Average	
C-1	3.0	3.1	3.1	3.1	3.1	3.1	-
C-20-25	3.9	4.1	4.0	6.3	6.4	6.4	1.29
C-20-30	4.7	5.1	4.9	6.9	7.0	7.0	1.58
C-20-35	4.9	5.1	5.0	7.4	7.6	7.5	1.61

3.2. Failure Modes and Cracking Behavior

The typical failure patterns for each group of tested specimens are presented in Figure 5. The failure of the cementitious composite tube-confined specimens, shown in Figure 5b–d, was similarly indicated by the formation of vertical cracks throughout the height of the entire outward surface; the width of the cracks increased with the increase of applied load, and ultimately failed. Bigger cracks appeared in all of the tube-confined specimens at the time of failure, which can be seen in Figure 5. The cracks in the cementitious tubes all had a width of more than 5 mm. However, during the failure of the non-reinforced specimens (Figure 5a) micro-cracks were observed in the specimens due to the application of load, and suddenly failed at the ultimate load. Furthermore, all of the non-reinforced specimen columns show a catastrophic collapse with the disintegration of structural integrity. The failure of all of the tube-confined specimens was due to the failure of the cementitious tube, where surface peeling in the tube was also observed.

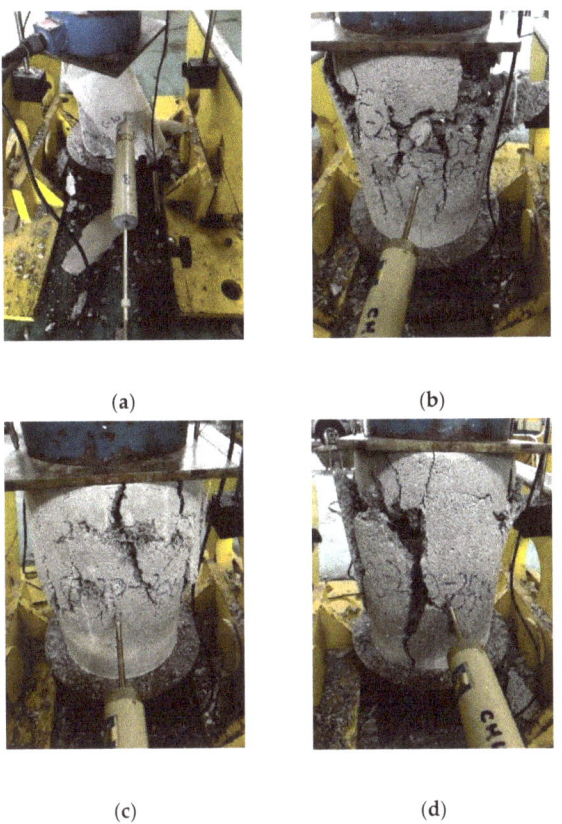

Figure 5. Typical failure mode of tested specimens. (**a**) C-1; (**b**) C-20-25; (**c**) C-20-30; (**d**) C-20-35.

For the CFPCCT specimens and mortar tube specimens, it is exhibited that cracking first occurred at the mid-height region, when the axial load reached about 80% of ultimate strength. Multiple cracks prominently extended following these initial cracks. The concrete cover peeled off at the same moment and the hoops were partially exposed at failure, as shown in Figure 5. For the control specimens, the first cracking point occurred throughout the wall surface when the axial load reached about 95% of ultimate value; then, a significant drop of loading was observed with an explosive failure of the specimens. From Figure 5, it can be seen that the 25 mm- and 35 mm-thick tube-confined specimens disintegrated upon failure. However, the 30 mm-thick tube-confined specimen showed a better performance in terms of disintegration. The disintegration of the specimens might be due to the fact that the tube also carried the axial load directly from the loading device. However, surprisingly, the spiral reinforcement was not yielded. This might be due to the fact that the reinforcement was not tied with any vertical reinforcement. In addition, the direct load applied to the tube was higher than that of the outward passive bursting pressure [6–8].

3.3. Stress-Strain Response

Figure 6 presents the stress-strain responses of the tested specimens. The stress-strain curves were established from thousands of original data points, obtained from the experimental test. Theoretically, the yield displacement is the yield point of an equivalent bilinear response curve that provides an equal area to that of the response curve [23]. In this study, the ultimate strain ε_c is considered to be the axial strain at the failure point when the load drops 20% from the peak load [23]. However, the yield point of all specimens in this study refers to the first cracking load.

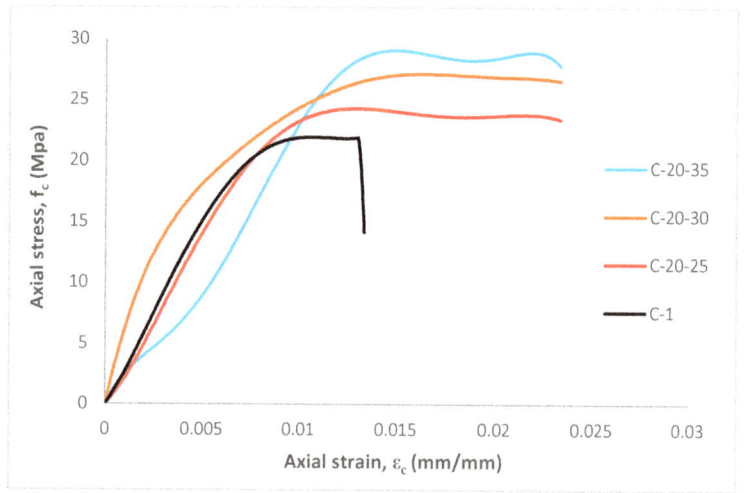

Figure 6. Stress-strain curves of specimens.

A descending branch, after the peak in the stress-strain curve, was expected for the tube confined specimens. However, the curve becomes flat to nearly flat after the peak strength and the compressive strength is reached, before the rupture of tube confinements. This behavior might be due to the fact that the external tube acted as an integrated part of the composite system, which carries both lateral and axial loads simultaneously. The stress-strain curves of the tube-confined specimens are quite similar. However, sudden failure was observed for unconfined concrete specimens. If the stress-strain curve terminates at a concrete stress higher than the compressive strength of non-reinforced specimens, the tube confinement is still working to enhance the strength [12,16]. It must be mentioned that the confining tube also carried the direct axial load, which contributes to the enhanced stress-strain responses.

It can be seen that the axial stress-strain curve of the C-20-30 specimens show the smoothest changing process, from load applying to failure. Conversely, the C-1 specimens show that there is a clear turning point, after the stress goes through the peak and drops sharply. This is also reflected in the testing procedure when there is an explosion occurred on C-1 specimens due to the limit of ductility.

The performance of the C-20-35 specimens is slightly different to the others around the yield point. Furthermore, they all pass through the linear stage before yield. In terms of its mechanical performance, the C-20-35 shows the higher compressive strength and better axial deformation, which means the higher wall-thickness of CFPCCT columns are functioning. However, it does not continue to 81.19% of ultimate compressive strength. In contrast to the high performance confined concrete tube, the wall-thickness of the cementitious composite tube is more sensitive to the load carrying capacity. Thus, it can be said that the loading carrying capacity will increase with an increasing wall-thickness of CFPCCT columns.

3.4. Ductility of Confinement

The tendency of a structural member to deform under large strains without fracturing is referred to as the ductility of that member, which is a desirable property of any structural element that provides warning before failure. On the one hand, the ductility of any specimens tested under compression can be computed in terms of displacement ductility, a ratio of the axial displacements corresponding to the yield and 0.85 ultimate loads [6]. On the other hand, Wang et al. [24] state that the ductility ratio of axial displacements is from the yield to 0.8 ultimate loads, as shown in Figure 7. This study adopts yield to 0.8 ultimate due to the fact that Kaish et al. [6] proposed the method for ferrocement, which is a well-known cementitious composite.

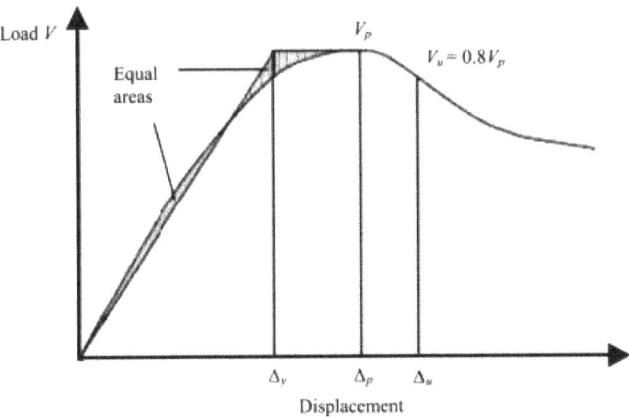

Figure 7. Definition of the ductility ratio.

The evaluated specimen's ductility is presented in Table 4. However, specimens that failed before even reaching 0.8 of ultimate stress on the descending branch are judged as the failure point directly, rather than theoretically calculating to achieve the perfect model, against the reality of the C-1 groups. The result shows that all tube reinforced specimens have a better performance of ductility ratio than non-reinforced specimens. The ductility values of the specimens are compared with the unconfined specimens. For the 25 mm, 30 mm and 35 mm wall-thickness cementitious composite tube-confined specimens groups, the ductility ratios are 3.1971, 3.3768 and 3.4247, respectively. Therefore, the results lead to the conclusion that higher wall-thickness causes better performance on ductility.

Table 4. Ductility of the tested specimens.

Specimen	Strain at Yield Stress (mm/mm)	Strain at 0.8 of Ultimate Stress (mm/mm)	Ductility Ratio
	Average	Calculated	
C-1	0.0052	0.0103	1.98
C-20-25	0.0067	0.0213	3.1791
C-20-30	0.0069	0.0233	3.3768
C-20-35	0.0073	0.0250	3.4247

4. Conclusions

In this investigative study, concrete-filled prefabricated cementitious composites tube (CFPCCT) columns with different tube wall-thickness were tested under axial compression. Based on the obtained test results, the following conclusions are drawn:

1. It is observed that the concrete-filled prefabricated cementitious composite tube columns performed a better load carrying and ductility capacity over unconfined concrete cylindrical specimens.

2. For the cementitious tube confined composite column specimens, the first crack occurred at the mid-height region when the axial load reached approximately 80% of the ultimate strength. After that, multiple cracks prominently extended with surface peeling. The 25 mm and 35 mm thick tubes showed complete disintegration upon failure. The cementitious composite tube with 30 mm wall thickness performed better in terms of failure behavior.
3. The strength and stress-strain behavior of CFPCCT columns enhanced with increased wall-thickness of the cementitious composite tube. This behavior is prominent as the external tube also directly carried the axial load.
4. Further investigation is recommended on this aspect to optimize the tube parameters for optimized performance of CFPCCT columns, together with suitable analytical models, to establish its practical application.

Author Contributions: Conceptualization, B.K. and A.B.M.A.K.; methodology, B.K. and A.B.M.A.K.; investigation, N.N. and A.B.M.A.K.; resources, N.N. and A.B.M.A.K.; data curation, B.K. and A.B.M.A.K.; writing—original draft preparation, B.K. and A.B.M.A.K.; writing—review and editing, N.N. and A.B.M.A.K.; visualization, N.N. and A.B.M.A.K.; supervision, N.N. and A.B.M.A.K.; project administration, N.N. and A.B.M.A.K.; funding acquisition, N.N. and A.B.M.A.K. All authors have read and agreed to the published version of the manuscript.

Funding: This research and APC were funded by Centre for Research and Instrumentation Management (CRIM), Universiti Kebangsaan Malaysia, grant number GGPM-2021-004.

Institutional Review Board Statement: Not applicable.

Informed Consent Statement: Not applicable.

Data Availability Statement: Not applicable.

Acknowledgments: The authors acknowledge the support from the Department of Civil Engineering, Universiti Kebangsaan Malaysia.

Conflicts of Interest: The authors declare no conflict of interest. The funders had no role in the design of the study; in the collection, analyses, or interpretation of data; in the writing of the manuscript; or in the decision to publish the results.

References

1. Afroughsabet, V.; Ozbakkaloglu, T. Mechanical and durability properties of high-strength concrete containing steel and polypropylene fibers. *Constr. Build. Mater.* **2015**, *94*, 73–82. [CrossRef]
2. Kaish, A.B.M.A.; Alam, M.R.; Jamil, M.; Zain, M.F.M.; Wahed, M.A. Improved ferrocement jacketing for restrengthening of square RC short column. *Constr. Build. Mater.* **2012**, *36*, 228–237. [CrossRef]
3. Kaish, A.B.M.A.; Alam, M.R.; Jamil, M.; Wahed, M.A. Ferrocement jacketing for restrengthening of square reinforced concrete column under concentric compressive load. *Procedia Eng.* **2013**, *54*, 720–728. [CrossRef]
4. Shan, B.; Lai, D.D.; Xiao, Y.; Luo, X.B. Experimental research on concrete-filled RPC tubes under axial compression load. *Eng. Struct.* **2018**, *155*, 358–370. [CrossRef]
5. Mollah, M.Y.A.; Adams, W.J.; Schennach, R.; Cocke, D.L. A review of cement–superplasticizer interactions and their models. *Adv. Cem. Res.* **2000**, *12*, 153–161. [CrossRef]
6. Kaish, A.B.M.A.; Jamil, M.; Raman, S.N.; Zain, M.F.M. Axial behavior of ferrocement confined cylindrical concrete specimens with different sizes. *Constr. Build. Mater.* **2015**, *78*, 50–59. [CrossRef]
7. Kaish, A.B.M.A.; Jamil, M.; Raman, S.N.; Zain, M.F.M.; Alam, M.R. An approach to improve conventional square ferrocement jacket for strengthening application of short square RC column. *Mater. Struct.* **2016**, *49*, 1025–1037. [CrossRef]
8. Kaish, A.B.M.A.; Jamil, M.; Raman, S.N.; Zain, M.F.M.; Nahar, L. Ferrocement composites for strengthening of concrete columns: A review. *Constr. Build. Mater.* **2018**, *160*, 326–340. [CrossRef]
9. Kaish, A.B.M.A.; Nahar, L.; Jaafar, A.; Ahmed, Y. Prospects of using Prefabricated Ferrocement Jacket for Semi-Automated Strengthening of RC Column. *Electron. J. Struct. Eng.* **2018**, *18*, 52–57. [CrossRef]
10. Lim, W.Y.; Park, H.G.; Oh, J.K.; Kim, C.S. Seismic resistance of cast-in-place concrete-filled hollow PC columns. *J. Korea Concr. Inst.* **2014**, *26*, 35–46. [CrossRef]
11. Hosoya, H.; Asano, Y. Seismic behavior of R/C column members using precast concrete shell under high axial load. In Proceedings of the 12th World Conference on Earthquake Engineering, Auckland, New Zealand, 30 January–4 February 2000.

12. Xu, L.-j.; Wang, Y.-b.; Zhang, Z.-g.; Lin, X.; Zhang, C. Quasi-static test study on precast ECC concrete-filled tubular bridge piers. *Eng. Mech.* **2021**, *38*, 229–238.
13. Pan, Z.; Zhu, Y.; Qiao, Z.; Meng, S. Seismic behavior of composite columns with steel reinforced ECC permanent formwork and infilled concrete. *Eng. Struct.* **2020**, *212*, 110541. [CrossRef]
14. Meng, W.; Khayat, K.H. Development of Stay-In-Place Formwork Using GFRP Reinforced UHPC Elements. In *International Interactive Symposium on Ultra-High Performance Concrete*; Iowa State University Digital Press: Ames, IA, USA, 2016; Volume 1.
15. Xiao, Y.; Ma, R. Seismic retrofit of RC circular columns using prefabricated composite jacketing. *J. Struct. Eng.* **1997**, *123*, 1357–1364. [CrossRef]
16. Shao, Y.; Kuo, C.W.; Hung, C.C. Seismic performance of full-scale UHPC-jacket-strengthened RC columns under high axial loads. *Eng. Struct.* **2021**, *243*, 112657. [CrossRef]
17. Hung, C.C.; Kuo, C.W.; Shao, Y. Cast-in-place and prefabricated UHPC jackets for retrofitting shear-deficient RC columns with different axial load levels. *J. Build. Eng.* **2021**, *44*, 103305. [CrossRef]
18. Guan, D.; Chen, Z.; Liu, J.; Lin, Z.; Guo, Z. Seismic performance of precast concrete columns with prefabricated UHPC jackets in plastic hinge zone. *Eng. Struct.* **2021**, *245*, 112776. [CrossRef]
19. Tian, H.; Zhou, Z.; Wei, Y.; Zhang, L. Experimental and numerical investigation on the seismic performance of concrete-filled UHPC tubular columns. *J. Build. Eng.* **2021**, *43*, 103118. [CrossRef]
20. Zhu, Y.; Zhang, Y.; Xu, Z. Analytical investigation of long-term behavior of normal concrete filled UHPC tube composite column. *Case Stud. Constr. Mater.* **2022**, *17*, e01435. [CrossRef]
21. Zhang, Y.; Zhu, Y.; Xu, Z.; Shao, X. Long-term creep behavior of NC filled UHPC tube composite column. *Eng. Struct.* **2022**, *259*, 114214. [CrossRef]
22. Shan, B.; Liu, G.; Li, T.Y.; Liu, F.C.; Liu, Z.; Xiao, Y. Experimental research on seismic behavior of concrete-filled reactive powder concrete tubular columns. *Eng. Struct.* **2021**, *233*, 111921. [CrossRef]
23. Mander, J.B.; Priestley, M.J.N.; Park, R. Observed stress-strain behavior of confined concrete. *J. Struct. Eng.* **1988**, *114*, 1827–1849. [CrossRef]
24. Wang, L.M.; Wu, Y.F. Effect of corner radius on the performance of CFRP-confined square concrete columns: Test. *Eng. Struct.* **2008**, *30*, 493–505. [CrossRef]

Article

Improved Shear Strength Equation for Reinforced Concrete Columns Retrofitted with Hybrid Concrete Jackets

Kyong Min Ro, Min Sook Kim and Young Hak Lee *

Department of Architectural Engineering, Kyung Hee University, Deogyeong-Daero 1732, Yongin 17104, Republic of Korea; kyongmin@khu.ac.kr (K.M.R.); kimminsook@khu.ac.kr (M.S.K.)
* Correspondence: leeyh@khu.ac.kr; Tel.: +82-31-201-3815

Abstract: The adequacy of retrofitting with concrete jacketing is influenced by the bonding between the old section and jacketing section. In this study, five specimens were fabricated, and cyclic loading tests were performed to investigate the integration behavior of the hybrid concrete jacketing method under combined loads. The experimental results showed that the strength of the proposed retrofitting method increased approximately three times compared to the old column, and bonding capacity was also improved. This paper proposed a shear strength equation that considers the slip between the jacketed section and the old section. Moreover, a factor was proposed for considering the reduction in the shear capacity of the stirrup resulting from the slippage between the mortar and stirrup utilized on the jacketing section. The accuracy and validity of the proposed equations were examined through a comparison with the ACI 318-19 design criteria and test results.

Keywords: seismic retrofitting; concrete jacketing; cyclic loading; shear equation

1. Introduction

Concrete jacketing is an effective seismic retrofit method to improve strength and rigidity by enlarging the cross-section of reinforced concrete columns. The seismic performance of a retrofitted member with a concrete jacket is affected by the reinforcement in the jacketing section, the compressive strength of concrete, and the bonding between the old section and the jacketing section. Therefore, recent studies have been conducted on seismic hooks and steel wire mesh (SWM) to improve the shear performance of the jacketing section and on the use of dowels to constrain the jacketing section and the old section [1–3]. Many studies focused on improving the strength of the jacketing section using ultra-high-performance fiber-reinforced concrete (UHPFRC) have been conducted [4–7]. UHPFRC can show excellent strength, ductility and durability by lowering the water-binder ratio (W/B) by 20% and mixing high-powder admixtures and high-strength steel fibers. The thickness of the jacketing section can be reduced when using such a high-performance material, and the strength and ductility of the retrofitted members were effectively improved. However, UHPFRC, with less water, has a large amount of admixture and no coarse aggregate compared to conventional concrete, which results in high self-shrinkage and a high risk of shrinkage cracking. In a previous study [8], a new hybrid concrete jacketing method was proposed with non-shrinkage mortar in which shrinkage was suppressed by adding an anti-shrinkage admixture. Steel fiber was mixed into the non-shrinkage mortar to enhance the strength of the jacketing section. Conventional concrete jacketing methods use a lot of reinforcement (such as dowels or cross ties) to improve bond capacity, but this reduces workability. Hybrid concrete jacketing methods involve steel wire mesh (SWM) and steel grid reinforcement (SGR). The welded SWM is attached to the surface of the old section to improve the adhesion between the old section and the jacketing section. This makes it possible to omit the process of chipping the concrete surface, thereby improving workability compared to the conventional concrete jacketing method.

Citation: Ro, K.M.; Kim, M.S.; Lee, Y.H. Improved Shear Strength Equation for Reinforced Concrete Columns Retrofitted with Hybrid Concrete Jackets. *Materials* 2023, 16, 3734. https://doi.org/10.3390/ma16103734

Academic Editor: Andreas Lampropoulos

Received: 26 April 2023
Revised: 11 May 2023
Accepted: 12 May 2023
Published: 15 May 2023

Copyright: © 2023 by the authors. Licensee MDPI, Basel, Switzerland. This article is an open access article distributed under the terms and conditions of the Creative Commons Attribution (CC BY) license (https://creativecommons.org/licenses/by/4.0/).

The hybrid concrete jacketing method is divided into two types according to the reinforcement in the jacketing method, as shown in Figure 1. Type 1 is easy to manufacture as welded reinforcing bar grids, but the confinement is reduced, so a small number of dowels are used to improve the bonding capacity between the old section and the jacketing section. Type 2 is a method involving making a hook at the end of the SGR. As compared to seismically designed transverse reinforcement with 135-degree hooks proposed in the ACI 318-19 standard [9], this construction is simpler and less hook loosening occurs, so excellent seismic performance can be expected. When seismic retrofitting a reinforced concrete structure with concrete jacketing, the most important factor is whether a slip occurs at the interface between the old section and jacketing section. Slip occurs at the interface between the core and the jacket concrete if the bonding is not properly secured when a load is applied to the reinforced concrete columns, so the seismic performance of the jacketing section is expected to be unacceptable. There are limited published studies on the contact surfaces of reinforced concrete members and jacketing [6]. Furthermore, when dealing with members having connections, it is important to conduct investigations to identify potential factors that could impact the monolithic behavior, such as slippage [10–13]. Psycharis and Mouzakis [10] examined the effect of dowel diameter, the number of dowels, and the placement of dowels from the edge of the section on the shear behavior of precast members under different loading patterns. The experimental results indicated that the resistance of the connection under cyclic loading was only half of that under monotonic loading, and the thickness of the cover concrete in the dowel installation direction was found to be related to dowel slippage, which was identified as a factor affecting the shear performance. Therefore, this study experimentally analyzed whether the hybrid concrete jacketing method proposed in a previous study [8] can ensure the appropriate bonding capacity and proposed a shear strength equation for reinforced concrete columns retrofitted with a hybrid concrete jacket.

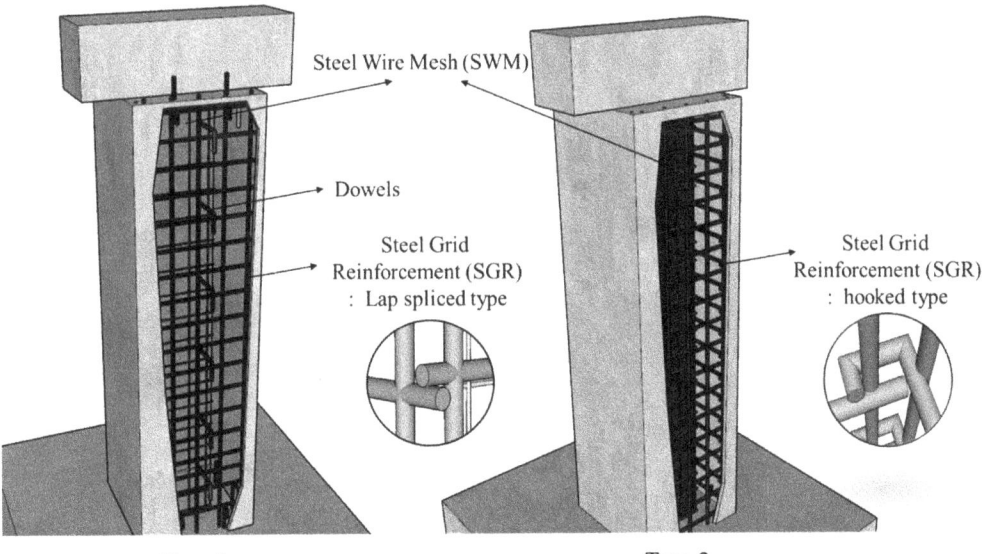

Figure 1. Types of hybrid concrete jacketing methods.

2. Experimental Program

To analyze the seismic performance of the hybrid concrete jacketing method, two test specimens of reinforced concrete columns without seismic design and three specimens retrofitted with hybrid concrete jackets were fabricated and subjected to a cyclic loading test. The details of the specimens are shown in Figure 2 and Table 1. The cross-section of the reinforced concrete column without seismic design was 250 mm × 250 mm, and the height was 1800 mm. For the jacketing section, 4 sides were retrofitted with a thickness of 125 mm by referring to the design method for the concrete jacketing section presented in Pennelis and Kappos [14]. The upper beam is for applying the axial and lateral load, the cross-section was 250 mm × 250 mm, and the length was 800 mm. The specimen was cast on a foundation that was 1400 mm × 1270 mm × 450 mm. The foundation was fastened to a strong floor through high-tensile bolts. The reinforcement was designed in accordance with the ACI 318-19 design standard. Four deformed bars with diameters of 22 mm and 90-degree closed external stirrups with diameters of 10 mm were placed at spacings of 125 mm in the column. The SWM was placed on the jacketing section using steel wire with a diameter of 10 mm, and the SGR, which served as the longitudinal bars and hoops of the column, was manufactured off-site using deformed reinforcing bars with diameters of 13 mm. The compressive strength of the concrete and the non-shrinkage mortar cast on the jacketing section were 24 MPa and 30 MPa, respectively. The yield strength of reinforcing bars placed in the old column and jacketing section was 400 MPa. In the hybrid concrete jacketing method, a non-shrinkage mortar was mixed with steel fibers to improve the structural performance of retrofitted members. The steel fiber was developed for non-shrinkage mortar and had a diameter of 0.34 mm, a length of 18 mm, and a tensile strength of 1250 MPa. Steel fiber content was designed to be 1.5% to ensure excellent performance and workability according to studies of the effect of steel fiber content on structural performance [15–18]. Table 2 summarizes the material properties used in the manufacture of the specimens.

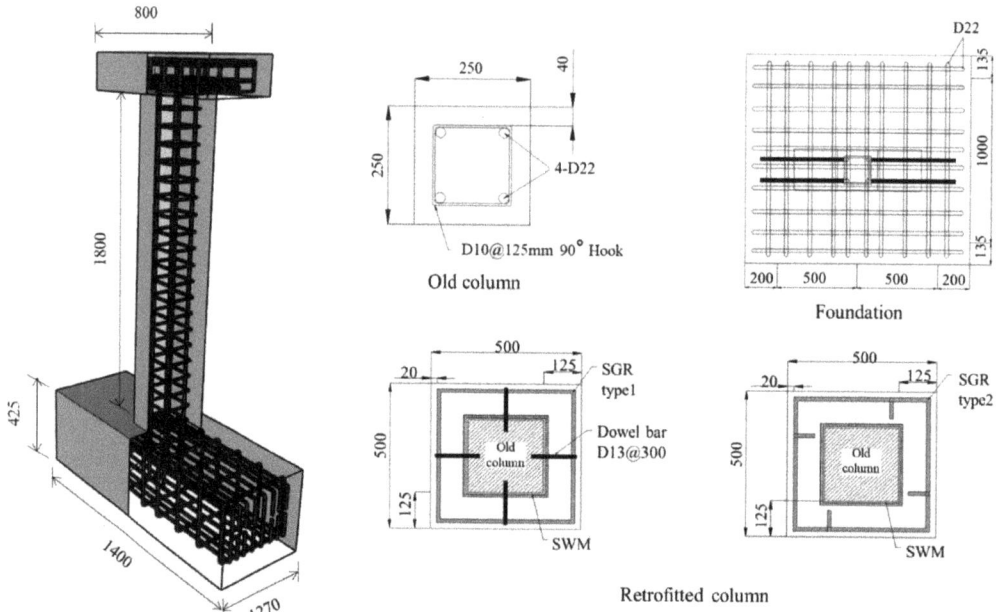

Figure 2. Specimen details (units: mm).

Table 1. Details of specimens.

Specimen	Retrofit Method	Loading Scheme	Cross Section [mm]
RC-U	–	Unidirectional	250×250
RC-B	–	Bidirectional	
HCJ-1U	Type 1	Unidirectional	500×500
HCJ-2U	Type 2	Unidirectional	
HCJ-1B	Type 1	Bidirectional	

Table 2. Mechanical properties of specimens.

	Concrete	Non-Shrinkage Mortar	SFNM	Reinforcement Steel
Compressive strength (MPa)	24	30	40.1	-
Yield strength (MPa)	-	-	-	400
Material	Diameter [mm]	Length [mm]	Aspect Ratio	Tensile Strength [MPa]
Steel fiber	0.34	18	0.019	1250

The manufacturing process of the specimens retrofitted with the hybrid concrete jacketing method is presented in Figure 3. The first step involves wrapping SWM around all four sides of the old column to enhance the bonding capacity between the old column and the jacketing section. The concrete core was fully confined using SWM, and strips of SWM were tightly fastened with steel wires. Subsequently, chemical anchors were installed by drilling the old column to place the SGR or dowel bar. Dowel bars were not used in the case of the specimen with Type 2 detail. Finally, a formwork was constructed on the jacketing section and non-shrinkage mortar mixed with steel fibers was poured. Chipping is a crucial process in conventional concrete jacketing that roughens the section of the old column. However, the hybrid concrete jacketing method improved adhesion performance by bonding SWM to the old column. Therefore, it was possible to omit the chipping process, which generates dust and hinders workability.

Figure 3. Retrofitting process of a column with the hybrid concrete jacketing method. (a) Placing SWM and drilling; (b) Installation of reinforcement in jacketing section; (c) Formwork of jacketing section.

In this study, a cyclic loading test considering axial load, lateral load, and torsional load was conducted to simulate actual seismic load. The test setup and the quasi-static loading protocol are shown in Figure 4. Details related to the loading setup and protocols are provided in a previous study [8]. The cyclic lateral load was applied through the horizontal actuator, and it was increased gradually from a drift ratio of 0.2% until the test was terminated. The definition of drift ratio is the ratio of the lateral displacement to the height from the bottom of the column to the loading point. A constant axial load of 255 kN, which is 17% of the axial load capacity, was applied. An eccentric load was applied to generate torsion with single (unidirectional) or multi-directional (bidirectional) loads. When a compressive force is applied at a location beyond the core of a section, tensile stress is induced in addition to compressive stress, and concrete is especially vulnerable to tensile stress. This study induced the tensile stress on the specimens to simulate extreme conditions during an actual earthquake by applying a load at a location beyond the core of a section. An eccentricity of 65 mm was set, considering the core of a section (1/6 of the section dimension for a rectangular section). To measure the strain of steel reinforcement and concrete, strain gauges were installed near the plastic hinge of the column where the damage is expected to be concentrated, as shown in Figure 5. As shown in Figure 5, the front sides of the specimen were designated Side 1, and the elevations were classified by naming Sides 2, 3, and 4 in the counterclockwise direction to identify the direction.

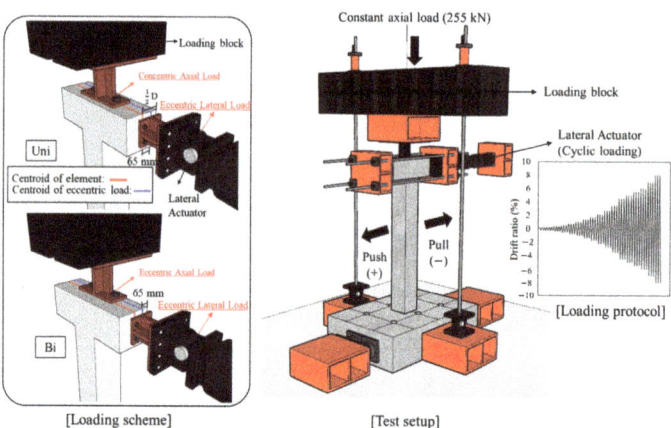

Figure 4. Test setup and loading protocol.

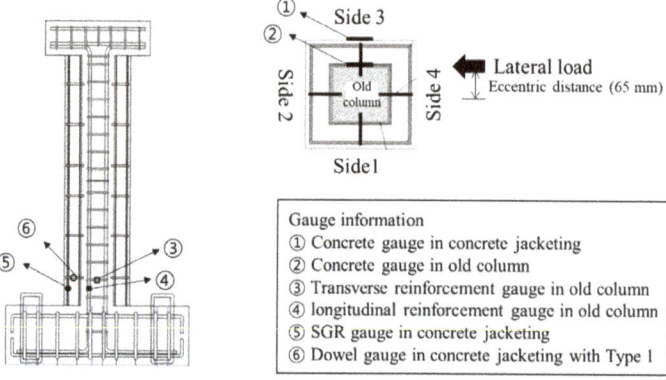

Figure 5. Locations of gauges.

3. Experimental Results

3.1. Load-Displacement Relationships

The load-displacement curves for each specimen are shown in Figure 6. Torsion was induced in all specimens, and shear cracks occurred at the bottom of the column, indicating shear failure. To confirm the seismic performance of hybrid concrete jacketing, the load at the occurrence of significant cracking and maximum load are summarized in Table 3. The maximum load of the reinforced concrete column without seismic design and the specimen retrofitted with hybrid concrete jacketing were compared under the same loading scheme. When unidirectional loading was applied, the maximum load of HCJ-1U was 3.9 times that of RC-U, and the maximum load of HCJ-1B was approximately 3.6 times that of RC-B when bidirectional loading was applied. This study confirmed that the hybrid concrete jacketing method is effective for seismic retrofit regardless of the loading scheme. When unidirectional loading was applied, the torsion induced in the specimen increased compared to bidirectional loading because the axes of axial force and lateral force did not coincide. If torsion is induced in reinforced concrete columns, shear cracking and concrete spalling are generally observed at the bottom of the column, and brittle failure occurs under extreme torsion. The hybrid concrete jacketing method can increase the strength and rigidity by enlarging the cross-section of the column, effectively resisting torsion because the steel fiber suppresses the propagation of shear cracks. Therefore, the difference in maximum load under unidirectional loading and bidirectional loading was insignificant, and there were no significant shear cracks or brittle failures.

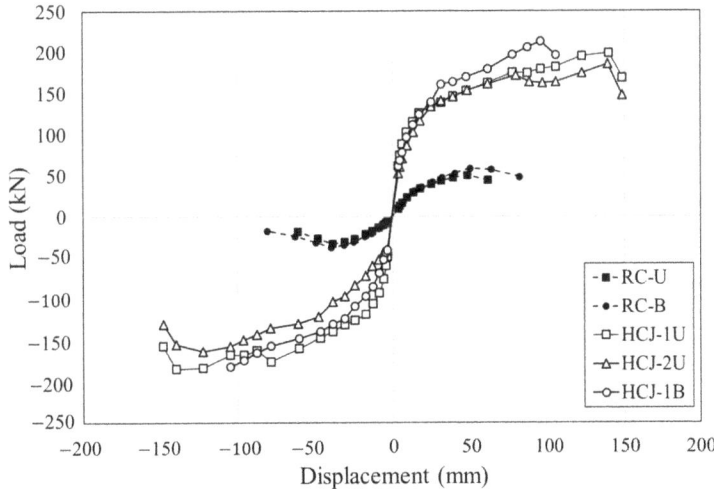

Figure 6. Backbone curve for specimens.

Table 3. Test results.

Specimen	Initial Crack		Shear Crack		Maximum	
	Load (kN)	Drift Ratio (%)	Load (kN)	Drift Ratio (%)	Load (kN)	Drift Ratio (%)
RC-U	15.21	1	43.53	2.2	50.61	3.5
RC-B	13.15	0.75	47.35	1.4	59.48	4.5
HCJ-1U	62.05	0.2	125.96	1	198.32	8.5
HCJ-2U	60.76	0.2	116.54	1	185.23	8.5
HCJ-1B	61.40	0.2	111.47	0.75	202.45	6

HCJ-1U and HCJ-1B retrofitted with Type 1 reached the maximum loads at 8.5% and 6% of the drift ratio, respectively, and the maximum loads were 198.32 kN and 202.45 kN, respectively. The maximum load of HCJ-2U retrofitted with Type 2 was 7% lower than that of HCJ-1U retrofitted with Type 2 under the same load scheme. Unlike Type 1, in which dowels were placed to improve the bonding performance between the old section and the jacketing section, Type 2 omitted additional reinforcement by placing SGR with hooked details in the jacketing section. The bonding capacity between the old section and the jacketing section was reduced in Type 2 compared to Type 1, resulting in relatively more slip. However, since similar seismic performances were observed compared to the simple construction process, this study confirmed that if Type 2 details are used, seismic performance can be secured without using additional shear reinforcement such as dowels. These results can be observed through the failure patterns of each specimen shown in Figure 7.

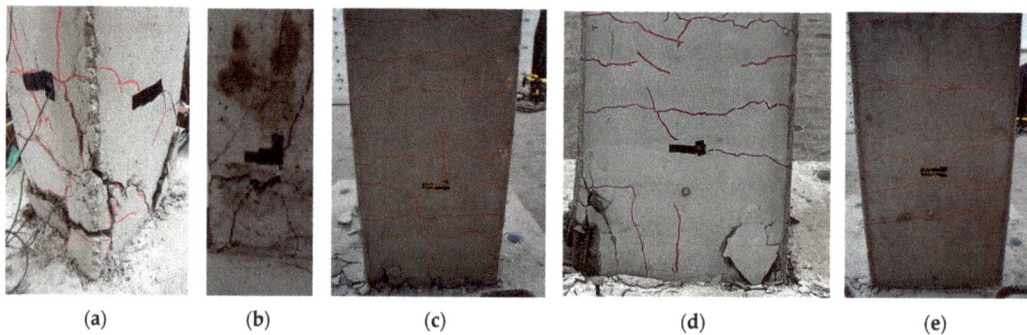

(a) (b) (c) (d) (e)

Figure 7. Failure of specimens at the bottom of the column: (**a**) RC-U, (**b**) RC-B, (**c**) HCJ-1U, (**d**) HCJ-2U, and (**e**) HCJ-1B.

3.2. Torsional Moment versus Twist Response

The torsional moment and twist were obtained from LVDTs and strain gauges, as shown in Figure 8. These were calculated as shown in Equations (1) and (2).

$$M_i = P \times \ell \times \cos \theta_i \tag{1}$$

$$\theta_i = \tan^{-1}\left(\frac{\Delta_2 - \Delta_1}{d}\right) \tag{2}$$

Here, M_i is the torsional moment generated to the specimen at the i-th drift ratio, P is the maximum load at the i-th drift ratio, ℓ is the eccentric distance, θ_i the twist of the column cross-section, and d is the distance between LVDTs.

Torsional moment-twist envelopes are shown in Figure 9. The torsional moment tended to increase gradually until shear cracks occurred in all specimens. The torsional moment decreased dramatically as the shear crack extended and failed. In the case of non-retrofitted specimens, the maximum torsional moment of RC-U under unidirectional loading was 2.13 kN·m, and the maximum torsional moment of RC-B under bidirectional loading was 1.41 kN·m. This result indicated that a degradation in seismic performance of about 60% was observed when a larger torsion was applied. This is a general tendency observed in reinforced concrete columns [19]. Comparing retrofitting with hybrid concrete jacketing and non-retrofitted specimens, the seismic performance improved by 7.8 times when unidirectional loading was applied and 5.8 times when bidirectional loading was applied. HCJ-2U with Type 2 reinforcement was about 80% of the maximum torsional moment of HCJ-2B with Type 1, and the twist increased. This means that the details of the reinforcement in the jacketing section can affect the contact surfaces between the old section and the jacketing section.

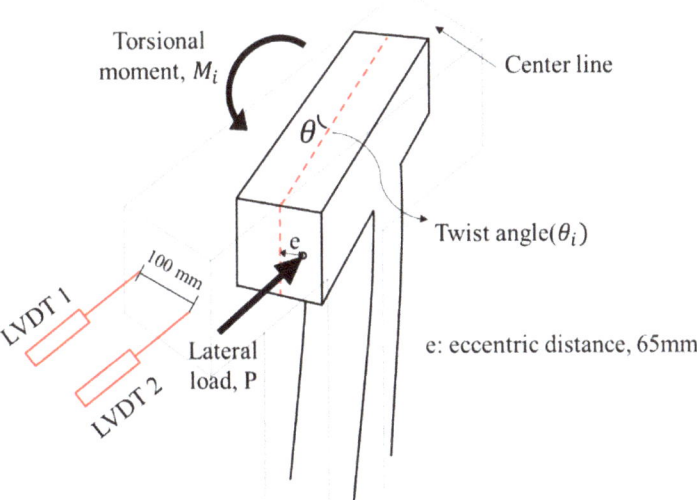

Figure 8. Torsional moment and twist.

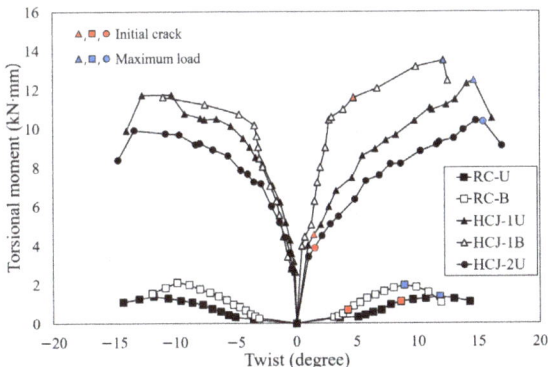

Figure 9. Torsional moment-twist relationships in the specimens.

3.3. Strain

Strains of the reinforcing bars in the old column and the jacketing section were compared to analyze the integration behavior of the hybrid concrete jacketing method. Figure 10 shows the load-strain relationship of the reinforcing bars of the retrofitted specimens. In HCJ-1U and HCJ-1B, the strain of the longitudinal bars of the old column and the vertical steel bars of the SGR (which act like longitudinal bars) showed a similar tendency until the maximum load. The dowels (which transmit stress to the SGR) also yielded increasing strain until the maximum load, confirming that the SGR and dowels placed in the jacketing section effectively resisted the load regardless of the loading scheme. In HCJ-1U and HCJ-2U, where the reinforcement details varied, the strain distribution of longitudinal reinforcing bars in the old column and the SGR was similar until yielding, and yielding occurred at the point of maximum load. The strain of transverse reinforcement of SGR in HCJ-1U increased gradually as the load increased, and it yielded at the maximum load. However, the strain of dowels increased non-linearly, and it did not yield until the maximum load. This indicated that dowels in Type 1 effectively transferred shear stress to the SGR. The strain of the transverse reinforcing bars of SGR in HCJ-2U also showed a gradual increase and then yielded at the point of maximum load. This means that the hooked end detail of SGR in Type 2 acted as a constraint and helped prevent buckling and shear crack control. Test results showed that columns retrofitted with Type 1 and Type 2 behaved similarly to the monolithic reinforced concrete column under the combined load considering torsion.

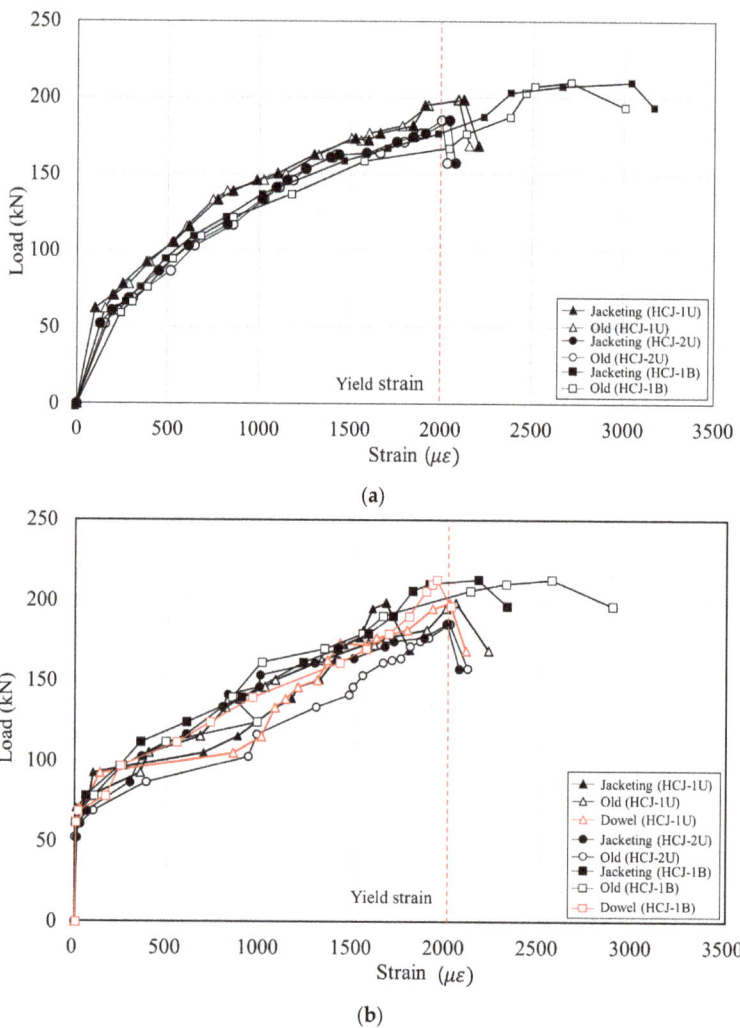

Figure 10. Strain of reinforcement: (**a**) longitudinal reinforcement and (**b**) stirrup.

When a reinforced concrete column retrofitted with concrete jacketing behaves like a monolithic column, the difference in concrete strain between the old section and the jacketing section tends to increase proportionally until the ultimate strain [6,20]. However, when slip occurs at the surface between the old section and the jacketing section, the concrete jacket does not resist deformation, and concrete strain at the jacketing section does not increase. Therefore, the bonding capacity of the hybrid concrete jacket was evaluated in this study by analyzing the concrete strain of the surface of the old column and the jacketing section. Figure 11 shows the load-strain relationship of concrete. HCJ-1U and

HCJ-2U subjected to unidirectional loading exceeded the ultimate strain of the concrete (0.003) when reaching the maximum load. However, the strain of the concrete in HCJ-1B did not reach the ultimate strain until the experiment terminated. This is because the torsion induced in the specimen was small, and the damage to the column until the failure was not significant. As the drift ratio increased in all retrofitted specimens, the difference in concrete strain between the old section and the jacketing section gradually increased. Compared to the increase in concrete strain in the old section, the increase in concrete strain in the jacketing section was lower. This means that a slip occurred at the interface.

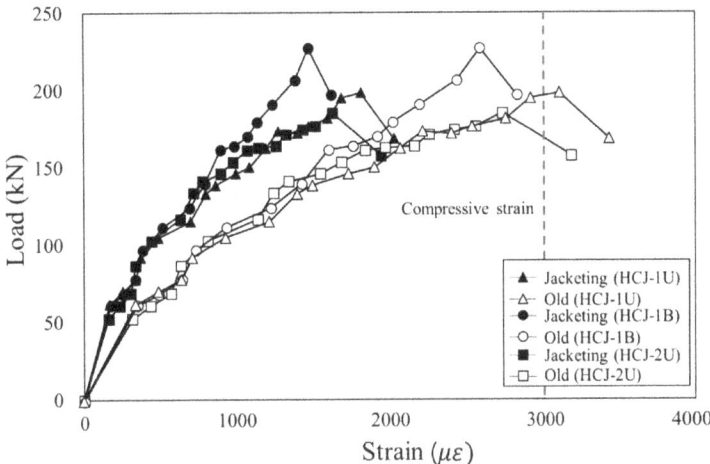

Figure 11. Strain of concrete.

4. Proposed Shear Strength Equation

The seismic performance of a concrete jacketed column depends on the bonding of the old concrete and added concrete. If appropriate shear reinforcement is designed in the jacketed section, a column with a concrete jacket should act monolithically under seismic load. Therefore, the shear strength and behavior of a reinforced concrete column retrofitted with concrete jacketing are predicted, considering the jacketed section to be an equivalent monolithic section [21]. The experimental results confirmed that the SGR of the hybrid concrete jacketing method performed the role of longitudinal reinforcements and hoops. Therefore, the reinforced concrete column retrofitted with hybrid concrete jacketing was an adequately designed shear reinforcement. In the past few decades, the bond-slip model was mainly considered for reinforced concrete strengthened with fiber-reinforced polymer (FRP) to avoid debonding failure. However, recent studies have confirmed that reinforced concrete columns retrofitted with concrete jackets should consider the slip when it takes place along the interfaces between the old section and the jacketing section [20,22,23]. The mechanics of reinforced concrete members retrofitted with a concrete jacket are quite complex. In particular, it is difficult to consider the behavior of the interface between the existing member and the jacket [23]. In this study, the slip coefficient was experimentally determined as a measure of the bonding capacity between the two sections to assess the behavior of the interface between the concrete jacket and the existing members. It is common to measure the slip coefficient experimentally considering various factors such as the strain, material of the jacket and existing member, friction coefficient, and angle. Figure 12 shows the strain profile of the jacketed cross-section. As shown in Figure 12, the slip coefficient, η, is a value that measures the frictional force between the jacketing section and the existing member. It is an indicator of the strength of the bonding capacity between the two sections. The slip coefficient is one of the most important factors in ensuring safe attachment between the jacket and the existing member in the concrete jacketing method.

If the old column is considered to be fully confined due to the confinement provided by the jacket, there is no sliding at the interface between the column and jacket. Therefore, the slip coefficient equals 1.0. On the other hand, in a partially confined column, the slip coefficient is less than 1.0, and the concrete strain of the jacketed section decreases by the ratio of the slip coefficient. This value is usually in the range of 0.8 to 1.0, with a higher value indicating a stronger friction force between the jacket and the existing member.

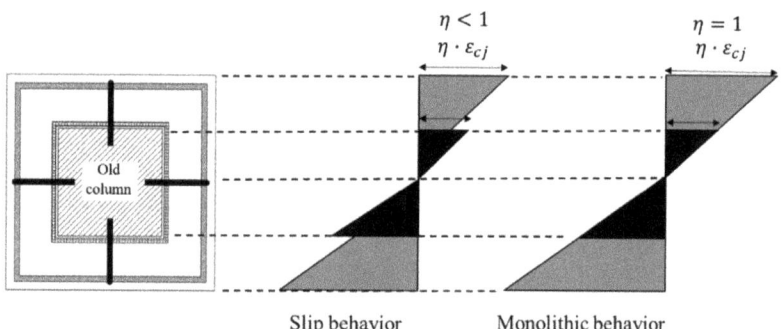

Figure 12. Strain profile.

To obtain the slip coefficient, gauges were used to measure the concrete strain in both the old section and the jacketing section, as shown in Figure 5. The gauges were placed at the same location in the cross-section of the existing member and the jacketing concrete, as illustrated in Figure 5. The slip coefficient was calculated based on the difference in concrete strain between the two sections, with a higher value indicating a smaller difference in strain and stronger bonding between the old section and the jacketed concrete section. Figure 13 depicts the slip coefficients obtained by calculating the difference in strain between the old section and the jacketing section for each drift ratio. At the onset of the experiment, the slip coefficient is approximately 1.0. However, all specimens exhibited a trend of decreasing slip coefficient as the drift ratio increased. This is because the torsion acting on the specimen increased according to the drift ratio. In addition, the slip coefficient decreased rapidly from the initial crack occurrence to the point of shear crack occurrence, and the slip coefficient converged after shear cracking. Bonding between the jacketing section and the old section decreased because torsion causes shear cracks in columns. The slip coefficient did not decrease beyond a certain level since the SWM attached to the old section secured the bonding capacity with the jacketing section. In this paper, the smallest value among the slip coefficients measured in the experiment was considered when calculating shear strength to ensure a conservative design. The slip coefficients for the HCJ-1U, HCJ-1B, and HCJ-2U specimens were verified to be 0.86, 0.88, and 0.82, respectively. Due to the misaligned axes of axial and lateral forces during unidirectional loading, HCJ-1U exhibited a smaller slip coefficient than HCJ-1B because the torsional forces were greater. Nevertheless, the discrepancy in slip coefficient between loading schemes was insignificant at approximately 2%. Since it is not feasible to anticipate the loading scheme under actual seismic loads, a conservative slip coefficient of 0.86 was assigned to Type 1. It should be noted that Type 2 had no dowel (unlike Type 1), and it ensured bonding capacity through the hooked details in SGR. As a result, the slip coefficient of Type 2 decreased compared to that of Type 1, with a value of approximately 95% of Type 1.

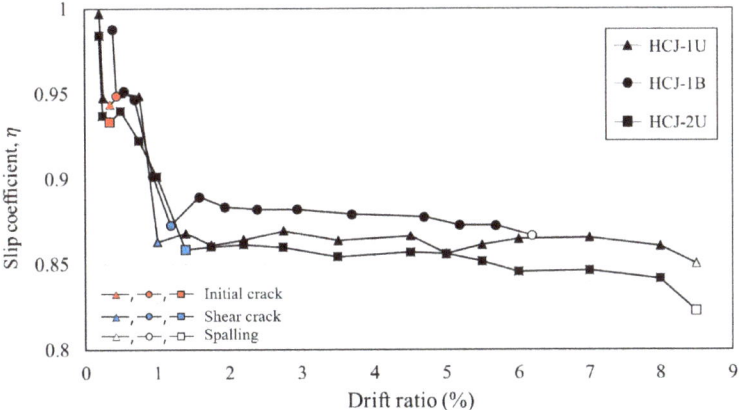

Figure 13. Slip coefficient from concrete strain.

This study proposed a novel concrete jacketing method that employs non-shrinkage mortar for concrete jacketing, with the objective of enhancing workability and alleviating the problem of high self-shrinkage that is typically associated with Ultra-High-Performance Fiber-Reinforced Concrete (UHPFRC). It is important to note that the slip of stirrups in concrete can have a significant impact on the strength and durability of columns, particularly in columns that use mortar. This is due to the relatively smooth surface of mortar, which can lead to increased stirrup slippage compared to concrete columns. As depicted in Figure 10, the strain patterns of the stirrups in the old and jacketing sections were similar. Nonetheless, slippage between the mortar and stirrup could potentially occur in the jacketing section, thereby compromising the shear performance of the stirrup. Consequently, the disparity in strains between the old and jacketing sections of the stirrup was analyzed, and this factor was accounted for in the evaluation of the shear strength of the stirrup in the jacketing section. The monolithicity factor (K) was introduced as a measure of bonding performance in the concrete jacketing method to investigate the reduction in shear capacity resulting from stirrup slippage within the jacketing section. The K factor is defined as the ratio of the response index of composite members to the response index of monolithic members with an identical geometry [23]. In this study, a comparison between a monolithic concrete column and a jacketed concrete column having identical geometries was not carried out. Nevertheless, the stirrup of the existing concrete column and that of the concrete jacket were designed to exhibit equivalent shear performance. Hence, the monolithicity factor was determined by computing the ratio of the stirrup strain in the jacketing section to that in the old section. The monolithicity factor was subsequently normalized to a maximum value of 1, and the results are presented in Figure 14.

During the initial stage of the experiment, the load was carried by the stirrup in the old column until yielding occurred, after which the stirrup within the jackets began to carry the load with an increase in load. Consequently, the monolithicity factor K, which denotes the ratio of stirrup strain in the jacketing section to that in the old section, gradually increased and approached unity in the early stages of the experiment. However, the load carried by the stirrup within the jacketing section increased as the torsional load increased, causing a reduction in friction between the stirrup and mortar in the jacket, resulting in a decrease in K. This phenomenon was observed consistently in all specimens, and K approached a constant value at the maximum load. The loading scheme was closely related to K, and the values of K at the failure for HCJ-1U, HCJ-2U, and HCJ-1B were 0.9, 0.82, and 0.97, respectively. Unidirectional loading led to an increase in torsional load, a reduction in bonding capacity between the stirrup and mortar within the jacketing section, and an increase in shear stress in the jacketing section. Moreover, Type 2 showed a greater decrease in bonding capacity between the stirrup and mortar within the jacketing section than Type 1 because no dowel was used to connect the old and jacketing sections, resulting in more sliding between the two sections and an increase in shear strength within the jacket.

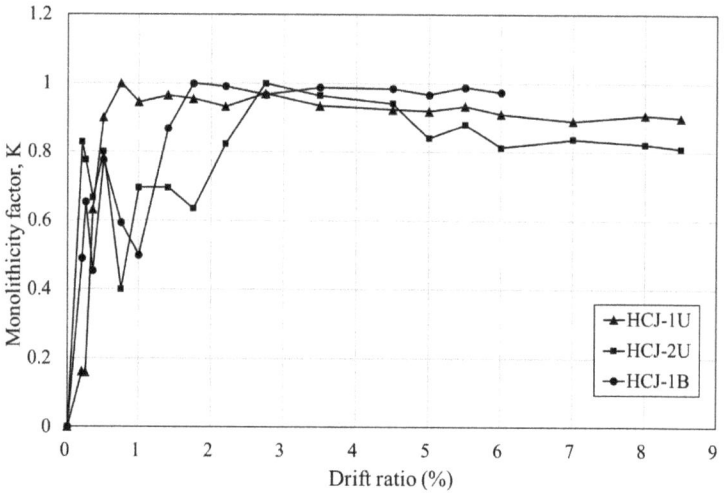

Figure 14. The monolithicity factor from stirrup strain.

In this study, the shear strength of concrete and stirrup was estimated using the ACI 318-19 design criteria as shown in Equations (3)–(5).

$$V_n = V_c + V_s \tag{3}$$

$$V_c = \frac{1}{6}(1 + \frac{N_u}{14A_g})\lambda\sqrt{f'_c}b_w d \tag{4}$$

$$V_s = \frac{A_v f_{yt} d}{s} \tag{5}$$

Here, A_g is the gross area of the concrete section, A_v and s are the area and spacing of shear reinforcement, b_w is the web width of the cross-section, d is the distance from extreme compression fiber to the centroid of longitudinal reinforcement, f'_c is the compressive strength of the concrete, f_{yt} is the yield strength of the stirrup, N_u is the factored axial force acting on the cross-section of the column, and λ represents the influence of lightweight concrete.

The shear strength equation of the reinforced concrete column retrofitted with hybrid concrete jacketing was proposed as shown in Equations (6)–(8). If slippage occurs between the old and jacketing sections, both sections may not fully exhibit their shear performance under seismic loads. To mitigate this issue, this study proposed a slip coefficient for each type of hybrid concrete jacketing method that considers the level of slippage between the old and jacketing sections based on experimental results. As a result, the contribution of concrete was estimated by multiplying the sum of the shear contributions of the old and jacketed sections by the slip coefficient to evaluate the shear strength of a reinforced concrete column retrofitted with hybrid concrete jacketing. The hybrid concrete jacketing method used mortar instead of concrete in the jacketing section, which can result in increased slip between the reinforcement steel and mortar due to the absence of coarse aggregate. Therefore, the reduction in stirrup shear capacity due to a slip between the mortar and the stirrup should be considered. The shear strength of the stirrup in the jacketing section was determined by multiplying the shear strength of the stirrup in the existing reinforced

concrete column by a monolithicity factor, which is a reduction factor. The variable n in Equation (8) indicates the number of pertinent entities drawn from the subsequent options:
(1) When a unidirectional load is applied.
(2) When no additional steel, such as dowels, is provided to enhance the bonding performance between the old section and the jacketing section.

$$V_{n,proposed} = \eta \left(V_{c,old} + V_{c,jacket} \right) + V_{s,old} + K \cdot V_{s,jacket} \tag{6}$$

$$\eta = 0.86 \text{ (for Type 1), } 0.82 \text{ (for Type 2)} \tag{7}$$

$$K = 0.9^n \tag{8}$$

The experimentally obtained shear strengths were compared with those predicted by the ACI 318 design code and the proposed equation presented in Table 4. The shear strength evaluated based on the ACI 318-19 and the proposed equations neglect the loading scheme applied to the reinforced concrete column. Huang et al. [19] conducted cyclic loading tests on nine reinforced concrete columns with variable load patterns and found that the shear strength of the columns decreased by approximately 60% when subjected to an eccentric lateral force as compared to the case where no eccentricity was present. Accordingly, this study compared the experimental results and the shear strength determined by the design criteria and proposed equations, with the latter being reduced to 60% of their calculated values. The ACI 318 design code, which does not consider slip, was found to overestimate the experimental results by approximately 23%. This was attributed to excessive deformation caused by seismic loads acting on the column, which led to a slip between the old and jacketing sections of the column. However, the proposed equation calculated the shear strength of reinforced concrete columns retrofitted with concrete jackets in detail by considering slip as a factor. The slip coefficient and monolithicity factor were derived by measuring the strain of the concrete and stirrups in the old and jacketing sections and considering the strength reduction due to slip. The proposed equation yielded a shear strength ratio of approximately 1.1 to that obtained from the experiment for all specimens, indicating a conservative prediction. Therefore, the proposed equation was deemed capable of accurately predicting the shear strength of reinforced concrete columns retrofitted with concrete jackets, mitigating the issue of overestimation observed in the existing design code.

Table 4. Validity of the proposed equation.

Specimen	V_{test} [kN]	$V_{n,ACI}$ [kN]	$V_{n,proposed}$ [kN]	$V_{test}/V_{n,ACI}$	$V_{test}/V_{n,Proposed}$
HCJ-1U	198.32		191.43	0.94	1.04
HCJ-2U	185.23	210.45	181.53	0.88	1.02
HCJ-1B	202.45		197.70	0.96	1.02

5. Conclusions

This study investigated the bonding capacity of a reinforced concrete column retrofitted with hybrid concrete jacketing under combined loading. Five specimens were fabricated, and cyclic loading tests were conducted. Based on the test results, the authors proposed a shear strength equation that considers the slip between the jacketed section and the old section. The following conclusions have been drawn:

(1) The hybrid concrete jacketing used non-shrinkage mortar and steel fiber to suppress shrinkage and enhance strength. This retrofitting method improved bond capacity with steel wire mesh and steel grid reinforcement, and this method is divided into two types, Type 1 and Type 2, according to the reinforcement details in the jacketing section. Additional dowels which can improve the bonding capacity were used only in Type 1;

(2) The cyclic loading test results demonstrated that the torsional moment of the jacketed column with Type 2 was approximately 80% of the jacketed column with Type 1. Additionally, the difference in strain between the old section and the jacketed section increased gradually, indicating that a slip occurred between the two sections. Therefore, when applying hybrid concrete jacketing for retrofitting old columns, it should be considered the slippage between two sections and shear resistance capacity according to the types of hybrid concrete jacketing methods used;

(3) The slip coefficient and the monolithicity factor were proposed in this study for reinforced concrete columns retrofitted with hybrid concrete jacketing. The slip coefficient accounts for the bonding capacity between the old and jacketed sections; it was dependent on the type of hybrid concrete jacketing, with Type 1 exhibiting a value of 0.86, while Type 2 had a coefficient of 0.82. The monolithicity factor was proposed as a parameter that accounts for the reduction of the stirrup shear strength resulting from the slip between the mortar and the stirrup in the jacketing section. This factor varies based on the presence of dowel bars and the loading pattern;

(4) In this study, the shear strength equation for reinforced concrete columns retrofitted with the hybrid concrete jacketing was proposed by introducing the slip coefficient and monolithicity factor. In contrast to the ACI 318-19, which overestimates test results by approximately 23%, the proposed equation yielded conservative estimates, underestimating the results by approximately 3%. It was indicated that the proposed equation is more reliable and accurate for evaluating the shear strength of jacketed columns. Overall, the results of this study provided important insights into the use of the hybrid concrete jacketing method for retrofitting reinforced concrete columns, and the proposed shear strength equation could be useful for assessing the structural performance of jacketed columns.

Author Contributions: Conceptualization, K.M.R. and Y.H.L.; methodology, K.M.R. and M.S.K.; validation, Y.H.L.; formal analysis, K.M.R. and M.S.K.; investigation, K.M.R. and Y.H.L.; writing—original draft preparation, K.M.R.; writing—review and editing, Y.H.L. All authors have read and agreed to the published version of the manuscript.

Funding: This work was supported by a National Research Foundation of Korea (NRF) grant funded by the Korean government (MSIT) (No. 2020R1A2C2009351).

Institutional Review Board Statement: Not applicable.

Informed Consent Statement: Not applicable.

Data Availability Statement: Not applicable.

Conflicts of Interest: The authors declare no conflict of interest.

References

1. Kaish, A.B.M.A.; Alam, M.R.; Jamil, M.; Zain, M.F.M.; Wahed, M.A. Improved ferrocement jacketing for restrengthening of square RC short column. *Constr. Build. Mater.* **2012**, *36*, 228–237. [CrossRef]
2. Hung, C.C.; Chen, Y.S. Innovative ECC jacketing for retrofitting shear-deficient RC members. *Constr. Build. Mater.* **2016**, *111*, 408–418. [CrossRef]
3. Maraq, M.A.A.; Tayeh, B.A.; Ziara, M.M.; Alyousef, R. Flexural behavior of RC beams strengthened with steel wire mesh and self-compacting concrete jacketing—Experimental investigation and test results. *J. Mater. Res. Technol. JMRT* **2021**, *10*, 1002–1019. [CrossRef]
4. Dagenais, M.A.; Massicotte, B. Cyclic behavior of lap splices strengthened with ultrahigh performance fiber-reinforced concrete. *J. Struct. Eng.-ASCE* **2017**, *143*, 04016163. [CrossRef]
5. Bae, B.I.; Chung, J.H.; Choi, H.K.; Jung, H.S.; Choi, C.S. Experimental study on the cyclic behavior of steel fiber reinforced high strength concrete columns and evaluation of shear strength. *Eng. Struct.* **2018**, *157*, 250–267. [CrossRef]
6. Sakr, M.A.; El Korany, T.M.; Osama, B. Analysis of RC columns strengthened with ultra-high performance fiber reinforced concrete jackets under eccentric loading. *Eng. Struct.* **2020**, *220*, 111016. [CrossRef]
7. Hong, S.G.; Lee, J.H.; Choi, Y.; Gu, I.Y. Seismic strengthening of concrete columns by ultrahigh-performance fiber-reinforced concrete jacketing. *J. Struct. Eng.-ASCE* **2021**, *147*, 04021157. [CrossRef]

8. Kim, M.S.; Lee, Y.H. Seismic Performance of Reinforced Concrete Columns Retrofitted with Hybrid Concrete Jackets Subjected to Combined Loadings. *Materials* **2022**, *15*, 6213. [CrossRef]
9. ACI Committee 318. *Building Code Requirements for Reinforced Concrete and Commentary (ACI 318-19)*; American Concrete Institute: Farmington Hills, MI, USA, 2019.
10. Psycharis, I.N.; Mouzakis, H.P. Shear resistance of pinned connections of precast members to monotonic and cyclic loading. *Eng. Struct.* **2012**, *41*, 413–427. [CrossRef]
11. Mou, B.; Bai, Y. Experimental investigation on shear behavior of steel beam-to-CFST column connections with irregular panel zone. *Eng. Struct.* **2018**, *168*, 487–504. [CrossRef]
12. Chen, Y.; Zhang, Q.; Feng, J.; Zhang, Z. Experimental study on shear resistance of precast RC shear walls with novel bundled connections. *J. Earthq. Tsunami* **2019**, *13*, 1940002. [CrossRef]
13. Fan, L.; Wei, J.; Chen, Y.; Feng, J.; Sareh, P. Shear Performance of Large-Thickness Precast Shear Walls with Cast-in-Place Belts and Grouting Sleeves. *ASCE-ASME J. Risk Uncertain. Eng. Syst. Part A Civ. Eng.* **2023**, *9*, 04023005. [CrossRef]
14. Penelis, G.G.; Kappos, A. *Earthquake Resistant Concrete Structures*; CRC Press: Boca Raton, FL, USA, 2014.
15. Kong, X.; Yao, Y.; Wu, B.; Zhang, W.; He, W.; Fu, Y. The Impact Resistance and Mechanical Properties of Recycled Aggregate Concrete with Hooked-End and Crimped Steel Fiber. *Materials* **2022**, *15*, 7029. [CrossRef] [PubMed]
16. Ali, B.; Kurda, R.; Herki, B.; Alyousef, R.; Mustafa, R.; Mohammed, A.; Raza, A.; Ahmed, H.; Ul-Haq, M.F. Effect of varying steel fiber content on strength and permeability characteristics of high strength concrete with micro silica. *Materials* **2020**, *13*, 5739. [CrossRef] [PubMed]
17. Rezakhani, R.; Scott, D.A.; Bousikhane, F.; Pathirage, M.; Moser, R.D.; Green, B.H.; Cusatis, G. Influence of steel fiber size, shape, and strength on the quasi-static properties of ultra-high performance concrete: Experimental investigation and numerical modeling. *Constr. Build. Mater.* **2021**, *296*, 123532. [CrossRef]
18. Fang, C.; Ali, M.; Xie, T.; Visintin, P.; Sheikh, A.H. The influence of steel fibre properties on the shrinkage of ultra-high performance fibre reinforced concrete. *Constr. Build. Mater.* **2020**, *242*, 117993. [CrossRef]
19. Huang, H.; Hao, R.; Zhang, W.; Huang, M. Experimental study on seismic performance of square RC columns subjected to combined loadings. *Eng. Struct.* **2019**, *184*, 194–204. [CrossRef]
20. Thermou, G.E.; Papanikolaou, V.K.; Kappos, A.J. Flexural behaviour of reinforced concrete jacketed columns under reversed cyclic loading. *Eng. Struct.* **2014**, *76*, 270–282. [CrossRef]
21. Murugan, K.; Sengupta, A.K. Seismic performance of strengthened reinforced concrete columns. *Structures* **2020**, *27*, 487–505. [CrossRef]
22. Alhadid, M.M.A.; Youssef, M.A. Analysis of reinforced concrete beams strengthened using concrete jackets. *Eng. Struct.* **2017**, *132*, 172–187. [CrossRef]
23. Thermou, G.E.; Kappos, A.J. Background to the monolithicity factors for the assessment of jacketed reinforced concrete columns. *Buildings* **2022**, *12*, 55. [CrossRef]

Disclaimer/Publisher's Note: The statements, opinions and data contained in all publications are solely those of the individual author(s) and contributor(s) and not of MDPI and/or the editor(s). MDPI and/or the editor(s) disclaim responsibility for any injury to people or property resulting from any ideas, methods, instructions or products referred to in the content.

Article

Experimental and Numerical Investigation on the Size Effect of Ultrahigh-Performance Fibre-Reinforced Concrete (UHFRC)

Andreas Lampropoulos [1,*], Demetris Nicolaides [2], Spyridon Paschalis [3] and Ourania Tsioulou [1]

1 School of Architecture, Technology and Engineering, University of Brighton, Cockcroft Building, Lewes Road, Brighton BN2 4GJ, UK; o.tsioulou@brighton.ac.uk
2 School of Engineering, Department of Civil Engineering, Frederick University, 1036 Nicosia, Cyprus; d.nicolaides@frederick.ac.cy
3 School of Engineering and Computing, University of West London, Lady Byron Building, St Mary's Road, London W5 5RF, UK; Spyros.Paschalis@uwl.ac.uk
* Correspondence: a.lampropoulos@brighton.ac.uk

Abstract: In the last few years, there has been increasing interest in the use of Ultrahigh-Performance Fibre-Reinforced Concrete (UHPFRC) layers or jackets, which have been proved to be quite effective in strengthening applications. However, to facilitate the extensive use of UHPFRC in strengthening applications, reliable numerical models need to be developed. In the case of UHPFRC, it is common practice to perform either direct tensile or flexural tests to determine the UHPFRC tensile stress–strain models. However, the geometry of the specimens used for the material characterization is, in most cases, significantly different to the geometry of the layers used in strengthening applications which are normally of quite small thickness. Therefore, and since the material properties of UHPFRC are highly dependent on the dimensions of the examined specimens, the so called "size effect" needs to be considered for the development of an improved modelling approach. In this study, direct tensile tests have been used and a constitutive model for the tensile behaviour of UHPFRC is proposed, taking into consideration the size of the finite elements. The efficiency and reliability of the proposed approach has been validated using experimental data on prisms with different geometries, tested in flexure and in direct tension.

Keywords: UHPFRC; strengthening; flexural strength; size effect; constitutive stress–strain model; numerical modelling

Citation: Lampropoulos, A.; Nicolaides, D.; Paschalis, S.; Tsioulou, O. Experimental and Numerical Investigation on the Size Effect of Ultrahigh-Performance Fibre-Reinforced Concrete (UHFRC). Materials 2021, 14, 5714. https://doi.org/10.3390/ma14195714

Academic Editor: Dario De Domenico

Received: 14 July 2021
Accepted: 28 September 2021
Published: 30 September 2021

Publisher's Note: MDPI stays neutral with regard to jurisdictional claims in published maps and institutional affiliations.

Copyright: © 2021 by the authors. Licensee MDPI, Basel, Switzerland. This article is an open access article distributed under the terms and conditions of the Creative Commons Attribution (CC BY) license (https://creativecommons.org/licenses/by/4.0/).

1. Introduction

The majority of the existing Reinforced Concrete (RC) structures need to be upgraded, either because they are designed with old seismic code provisions or without them entirely, or because of existing damages. Nowadays, there is a wide range of techniques for the structural upgrade of existing RC elements, and the use of novel high-performance materials has been shown to offer enhanced structural performance and durability. Ultrahigh-Performance Fibre-Reinforced Concrete (UHPFRC) is a relatively new construction material with superior mechanical characteristics. It is characterised by significantly enhanced compressive and tensile strength, and exceptional ductility and energy absorption capacity. These characteristics are directly linked to the mix design, and there are numerous published studies where the effect of the mix design, the type and size of aggregates, and more significantly the dosage and characteristics of steel fibres have been examined [1–20].

In most cases, the behaviour of UHPFRC is significantly affected by the microstructure of the cementitious matrix and the characteristics of the fibres, since the strain-hardening characteristics are attributed to the bond between the fibres and the matrix, which is directly linked to the bridging effect of the fibres.

The effect of different type, length and volume fraction of fibres on the mechanical properties of UHPFRC has been examined by a number of researchers [5–11]. Paschalis and Lampropoulos [5] found that an increase in the steel fibre content from 1 to 6 Vol.-% enhanced the tensile strength by 92% and the compressive strength by 72%. Hannawi et al. [6] examined various types of fibres, and for a volume fraction of 1%, they found that the effect of the fibres on the compressive strength and the elastic modulus of UHPFRC specimens was negligible. Abbas et al. [7] explored the effects of the steel fibre length and volume fraction on the mechanical properties and durability of the UHPFRC. Based on this study [7], the addition of steel fibres significantly enhanced the tensile and flexural strength, while the compressive strength was only slightly increased. It was also observed that the addition of fibres altered the failure pattern from sudden and explosive to ductile behaviour. The length of the fibres had negligible effect on the compressive strength, but they considerably affected the peak load carrying capacity and load-deflection behaviour. Gesoglu et al. [8], tested the effect of microsteel, hooked steel and microglass fibres vol % up to 2% on the properties of the UHPFRC. They observed that an increase in fibre content led to increased compressive, tensile and flexural strength and increased modulus of elasticity of the UHPFRC regardless the fibre type. However, the strength values began to decrease after 1.5% volume of glass fibres, while 2% hooked steel fibres led to UHPFRC enhanced ductility. Kazemi et al. [9] examined the mechanical properties of UHPFRC containing up to 5% volume fraction of smooth steel fibres. The key findings were that an increase in the fraction of steel fibres led to significant increases in flexural and shear strength. Wu et al. [10,11] investigated the influence of straight, corrugated, and hooked fibres on fibre–matrix bond properties, and compressive and flexural properties of UHPFRC. It was found that the compressive and ultimate flexural strengths increased with the increase in fibre content and age. Additionally, pull-out bond strength and toughness of embedded hooked fibres were much higher compared with those with straight and corrugated fibres. Yoo et al. [12], conducted four-point bending tests on UHPFRC beams with smooth steel fibres of different length. Based on this study [12], fibre length significantly increased the load and toughness of the beams after the limit of proportionality, due to the improved fibre bridging capacity. In addition, beams with a longer fibre length exhibited a higher number of microcracks.

Another important parameter is the orientation of the fibres, which is affected by the method of pouring and by the dimensions of the examined specimens. There are a few studies on the so-called "size effect" of UHPFRC [9,13,14], which prove that the size of the examined specimens is important for both the compressive and flexural strength characteristics. Kazemi et al. [9] observed that smaller samples tend to show higher compressive and direct shear strength. Mahmud et al. [13] investigated the size effect on the flexural strength of UHPFRC beams tested under three-point bending tests. Results showed that the size effect on the flexural strength of UHPFRC is negligible and follows the yield criterion because of its high ductility. An et al. [14] examined the size effect on the compressive strength of UHPFRC cubes with different sizes, and they found that the larger specimens had lower compressive strength when compared with the smaller ones. Even if the existing studies present some useful experimental data on the performance of UHPFRC specimens with carrying sizes, the so-called "size effect" has not been sufficiently explained and there is not any available methodology on how to take this size effect into consideration for practical applications.

The addition of fibres to the concrete matrix can dramatically improve the overall mechanical performance and fracture behaviour of the composite and can also impart it with additional strength in tension, shear and flexure [5–11]. It is, however, extremely difficult to achieve an even distribution of fibres within the mix, especially when a large quantity of fibres is being used. Failure to attain this goal may result in low mechanical properties, whereas a proper and even fibre distribution can guarantee considerably higher values of properties. This is more evident in the case of UHPFRC, where (i) the interfacial bond between the fibres and the matrix is particularly strong, due to the dense structure of the material and (ii) unreinforced matrices are extremely brittle due to the absence of coarse aggregates in these types of materials. The mechanical and fracture properties of any UHPFRC depend to a high degree on the uniform distribution of fibres in the bulk of the material. Any regions with a low concentration of fibres, or with no fibres, are potential sites of weakness. The distribution of fibres in the mix depends on a number of factors, such as how the fibres were introduced into the mix, on the vibration frequency during compaction, and on the size and shape of the object cast from UHPFRC [1,2,4]. The difficulty in achieving an even distribution of fibres is more pronounced in thicker specimens (e.g., 100 mm), whereas an even distribution can be achieved without difficulty in specimens with a relatively small thickness (e.g., 5 mm–30 mm) [1–3]. This particular observation is important, as it can be a crucial factor towards the discrepancy of the experimental results between specimens of several sizes [1–3]. The energy absorbing mechanisms (expressed through the size of the Fracture Process Zone (FPZ)) are reduced in the case of larger UHPFRCC specimens because of the difficulty to achieve a uniform fibre distribution, and thus engage more mechanisms in the energy absorbing process. On the other hand, specimens of smaller thicknesses exhibit an even distribution of fibres which apply substantial closure pressure, thus increasing the flexural capacity of the beams [1,2]. Previous research [1–3] that thoroughly examined the failure surfaces of three-point bend-tested beams of several sizes confirmed the aforementioned remark. Awinda et al. [15] performed experimental and numerical investigations on UHPFRC prisms with various geometries, where a sensitivity analysis was conducted. According to this study [15], the fibres' orientations and the alignment of the fibres seem to be quite prominent for specimens with depths of 50 mm or less. It was also highlighted that further work is required to consider the effect of parameters such as the fibre content and length on the numerical modelling of UHPFRC [15]. This is a particularly important aspect, since the application of UHPFRC elements with small thickness, such as bridge decks and strengthening layers, has been extensively used in the last few years. The use of additional UHPFRC layers or jackets has been shown to be quite effective for the enhancement of the flexural and shear capacity of Reinforced Concrete (RC) structures [17–20]. Additionally, the application of UHPFRC layers in connection with existing RC slabs has been found to offer significant improvement of the punching shear resistance [21], and the enhancement of the punching shear of the composite/strengthened elements can be calculated using an analytical model [22]. In all these applications, thin UHPFRC are used, however the size effect is not taken into consideration and the tensile characteristics are derived using either prisms tested under flexural loading or from the direct tensile testing of dog-bone shaped specimens with various geometries. The current study aims to examine the effect of steel fibres volume fraction and the effect of the dimensions of the examined specimens on the flexural strength characteristics of UHPFRC, and to propose a suitable methodology for the numerical modelling of UHPFRC elements with various geometries and thicknesses.

2. Experimental Investigation

In this paper the results of two different experimental studies have been combined and used to investigate a number of different depths of UHPFRC prisms. Flexural tests have been conducted on prisms with various geometries and these results have been used for the validation of the numerical model. Information related to these experimental works (e.g., materials' mix designs, manufacturing processes, experimental results, etc.) are presented in the following sections.

2.1. Material Preparation and Geometry of the Examined Specimens

In the present paper, two different UHPFRC mix designs (i.e., UHPFRC-1 and UHPFRC-2) have been selected and used. The selection of the particular mixtures was made in order to accumulate a sufficient number of different depths of prisms, thus allowing for investigation on the size effect on the performance of UHPFRC layers. UHPFRC-1 was developed at Cardiff University (Cardiff, UK), as is described by Nicolaides [1], whereas UHPFRC-2 is based on the experimental investigation of Hassan et al. [16]. The mix proportions for the two mixtures are provided in Table 1.

Table 1. UHPFRC 1 and 2 mix designs.

Material	Mix Proportions (kg/m^3)	
	UHPFRC-1	UHPFRC-2
Cement (52.5N)	855	657
GGBS		418
Silica Fume	214	119
Silica Sand	940	1051
Superplasticizers	28	59
Water	188	185
Steel Fibres	468	234

Both mixtures are characterized by the use of high cement content 52.5R, along with the use of microsilica and low water–binder ratios, which were achieved by the use of superplasticizers. The maximum particle sand size for UHPFRC-1 was 0.6 mm, whereas the corresponding value for UHPFRC-2 mix was 0.5 mm. For the development of UHPFRC-2 mix a considerable amount of Ground Granulated Blast Furnace Slag (GGBS) was also utilized. Large volumes of steel fibres were also used in each mixture, i.e., 6% (468 kg/m^3) and 3% per volume (234 kg/m^3) for mixtures 1 and 2, respectively. For the development of UHPFRC-1, a combination of shorter (i.e., 6 mm) and longer (i.e., 13 mm) brass coated steel fibres were added, whereas for the development of UHPFRC-2, only one length of fibres (13 mm) was incorporated in the mixture. All fibres had a diameter of 0.16 mm, a tensile strength of 3000 MPa and Modulus of Elasticity 200 GPa.

Both mixtures were produced in the labs by applying dry mixing process, i.e., mixing of the dry materials (sand, silica fume, cement and GGBS) first, before the addition of any liquid material. In UHPFRC-1, the steel fibres were also added into the dry mixture, just before the addition of water and superplasticizer. In contrast, in UHPFRC-2 steel fibres were incorporated in the mix right after the addition of the liquid constituents of the mixture. For the production of both materials, high-shear pan mixers were used.

The moulds of UHPFRC-1 were left in environmental conditions for 24 h, and then the demoulded specimens were placed into a hot curing tank, filled with water controlled at 90 °C. The specimens were left in the tank for 9 days. On the first day the temperature of the curing tank was increased (20–90 °C), and on the ninth day it was decreased (90–20 °C) gradually, in order to prevent thermal shock of the specimens. The hot curing regime was applied in order to minimize the curing period of the material. The same curing procedure was also followed for the UHPFRC-2, and after demoulding the specimens were placed into a hot curing tank at 90 °C and tested after 14 days.

From mix UHPFRC-1, two different layer depths were investigated, namely 35 mm and 100 mm, whereas from mix UHPFRC-2, four different layer depths were investigated, namely 25 mm, 50 mm, 75 mm and 100 mm. At least three beams were prepared for each layer depth. In addition, six dog-bone specimens were also prepared and tested in order to determine the direct tensile strength of UHPFRC-2. The corresponding value of the direct tensile strength of UHPFRC-1 was determined in an earlier study by Benson and Karihaloo [23]. In Figure 1, pictures taken during the preparation of beams with different layer depths are presented.

2.2. Flexural Prism Tests

In this section, the experimental results of the flexural testing of prisms with different section depths are presented. At least three identical specimens were tested for each different depth. Figure 2 illustrates schematically the experimental setups of (a) the three-point bending test (UHPFRC-1) and (b) the four-point bending test (UHPFRC-2), including the dimensions of both the overall and the testing spans and the location of the applied loads (P). For the determination of the tensile/flexural strength of UHPFRC-1, prisms were tested by three-point bending (Figure 3a) for deformation control. Two types of measurement were recorded for each beam: (1) the load from the load cell of the testing machine; (2) the vertical deflection at the centre point. The vertical deflection was measured by a single LVDT placed underneath the testing beam at the centre point. The tests were performed in a stiff self-straining testing frame. Samples from UHPFRC-2 were tested under four-point loading (Figure 3b). As can be seen in Figure 3b, for the testing of UHPFRC-2, an external yoke was used together with two LVDTs which were attached on both sides of the specimens to record the average deflections of the beams. All tests were conducted under a constant displacement rate of 0.001 mm/sec according to JSCE [24]. The flexural strengths of all specimens were calculated from the recorded data using Equations (2) and (3), depending on the loading conditions.

Figure 1. Preparation of prisms with different depths.

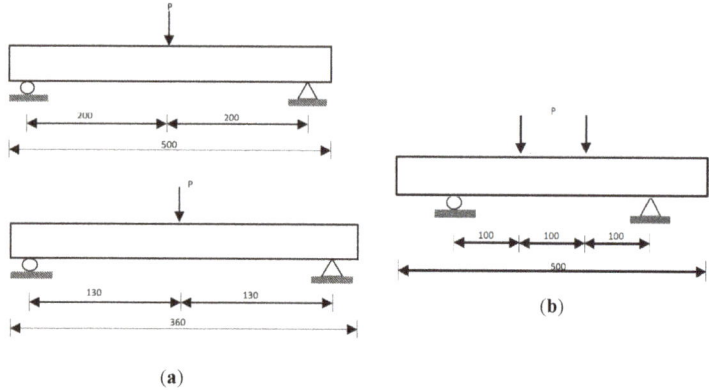

Figure 2. Schematic presentation of (**a**) the three-point bending test (UHPFRC-1) and (**b**) the four-point bending test (UHPFRC-2) (dimensions in mm).

 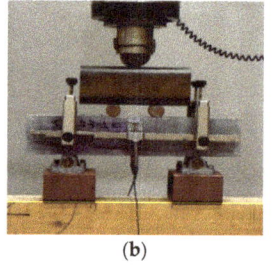

(**a**) (**b**)

Figure 3. Experimental setup for the testing of (**a**) UHPFRC-1 and (**b**) UHPFRC-2 beams.

The typical failure in all the examined cases was formed with a main crack in the middle of the span. The crack patterns for selected typical prisms with different depths are presented in Figure 4. The load-deflection results for all the examined specimens for UHPFRC-1 and UHPFRC-2 are presented in Figure 5a,b, respectively.

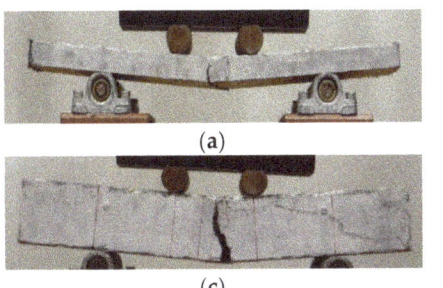
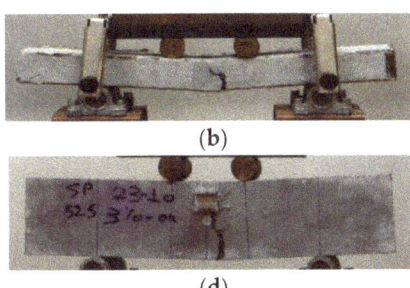

Figure 4. Characteristic failures of selected typical prisms for (**a**) 25 mm, (**b**) 50 mm, (**c**) 75 mm and (**d**) 100 mm depths.

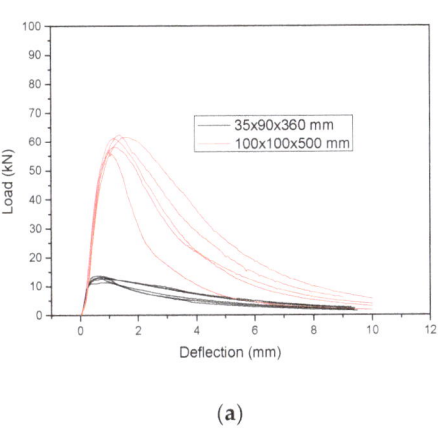
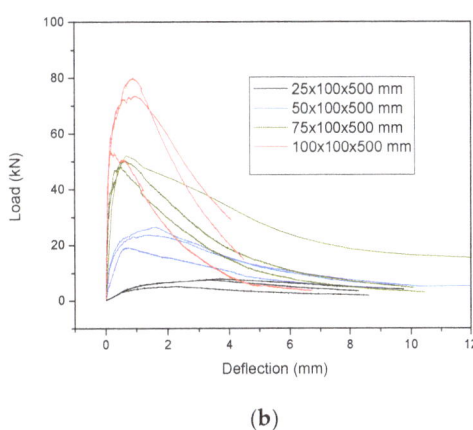

Figure 5. Load-deflection results for all the examined specimens of (**a**) UHPFRC-1 and (**b**) UHPFRC-2.

The values of Figure 5 have been used for the calculation of the flexural strength for both UHPFRC-1 and UHPFRC-2.

The flexural strength σ_t is calculated using Equation (1):

$$\sigma_t = \frac{M \cdot y}{I} \qquad (1)$$

where:
- M is the bending moment;
- I is the moment of inertia;
- y is the distance of the centroid from the extreme fibre.

Using Equation (1), the following models are derived for the three-point Equation (2) and four-point Equation (3) bending testing, respectively, of UHPFRC-1 and UHPFRC-2.

$$\sigma_{t-3p} = \frac{3 \cdot P \cdot L}{2 \cdot b \cdot d^2} \qquad (2)$$

$$\sigma_{t-4p} = \frac{P \cdot L}{b \cdot d^2} \qquad (3)$$

where:

σ_{t-3p} and σ_{t-4p}	are the flexural strength values calculated from the three-point and four-point bending tests (MPa);
P	is the peak load (N);
L	is the effective span length (mm);
b	is the width of specimen (mm);
d	is the depth of the specimens (mm).

The flexural strength results for all the different examined thicknesses have been calculated and the average results together with the scatter plot are presented in Figure 6.

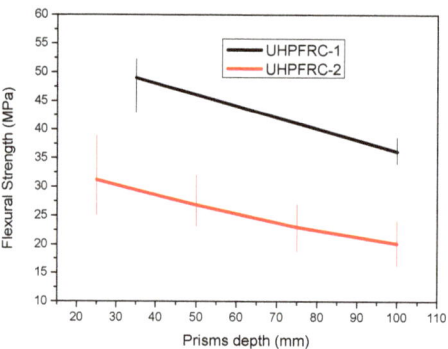

Figure 6. Flexural strength results for UHPFRC-1 and UHPFRC-2 for all the examined prism depths.

From the results presented in Figure 6, it can be clearly observed that there is a reduction in the flexural strength as the depth of the specimens is increased, which confirms the so called "size effect". This reduction is attributed to the uneven distribution of fibres in thicker elements (e.g., 100 mm) as opposed to specimens with smaller thicknesses where there is a more even distribution of the fibres and therefore increased flexural strength is achieved. Additionally, the results of Figure 4 show a similar trend in the reduction in the flexural strength values, with a slightly more pronounced rate of reduction in the case of UHPFRC-1 which is linked to the higher percentage of steel fibres (UHPFFRC-1 has 6% steel fibres while UHPFRC-2 has 3%). Additionally, the overall flexural strength of UHPFRC-1 is higher than the respective values of UHPFRC-2 due to the higher percentage of steel fibres.

3. Constitutive Modelling and Numerical Analysis

3.1. Direct Tensile Tests

Direct tensile tests have been conducted for both UHPFRC-1 and UHPFRC-2. The direct tensile test results of UHPFRC-2 have been used for the constitutive modelling of UHPFRC and the values of the model have been accordingly adjusted to consider the different Youngs modulus and tensile strength of the two mixes. In total, six UHPFRC-2 dog-bone specimens were tested under direct tensile testing [17]. A pair of steel grips was used to apply the tensile load and all the specimens were tested under displacement control with a rate equal to 0.007 mm/s. The extension and the respective strain values were calculates using the measurements of Linear Variable Differential Transformers (LVDTs) using the setup presented in Figure 7 [17].

The direct tensile test results for the UHPFRC-2 specimens are presented in Figure 8. The results of Figure 8 present the distribution of the stress with the total strain for all the individual specimens together with the average results [17]. According to the direct tensile test results of Figure 8, the tensile strength of UHPFRC-2 was found to be in the range of 11.74 MPa to 14.20 MPa while the respective strength of the average stress–strain curve was calculated equal to 12.15 MPa.

Figure 7. Direct tensile testing of UHPFRC-2 dog-bone specimens.

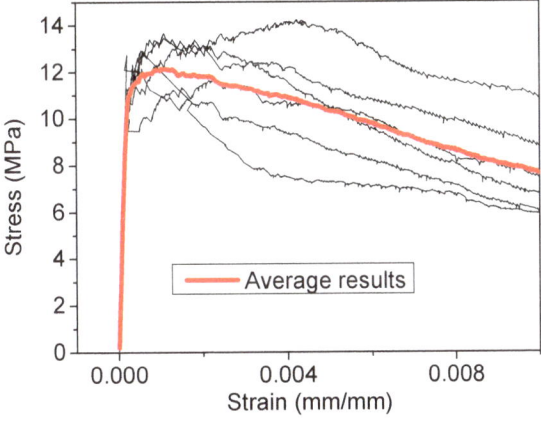

Figure 8. Direct tensile stress (MPa)–strain (mm/mm) results for UHPFRC-2.

3.2. UHPFRC Constitutive Modelling

The experimental results of Figure 8 were used for the development of the constitutive model presented in Figure 9 which represents the stress–strain distribution after the end of the elastic part. This model consists of a linear part up to the maximum tensile stress value (f_t) followed by a bilinear descending branch (Figure 9). This model will be used for the numerical modelling of both UHPFRC-1 and UHPFRC-2 to evaluate the reliability of the model for these two different types of UHPFRC.

ATENA software was used for the numerical simulations, and experimental values for the tensile strength and the young modulus were used for the modelling of the tensile stress–strain behaviour using the model of Figure 9. For UHPFRC-1, 16 MPa tensile strength was obtained experimentally, while the Youngs Modules and the Compressive strength were equal to 48 GPa and 193.6 Mpa, respectively [1]. Regarding UHPFRC-2, tensile strength of 11.5 Mpa and Youngs modulus equal to 57.5 Gpa were used, while the compressive strength was found equal to 164 MPa.

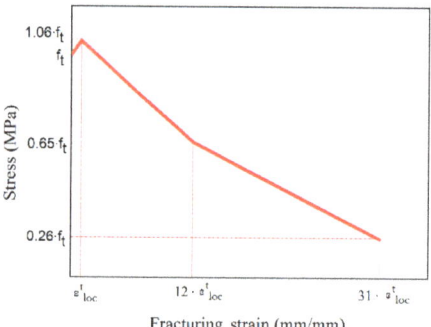

Figure 9. Constitutive modelling for UHPFRC.

To determine the strain values for the characteristics points of the model in Figure 9, a strain equal to 0.042 was taken as ε_{loc}^{t} using characteristic size (l_{ch}) equal to 2 mm and finite elements size (l_t) equal to 65 mm [17]. This model has been found to be able to accurately predict the behaviour of UHPFRC-2, however the reliability of this model is highly dependent on the values of the characteristic size and the mesh size of the elements of the numerical models, which significantly affect the results in the post-crack region. Therefore, it is important to develop models which can accurately predict the behaviour of UHPFRC independently of the size of the Finite Element Models. This crucial aspect is addressed in this study with the development of a model which takes into consideration the size of the elements and can be used to accurately predict the behaviour of various geometries of UHPFRC specimens. The proposed tensile stress-strain characteristics are defined each time depending on the size of the finite elements (l_t), following the model presented in Figure 10.

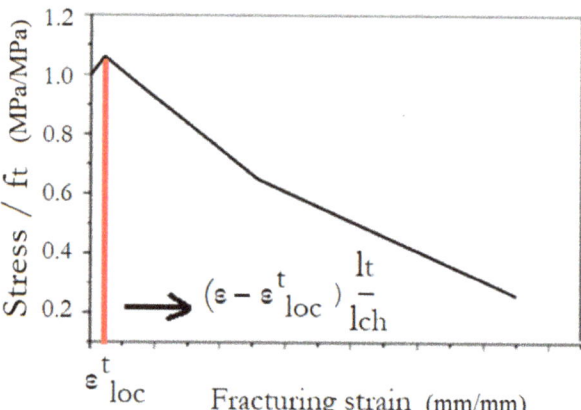

Figure 10. Constitutive modelling for UHPFRC depending on the FEA mesh size.

This approach has been examined for both UHPFRC-1 and UHPFRC-2 and the numerical models presented in Figures 11 and 12 have been analysed.

For the specimens of Figure 12b–d, the finite element size ($l_{t\ new}$) was 14.7 mm while the respective value for the specimen of Figure 12a was 15 mm. Following the procedure described in Figure 10, the UHPFRC constitutive model was altered by adjusting the characteristic size value (l_{ch} = 2 mm) which was initially proposed for l_t = 65 mm, multiplying it with the ratio $\frac{l_{t\ new}}{l_t}$ (i.e., for $l_{t\ new}$ = 14.7 mm, $\frac{l_{t\ new}}{l_t}$ = 0.2 and $l_{ch\ new}$ = 0.2 · l_{ch} = 0.4 mm).

Figure 11. Numerical models for UHPFRC-1 prisms (**a**) 25 × 90 × 360 mm and (**b**) 100 × 100 × 500 mm.

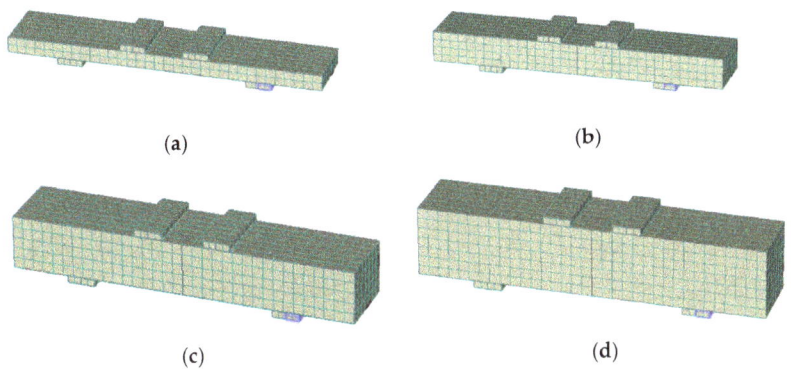

Figure 12. Numerical models for UHPFRC-1 prisms (**a**) 25 × 100 × 500 mm, (**b**) 50 × 100 × 500 mm, (**c**) 75 × 100 × 500 mm, and (**d**) 100 × 100 × 500 mm.

Simply supported conditions were applied to all the examined specimens and a monotonically increasing displacement was applied to the middle of the span, reproducing the conditions of the experimental tests. The comparisons between the numerical and analytical results are presented in Section 4.

4. Results and Discussion

The numerical results are compared with the respective experimental results, and the results for UHPFRC-1 are presented in Figure 13 while the respective results for UHPFRC-2 are presented in Figure 14. The results of Figures 13 and 14 show that the numerical modelling results are in a good agreement with all the experimental results for both UHPFRC-1 and UHPFRC-2 for all the examined prism dimensions.

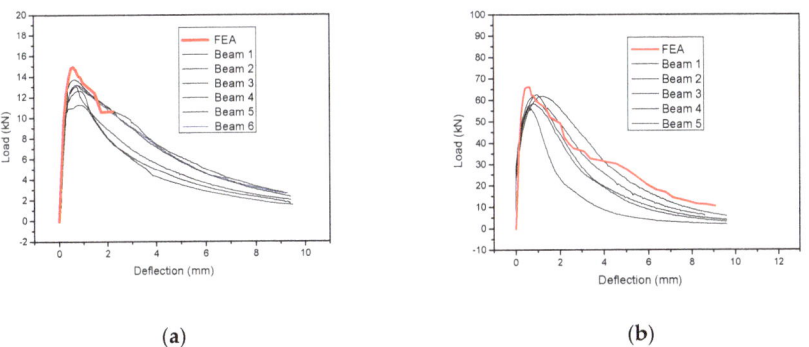

Figure 13. Load deflection results: Numerical vs. Experimental for UHPFRC-1 prisms with (**a**) 25 and (**b**) 100 mm depth values.

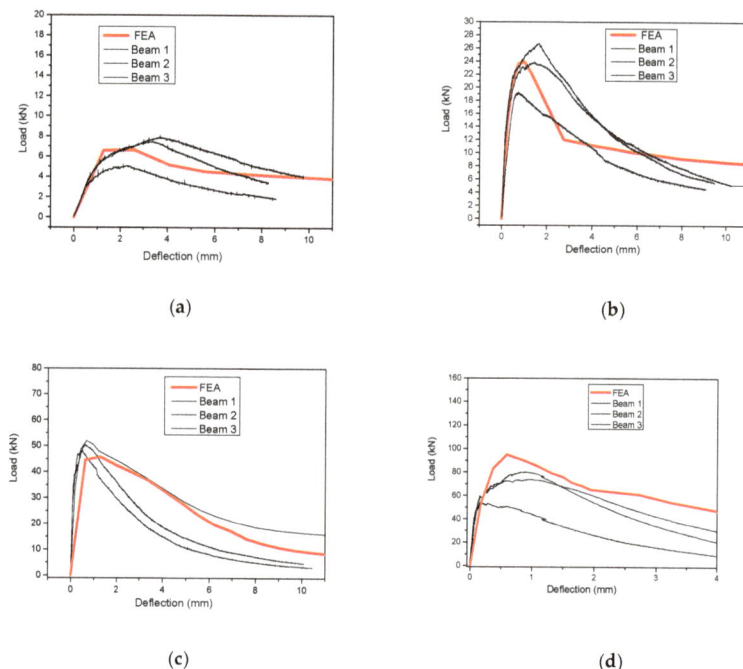

Figure 14. Load deflection results: Numerical vs. Experimental for UHPFRC-2 prisms with (**a**) 25 mm, (**b**) 50 mm, (**c**) 75 mm and (**d**) 100 mm depth values.

More specifically, the numerical models were found to successfully simulate the initial stiffness, the maximum load and the post-cracking behaviour of most of the examined samples. It should be highlighted here that there is a significant deviation between the experimental results of the multiple samples of each of the examined mixes, and the geometry which is attributed to the distribution of the fibres of the experimental samples. In most of the examined cases the numerical results are approaching the experimental results of the specimens with the highest load capacity (among the numerous specimens examined for each mix/type). Additionally, it is worth mentioning that in the case of specimens with relatively small depth (i.e., 25 mm, 50 mm and 75 mm), the numerical model results are near the average of the experimental results, while in case of specimens with 100 mm the numerical results are slightly higher than the highest values of the experimental results. This is due to the fact that the scatter of the experimental results is linked to the non-uniform distribution and orientation of the fibres, which is more important in the case of specimens with high thickness (i.e., 100 mm). Therefore, the proposed methodology can be used to accurately simulate the response of relatively thin UHPFRC layers and with thicknesses no more than 100 mm. These results prove the applicability of the proposed method for the simulation of UHPFRC specimens with different dimensions, eliminating the impact of the size effect.

5. Conclusions

In this study, extensive experimental work on prisms with different fibre volume fractions and different geometries has been conducted, and a methodology for the numerical modelling of UHPFRC has been proposed. More specifically, the development of a widely applicably model is proposed for the simulation of UHPFRC independently of the size of the Finite Element Models. The proposed model takes into consideration the size of the elements and can be used to accurately predict the behaviour of various geometries of UHPFRC specimens.

The following conclusions can be drawn:

- The flexural strength of the examined prisms is reduced as the depth of the specimens is increased, which confirms the so called "size effect".
- This reduction is attributed to the uneven distribution of fibres in thicker elements (e.g., 100 mm), as opposed to specimens with smaller thickness where there is a more even distribution of the fibres and therefore increased flexural strength is achieved.
- The reduction rate of the flexural strength values is more pronounced in case of UHPFRC-1, which is linked to its higher percentage of steel fibres (UHPFFRC-1 has 6% steel fibres while UHPFRC-2 has 3%).
- The proposed numerical modelling approach can accurately predict all the examined types (with different fibre volume fractions and different geometries), confirming the applicability of the proposed method for the simulation of UHPFRC specimens with different dimensions, eliminating the impact of the size effect. The proposed methodology can be used to accurately simulate the response of relatively thin UHPFRC layers and with thicknesses no more than 100 mm. Further research is required for the simulation of UHPFRC specimens with thickness higher than 100 mm.

Author Contributions: Conceptualization, A.L. and D.N.; Methodology, A.L., D.N., S.P. and O.T.; Software, A.L. and O.T.; Validation, A.L., D.N. and S.P.; Formal analysis, A.L. and O.T.; Investigation, A.L., D.N., S.P. and O.T.; Resources, A.L., D.N. and S.P.; Data curation, A.L. and S.P.; Writing—original draft preparation, A.L., D.N., S.P. and O.T.; Writing—review and editing, A.L., D.N., S.P. and O.T.; Visualization, A.L. and O.T. All authors have read and agreed to the published version of the manuscript.

Funding: This research received no external funding.

Institutional Review Board Statement: Not applicable.

Informed Consent Statement: Not applicable.

Data Availability Statement: Data is contained within the article.

Conflicts of Interest: The authors declare no conflict of interest.

References

1. Nicolaides, D. Fracture and Fatigue of CARDIFRC. Ph.D. Thesis, Cardiff University, Wales, UK, 2004.
2. Nicolaides, D.; Kanellopoulos, A.; Karihaloo, B.L. Investigation of the effect of fibre distribution on the fatigue performance and the autogenous shrinkage of CAR-DIFRC®. In *Measuring, Monitoring and Modelling Concrete Properties*; Konsta-Gdoutos, M.S., Ed.; Springer: Dordrecht, The Netherlands, 2006; pp. 3–16.
3. Farhat, F.A.; Nicolaides, D.; Kanellopoulos, A.; Karihaloo, B.L. CARDIFRC®—Performance and application to retrofitting. *J. Eng. Fract. Mech.* **2007**, *74*, 151–167. [CrossRef]
4. Nicolaides, D.; Kanellopoulos, A.; Petrou, M.; Savva, P.; Mina, A. Development of a new Ultra High Performance Fibre Reinforced Cementitious Composite (UHPFRCC) for impact and blast protection of structures. *Constr. Build. Mater.* **2015**, *95*, 667–674. [CrossRef]
5. Paschalis, S.; Lampropoulos, A. Fiber content and curing time effect on the ten-sile characteristics of ultra high performance fiber reinforced concrete. *Struct. Concr.* **2017**, *18*, 577–588. [CrossRef]
6. Hannawi, K.; Bian, H.; Prince-Agbodjan, W.; Raghavan, B. Effect of different types of fibers on the microstructure and the mechanical behavior of ultra-high performance fiber-reinforced concretes. *Compos. Part B Eng.* **2016**, *86*, 214–220. [CrossRef]
7. Abbas, S.; Soliman, A.M.; Nehdi, M.L. Exploring mechanical and durability properties of ultra-high performance concrete incorporating various steel fiber lengths and dosages. *Constr. Build. Mater.* **2015**, *75*, 429–441. [CrossRef]
8. Gesoglu, M.; Güneyisi, E.; Muhyaddin, G.F.; Asaad, D.S. Strain hardening ultrahigh performance fiber reinforced cementitious composites: Effect of fiber type and concentration. *Compos. B Eng.* **2016**, *103*, 74–83. [CrossRef]
9. Kazemi, S.; Lubell, A.S. Influence of specimen size and fiber content on mechanical properties of ultra-high-performance fiber-reinforced concrete. *ACI Mater. J.* **2012**, *109*, 675–684.
10. Wu, Z.; Shi, C.; He, W.; Wu, L. Effects of steel fiber content and shape on mechanical properties of ultra high performance concrete. *Constr. Build. Mater.* **2016**, *103*, 8–14. [CrossRef]

11. Wu, Z.; Khayat, K.H.; Shi, C. How do fiber shape and matrix composition affect fiber pullout behavior and flexural properties of UHPC? *Cem. Concr. Compos.* **2018**, *90*, 193–201. [CrossRef]
12. Yoo, D.Y.; Kang, S.T.; Yoon, Y.S. Enhancing the flexural performance of ultrahigh-performance concrete using long steel fibers. *Compos. Struct.* **2016**, *147*, 220–230. [CrossRef]
13. Mahmud, G.; Yang, Z.; Hassan, A. Experimental and numerical studies of size effects of ultra high performance steel fibre reinforced concrete (UHPFRC) beams. *Constr. Build. Mater.* **2013**, *48*, 1027–1034. [CrossRef]
14. An, M.; Zhang, L.; Yi, Q. Size effect on compressive strength of reactive powder con-crete. *J. China Univ. Min. Technol.* **2008**, *18*, 279–282. [CrossRef]
15. Awinda, K.; Chen, J.; Barnett, S. Investigating geometrical size effect on the flexural strength of the ultra high performance fibre reinforced concrete using the cohesive crack model. *Constr. Build. Mater.* **2015**, *105*, 123–131. [CrossRef]
16. Hassan, A.M.T.; Jones, S.W.; Mahmud, G.H. Experimental test methods to determine the uniaxial tensile and compressive behaviour of ultra high performance fibre reinforced concrete (UHPFRC). *Constr. Build. Mater.* **2012**, *37*, 874–882. [CrossRef]
17. Lampropoulos, A.; Paschalis, S.; Tsioulou, O.; Dritsos, S. Strengthening of rein-forced concrete beams using ultra high performance fibre reinforced concrete (UHPFRC). *Eng. Struct.* **2016**, *106*, 370–384. [CrossRef]
18. Bastien-Masse, M.; Brühwiler, E. Contribution of R-UHPFRC Strengthening layers to the shear resistance of RC elements. *Struct. Eng. Int.* **2016**, *4*, 365–374. [CrossRef]
19. Paschalis, S.; Lampropoulos, A.; Tsioulou, O. Experimental and numerical study of the performance of ultra high performance fiber reinforced concrete for the flexural strengthening of full scale reinforced concrete members. *Constr. Build. Mater.* **2018**, *186*, 351–366. [CrossRef]
20. Paschalis, S.; Lampropoulos, A. Developments in the use of Ultra High Perfor-mance Fiber Reinforced Concrete as strengthening material. *Constr. Build. Mater.* **2021**, *233*, 111914.
21. Bastien-Masse, M.; Brühwiler, E. Experimental investigation on punching re-sistance of R-UHPFRC–RC composite slabs. *Mat. Struct.* **2016**, *49*, 1573–1590. [CrossRef]
22. Bastien-Masse, M.; Brühwiler, E. Composite model for predicting the punching re-sistance of R-UHPFRC–RC composite slabs. *Eng. Struct.* **2016**, *117*, 603–616. [CrossRef]
23. Benson, S.D.P.; Karihaloo, B.L. CARDIFRC®—Development and mechanical properties. Part III: Uniaxial tensile response and other mechanical properties. *Mag. Conc. Res.* **2005**, *57*, 433–443. [CrossRef]
24. JSCE-SF4 III P. *Method of Tests for Steel Fiber Reinforced Concrete*; Concrete library of JSCE, The Japan Society of Civil Engineering: Tokyo, Japan, 1984.

Article

Tensile Experiments and Numerical Analysis of Textile-Reinforced Lightweight Engineered Cementitious Composites

Mingzhao Chen [1], Xudong Deng [1,2], Rongxin Guo [1,*], Chaoshu Fu [1] and Jiuchang Zhang [1,3]

1. Yunnan Key Laboratory of Disaster Reduction in Civil Engineering, Faculty of Civil Engineering and Mechanics, Kunming University of Science and Technology, Kunming 650500, China
2. Yunnan Highway Science and Technology Research Institute, Kunming 650051, China
3. Department of Civil Engineering, Yunnan Minzu University, Kunming 650504, China
* Correspondence: guorx@kmust.edu.cn

Abstract: Despite many cases of textile-reinforced engineered cementitious composites (TR-ECCs) for repairing and strengthening concrete structures in the literature, research on lightweight engineered cementitious composites (LECC) combined with large rupture strain (LRS) textile and the effect of textile arrangement on tensile properties is still lacking. Therefore, this paper develops textile-reinforced lightweight engineered cementitious composites (TR-LECCs) with high strain characteristics through reinforcement ratio, arrangement form, and textile type. The study revealed that, by combining an LRS polypropylene (PP) textile and LECC, TR-LECCs with an ultimate strain of more than 8.0% (3–4 times that of traditional TR-ECCs) could be developed, and the PP textile's utilization rate seemed insensitive to the enhancement rate. The basalt fiber-reinforced polymer (BFRP) textile without epoxy resin coating had no noticeable reinforcement effect because of bond slip; in contrast, the BFRP grid with epoxy resin coating had an apparent improvement in bond performance with the matrix and a better reinforcement effect. The finite element method (FEM) verified that a concentrated arrangement increased the stress concentration in the TR-LECC, as well as the stress value. In contrast, a multilayer arrangement enabled uniform distribution of the stress value and revealed that the weft yarn could help the warp yarn to bear additional tensile loads.

Keywords: fiber-reinforced polymer; lightweight engineered cementitious composites; numerical analysis; textile grid; repair and reinforcement

1. Introduction

Fiber-reinforced polymer (FRP) is extensively applied in the repair and strengthening of reinforced concrete structures [1–4] following numerous theoretical designs on the strength of FRP materials [5–8], fiber-reinforced concrete [9], and slope and foundation reinforcement [10,11]. Its advantages include light weight, high strength, corrosion resistance, and durability [12,13]. Epoxy resin is generally employed as the binder when FRP is used to strengthen structures since it works well in concert with existing structures. However, epoxy resin still has flaws, including easy aging [14], low adaptability [15], brittle damage causing interface peeling [16], and the emission of irritating poisonous gases, which significantly reduces the effectiveness of FRP reinforcement [17]. Therefore, the combination of epoxy resin and FRP cannot fully exert the effect of strengthening the structure.

To address the shortcomings of organic resins, efforts have been undertaken to substitute organic binders such as epoxy resins with inorganic bonding materials, e.g., cement-based matrices. The high-performance ferrocement laminate (HPFL) [18] and the textile-reinforced mortar/concrete (TRM/TRC) approach have also been proposed [19]. TRM is a cementitious composite material composed of a fine-grained mortar matrix reinforced by textiles. It has high tensile strength, as well as multiple cracking behaviors under loading, and it can reduce the crack width [20]. It is typically used for flexural [21] and shear reinforcement of concrete [22], as well as repair reinforcement of concrete columns and thin shell structures [23,24]. Although the crack width can remain small, the multiple cracking stages of TRM structures are typically brief, lasting only until roughly 0.5% of the tensile strain [25]. The textile near the crack is vulnerable to debonding from the matrix section due to stress concentration, and this is also when the matrix at the cracking spot can no longer transmit loads. As a result, the reinforcing system fails before the FRP reaches the maximum damage state [26,27]. Although TRM and HPFL can improve the mechanical properties of concrete components, inorganic bonding materials still suffer from interfacial peeling, poor compatibility with FRP materials, low elongation, and brittleness.

In recent years, high-ductility engineered cementitious composites (ECCs) have become an alternative matrix because the tensile properties of the ECC materials and the bridging ability of short fibers to the matrix compensate for the brittleness and interfacial peeling problems of cementitious materials. ECCs are inorganic cementitious materials reinforced by short fibers with ultimate strains of more than 3%, multiple cracking, and reasonable ductility and toughness [28]. The crack width in ECCs can be less than 100 um during the strain-hardening stage [29]. Therefore, ECC materials have higher ductility and durability compared with TRM. Yang et al. [30] and Zheng et al. [31] used FRP-reinforced ECCs for flexural strengthening of RC beams. They found that the resulting TR-ECC system was very effective because the ECC's strain-hardening and multiple cracking behaviors alleviated the stress concentration at the interface between TR-ECCs and concrete, thereby suppressing interfacial peel failure. According to Chen's [32] study, BFRP textile-reinforced ECC-constrained columns outperformed TRM-deprived columns in terms of ultimate load capacity. The ECC's tensile properties and the textile's outstanding bonding with the ECCs significantly delayed the onset of cracks and maintained structural integrity. Numerous studies have demonstrated the many benefits of TR-ECCs for structural reinforcement. It has been confirmed that textiles can rely on the bridging action in the ECC matrix to transfer loads and maintain good bonding properties because of the strain-hardening ability, multi-cracking behavior, and ultrahigh-ductility properties of the ECC materials [33]. TR-ECCs can repair reinforcement protection for building systems [31] and engineering applications for structural seismic energy resistance [34]. Compared with TRM, TR-ECC can not only prevent interfacial peeling damage but also improve the use efficiency of textiles. Therefore, ECC can replace cement mortar as a potential inorganic bonding material.

There was no relative slip at the interface between the BFRP grid and the ECC matrix when Zheng et al. [35] applied BFRP-reinforced ECC, according to numerous investigations on the tensile mechanical properties of TR-ECCs. At different BFRP grid reinforcement rates, the load-carrying capacity of TR-ECCs specimens rose from 42% to 172%. Li et al. [36] studied the impact of textile volumetric ratios, textile geometry, and matrix thicknesses on tensile mechanical characteristics and demonstrated that TR-ECC composites could enhance textile reinforcing to a significantly greater extent than TRM materials. The textile spacing impacts how the matrix is impregnated onto the textile, which impacts the ability of the matrix and textile to link together. In contrast, the matrix thickness impacts the bridging mechanism of short fibers. Zhang et al. [37] demonstrated that the volume proportion of short fibers and the number of textile layers had a dominant influence on the tensile behavior of TR-ECCs by examining the matrix type, volume fraction of short fibers, and number of textile layers. The matrix type, however, had a negligible effect. The textile also demonstrated multi-seam cracking, strain hardening, and good crack width management, greatly increasing its tensile qualities. Obviously, the study of the tensile mechanical properties of TR-ECC has demonstrated the superiority of ECC as an inorganic binder material.

These investigations often used materials with high modulus and low elongation (typically less than 3%), such as BFRP and carbon fiber-reinforced polymer (CFRP), and they produced TR-ECC tensile properties less than 2%. However, the tensile properties of ECCs are typically larger than 3%, especially recently created ECCs with 8% tensile strain [36]. The high-ductility properties of ECCs are not entirely utilized when using traditional small-strain BFRP, and CFRP is reinforced with large-strain ECCs. For some buildings with high seismic requirements, LRS-FRP shows a competitive advantage [38,39] because various components of concrete restrained with LRS-FRP show a great energy dissipation capacity [40,41]. Therefore, it is necessary to study the tensile mechanical properties of the LRS textile combined with the ECC matrix.

To increase the tensile deformation capacity of TR-ECC, it is, therefore, essential to produce reinforced materials with LRS; however, there are relatively few reports on this aspect. Additionally, ECCs with lightweight characteristics have recently been created with densities of 1300–650 kg/m^3. In large-span, weight-sensitive buildings, these ECCs have demonstrated more significant advantages; nevertheless, the usage of reinforced material in conjunction with these ECCs has not been reported. Therefore, we take the combination of LECC and LRS materials in the paper and investigate the effects of textile type, reinforcement rate, and arrangement form on the tensile properties of TR-LECC, aiming to explore the feasibility of developing TR-LECC with LRS materials and LECC. Overall, we found that the combination of LRS textiles and LECC could develop TR-LECC with similar strength to conventional TR-ECC but 8.0% ultimate strain (3–4 times that of conventional TR-ECC).

2. Test Program

2.1. Performance of Materials

Polypropylene (PP) is a thermoplastic resin made from the polymerization of propylene, which has good resistance to acid, alkali, corrosion, and aging, as well as high tensile strength, lightweight properties, high ductility, and low cost. The ductility of PP textile can reach 10% when strength is considered [42], and the tensile stress of PP textile increases with strain under tensile stress. Therefore, combining a high-ductility LECC matrix and LRS-FRP can not only bring out the reinforcing properties but can also give full play to the ductility properties of the LECC matrix.

Continuous basalt fiber is a new type of high-performance inorganic material, with the advantages of good stability, corrosion resistance, anti-combustion, and high-temperature resistance. The raw material is natural, environmentally friendly, and low-cost. Continuous basalt fiber is divided into two forms: equally spaced dry fiber without epoxy resin impregnation, called BFRP textile (BFRP-F), and epoxy resin-impregnated basalt textile forming a BFRP grid (BFRP-T) after the epoxy resin is cured. Detailed information on the three textile types is shown in Figure 1 and Table 1.

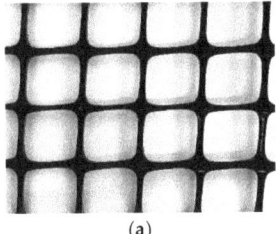 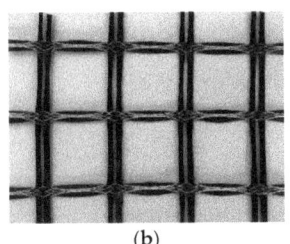 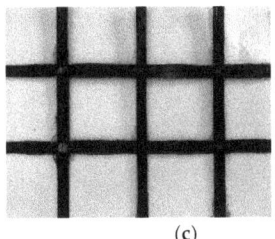

(a) (b) (c)

Figure 1. The specific shape of the textile (**a**) PP, (**b**) BFRP-F, (**c**) BFRP-T.

Table 1. The details of textile types.

Types	PP-1	PP-2	BFRP-F	BFRP-T-1	BFRP-T-2	BFRP-T-3
Textile size	30 mm	30 mm	25 mm	50 mm	50 mm	50 mm
Cross-section area	4.76 mm^2	1.59 mm^2	1.47 mm^2	7.95 mm^2	13.86 mm^2	23.32 mm^2

2.2. Design of the Experiment

To investigate the tensile properties of the LECC matrix according to textile type, enhancement rate, and arrangement form, a total of 11 groups of tests were designed; each group contained three identical specimens, with a total of 33 axial tensile specimens, including five groups of PP textile-reinforced LECC composite specimens, three groups of BFRP textile-reinforced LECC composite specimens, and three groups of BFRP grid-reinforced LECC composite specimens. The detailed test parameters are shown in Table 2. The length of the TR-LECC specimens was 200 mm, the width was 70 mm, and the casting thickness was 15 mm.

Table 2. The design of the TR-LECC specimens.

Specimen ID [a]	Reinforcement Rate (%)	Material Types	Textile Grid Plies
PP-1-1.30	1.30	PP textile	1
PP-2-1.30	1.30	PP textile	2
PP-3-1.30	1.30	PP textile	3
PP-2-2.41	2.41	PP textile	2
PP-3-4.82	4.82	PP textile	3
BF-1-0.42	0.42	BFRP textile	1
BF-2-0.84	0.84	BFRP textile	2
BF-3-1.29	1.29	BFRP textile	3
BT-1-1.30	1.30	BFRP grid	1
BT-1-2.70	2.70	BFRP grid	1
BT-1-4.21	4.21	BFRP grid	1

Note: [a] Taking "PP-1-1.30" as an example, the first two letters are the type of the material in Table 2, the first number represents textile grid plies, and the second number denotes the reinforcement rate.

2.3. TR-LECC Specimen Preparation

Ordinary silicate P.O.52.5 grade cement was used as the main cementitious material, and F-type fly ash (FA) was used as the supplementary cementitious material. Fly ash cenospheres (FACs) used as fine aggregate are waste collected in coal-fired power plants with a diameter of 0.01–0.5 mm and a density of 530 kg/m^3. The addition of nano-silica with a particle size of 40 nm improved the mechanical properties of LECC. The LECC matrix was stirred by a mixing pot, and we adopted a layered casting process [36,37], as shown in Figure 2. The steps for making standard specimens were as follows: (1) pour a layer of the LECC matrix, and then vibrate the LECC on the vibrating table; (2) flatten the matrix, and paste the cut reinforcement textile on the LECC matrix; (3) fix the position of the reinforcement textile, and continue to pour the next layer of the LECC; (4) cover the surface of the specimen with a film to prevent moisture evaporation after the vibrating process is finished. After the cast specimens were cured at room temperature for 24 h, the specimens in the mold were removed and maintained at a temperature of 20 ± 2 °C and 95% relative humidity for 27 days.

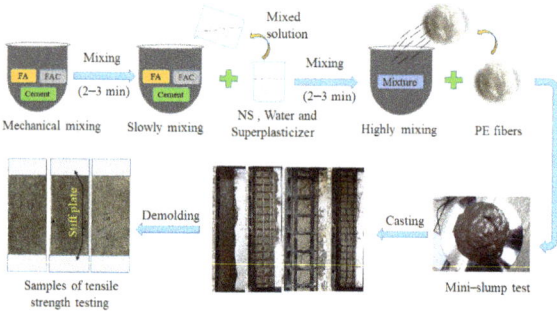

Figure 2. Flowchart of specimen preparation. Note: FA, fly ash; FAC, fly ash cenospheres; NS, nano-silica.

2.4. FRP and TR-LECC Tensile Test

The load was controlled at a constant displacement of 0.5 mm/min under a static tensile load. The testing machine executed the test procedure, and the test was terminated after the specimen was damaged [36]. The specimen strains were collected with a linear variable differential transformer (LVDT) of a 100 mm scale length, and the ends of the gauge were fixed in the test length interval, as shown in Figure 3. All the tests were carried out to ensure that the damage occurred in the middle position (within the measured length of the specimen) away from the aluminum sheet, and accurate tensile test results were obtained.

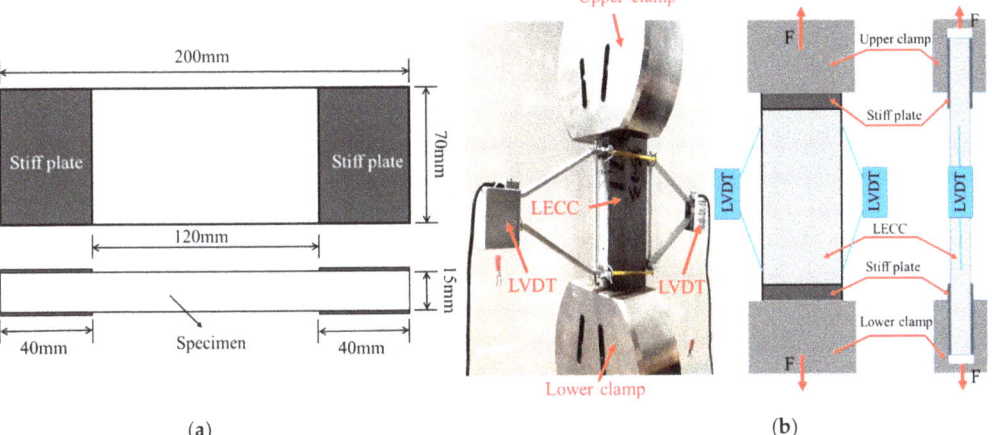

Figure 3. (a) Dimensions of the TR-LECC specimen and (b) test setup.

3. Results and Discussion

3.1. Matrix Material

The design of LECCs as a bonded matrix was based on the study of Fu et al. [43,44]. The uniaxial tensile behavior of the matrix was tested using "dog-bone" specimens according to the recommendation of JSCE [45], and the obtained tensile stress–strain curves are shown in Figure 4. The specific mechanical properties and material mix ratios are shown in Tables 3 and 4, respectively.

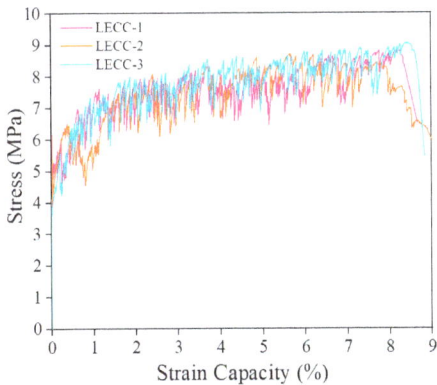

Figure 4. Stress–strain curve of the LECC.

Table 3. Mechanical properties of LECC.

Cracking Stress (MPa)	Cracking Strain (%)	Ultimate Stress (MPa)	Ultimate Strain (%)
5.5 MPa	0.037%	8.8 MPa	7.8%

Table 4. Mix proportions of LECC mixtures (kg/m^3).

Cement	Fly Ash	Nanosilica	Fly Ash Cenospheres	Water	Superplasticizer	PE Fiber
874	391.5	39	195.8	238.7	91.7	20

3.2. Tensile Behavior of Reinforced Materials

The PP textile used in the test was provided by Shandong Lianshun Engineering Materials Co. BFRP textile and BFRP grid materials were provided by Shandong Dalu Engineering Materials Co and Jiangsu Green Material Valley New Material Technology Development Company, respectively. The key mechanical properties of the reinforcement textile were measured in the laboratory, as shown in Table 5. The mechanical properties of the reinforced textile were obtained by tensile test, as shown in Figure 5a. The tensile test was carried out on a 300 KN MTS testing machine, and a 200 mm length single yarn sample was tested with a displacement loading method of 0.5 mm/min.

The stress–strain curve of the PP textile during the entire tensile stage was nonlinear, and the growth rate of tensile stress decreased gradually with the increase in strain. In the tensile test, the PP textile showed a phenomenon similar to the "necking" of steel bars, where the stress and strain continuously increased. Finally, the test ended with a sudden fracture at the nodes of the PP textile, during which the PP textile showed excellent ductility performance, as shown in Figure 5b.

The uniaxial tensile tests yielded the primary failure modes of two different structures of basaltic materials: (1) the BFRP textile not impregnated with epoxy resin had an apparent nonlinear behavior observed at the initial stage of loading, and the nonlinear behavior gradually changed to a linear behavior when a specific load was reached, with a gradual loss of load-bearing capacity as the number of fractured slender filaments increased; (2) the stress–strain curves of the epoxy resin-impregnated BFRP grids were linearly correlated, and the damage behavior was that of an abrupt fracture. The Young's modulus was obtained by extracting the area of the linear section of the BFRP textile and the BFRP grid; the maximum stress in the linear section was used as the fracture strength, and the maximum strain was used as the fracture strain. The stress–strain curves of the three materials are shown in Figure 5b.

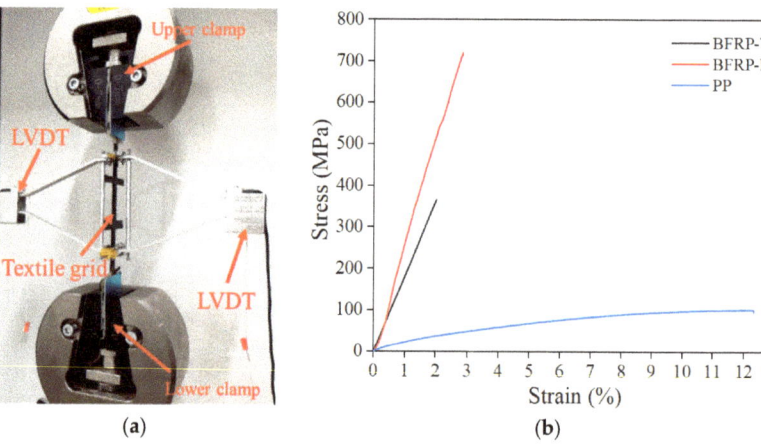

Figure 5. (a) Test setup; (b) stress–strain curves of different materials.

Table 5. Key mechanical property parameters of reinforced materials.

Material Types	Ultimate Tension (KN)	Young's Elastic Modulus (GPa)	Ultimate Stress (MPa)	Ultimate Strain (%)
PP	0.646	-	102.17	12.14
BFRP-T	5.068	18.2	365.65	2.01
BFRP-F	1.061	25.5	719.56	2.82

3.3. Tensile Stress–Strain Curve of TR-LECC

The experimental results showed that two different stress–strain curves characterized the tensile behavior of the TR-LECC. The PP textile and the BFRP grid-reinforced LECC matrix had a two-stage characteristic curve. In contrast, the BFRP textile had a three-stage characteristic curve, and similar characteristics have been observed in the literature [35], as shown in Figure 6. The first two stages were characterized similarly for all three materials. The first stage was the behavior of linear elasticity, where stress and strain were linearly related until the first crack appeared in the matrix. The second stage was strain-hardening behavior following matrix cracking. The matrix exhibited the characteristics of multi-slit cracking, and the stage ended when the increase in the stresses reached its ultimate value. The stiffness of the composite material contributed less to this stage due to matrix cracking; hence, the stiffness of this stage decreased significantly, a phenomenon that was also observed in [46]. A comparison with the stress–strain curve of the LECC matrix showed that the fluctuation was minor because the reinforcing material could contribute to the tensile strength in the matrix, and a larger material tensile strength resulted in a smaller fluctuation. This indicates that the reinforcing material not only improved the strength of the matrix but also had a suppressive effect on the number and width of cracks in the matrix [47]. After this, the BFRP textile-reinforced LECC matrix entered the third stage, i.e., the bonding slip process of the BFRP textile in the matrix. Due to the unimpregnated epoxy resin, the bonding ability between the BFRP textile and the LECC matrix decreased; thus, the BFRP textile appeared to debond after the peak bonding strength. The tensile stress was shared by the bridging force of the polyethylene (PE) fibers and the frictional force between the BFRP textile and the matrix. Finally, the damage of the TR-LECC ended with fracture or debonding failure of the reinforcement and PE fibers that could not continue to bear the load.

Typical stress–strain curves for the textile types at different reinforcement rates are shown in Figure 7a. Similar to the findings in the literature [30], in the elastic stage, the reinforced material did not bear additional tensile stress, and the matrix mainly bore the tensile load. Therefore, the reinforcement rate had little influence on the elastic stage, and the cracking strength of the matrix was not improved. In the multi-slit cracking stage, the slope of the curve increased with the material reinforcement rate, which indicates that the material reinforcement rate could improve the tensile stiffness of the specimen after cracking. The different textile materials had a significant effect on the slope of the curve, which was determined by the materials' inherent characteristics; high tensile strength and good bonding performance exhibited a more significant slope of the curve. Additionally, it can be seen from the curves of the different textile materials that the increase in the enhancement rate reduced the fluctuation range of the curve, indicating that the enhancement rate had an inhibitory effect on cracks [48]. Because the tensile stress after the cracking of the matrix was mainly provided by the reinforcement material, which replaced the PE fibers in the matrix to assume the bridging role between the cracks to inhibit the further development of the cracks, this resulted in a significant reduction in the number of cracks in the specimens [49,50].

The stress–strain curves of PP textile under different arrangements are shown in Figure 7b. The difference in the arrangement form had little effect on the elastic stage of the specimen. In the strain-hardening stage, it can be seen that the increase in the number of arrangement layers improved the reinforcement effect, but there was no difference in the final ultimate stress and ultimate strain. After the failure of the specimen, the stress of the textiles arranged in multiple layers decreased gradually, and the textile arranged in one layer dropped rapidly and soon lost their bearing capacity.

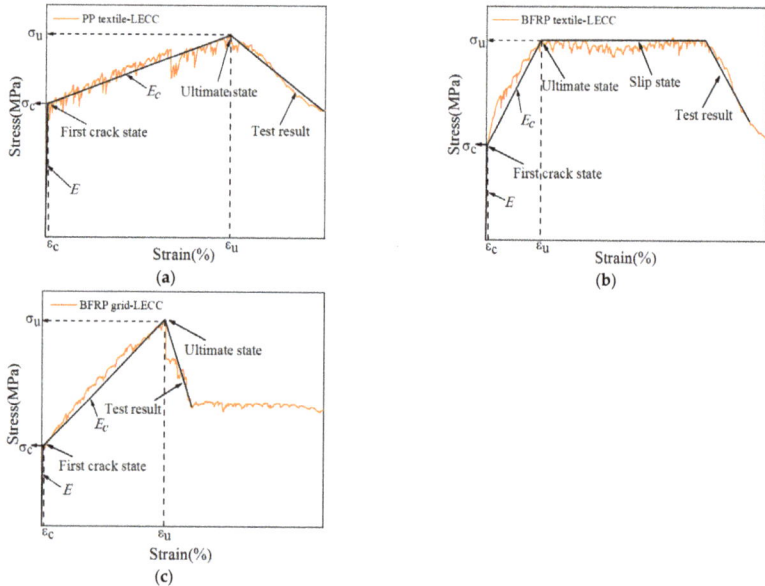

Figure 6. Stress–strain curve of reinforced textile type: (**a**) PP textile, (**b**) BFRP textile, and (**c**) BFRP grid.

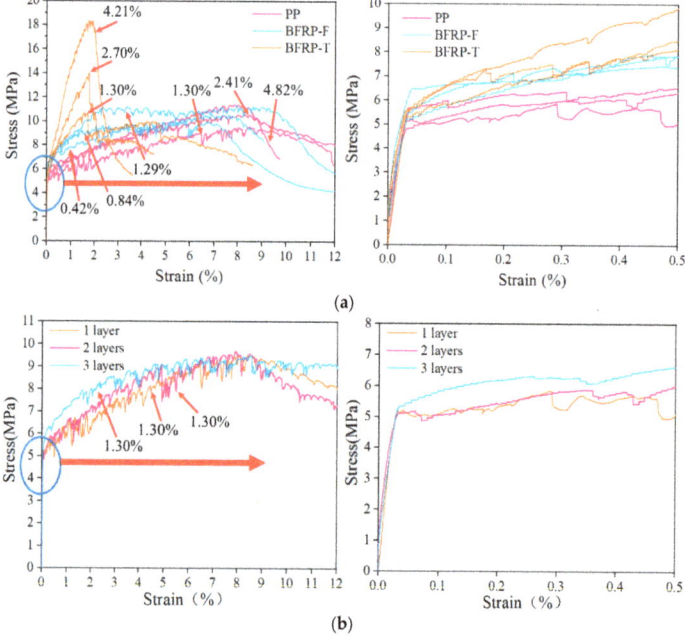

Figure 7. (**a**) Comparison of enhancement ratios of different reinforcement materials and (**b**) comparison of arrangement forms.

3.4. Key Mechanical Parameters

The stress of the TR-LECC was obtained as the ratio of the tensile load to the average cross-sectional area of the TR-LECC panel; the average strain was obtained by dividing the average displacement (average of displacement recorded by two LVDTs) by the length of the LVDT measurement area (100 mm) for calculation. The typical curve of the TR-LECC composite material is shown in Figure 8. The critical parameters in the curve are the cracking stress, cracking strain, and the ultimate stress and ultimate strain of the specimen at the maximum tensile force, as well as the elastic stage and multiplicity's elastic modulus [51]. This experiment investigated the effect on critical parameters according to textile type, enhancement rate, and arrangement form.

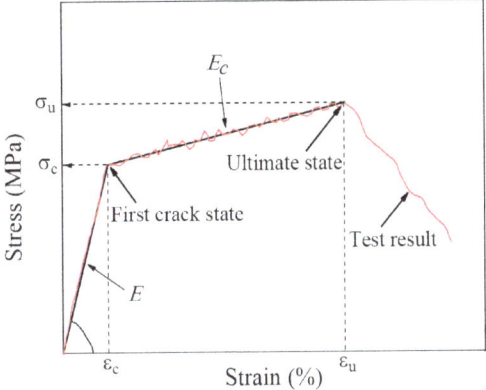

Figure 8. Stress–strain relationship of TR-LECC composites.

3.5. Effect of Arrangement Form on LECC Matrix

Three different layouts (one layer, two layers, and three layers) of the specimens with the same enhancement rate were evaluated, and the stress–strain values, as shown in Figure 9, were obtained from the uniaxial tensile test. The average ultimate stresses were 9.46 MPa, 9.52 MPa, and 9.53 MPa, and the average ultimate strains were 8.27%, 8.33%, and 8.31% for the one-layer, two-layer, and three-layer textile-reinforced LECC, respectively. Therefore, the arrangement form had little effect on the reinforcement and ductility properties of the LECC matrix at the same enhancement rate. Nevertheless, the change in the arrangement form affected the crack distribution and the number of cracks in the composite. Figure 9 shows the crack morphology under the uniaxial tensile tests for specimens with different arrangement forms, as also observed in [52]. The increasing number of textile layers led to apparent multiple cracking behaviors, which suppressed the development of crack width and improved the concentration of tensile stresses [53]. When one layer of the textile was arranged, the specimen cracked with increasing stress accompanied by crack generation, but the number of cracks was relatively small. When the specimen reached the ultimate tensile stress, the matrix crack was no longer generated, one of the cracks gradually widened into the main crack, and the specimen failed, as shown in Figure 9a. The matrix microcracks increased significantly with the increased layers in the arrangement. They then closed automatically after the load was removed, and the damage form of the TR-LECC changed from main crack damage to multiple crack damage, as shown in Figure 9b,c. This indicates that the formation of cracks in the one-layer arrangement is more likely to lead to stress concentration. The bridging effect of the textile in high-stress conditions is weakened, whereby it is difficult to transfer the stress to the matrix through textile; hence, the cracking of the matrix is not inhibited, the crack width is increased, and the number of cracks is reduced. However, the tensile stresses transferred between the matrix by the multilayer arrangement of the textile enhance the control of cracks such that the stresses are more uniformly distributed within the matrix, and the crack width decreases when the number of cracks increases. In addition, the problem of stress concentration in the reinforcing textile can be avoided when multiple layers of textile are arranged due to the coupling effect between the textile layers.

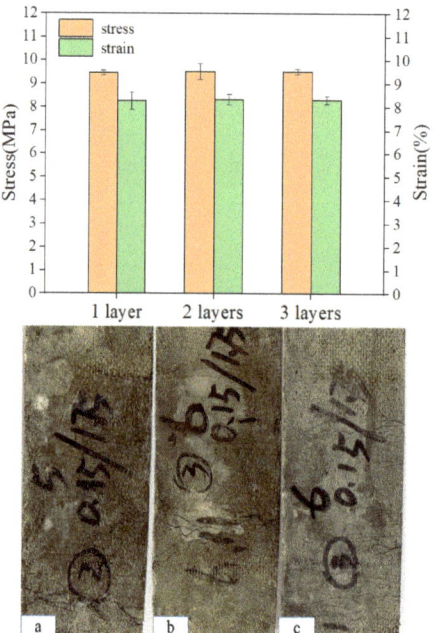

Figure 9. Stress–strain values and crack distribution in the tensile region with different arrangements: (**a**) one-layer textile, (**b**) two-layer textile, and (**c**) three-layer textile.

3.6. Effects of Enhanced Textile Type

In Figure 10, the strengthening effect of the three reinforcing materials on the LECC matrix, BFRP textile, and BFRP grid shows the strengthening effect after the matrix cracking. The strengthening result of the PP textile had a lag phenomenon after the matrix cracking. The BFRP grid-reinforced LECC matrix showed the best strengthening ability; the curve's fluctuation in the multi-slit cracking stage was slight. The sound of a BFRP grid fracture was heard when the tensile stress reaches the limit, followed by a rapid decrease in the bearing capacity of the specimen and damage. The BFRP textile was affected by the production process, and the strengthening effect was slightly weaker after the multi-slit cracking; finally, the bonding failure occurred due to a slip, and the gradual loss of friction apparently reduced the trend of the curve, with a somewhat complementary ductility performance. Because its tensile strength was not high, the PP textile could only enhance the rate of a more significant case. The BFRP grid had a comparable reinforcement effect, but the ductility performance was excellent.

Because of the differences in the materials, each reinforcing material exhibited different reinforcing effects, ductility properties, and the ability to control cracks. The epoxy-impregnated BFRP grid bonded better to the matrix interface. No significant slip between the BFRP grid and the matrix was observed throughout the tests, and this conclusion was fully verified by Dvorkin et al. [54] and Hegger [21]. However, the bonding effect of BFRP textile was mainly provided by the external basalt fiber bundle. In the case of bonding failure, it was provided by the bridging action of PE fibers and the frictional force of the inner and outer basalt fibers [51]. The PP textile had a good bonding effect and ductile deformation, and the minimum tensile strength was the reason for the insignificant strengthening effect of the PP textile. However, the combination with the LECC matrix could fully make use of the ultrahigh-ductility performance. The mechanism of action of the TR-LECC is mainly determined by the interfacial bonding properties between the fiber bundles and matrices. Good interfacial bonding properties can enable the materials to be combined and then synergistically stressed to form an excellent structure with integral properties. Therefore, the BFRP grid is more suitable as a reinforcing material but at the cost of less ductility.

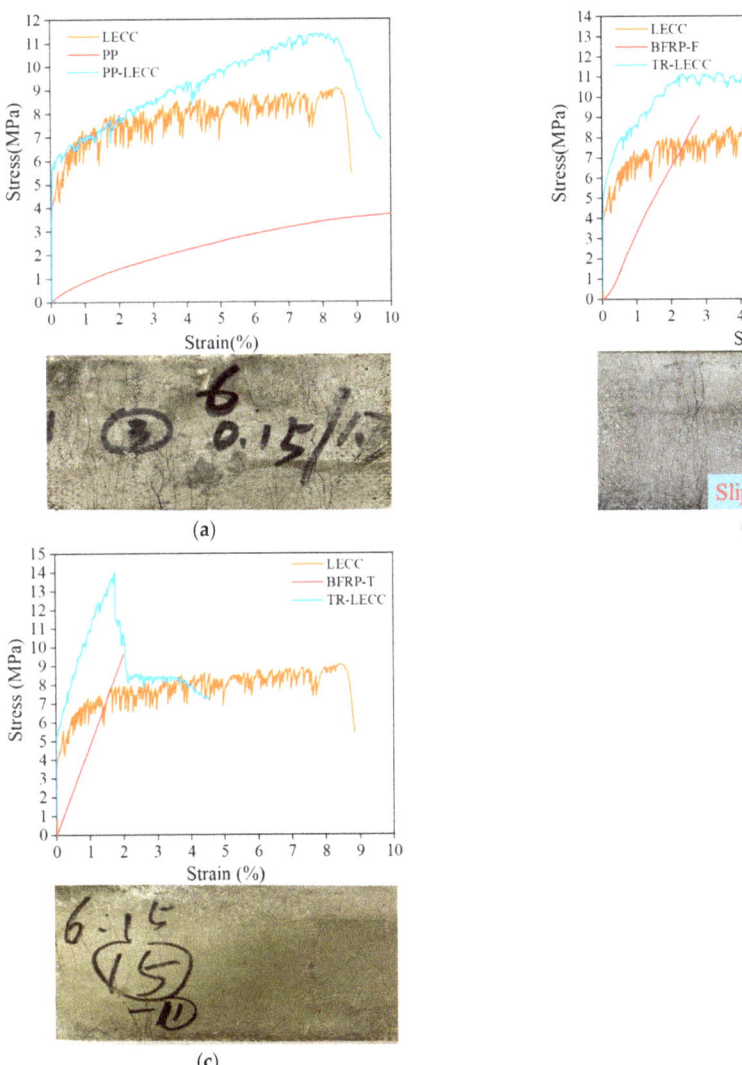

Figure 10. Reinforcement effect of reinforcement textile types and crack diagram: (**a**) PP textile, (**b**) BFRP textile, and (**c**) BFRP grid.

From Figure 11, it can be seen that increasing the enhancement rate of different materials could increase the ultimate stress. Furthermore, the reinforcements rely on the adequate bonding of the reinforcing textile to the LECC matrix to exert its effect. The studies in [55,56] gave the average bond strength of the textile within the gelling matrix along the embedding length to compare the bonding efficiency, which is mainly related to the external perimeter of the warp and the embedding length, with the bond stress expressed as

$$\tau = \frac{F}{l \cdot l_y},\qquad(1)$$

where F is the pullout force, and l_y and l are the warp perimeter and embedding length, respectively.

The concept of material utilization was proposed for the effect of the external perimeter and cross-sectional area of the yarn on the reinforcement effect [21]. This was expressed as the ratio of the maximum stress of the TR-LECC to the stress of the reinforcement textile and LECC.

$$\Delta = \sigma_{\max,\,TR-LECC} / (\sigma_{LECC} + \sigma_m),\qquad(2)$$

where $\sigma_{\max,\,TR-LECC}$ is the ultimate stress of the TR-LECC, σ_{LECC} is the stress of the linear interpolation of the LECC matrix, and σ_m is the ultimate stress of the reinforcement textile. Therefore, the textile warp perimeter and cross-sectional area related to the efficiency of reinforcement utilization were used for the calculation. Because of the same embedding length of the textile in the matrix, its effect on the bond stresses was not considered. The utilization factor β is defined as the warp circumference ratio to the reinforced material's cross-sectional area. Both the warp circumference and the cross-sectional area were obtained from SEM.

Figure 11a,b show the PP and BFRP textile material utilization rates for the different numbers of layers (enhancement rate). The utilization rate of the PP textile decreased from 99.35% to 90.72%, while the enhancement rate increased from 1.3% to 4.82%. The PP textile utilization rate seemed insensitive to the enhancement rate, but the LECC textile reinforced by BFRP decreased from 84.11% to 65.21%, while the enhancement rate increased from 0.43% to 1.29%. From Equation (1), it can be seen that the enhancement rate has no effect on the bond strength; hence, when the utilization factor is the same, the material utilization decreases gradually as the number of layers increases. This is mainly because the textile is not fully utilized, and the utilization efficiency decreases as the number of layers increases. It was also found that the BFRP textile material utilization rate was lower when the utilization factor was significantly higher than that of the PP textile. This indicates that the material utilization rate was also related to the bonding performance of the material. Although increasing the perimeter of the warp improved the bond strength, due to the poor bonding performance, the material utilization was lower because the BFRP textile could not fully utilize the tensile stress. Figure 11c shows the material utilization rate of the BFRP grid in a one-layer arrangement; the BFRP grid had an enhancement rate from 1.3% to 4.21%, and its utilization rate decreased from 81.19% to 71.32%. It can be observed that the utilization rate decreased with the decrease in the utilization factor. The main reason for this phenomenon is that a larger utilization factor denoted a larger external perimeter of the yarn, which increased the contact area between the textile and the matrix, and the increased friction between the yarn and the matrix increased the bond strength. This means that the material utilization rate is not only controlled by the enhancement rate but is also related to the material utilization factor, which means that the ratio of the material's circumference to the cross-sectional area positively affects the material utilization rate.

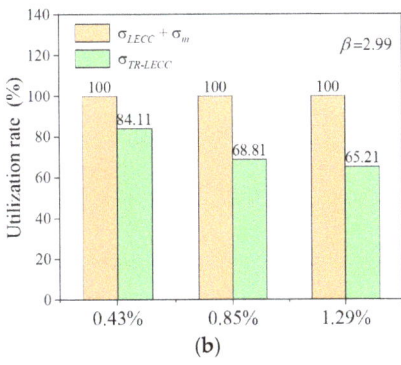

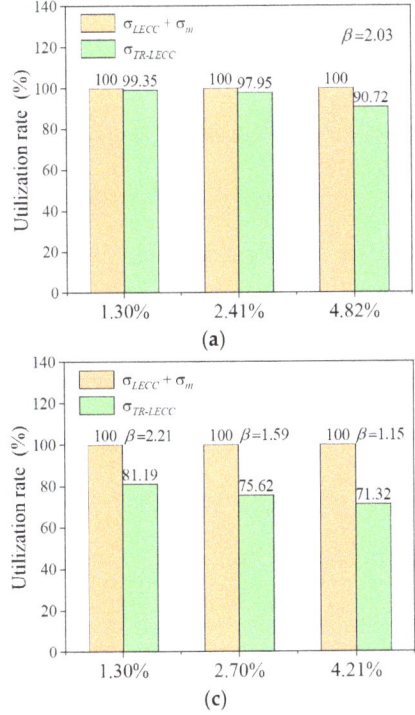

Figure 11. Material utilization of different textile types: (**a**) PP textile, (**b**) BFRP textile, and (**c**) BFRP grid.

3.7. Effect of Enhancement Rate on LECC

In Section 3.5, it was noted that the arrangement form of the reinforcement material had a more negligible effect on the reinforcement effect of the LECC matrix. Thus, the effect caused by the change in the arrangement form was not considered in the TR-LECC. Moreover, on the basis of this conclusion, the effect of the reinforcement rate on the LECC matrix was analyzed again. Figure 12 shows the comparison diagram of the ultimate stress and ultimate strain under different reinforcement rates. With the increase in reinforcement rate, the ultimate stress of specimens was increased to a certain extent. The ultimate strain of the PP textile rose continuously with the addition of a reinforcement rate, while the ultimate strain of the BFRP textile and BFRP grid remained stable with the rise in the enhancement rate. It is not difficult to understand that the ultimate stress changes with the enhancement rate change, under the influence of many factors. For example, the ultimate strain of the reinforcing material plays a critical role. The PP textile-reinforced LECC exhibited a higher strain than the matrix due to its ultrahigh ductility, while the BFRP grid failed early due to the slight ultimate strain. In addition, the bond-slip effect can also increase the ultimate strain. The frictional force between the BFRP textile and the matrix after debonding and the bridging force of the PE fibers together maintain the ductility performance.

For the PP textile, increases in ultimate stress values of 8.3%, 21.1%, and 33.3% were achieved at the enhancement rates (1.30%, 2.41%, and 4.82%), while the ultimate strains increased by 1.5%, 8.5%, and 8.7%, respectively. On the other hand, the BFRP textile increased the ultimate stress by 17.5%, 21.6%, and 29.3% at its enhancement rates, and the ultimate strain was maintained by friction and the PE fiber bridging force as in the LECC matrix. The increases in ultimate stress for BFRP grids at the enhancement rate were 40.2%, 79.2%, and 133.5%, respectively, but the ultimate strain was maintained at only 1.8%.

Combined with the study of Peled [57], it can be found that the strengthening effect is closely related to the textile's material type, and the strengthening effect is more evident if the axial tensile strength is more significant. Meanwhile, the bonding performance of the textile material and the matrix is also a key factor because the tensile strength of the BFRP textile is about twice that of the BFRP grid. Nevertheless, the strengthening effect of the BFRP textile accounted for 83% of the BFRP grid; thus, the bonding performance also significantly influenced the strengthening effect.

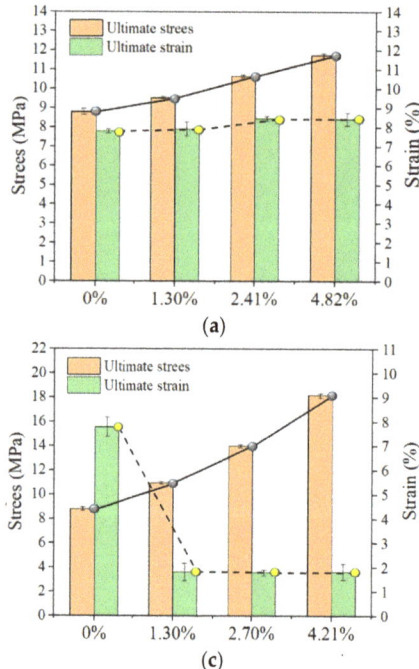
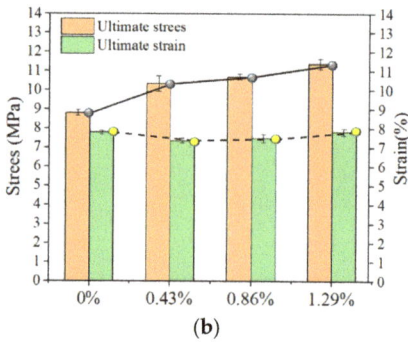

Figure 12. Comparison of ultimate stress and ultimate strain for different enhancement rates: (a) PP textile, (b) BFRP textile, and (c) BFRP grid.

4. Finite Element Simulation

Accurate finite element simulation is a reliable analytical tool that not only simulates the overall response of structure but also enables the study of properties and effects that are difficult or impossible to determine experimentally, such as stress or strain distributions within the material. However, this is essential to understand the macroscopic structural behavior, which is based on the mechanical behavior at a much smaller scale. The authors of [26,58] provided a constitutive model of the ECC matrix and a constitutive relationship of the BFRP material. On the basis of these studies, the LECC matrix model used a C3D8R unit in the solid form. The BFRP material used a three-dimensional truss model (T3D2) for rod members that could only withstand tensile loads but not bending moments, as shown in Figure 13a.

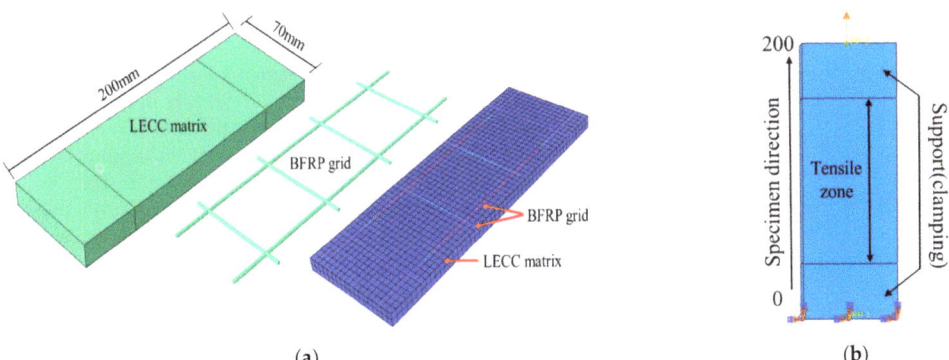

Figure 13. Finite element model. (**a**) Details of the LECC matrix and BFRP grid, (**b**) Schematic diagram of tensile test.

4.1. Material Model

Since no relative slip was observed between the BFRP grid and the LECC matrix with perfective bonding properties, the interaction between the BFRP and the LECC matrix was simulated using the "embedded element" in ABAQUS. The interface effect between the LECC matrix and BFRP was achieved by defining "tensile stiffing" in the concrete damaged plasticity model.

We established the reference point (RP) and the test clamping part to achieve the coupling through interaction and established a wholly fixed restraint method at the RP position; we continued to set the RP at the tensile end. Moreover, we continued to establish the coupling role of the clamping surface with the RP, and we finally completed the test using the displacement loading method, as shown in Figure 13b.

4.2. Results and Discussion

To validate the finite element model, the numerical model and the experimental results were compared in terms of both the tensile stress–strain curve and the failure mode of the specimen. The numerical behavior of the tensile stress–strain curves obtained from the finite element analysis with the experimental results is shown in Figure 14. The slope of the curve obtained from the finite element model was close to that of the experimental results; the elastic stage in the simulation almost coincided with that of the experimental curve, and the elastic modulus, peak stress, and ductility of the multi-slit cracking stage were highly consistent with the experimental results. This indicates that the modeling method can reflect the mechanical properties of the BFRP grid-reinforced LECC composite under axial tensile loading.

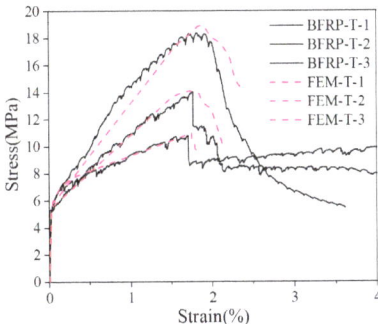

Figure 14. Stress–strain curves of the experimental results and numerical simulation results.

The simulations revealed the tensile failure modes of the TR-LECC for different textile enhancement rates, as shown in Figure 15. It is noteworthy that the simulation test results for tensile damage distribution and crack development patterns were the same as the experimental results. As shown in Figure 15a, the middle region of the specimen suffered more tensile damage than other locations. It can be seen that there were more fine cracks in the central part of the specimen compared with the test, indicating that the tensile damage was also more significant in the central location in the test. Figure 15b,c show that the maximum tensile damage from the simulations was produced at the two ends, with relatively minor tensile damage in the middle region. Furthermore, the test specimens also had fewer cracks and significantly smaller crack widths in the middle before finally failing at the end of the tensile area specimen. From the finite element results of the three different reinforcement rates, it can be concluded that the tensile damage cracking conditions of the simulation and test are the same. The damage produced by the specimen during the tensile process was not severe, which explains why the number of cracks and crack width decreased in the test, indicating that the BFRP grid bore the main load in the tensile state, which not only fully shows the reinforcing effect and the inhibition of the development of cracks, but also shows the improvement to the durability performance of the TR-LECC composite.

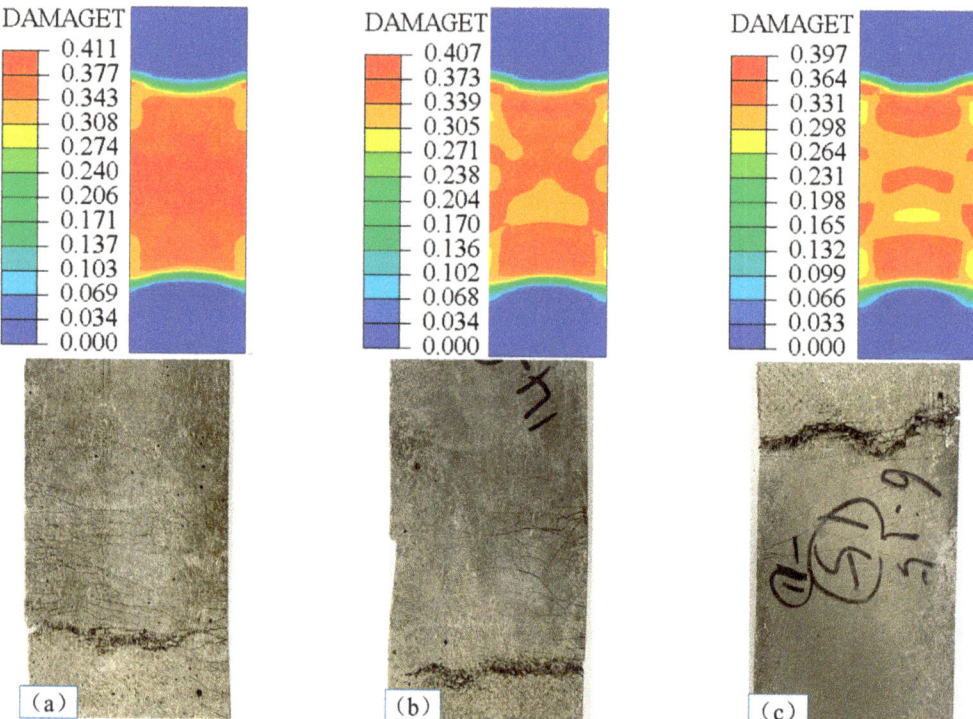

Figure 15. Failure modes for different enhancement rates in finite element simulations and tests: (**a**) 1.30%, (**b**) 2.70%, and (**c**) 4.21%.

4.3. The Impact of Arrangement Form

Figure 16 shows the effects of three different arrangements on the stress distribution of the LECC matrix. When using a one-layer BFRP grid, the LECC matrix in the tensile region exhibited a nonuniform stress distribution, with a prominent stress concentration area in the middle of the model and gradually decreasing to the sides; the arrangement of the two-layer BFRP grid exhibited a better stress distribution, and the tensile region improved the problem of concentrated stress; the three-layer BFRP grid arrangement not only showed a uniform stress distribution in the whole tensile area but also a reduced stress value. From the finite element simulation, we can see that the internal BFRP grid arrangement mainly influenced the stress distribution of the LECC matrix. When the one-layer arrangement had a smaller bridging effect on the BFRP grid in the matrix, the stress concentration was formed after the matrix cracked. Finally, cracks gradually developed under a high-stress state and created major crack damage. However, the multi-layer grid arrangement provided a greater fiber-bridging effect. Thus, the tensile stress of the BFRP grid was uniformly transferred to the matrix material, avoiding the stress concentration in the reinforced material and improving the stress form of the matrix. In addition, the stress value of the matrix at the location of the weft yarn in the tensile area was significantly lower than that of the other parties, indicating that the presence of the weft yarn also changed the stress distribution and the magnitude of the stress value of the matrix.

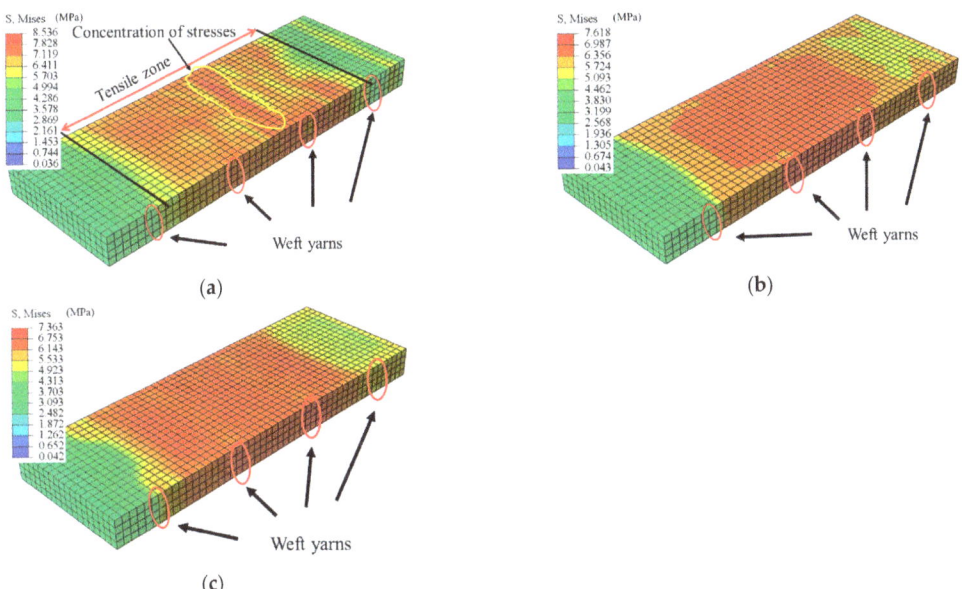

Figure 16. Stress distribution for different BFRP grid layers: (a) one-layer grid, (b) two-layer grid, and (c) three-layer grid.

We further investigated the effect of the arrangement form on the stress and strain distribution; Figure 17 shows the stress and strain values along with the length and width directions of the specimen obtained for different layers of the BFRP grid. The more significant stress and strain values at the tensile end of the specimen were caused by setting the displacement loading surface of the load boundary. Then, the displacement loading caused tensile stress and strain in the tensile end section under tension. It can be seen that there was an apparent stress concentration in the middle of the specimen with a one-layer grid arrangement, and there was also a stress–strain mutation at the location of the weft yarn. Moreover, the stress in the specimen direction of the BFRP grid decreased gradually as the number of arrangement layers increased. Therefore, a multilayer arrangement of the grid can avoid the stress concentration problem and improve the stress distribution of the BFRP grid in the matrix, and the presence of the weft yarn can take up the stress in the warp direction.

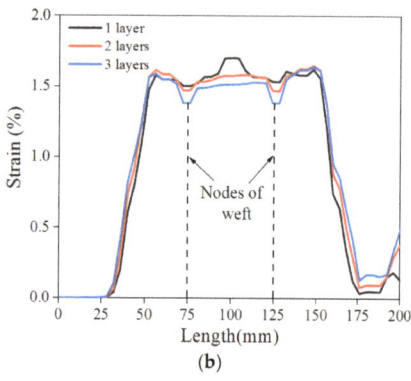
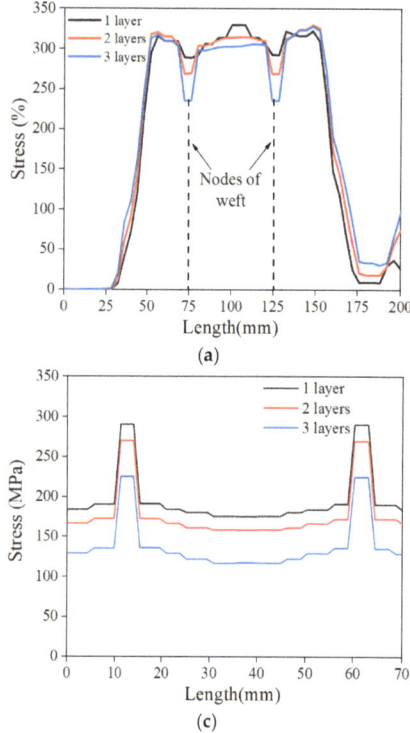

Figure 17. Stress–strain of BFRP grid: (**a**) stress in the length direction of the specimen, (**b**) strain in the length direction of the specimen, and (**c**) stress in the width direction of the specimen.

4.4. Influence of Weft Yarn

Figure 18a shows the stress distribution along the length of the specimen under tensile loading for three grids spacings, which contained no weft reinforcement (BS-0), a weft spacing of 25 mm (BS-25), and a weft spacing of 50 mm (BS-50), as shown in Figure 18b–d, respectively. It can be seen that the specimen without weft reinforcement reached the ultimate stress state at several locations, and the presence of weft reinforcement significantly reduced the ultimate stress and decreased the stress at the nodes of the weft and warp. With the reduction in the weft spacing, the stress concentration of the specimen was improved, which means that the presence of the weft not only helped the warp to bear part of the load but also improved the stress state by avoiding stress concentration. Due to the impregnation of the epoxy resin at the node, the weft and warp yarns formed a rigid node so that the weft yarns anchored to the matrix produced an interlocking effect on the warp yarns through the node. Finally, the interlocking effect could transfer the stress to the weft yarns through the node and help the warp yarns to bear part of the load, which was also confirmed by Lior [25] in his experiments.

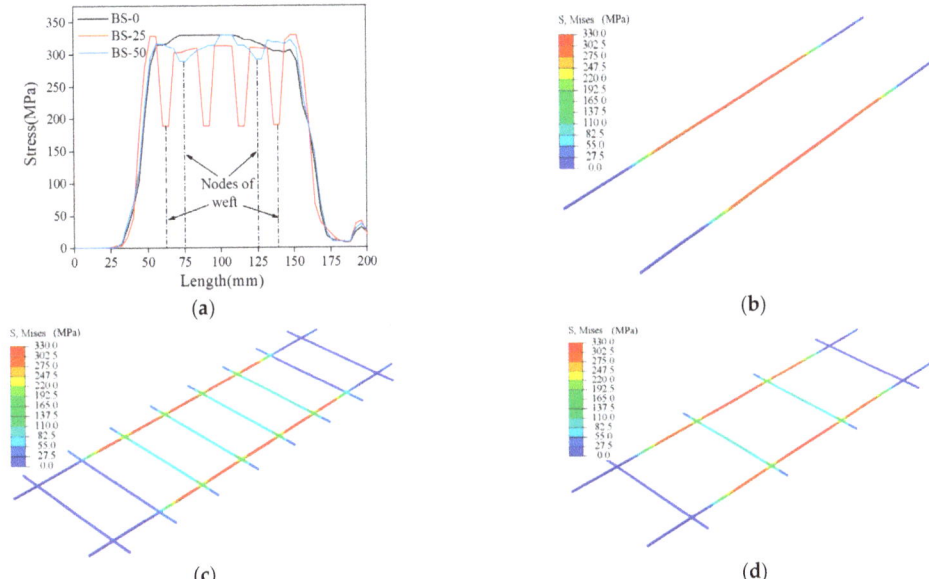

Figure 18. (a) Stress distribution pattern of weft yarn acting on the length direction of the specimen; (b) no weft yarn reinforcement, (c) grid weft yarn spacing of 25 mm, and (d) grid weft yarn spacing of 50 mm.

5. Conclusions

On the basis of the LRS textile and LECCs, TR-LECCs with lightweight and high-ductility properties were developed in this research. Through experiments and FEM, the impacts of variables including textile type, enhancement rate, and arrangement form on the performance of TR-LECCs were investigated. According to the results of this work, the following findings can be summarized:

(1) It is possible to develop TR-LECCs with strengths similar to conventional TR-ECCs and with ultimate strains of 8.0% (3–4 times those of traditional TR-ECCs) by combining the LRS textile and LECCs. This can improve their significant energy dissipation capacity when used to strengthening and repair structures.

(2) Although the tensile characteristics of TR-LECCs are essentially unaffected by the type of textile arrangement (a multilayer arrangement or concentrated arrangement), TR-LECC cracking patterns are nevertheless impacted. While the concentrated arrangement greatly reduces the number of cracks and increases the crack width due to the stress concentration, the multilayer arrangement is advantageous for the fine dispersion of cracks.

(3) The textile type significantly influences the tensile performance of TR-LECCs. Because PP textiles have a higher tensile strain capacity (>8%), TR-LECC reinforcement provides much better strain ductility. Although this does not contribute as much to stiffness as traditional BFRP, increasing the enhancement rate can compensate for it. Due to bond-slip failures, BFRP textiles cannot fully utilize its reinforcing effect. However, BFRP grids impregnated with epoxy resin efficiently utilize the BFRP material's reinforcing effect, increasing the LECC matrix's tensile strength by 40.2% to 133.5%.

(4) The ultimate tensile stress of TR-LECCs improves significantly with an increasing enhancement rate of the textile, but the increase in the enhancement rate decreases the material utilization rate. For instance, as the enhancement rate increased from 0.43% to 1.29%, the material utilization of LECC reinforced with BFRP textile decreased from 84.11% to 65.21%. Notably, the utilization rate of PP textile seems insensitive to the

enhancement rate, decreasing by just 8.63% as the enhancement rate increased from 1.3% to 4.82%, which is favorable for large-volume applications of PP textile.

(5) According to the FEM analysis, the arrangement forms considerably alter how the stress values are distributed in the TR-LECC. The centralized arrangement causes a stress concentration in the TR-LECC, and the stress value is large, while the multilayer arrangement facilitates uniform distribution of stress values in the TR-LECC. In addition, the presence of weft yarns has an important influence on the stress form in the TR-LECC; as the number of weft yarns increases, the stress concentration in the tensile region of the TR-LECC tends to weaken. On the other hand, adding additional weft yarns can help the warp yarns bear a higher axial load.

The current research demonstrates the feasibility of TR-LECC composites and provides an essential basis for the design of textile-reinforced LECC. However, more experiments and finite element simulations are needed to explore suitable reinforcing materials for concrete structure repair and strengthening and optimize contact interface properties for improvement.

Author Contributions: Conceptualization, M.C. and R.G.; methodology, M.C.; validation, M.C., R.G., C.F. and X.D. formal analysis, R.G.; investigation, M.C. and C.F.; resources, R.G. and J.Z.; data curation, R.G.; writing—original draft preparation, M.C.; writing—review and editing, M.C.; visualization, M.C.; supervision, R.G.; project administration, R.G.; funding acquisition, R.G. All authors have read and agreed to the published version of the manuscript.

Funding: This study was funded by the "National Natural Science Foundation of China (NSFC) (Grant No.11962009)". All the help and support are greatly appreciated.

Institutional Review Board Statement: Not applicable.

Informed Consent Statement: Not applicable.

Data Availability Statement: The data presented in this study are available from the corresponding author upon reasonable request.

Conflicts of Interest: The authors declare no conflict of interest.

References

1. Larbi, A.S.; Agbossou, A.; Hamelin, P. Experimental and numerical investigations about textile-reinforced concrete and hy-brid soluteions for repairing and/or strengthening reinforced concrete beams. *Compos. Struct.* **2013**, *99*, 152–162. [CrossRef]
2. Zeng, J.J.; Ye, Y.Y.; Guo, Y.C.; Lv, J.F.; Jiang, C. PET FRP-concrete-high strength steel hybrid solid columns with strain-hardening and ductile performance: Cyclic axial compressive behavior. *Compos. Part B* **2020**, *190*, 107903. [CrossRef]
3. Qu, W.; Zhang, X.; Huang, H. Flexural Behavior of Concrete Beams Reinforced with Hybrid (GFRP and Steel) Bars. *J. Compos. Constr.* **2009**, *13*, 350–359. [CrossRef]
4. Zhang, C.; Abedini, M. Development of PI model for FRP composite retrofitted RC columns subjected to high strain rate loads using LBE function. *Eng. Struct.* **2022**, *252*, 113580. [CrossRef]
5. Kolpakov, A.G.; Rakin, S.I. Homogenized strength criterion for composite reinforced with orthogonal systems of fibers. *Mech. Mater.* **2020**, *148*, 103489. [CrossRef]
6. Kolpakov, A.G.; Rakin, S.I. Local stresses in the reinforced plate with orthogonal sytems of fibers. *Compos. Struct.* **2021**, *265*, 113772. [CrossRef]
7. Kolpakov, A.A.; Kolpakov, A.G.; Rakin, S.I. "Skin" boundary layers and concept of representative model in inhomogeneous plates. *Eur. J. Mech.-A/Solids* **2022**, *93*, 104552. [CrossRef]
8. Kolpakov, A.G.; Rakin, S.I. Comparative analysis of local stresses in unidirectional and cross-reinforced composites. In *Mechanics and Physics of Structured Media*; Academic Press: Cambridge, MA, USA, 2022; pp. 395–416.
9. Wang, X.; Yang, Y.; Yang, R.H.; Liu, P. Experimental Analysis of Bearing Capacity of Basalt Fiber Reinforced Concrete Short Columns under Axial Compression. *Coatings* **2022**, *12*, 654. [CrossRef]
10. Xu, J.; Wu, Z.; Chen, H.; Shao, L.T.; Zhou, X.G.; Wang, S.H. Study on strength behavior of basalt fiber-reinforced loess by digital image technology (DIT) and scanning electron microscope (SEM). *Arab. J. Sci. Eng.* **2021**, *46*, 11319–11338. [CrossRef]
11. Wu, Z.; Xu, J.; Chen, H.; Shao, L.T.; Zhou, X.G.; Wang, S.H. Shear strength and mesoscopic characteristics of basalt fiber–reinforced loess after dry–wet cycles. *J. Mater. Civ. Eng.* **2022**, *34*, 04022083. [CrossRef]
12. Zhou, Y.; Zheng, Y.; Pan, J.; Sui, L.; Xing, F.; Sun, H.; Li, P. Experimental investigations on corrosion resistance of innovative steel-FRP composite bars using X-ray microcomputed tomography. *Compos. Part B* **2019**, *161*, 272–284. [CrossRef]

13. Yang, G.; Feng, X.; Wang, W.; Yang, Q.O.; Liu, L. Effective interlaminar reinforcing and delamination monitoring of carbon fibrous composites using a novel nano-carbon woven grid. *Compos. Sci. Technol.* **2021**, *213*, 108959. [CrossRef]
14. Ali, M.S.; Mirza, M.S.; Lessard, L. Durability assessment of hybrid FRP composite shell and its application to prestressed concrete girders. *Constr. Build. Mater.* **2017**, *150*, 114–122. [CrossRef]
15. Benzarti, K.; Chataigner, S.; Quiertant, M.; Marty, C.; Aubagnac, C. Accelerated ageing behaviour of the adhesive bond between concrete specimens and CFRP overlays. *Constr. Build. Mater.* **2011**, *25*, 523–538. [CrossRef]
16. Catherine, G.; Papanicolaou, T.C.; Triantafillou, M.P.K. Textile reinforced mortar (TRM) versus FRP as strengthening material of URM walls: Outof-plane cyclic loading. *Mater. Struct.* **2008**, *41*, 143–157.
17. Ali, A.; Abdalla, J.; Hawileh, R.; Galal, K. CFRP mechanical anchorage for externally strengthened RC beams under flexure. *Phys. Procedia* **2014**, *55*, 10–16. [CrossRef]
18. Huang, H.; Huang, M.; Zhang, W.; Yang, S.L. Experimental study of predamaged columns strengthened by HPFL and BSP under combined load cases. *Struct. Infras. Eng.* **2021**, *17*, 1210–1227. [CrossRef]
19. Triantafillou, T.C.; Papanicolaou, C.G. Shear strengthening of reinforced concrete members with textile reinforced mortar (TRM) jackets. *Mater. Struct.* **2006**, *39*, 93–103. [CrossRef]
20. Hegger, J.; Will, N.; Curbach, M.; Bruckermann, O. Loading-bearing behavior and simulation of textile reinforced concrete. *Mater. Struct.* **2006**, *39*, 765–776. [CrossRef]
21. Hegger, J.; Voss, S. Investigations on the bearing behaviour and application potential of textile reinforced concrete. *Eng. Struct.* **2008**, *30*, 2050–2056. [CrossRef]
22. Portal, N.W.; Thrane, L.N.; Lundgren, K. Flexural behaviour of textile reinforced concrete composites: Experimental and numerical evaluation. *Mater. Struct.* **2017**, *50*, 4. [CrossRef]
23. Sharei, E.; Scholzen, A.; Hegger, J.; Chudoba, R. Structural behavior of a lightweight, textile-reinforced concrete barrel vault shell. *Compos. Struct.* **2017**, *17*, 505–514. [CrossRef]
24. Scholzen, A.; Chudoba, R.; Hegger, J. Thin-walled shell structures made of textile-reinforced concrete: Part I: Structural design and construction. *Struct. Concr.* **2013**, *16*, 106–114. [CrossRef]
25. Lior, N.; Erez, G.; Alva, P. Tensile behavior of fabric-cement-based composites reinforced with non-continuous load bearing yarns. *Constr. Build. Mater.* **2020**, *236*, 117432. [CrossRef]
26. Larrinaga, P.; Chastre, C.; Biscaia, H.C.; San-Jose, J.T. Experimental and numerical modeling of basalt textile reinforced mortar behavior under uniaxial tensile stress. *Mater. Des.* **2014**, *55*, 66–74. [CrossRef]
27. Carozzi, F.G.; Poggi, C. Mechanical properties and debonding strength of Fabric Reinforced Cementitious Matrix (FRCM) systems for masonry strengthening. *Compos. Part B* **2015**, *70*, 215–230. [CrossRef]
28. Hung, C.C.; Su, Y.F.; Su, Y.M. Mechanical properties and self-healing evaluation of strain-hardening cementitious composites with high volumes of hybrid pozzolan materials. *Compos. Part B* **2018**, *133*, 15–25. [CrossRef]
29. Li, V.C. *Engineered Cementitious Composites (ECC) Material, Structural, and Durability Performance*; University of Michigan: Ann Arbor, MI, USA, 2008.
30. Yang, X.; Gao, W.Y.; Dai, J.G.; Lu, Z.D.; Yu, K.Q. Flexural strengthening of RC beams with CFRP grid-reinforced ECC matrix. *Compos. Struct.* **2018**, *189*, 9–26. [CrossRef]
31. Zheng, Y.Z.; Wang, W.W.; Brigham, J.C. Flexural behaviour of reinforced concrete beams strengthened with a composite reinforcement layer: BFRP grid and ECC. *Constr. Build. Mater.* **2016**, *115*, 424–437. [CrossRef]
32. Chen, X.; Zhu, G.Y.; Al-Gemeel, A.N.; Xiong, Z.M. Compressive behaviour of concrete column confined with basalt textile reinforced ECC. *Eng. Struct.* **2021**, *243*, 112651. [CrossRef]
33. Li, V.C.; Horii, H.; Kabele, P.; Kanda, T.; Lim, Y.M. Repair and retrofit with engineered cementitious composites. *Eng. Fract. Mech.* **2000**, *65*, 317–334. [CrossRef]
34. Fischer, G.; Li, V.C. Effect of matrix ductility on deformation behavior of steel reinforced ECC flexural members under reversed cyclic loading conditions. *ACI Struct. J.* **2002**, *99*, 781–790.
35. Zheng, Y.Z.; Wang, W.W.; Mosalam, K.M.; Zhu, Z.F. Mechanical behavior of ultra-high toughness cementitious composite strengthened with Fiber Reinforced Polymer grid. *Compos. Struct.* **2018**, *184*, 1–10. [CrossRef]
36. Li, B.B.; Xiong, H.B.; Jiang, J.F.; Dou, X.X. Tensile behavior of basalt textile grid reinforced Engineering Cementitious Composite. *Compos. Part B.* **2019**, *156*, 185–200. [CrossRef]
37. Zhang, W.; Deng, M.K.; Han, Y.G.; Li, R.Z.; Yang, S. Uniaxial tensile performance of high ductile fiber-reinforced concrete with built-in basalt textile grids. *Constr. Build. Mater.* **2021**, *315*, 125716. [CrossRef]
38. Dai, J.G.; Bai, Y.L.; Teng, J.G. Behavior and modeling of concrete confined with FRP composites of large deformability. *J. Compos. Constr.* **2011**, *15*, 963–973. [CrossRef]
39. Huang, L.; Zhang, S.S.; Yu, T.; Wang, Z.Y. Compressive behaviour of large rupture strain FRP-confined concrete-encased steel columns. *Constr. Build. Mater.* **2018**, *183*, 513–522. [CrossRef]
40. Han, Q.; Yuan, W.Y.; Ozbakkaloglu, T.; Bai, Y.L.; Du, X. Compressive behavior for recycled aggregate concrete confined with recycled polyethylene naphthalate/terephthalate composites. *Constr. Build. Mater.* **2020**, *261*, 120498. [CrossRef]
41. Dai, J.G.; Lam, L.; Ueda, T. Seismic retrofit of square RC columns with polyethylene terephthalate (PET) fibre reinforced polymer composites. *Constr. Build. Mater.* **2012**, *27*, 206–217. [CrossRef]

42. Yao, W.; Li, J.; Wu, K. Mechanical properties of hybrid fiber-reinforced concrete at low fiber volume fraction. *Cem. Concr. Res.* **2003**, *33*, 27–30. [CrossRef]
43. Fu, C.; Guo, R.; Lin, Z.; Xia, H.; Yang, Y.; Ma, Q. Effect of nanosilica and silica fume on the mechanical properties and microstructure of lightweight engineered cementitious composites. *Constr. Build. Mater.* **2021**, *298*, 123788. [CrossRef]
44. Fu, C.; Chen, M.; Guo, R.; Qi, R.Q. Green-Engineered Cementitious Composite Production with High-Strength Synthetic Fiber and Aggregate Replacement. *Materials* **2022**, *15*, 3047. [CrossRef] [PubMed]
45. Rokugo, K. *Recommendations for Design and Construction of High Performance Fiber Reinforced Cement Composites With Multiple Fine Cracks (HPFRCC)*; The Japan Society of Civil Engineering: Tokyo, Japan, 2008.
46. Bolster, E.D.; Cuypers, H.; Itterbeeck, P.V.; Wastiels, J. Use of hypar-shell structures with textile reinforced cement matrix composites in lightweight constructions. *Compos. Sci. Technol.* **2009**, *69*, 1341–1347. [CrossRef]
47. Peled, A.; Bentur, A. Fabric structure and its reinforcing efficiency in textile reinforced cement composites. *Compos. Part A* **2003**, *34*, 107–118. [CrossRef]
48. Cuypers, H.; Wastiels, J. A stochastic cracking theory for the introduction of matrix multiple cracking in textile reinforced concrete under tensile loading. In *Finds and Results from the Swedish Cyprus Expedition: A Gender Perspective at the Medelhavsmuseet*; RILEM: Paris, France, 2006; pp. 193–202.
49. Barhum, R.; Mechtcherine, V. Effect of short, dispersed glass and carbon fibres on the behaviour of textile-reinforced concrete under tensile loading. *Eng. Fract. Mech.* **2012**, *92*, 56–71. [CrossRef]
50. Barhum, R.; Mechtcherine, V. Influence of short dispersed and short integral glass fibres on the mechanical behaviour of textile-reinforced concrete. *Mater. Struct.* **2013**, *46*, 557–572. [CrossRef]
51. Contamine, R.; Larbi, A.S.; Hamelin, P. Contribution to direct tensile testing of textile reinforced concrete (TRC) composites. *Mater. Sci. Eng.* **2011**, *528*, 8589–8598. [CrossRef]
52. Liu, D.J.; Huang, H.W.; Zuo, J.P.; Duan, K.; Xue, Y.D.; Li, Y.J. Experimental and numerical study on short eccentric columns strengthened by textile-reinforced concrete under sustaining load. *J. Reinf. Plast. Compos.* **2017**, *36*, 1712–1726. [CrossRef]
53. Larrinaga, P.; Chastre, C.; San-José, J.T.; Garmendia, L. Non-linear analytical model of composites based on basalt textile reinforced mortar under uniaxial tension. *Compos. Part B* **2013**, *55*, 518–527. [CrossRef]
54. Dvorkin, D.; Poursaee, A.; Peled, A.; Weiss, W.J. Influence of bundle coating on the tensile behavior, bonding, cracking and fluid transport of fabric cement-based composites. *Cem. Concr. Compos.* **2013**, *42*, 9–19. [CrossRef]
55. Peled, A.; Bentur, A. Geometrical characteristics and efficiency of textile fabrics for reinforcing cement composites. *Cem. Concr. Res.* **2000**, *30*, 781–790. [CrossRef]
56. Portal, N.W.; Perez, I.F.; Thrane, L.N.; Lundgren, K. Pull-out of textile reinforcement in concrete. *Constr. Build. Mater.* **2014**, *71*, 63–71. [CrossRef]
57. Peled, A.; Cohen, Z.; Pasder, Y.; Roye, A.; Gries, T. Influences of textile characteristics on the tensile properties of warp knitted cement based composites. *Cem. Concr. Compos.* **2008**, *30*, 174–183. [CrossRef]
58. Meng, D.; Huang, T.; Zhang, Y.X.; Lee, C.K. Mechanical behaviour of a polyvinyl alcohol fibre reinforced engineered cementitious composite (PVA-ECC) using local ingredients. *Constr. Build. Mater.* **2017**, *141*, 259–270. [CrossRef]

Article

Investigation of the Failure Modes of Textile-Reinforced Concrete and Fiber/Textile-Reinforced Concrete under Uniaxial Tensile Tests

Giorgio Mattarollo [1,2,*], Norbert Randl [2] and Margherita Pauletta [1]

[1] Polytechnic Department of Engineering and Architecture, Università degli Studi di Udine, Via delle Scienze 206, 33100 Udine, Italy
[2] Faculty of Civil Engineering and Architecture, Carinthia University of Applied Sciences (CUAS), Villacher Straße 1, A-9800 Spittal an der Drau, 9800 Carinthia, Austria
* Correspondence: giorgio.mattarollo@phd.units.it

Abstract: Recently, innovations in textile-reinforced concrete (TRC), such as the use of basalt textile fabrics, the use of high-performance concrete (HPC) matrices, and the admixture of short fibers in a cementitious matrix, have led to a new material called fiber/textile-reinforced concrete (F/TRC), which represents a promising solution for TRC. Although these materials are used in retrofit applications, experimental investigations about the performance of basalt and carbon TRC and F/TRC with HPC matrices number, to the best of the authors' knowledge, only a few. Therefore, an experimental investigation was conducted on 24 specimens tested under the uniaxial tensile, in which the main variables studied were the use of HPC matrices, different materials of textile fabric (basalt and carbon), the presence or absence of short steel fibers, and the overlap length of the textile fabric. From the test results, it can be seen that the mode of failure of the specimens is mainly governed by the type of textile fabric. Carbon-retrofitted specimens showed higher post-elastic displacement compared with those retrofitted with basalt textile fabrics. Short steel fibers mainly affected the load level of first cracking and ultimate tensile strength.

Keywords: textile-reinforced concrete (TRC); fiber/textile-reinforced concrete (F/TRC); textile fabric; short fibers; carbon textile; basalt textile; uniaxial tensile test; overlap length; pull-out

1. Introduction

Textile-reinforced concrete (TRC) is a relatively new material, composed of a textile fabric and a cementitious matrix. In the literature, this material is also called fiber-reinforced concrete mortar (FRCM) and textile-reinforced mortar (TRM) [1]. Its main functions concern the retrofit of existing structures and the realization of new structural elements [2]. TRC became popular as an alternative solution to the well-known fiber-reinforced polymer (FRP) [3–5]. FRP is characterized by poor behavior above the glass transition temperature of epoxy resin, the high cost of epoxy resins, being hazardous for manual workers, non-applicability on wet surfaces and the incompatibility of epoxy resins with the substrate materials [3]. Such drawbacks are mainly related to the organic matrix [3–5]. Therefore, the use of a cement-based matrix instead of an organic one has led to the rise of TRC. Furthermore, TRC shares some advantages with FRP: the light weight to high strength ratio, the low impact on the original geometry and the corrosion resistance. For both components of the composite, the textile fabric and cementitious matrix, investigations to optimize the performance of TRC have been carried out in the last two decades. The effect of the matrix on the performance of TRC has been investigated by considering different types of cement-based matrices [6], high-performance concrete (HPC) [7,8] and ultra-high-performance concrete (UHPC) [9,10]. Furthermore, in order to improve the performance of cementitious matrices, lacking in terms of tensile strength, the admixture with short fibers has been studied. The use of short fibers in the TRC matrix has led to the creation of a new material called fiber/textile-reinforced concrete (F/TRC) [7,8]. The main advantages of admixing short fibers in the cementitious matrix are the increase in first crack stress and ultimate tensile strength and the improvement of crack control [8,10–13]. Concerning the textile fabric, different warps and materials have been studied, such as carbon, steel, glass, basalt, polypara-phenylene-benzobisoxazole (PBO) and aramid [14–19]. Among these materials, carbon is the most commonly used. The success of this material is mainly related to its high performance and light weight. However, among the materials available in the market, basalt is gaining increasing interest due to its mechanical performance and low environmental impact. In fact, the tensile stress and the elastic modulus of basalt fibers are generally higher than those of glass fibers and lower than those of carbon fibers [19]. Concerning the low environmental impact of basalt fibers, their production requires less energy than the production of carbon and glass fibers [20–22]. Furthermore, basalt fibers are characterized by high corrosion and temperature resistance. Basalt resists pH values up to 13 or 14, and shows good resistance to alkaline environments [19,23].

Despite the advantages of using a cementitious matrix, some drawbacks have to be considered. Due to the granularity of the mortar, the impregnation of the fibers of the rovings is very difficult to achieve [3]. This negatively affects the bond performance of the textile fabric with the matrix. The interaction between these two components, on TRC or F/TRC elements, is arduous to study if the properties of the textile fabric and the inorganic matrix are known separately. Furthermore, which of the different modes of failure that is most likely to happen cannot be easily estimated from the mechanical properties of the textile and cementitious matrix if studied separately. For this reason, tensile tests on composite elements are commonly executed to estimate their mechanical properties [9,24–27].

From the literature, it has been observed that, because of the large amount of materials available to realize textile fabrics and inorganic matrices, each combination of TRC system must be tested in order to characterize its tensile properties [27]. De Domenico et al. [6] conducted an experimental campaign of tensile tests on FRCM specimens. The materials used to realize the specimens were basalt, carbon and steel textile fabrics, short polymer fibers and three cementitious matrices characterized by compressive strengths of 22.03, 29.30 and 19.87 MPa. In this work, a comparison between the different configurations of the matrix and textile is carried out.

D'Antino and Papanicolaou [28] realized a mechanical characterization of TRM composites. They adopted a carbon textile fabric with and without coating, coated basalt textile fabric, coated glass textile fabric and galvanized steel textile fabric. Three different matrices were used, Matrix C, L and P, with compressive strengths of 16.40, 12.10 and 10.30 MPa, respectively.

De Felice et al. [29] executed an experimental campaign of uniaxial tensile tests and bond tests on brick and stone substrates. Carbon, basalt and steel textile fabrics were used with three different matrices: UNIRM3, a pozzolanic-cement mortar with a compressive strength of 37.0 MPa; UMINHO, a pozzolan lime-based mortar with a compressive strength of 13.0 MPa; and TECNALIA, a cementitious mortar with a compressive strength of 22.6 MPa.

Lignola et al. [30] conducted an extensive campaign of tensile and bond tests on FRCM specimens realized with three different basalt textile fabrics and four different inorganic matrices characterized by 25, 12, 15 and 15 MPa of compressive strength.

Larrinaga et al. [31], in order to propose an empirical non-linear approach to predict the stress–strain curve of TRM, executed an experimental test campaign on 31 specimens. The specimen utilized in this study was a basalt textile fabric covered by a bitumen coat. The matrix was a non-commercial cement-based mortar, characterized by a compressive strength of 19.8 MPa.

Hojdys and Krajewski [27] present the results of direct tensile tests on FRCM specimens characterized by three different textile fabrics (carbon, glass and PBO) and four types of inorganic matrix characterized by a compressive strength of 15.4, 14.9, 9.8 and 44.3 MPa, respectively. This study investigates the mechanical properties of FRCM systems, modes of failure and finding a bi- or tri-linear curve for the tensile stress–strain relationship.

Beßling et al. [32] developed a TRC/TRM system in which two inorganic matrices were investigated, a UHPC and a HPC matrix with a compressive strength of 83 ± 7 and 60 ± 5 MPa, respectively. Furthermore, carbon and basalt textile fabrics were used.

Zhou et al. [33] present an investigation about the effect of the reinforcement ratio, volume fraction of steel fibers and prestressing on Carbon TRM specimens under uniaxial tensile tests. The cementitious matrix adopted is a high-performed fine-grained mortar, characterized by a compressive strength of 76.7 MPa after 28 days. Short steel fibers characterized by a length of 12–15 mm and a diameter of 0.18–0.23 mm were admixed in the cementitious matrix. The textile fabric adopted is made by carbon fibers. A summary of the literature review presented above is listed in Table 1.

Table 1. Summary of the inorganic matrices and textile fabrics investigated in the literature.

Authors	Type of Inorganic Matrix	f_c	Short Fibers	Textile Fabric
De Domenico et al. [6]	Cementitious matrix	22.03 29.30 19.87	Polymer	Basalt Carbon Steel
D'Antino and Papanicolaou [28]	Tixotropic fiber-reinforced cement-based matrix Lime-based mortar with silica sand Lime-based mortar with pozzolanic binders, synthetic fibers and graded sand	16.40 12.10 10.30		Carbon Basalt Glass Steel
De Felice et al. [29]	Pozzolanic-cement mortar Pozzolan lime-based mortar Cementitious mortar	37.0 13.0 22.6		Steel Carbon Basalt
Lignola et al. [30]	Cement based Lime based	25 12 15 15		Basalt
Larrinaga et al. [31]	Cement-based mortar	19.8		Basalt
Hojdys and Krajewski [27]	Cement based one-component mortar Cement-free NHL-based one-component mortar Natural hydraulic lime-based mortar Cement-based mortar	15.4 14.9 9.8 44.3	Synthetic fibers Fiber-reinforced	Carbon Glass PBO
Beßling et al. [32]	UHPC HPC	83 ± 7 60 ± 5		Carbon Basalt
Zhou et al. [33]	HPC	76.7	Steel	Carbon

Among the research works presented above, in those of Beßling et al. [32] and Zhou et al. [33] similar solutions were used for the cementitious matrices of the tested specimens. In fact, Beßling et al. [32] used HPC and UHPC cementitious matrix and Zhou et al. [33] adopted a high-performed fine-grained mortar admixed with short steel fibers.

The inorganic matrices used by the authors differ in their composition and the presence or absence of short fibers, resulting in different mechanical performance, affecting aspects such as the compressive strength and the tensile strength. The textile fabrics differ from each other in the material of the constituent fibers, the shape of the rovings, the warping and the presence of a pre-coating. All of these aspects affect the behavior and performance of TRC, leading to each system being considered as unique.

2. Research Method, Content and Significance

By considering the state of the art in the TRC retrofit, the combination of high-performance concrete (HPC) admixed with short steel fibers with basalt and carbon textiles adopted in this work is not yet sufficiently investigated. For this reason, because the materials adopted here are commercial products, this work is an important contribution to the research and innovation of TRC and F/TRC. This is also a preliminary step of a larger experimental investigation on applications of these TRC and F/TRC systems on RC structural elements [34,35]. The results will be applied to refine in future research works the analytical formulations used to calculate the increase in performance of structural elements retrofitted with these TRC or F/TRC systems.

Finally, the main purpose of this study is to determine the tensile behavior and the mode of failure of these composite strengthening layers. The variables investigated are the material of the textile fabric (basalt and carbon), the influence of admixing short steel fibers in the cementitious matrix and the presence of an overlap of the textile fabric in the specimen. A total of 24 uniaxial tensile tests were conducted on specimens of Carbon-TRC (C-TRC), Basalt-TRC (B-TRC), Carbon-F/TRC (C-F/TRC) and Basalt-F/TRC (B-F/TRC) at the Science & Energy Labs of the Carinthia University of Applied Sciences in Villach (Austria).

3. Materials and Methods

The cementitious matrix of the specimens was obtained as a fine-grained premix approved in Germany to be used as a cementitious matrix in TRC retrofit [36]. Short steel fibers with a length l of 5 mm and a diameter d of 0.15 mm were admixed in the cementitious matrix in order to be used in F/TRC. An amount of 2.5 vol.-% of short steel fibers was admixed in the cementitious matrix. Through compressive tests on $100 \times 100 \times 100$ mm^3 cubes and a splitting tensile test on 100×200 mm^2 cylinders, the compressive and splitting tensile strength f_c and f_t of the cementitious matrix are approximately 93.6 and 3.6 MPa, respectively. In the fiber-reinforced matrix, the compressive and tensile strengths are, respectively, 105.2 and 10.9 MPa. The following geometrical and mechanical properties of the textile fabrics are declared by the producers [37–39]. The grid opening g of the basalt textile fabric is 20×20 mm^2 and the cross sectional area A per meter is 65 mm^2/m. The elastic modulus E, the tensile strength $\sigma_{t,tex}$ and the strain at the maximum load ε_u of basalt textile fabric are 92.7 GPa, 1495 MPa and 1.61% in both directions, respectively. The carbon textile fabric is characterized by a grid opening g of 22×22 mm^2, and the cross-sectional area A per meter is 71 mm^2/m. The tensile strength $\sigma_{t,tex}$ and strain at the maximum load are 2531 MPa and 1.71% lengthways and 2841 MPa and 1.47% crosswise, respectively. The properties of the cementitious matrices and the textile fabrics are presented in Table 2 and in Table 3, respectively.

Table 2. Mechanical properties of the cementitious matrices.

	f_c (MPa)	f_t (MPa)
Cement matrix with short steel fibers	105.2	10.9
Cement matrix without short steel fibers	93.6	3.6

Table 3. Mechanical and geometrical properties of textile fabrics.

	g (mm)	A (mm^2/m)	$\sigma_{t,tex}$ Lengthways (MPa)	$\sigma_{t,tex}$ Crossways (MPa)	ε_u Lengthways	ε_u Crossways	E (GPa)
Carbon textile fabric	22	71	2531	2841	1.71%	1.47%	
Basalt textile fabric	20	65	1495	1495	1.61%	1.61%	92.7

The experimental test campaign included 24 specimens divided into two series: A and B. Series A concerns specimens composed of one layer of basalt or carbon TRC and F/TRC to be tested under the uniaxial tensile test. The specimens are 120 mm wide, 15 mm thick and 600 mm long. The textile fabric is placed halfway through the thickness of the specimen (Figure 1). The design of the specimens is in accordance with the recommendation of RILEM TC 232-TDT [24]. The aim of testing specimens of series B is to investigate the resistance of the textile fabric overlap in one layer of TRC or F/TRC. In this series, the specimen's dimensions differ from those of series A in terms of length (810 mm for the specimen with basalt textile fabric and 710 mm with carbon textile fabric) and thickness (on carbon- and basalt-retrofitted specimens, the thickness is close to 18 and 16 mm, respectively). The overlap length is 150 mm and 250 mm on carbon and basalt textile fabric, respectively. Two pieces of textile fabrics, overlapped in the center of the specimens (see Figure 1), are placed in each specimen. The overlap is realized with two pieces of textile 430 mm and 530 mm long, overlapped in order to obtain an overlap length of 150 mm and 250 mm on carbon and basalt textile fabric, respectively. These values refer to the overlap length suggested by the producers, which is increased by 50 mm in order to avoid premature failure in the overlap zone. The reinforcement ratio $\rho = A_{textile}/A_{matrix}$ is 0.5% for C-TRC and C-F/TRC and 0.4% for B-TRC and B-F/TRC specimens (Figure 1).

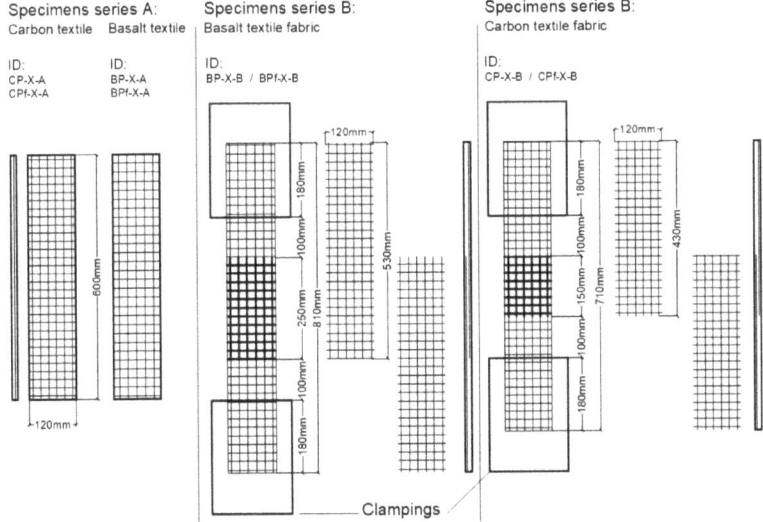

Figure 1. Details of the design for specimens of series A and series B.

The first step for the preparation of the specimens is the realization of the wooden formworks, which are 6 mm-thick rectangular frames suitable for the realization of a 6 mm-thick homogeneous layer of cementitious matrix, as shown in Figure 2a. The next step is to apply the textile fabric to the fresh matrix and fix it on the two opposite sides to the formwork in order to keep it tight and maintain the position during the application of the last layer of cementitious matrix. Finally, the last 6 mm-thick layer of cementitious matrix is applied as a cover (Figure 2b–e). The thickness of the textile and the clips used to fix it is approximately 3 mm. The fabrication process for specimens of series B differs only during the phase of application of the textile fabric. Two pieces of textile fabric are overlapped: once the first 6 mm thick layer of cementitious matrix is applied, the first piece of textile fabric is placed and fixed along the lateral sticks of wood with clips; subsequently, wood sticks are placed laterally to the specimen in order to act as a guide for the application of a 2 mm-thick layer of cementitious matrix; then, the last piece of textile fabric is applied and fixed. Finally, the last 6 mm of cementitious matrix is applied as a cover.

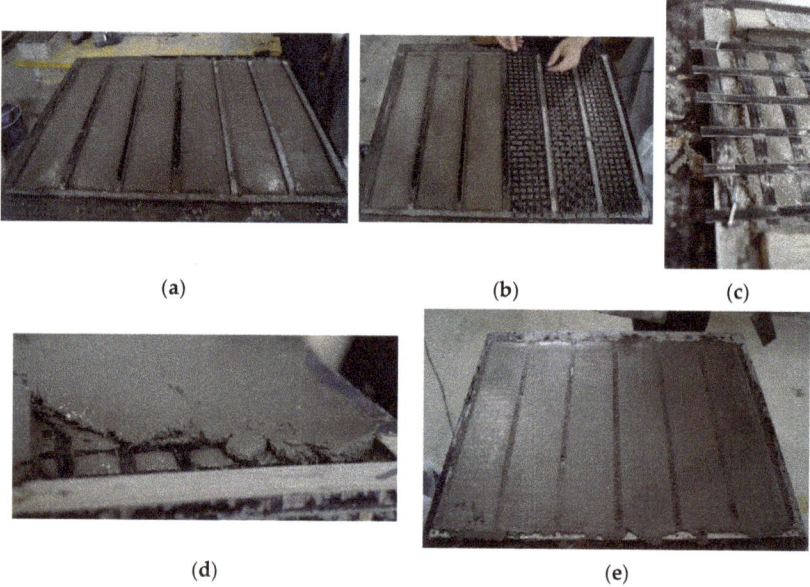

Figure 2. Details of the design for specimens of series A and series B. In (**a**), the first layer of cementitious matrix is applied; in (**b**), the basalt textile fabric is placed over the cementitious matrix; in (**c**), is shown a detail of how textile fabric is fixed on the wooden formwork; in (**d**), the last layer of cementitious matrix is applied; and in (**e**) is shown the specimens when the production process is concluded.

The test setup, shown in Figure 3a, refers to the recommendations of RILEM TC 232-TDT [24], Digital Image Correlation system was used to measure strains instead of the traditional strain transducers. To perform the tests, the two ends of the specimens are placed inside two steel clamps of the test machine (Figure 3b). The load is applied on specimens through friction by tightening the six bolts present for each clamp. A thin layer of rubber is placed between the steel clamps and the specimen to improve the adherence between steel and the cementitious matrix.

The steel clamps are 180 mm long, resulting in a free length of 240 mm on uniaxial specimens of series A. On specimens of series B, the free length is 350 and 450 mm for specimens with carbon and basalt textile fabric, respectively. The age of the cement-based matrix during the test phase is between 211 and 224 days on specimens with matrix not admixed with short steel fibers, and between 240 and 252 days for specimens with short steel fibers admixed in the matrix.

Specimens are aligned with the jaws in order to apply pure tension to the specimen. Tests are performed under displacement control with a speed of 0.5 mm/min. The tests are also measured with the Digital Image Correlation system (DIC), which is appropriate to survey the cracking behavior. Furthermore, the axial displacement of the machine was measured during the tests through the DIC. Through the analysis with the DIC software it is possible to select two points in the chosen area and to measure the displacement of these points during the test. These points together are referred to in this article as the "virtual extensometer". The axial displacement imposed by the test machine during the test is measured through a virtual extensometer by placing two points in the middle of the two steel clamps, one on the top and one on the bottom of the test machine. The measurement of crack opening and deformation are obtained by placing three virtual extensometers across the crack that propagate in the specimen. Therefore, the two virtual points are placed on the right and on the left of the measured crack. Crack deformation values are obtained by dividing the crack displacement by the initial displacement.

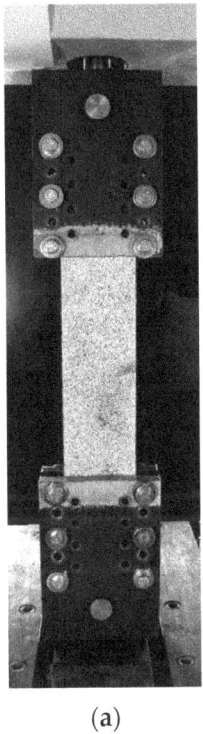

(a)

(b)

Figure 3. Specimen clamped in the testing machine (a), and a detail of the steel clamp (b).

4. Results

Table 4 summarizes the main results of the tests: maximum load, equivalent yielding displacement δ_y, ultimate displacement δ_u and index of inelastic displacement μ. The index of inelastic displacement is the ratio between the ultimate displacement and the equivalent yielding displacement, which expresses the capacity of the specimens to undergo increasing displacement after the first elastic branch to a certain extent before failing. The ultimate displacement is considered here as the displacement corresponding to a decrease of 20% in the ultimate load of the descendant curve [40]. The equivalent yielding displacement corresponds to the displacement occurring at the first transversal crack. In Table 4, values of the coefficient of variation (CV) are listed. While values of maximum load are characterized by low CV, the index of inelastic displacement is generally characterized by a higher CV. Firstly, μ is the ratio between δ_u and δ_y; hence, phenomena that influence these two parameters are taken into consideration. For TRC specimens in particular, the values of δ_y are low; therefore, even slight variations in δ_y have a significant impact on μ values and may increase the CV of μ. The study of crack propagation in the specimens is limited by the fact that, during the tests, the test machine steel clamps cover the ends of the specimens. These areas are not measured by the DIC systems; thus, the propagation of cracks cannot be observed there. The propagation of cracks within the steel clamps would affect the total displacement capacity of the specimens, and consequently the ultimate displacement δ_u

and μ. Finally, minimal geometric imperfections of the textile fabric and imprecision during the specimens' production process may increase the variability of the test results.

Table 4. Test results expressed in terms of average maximum load, equivalent yielding displacement, ultimate displacement and index of inelastic displacement.

Specimen ID	Maximum Load (N)	Average Maximum Load (N) (Stress (MPa))	CV Max Load	δ_y (mm)	δ_u (mm)	μ	μ Average	CV μ
BP-1-A	6364.5	6178.3 (792.1)	0.04	0.029	2.8	96.6	71.0	0.45
BP-2-A	6253.4			0.047	3.8	80.9		
BP-3-A	5916.9			0.090	3.2	35.6		
BPf-1-A	11,231	11,308 (1449.7)	0.01	0.65	2.2	3.38	4.4	0.21
BPf-2-A	11,386			0.45	2.2	4.9		
BPf-3-A	11,306			0.38	1.9	5		
CP-1-A	11,107	10,296.7 (1311.7)	0.07	0.06	3.76	62.7	71.5	0.40
CP-2-A	10,135			0.04	4.14	103.5		
CP-3-A	9648			0.07	3.38	48.3		
CPf-1-A	12,365	12,159 (1548.9)	0.03	0.57	4.43	7.77	5.65	0.38
CPf-2-A	12,315			0.49	2.78	5.67		
CPf-3-A	11,797			0.46	1.61	3.5		
BP-1-B	7848	7289.7 (934.6)	0.11	0.077	1.77	22.96	41.23	0.39
BP-2-B	7610			0.07	3.77	53.85		
BP-3-B	6411			0.07	3.28	46.89		
BPf-1-B	13,455	13,973.7 (1791.5)	0.06	0.78	1.96	2.51	8.46	1.28
BPf-2-B	14,944			0.63	1.22	1.94		
BPf-3-B	13,522			0.077	1.61	20.93		
CP-1-B	8708.9	8905.1 (1134.4)	0.12	0.18	6.37	35.39	18.18	0.84
CP-2-B	10,066			0.82	5.37	6.55		
CP-3-B	7940.6			0.17	2.14	12.59		
CPf-1-B	15,382	15,638.7 (1992.2)	0.02	0.88	1.54	1.75	3.2	0.47
CPf-2-B	15,482			0.69	2.14	3.11		
CPf-3-B	16,052			0.52	2.47	4.74		

The ID used to identify the specimens consists of a code in which the first letter refers to the material of the textile fabric (B for basalt and C for carbon), the second letter refers to the absence or presence of short steel fibers in the cementitious matrix (P is without fibers and Pf is with fibers), the number identifies one of the three identical specimens and the last letter is the name of the series (series A or series B). For example, the ID BPf-3-A refers to a specimen retrofitted with basalt textile embedded in a cement-based matrix admixed with short steel fibers, the third specimen of series A.

4.1. Series A

4.1.1. B-TRC vs. B-F/TRC

Force–displacement and stress–strain curves of B-TRC (in red, in Figure 4a,b) can be simplified as a bilinear curve characterized by a stiff linear first branch and a second branch characterized by a lower stiffness, which lasts until the maximum load. The specimens fail abruptly after the maximum load is reached. The average maximum load for the B-TRC specimens is 6178.3 N. The change in stiffness between the two linear branches is approximately situated at a load of 3000 N in the force–displacement curve. Alongside this changes of stiffness, the first crack appears, followed by a second crack that appears during the development of the second branch. The deformations of the specimens are mainly concentrated in the widening of one crack.

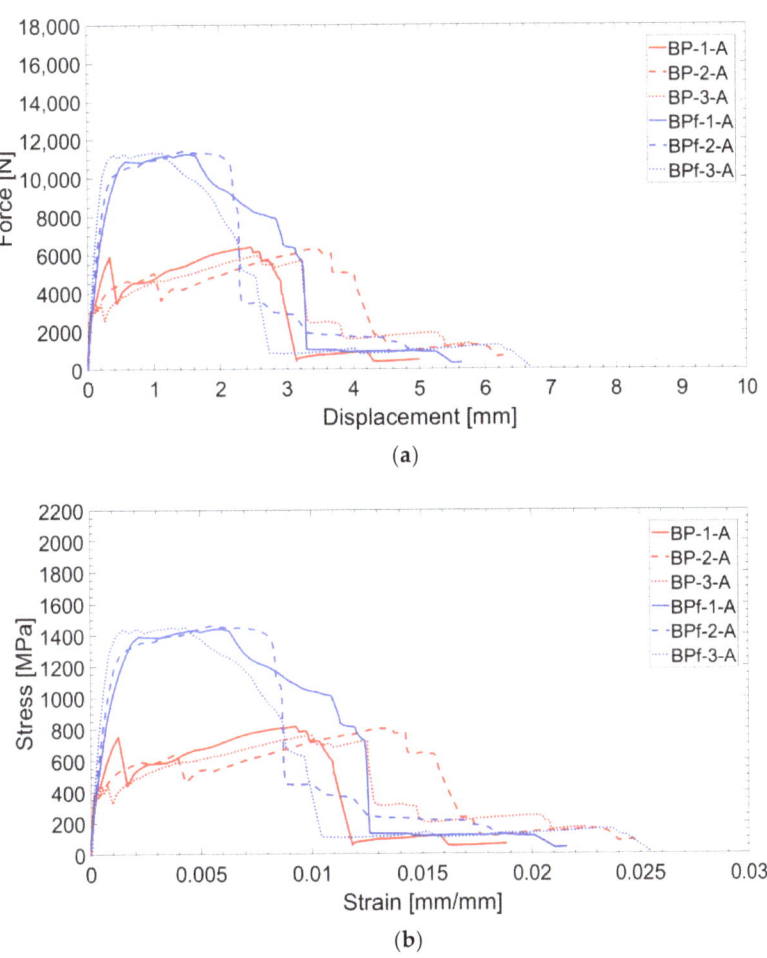

Figure 4. Force–displacement (**a**) and stress–strain (**b**) curves of B-TRC and B-F/TRC specimens.

From Figures 5 and 6, it is possible to observe that, at failure, the main crack propagates from one side to the other of the specimen. In these pictures, colors are scaled to represent the major principal strain values. Figure 7 shows six groups of graphs. In each group, the first two graphs refer to the measurement of the main crack opening, and the third is the force–displacement curve of the specimens during the test. Crack opening is expressed through three curves obtained by three virtual extensometers positioned to measure the crack opening; two are placed at the ends and one in the center of the specimens (see Figure 5). In the legend of Figure 7, E1, E2 and E3 represent the three virtual extensometers. In these graphs, the vertical axis corresponds to the displacement between the two points that define the virtual extensometer, while the horizontal axis corresponds to the axial displacement of the two steel clamps during the test. Finally, one last virtual extensometer is applied on the steel clamps, to measure the displacement imposed by the test machine. The force–displacement curve is obtained by displaying the force measured by the test machine versus the axial displacement measured by the virtual extensometer. In each group, graphs are vertically aligned to observe how the crack opening behaves compared with the force–displacement curves. In line with the maximum load, cracks tends to open abruptly, followed by a sudden decrease in load (see Figure 7a,c,e). From the curves of the virtual extensometers, it can be observed that the main crack tends to widen on one side and to close on the opposite side of the specimen. Therefore, in all specimens, the crack opening is not uniform along the transverse section. This produces specimen rotation, as can be seen in Figure 5.

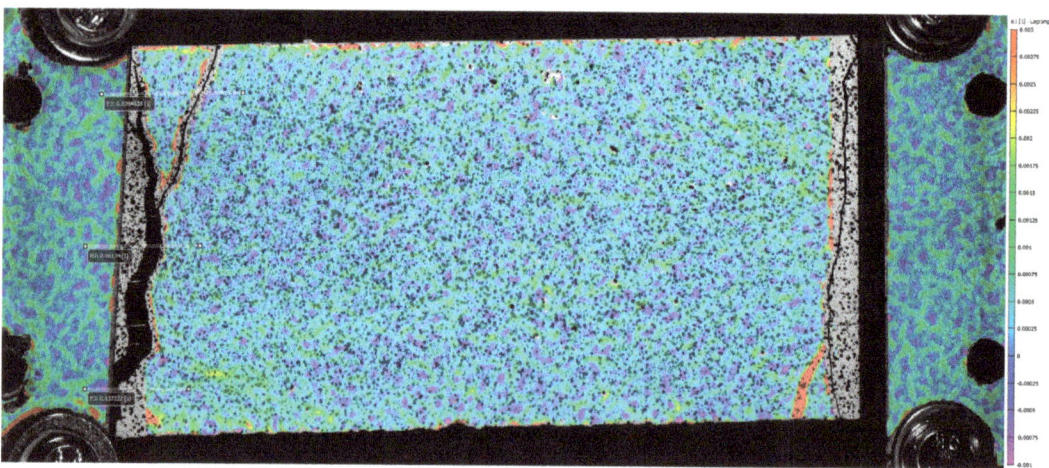

Figure 5. The virtual extensometer on the BP-A specimen. The color scale represents the grade of the major principal strains, from 0.003 units in red to −0.001 units in purple.

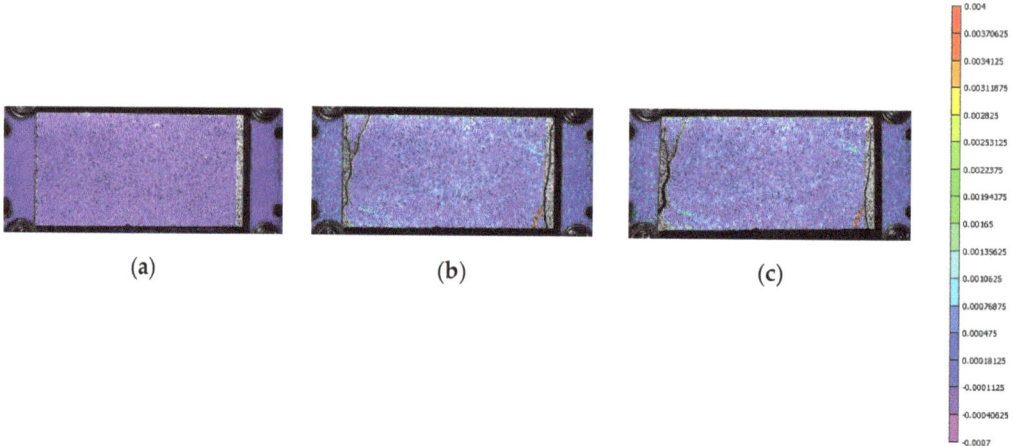

Figure 6. Pictures of specimen BP-2-A at the beginning of the test (**a**), maximum load (**b**) and after the drop in load (**c**). The color scale represents the grade of the major principal strains, from 0.004 units in red to −0.0007 units in purple.

By observing the failure conditions for the basalt textile B-TRC specimens, it appears that the failure of the specimens occurred due to the rupture of the textile after the principal crack (Figure 8a).

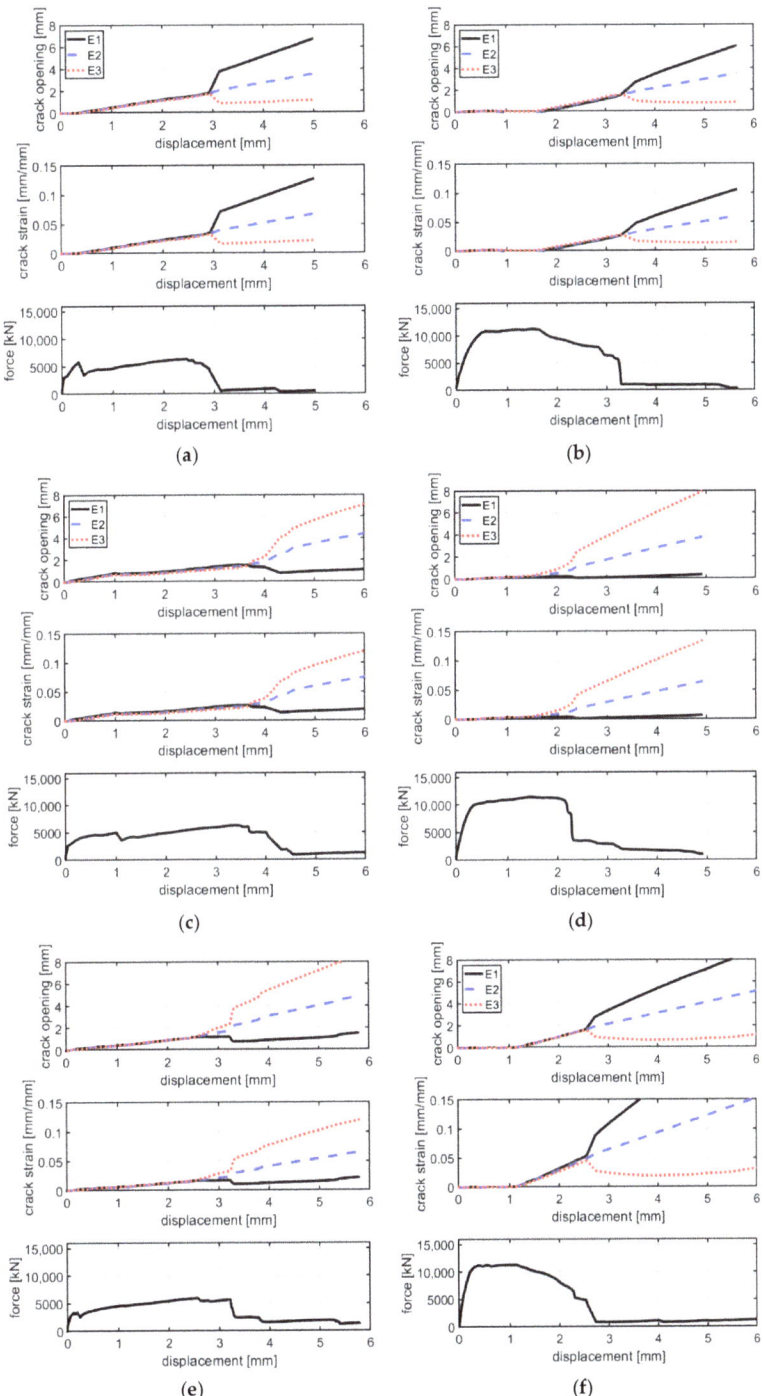

Figure 7. Comparison between crack opening, crack strain and axial force with the axial displacement of specimen BP-1-A (**a**), BP-2-A (**c**), BP-3-A (**e**), BPf-1-A (**b**), BPf-2-A (**d**) and BPf-3-A- (**f**).

Figure 8. Failed specimens: (**a**) BP-1-A specimens, textile rupture; (**b**) main crack on specimen BPf-1-A and detail of the textile at the two ends of the specimen (**c**,**d**).

The test results of B-F/TRC specimens are represented in Figure 4 in terms of force–displacement curves. The behavior of these curves can be simplified into three parts: a first, almost linear behavior, a plateau, and a descendant branch which leads to the failure of the specimen. The first branch is characterized by a linear increase in the load until approximately 9000–10,000 N. Once this phase is reached, a decrease in stiffness follows, leading to a plateau in the force–displacement curves. During the degradation of stiffness and until the end of the plateau, a progressive propagation of transversal cracks develops along the specimens (Figure 9a–c). At the end of the plateau, the progressive drop in load results in the concentration of the imposed displacement in the widening of one principal crack (Figure 9c). During the decrease in load, at approximately 6000 N, the main crack is approximately 2 mm wide and expands suddenly at failure. During the failure of the specimens, it has been observed that the main crack tends to widen more on one side than the other, which results in a "rotation" of the specimen during the failure. This observation is confirmed by the crack opening curves in Figure 7b,d,f, in which the curves of the virtual extensometers show an increase in displacement (or deformation) on one side of the specimen and a decrease in the other side.

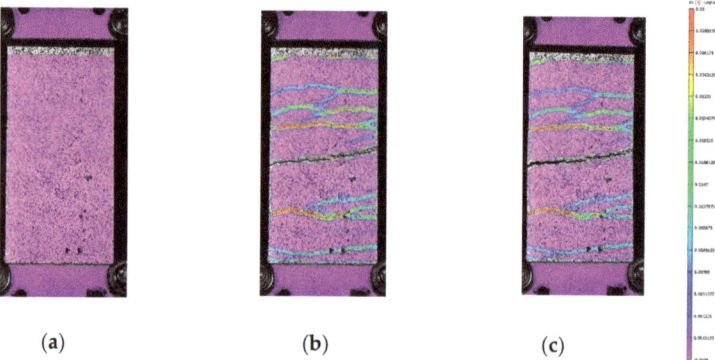

Figure 9. Crack pattern evolution of specimen BPf-3-A (**a**–**c**) during the tests. Pictures are taken at the beginning of the test (**a**), the end of the plateau (**b**) and immediately after the final drop in load (**c**). The scale of color represents the grade of the major principal strains, from 0.03 units in red to −0.0007 units in purple.

The mode of failure of B-F/TRC specimens is the same as that of B-TRC specimens: rupture of the basalt textile fabric. The possibility of the pull-out of the textile fabric is excluded by observing the failed specimens. Figure 8b–d show the main crack of specimen BPf-1-A and the details of the textile at the two ends, showing no apparent slippage of the rovings.

4.1.2. C-TRC vs. C-F/TRC

The results of the tests are represented by force–displacements and stress–strain curves in Figure 10a,b.

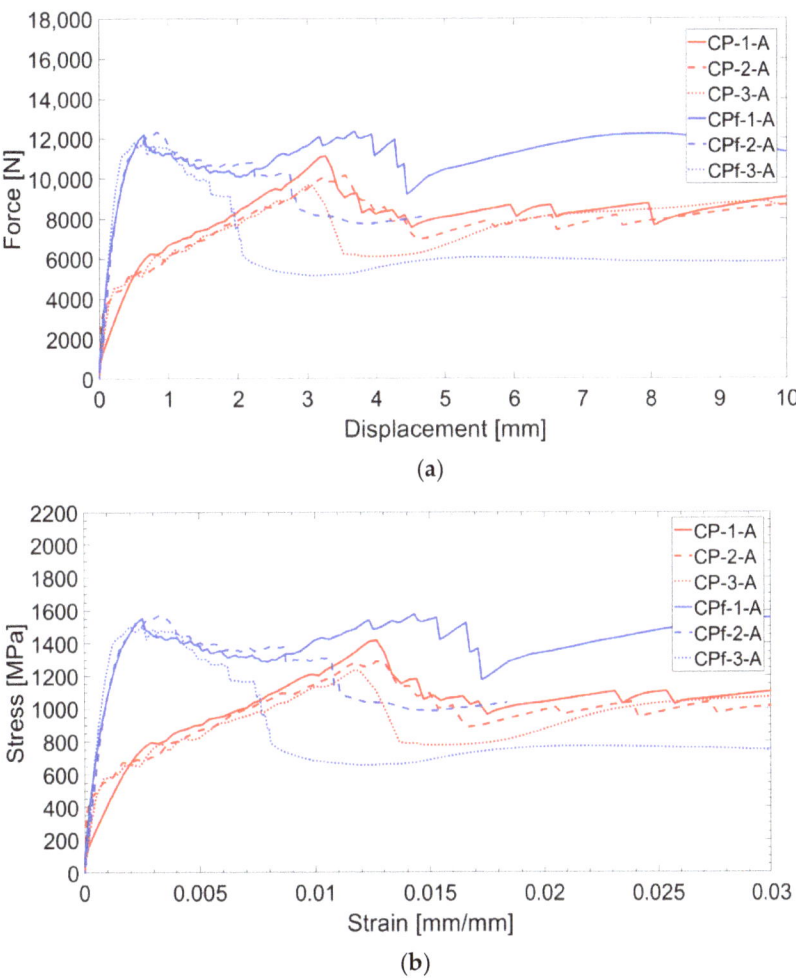

Figure 10. Force–displacement (a) and stress–strain (b) curves of C-TRC (in red) and C-F/TRC specimens (in blue).

The first stage is characterized by high stiffness, governed by the interaction between the matrix and the textile fabric. This first linear behavior lasts until the load reaches approximately 4000–5000 N, where the first transversal crack appears on the cementitious matrix. The second stage is characterized by lower stiffness and a behavior which could be approximated as linear in the first instance. During the increase in load, the displacement imposed in the test is mainly concentrated in the widening of the crack. Once the maximum load is reached, a drop in load of approximately 2000–3000 N characterizes the third stage. This leads to the fourth stage, in which the load slightly increases asymptotically to approximately 8000 N, resulting in a pull-out failure. This is confirmed by checking the details of the cracks in Figure 11 and the graphs in Figure 12a–c, which show that cracks, at the last stage of the force–displacement behavior, tend to open continuously at almost constant load. This is expressed by the crack opening curves in which the virtual extensometer, after the drop in load, proceeds almost linearly.

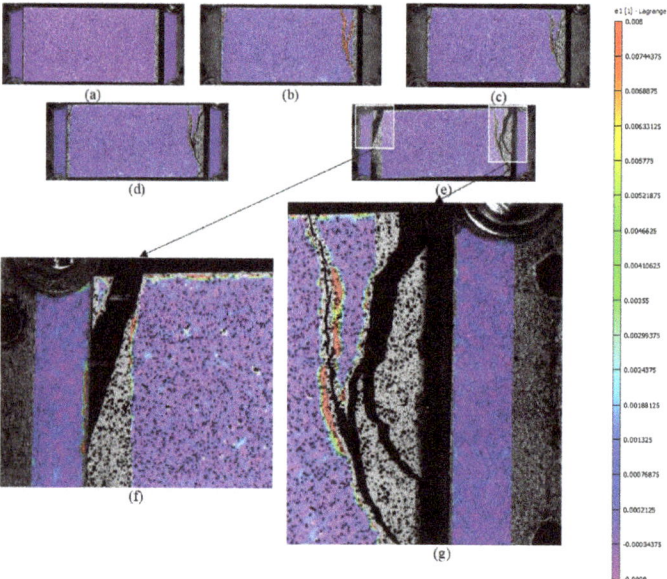

Figure 11. Pictures of specimen CP-1-A from DIC measurement during the various phases of the test: beginning of the test (**a**), appearance of the first crack (**b**), immediately before and after the first load drop (**c**,**d**), and when the test finished (**e**). Figures (**f**,**g**) are details of the cracks on the specimen. The color scale represents the grade of the major principal strains, from 0.008 units in red to −0.0009 units in purple.

The mode of failure of C-F/TRC specimens is governed by the pull-out of the textile fabric from the cementitious matrix; however, the shape of the force–displacement curve slightly differs from that of the C-TRC specimens in the first and second stage. As shown in Figure 10, the maximum load is generally reached in the first stage, while the second branch is descendant. The average maximum load is 12,159 N. In close proximity to the maximum load, a first transversal crack appears, whereas the stiffness is reducing. This event leads to the second stage, which is characterized by a decrease in the load and a progressive propagation of transversal cracks along the specimen. The propagation of transversal cracks along the specimen is shown in Figure 13. In this phase, the presence of the short steel fibers affects the performance of the layer by reaching higher loads in the various stages and by distributing stresses along the specimens through the effect of the short fibers. In the third stage, the drop in load also constitutes a quick widening of the crack, leading to the last part of the graph: an almost linear behavior in which the load slightly increases due to the friction between the rovings of the textile fabric and the cementitious matrix. This is also observed in Figure 12d–f: once the maximum load is reached, the displacement measured by the virtual extensometers increases almost linearly until the end of the test.

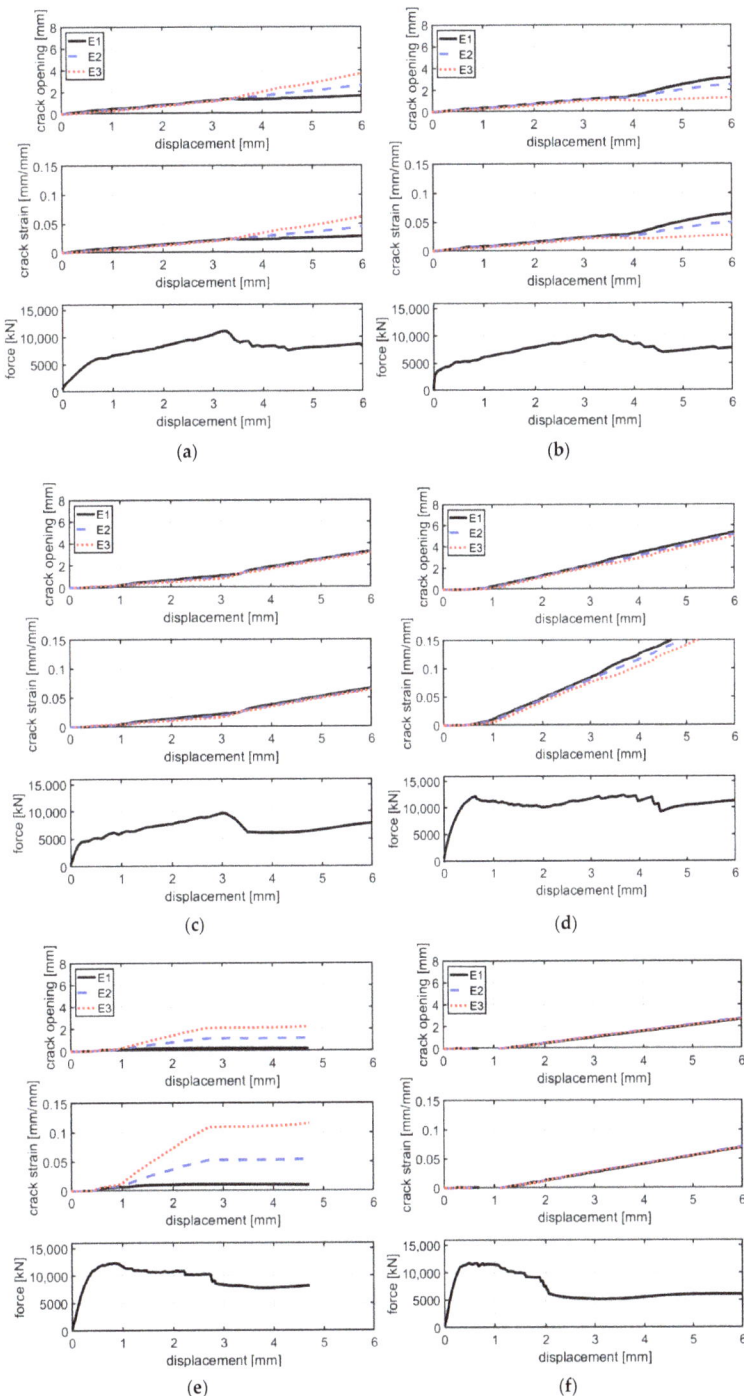

Figure 12. Crack opening/crack deformation/force–displacement curves for specimens CP-1-A (**a**), CP-2-A (**b**), CP-3-A (**c**), CPf-1-A (**d**), CPI-A (**e**) and CPf-3-A (**f**).

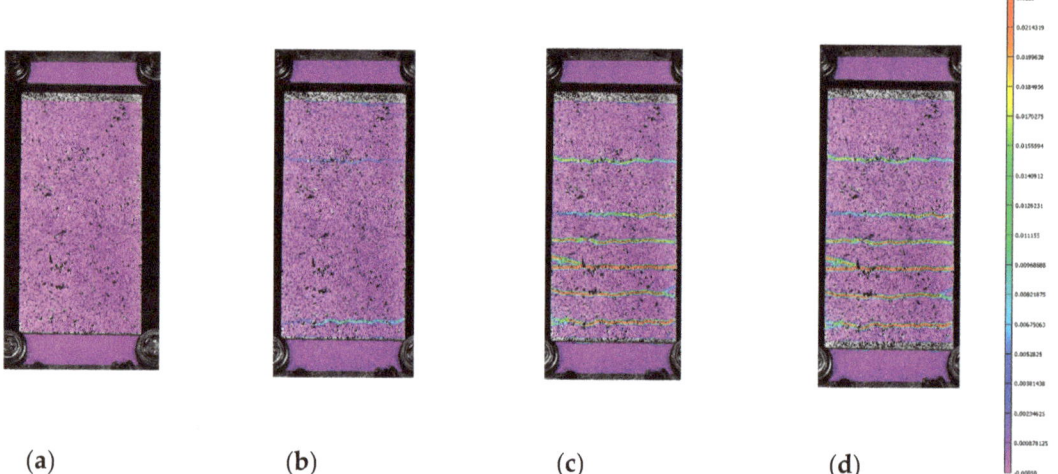

Figure 13. Frames of the DIC measurement of C-F/TRC specimen 3 at the beginning (**a**), the first crack (**b**), the end of stage 3 (**c**) and at the beginning of stage 4 (**d**). The color scale represents the grade of the major principal strains, from 0.0229 units in red to −0.00059 units in purple.

4.2. Series B

4.2.1. B-TRC vs. B-F/TRC

The test results of the B-TRC specimens of series B are represented by force–displacement and stress–strain curves in Figure 14a,b.

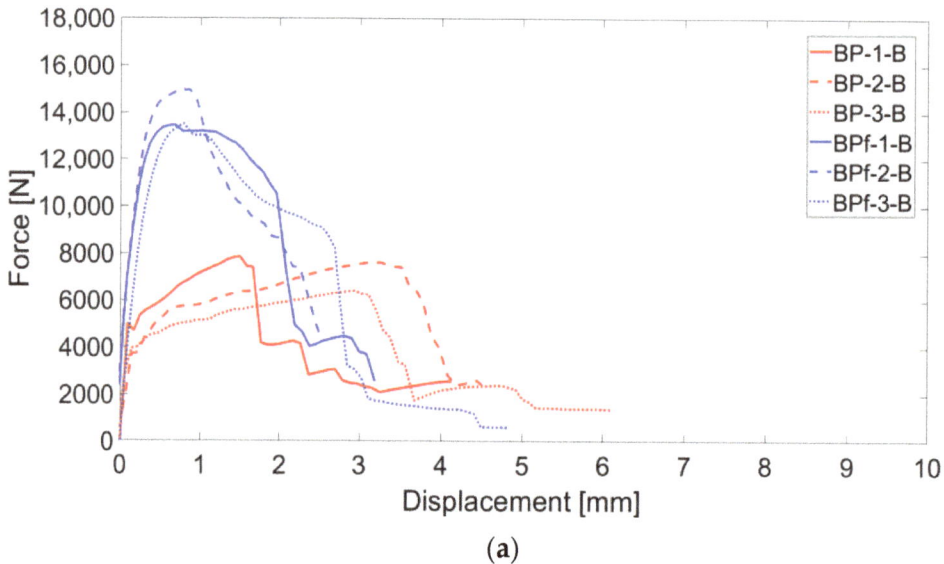

Figure 14. *Cont.*

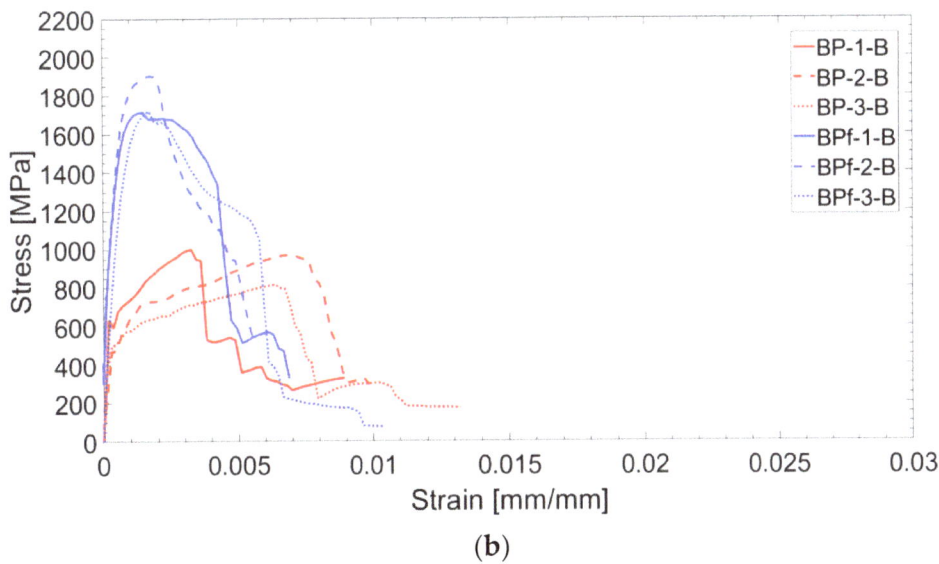

Figure 14. Force–displacement (**a**) and stress-strain (**b**) curves of B-TRC specimens (in red) and B-F/TRC specimens (in blue).

Generally, the behavior of these tests is similar to that of series A. In fact, the force–displacement curves are characterized by a similar bilinear behavior. The linear branch lasts until approximately 4000–5000 N. At the end of the first linear branch, the first crack appears and there is a sudden decrease in stiffness. In the second branch, the load increases—with a reduced stiffness—until the maximum is reached. The average maximum load is 7290 N, which is 18% higher compared with that of series A. The failure mode of specimens of series B is the same of those of series A: rupture of the textile fabric. Once the maximum load is reached, the force instantly decreases and the principal crack simultaneously widens. Regarding the mode of failure of B-TRC specimens in series A, the propagation of the principal crack starts from one side and propagates transversally through the specimen. In Figure 15, a comparison between the force–displacement curves and the crack opening–axial displacement of the main crack is presented for each B-TRC and B-F/TRC specimen. The behavior of the virtual extensometer's curves is similar to that of series A for both B-TRC and B-F/TRC specimens. Once the maximum load is reached, the displacement measured on one side of the specimen starts to increase, while on the other side it tends to decrease, resulting in a slight rotation of the specimens around the main crack.

The behavior of the force–displacement curves of B-F/TRC specimens of series B is similar to that of series A in terms of average maximum load and shape of the curves. Four stages characterize the force–displacement curves. Firstly, a linear branch lasts until 10,000–11,000 N. Due to the formation of the first transversal cracks, the stiffness decreases, leading to the second branch in which the load reaches its maximum values and begins to decrease. The third stage, related to the decrease in the load coincides with the widening of the principal crack. During the widening of the principal crack, there is the pull-out of the short steel fibers. During this phase, the load tends to decrease and to intercept the curves of B-TRC specimens. At approximately 7000–8000 N, the curves tend to intercept those of B-TRC specimens, leading to a final drop in load corresponding to the rupture of the basalt textile fabric. The mode of failure is still characterized by the non-homogeneous propagation of the crack.

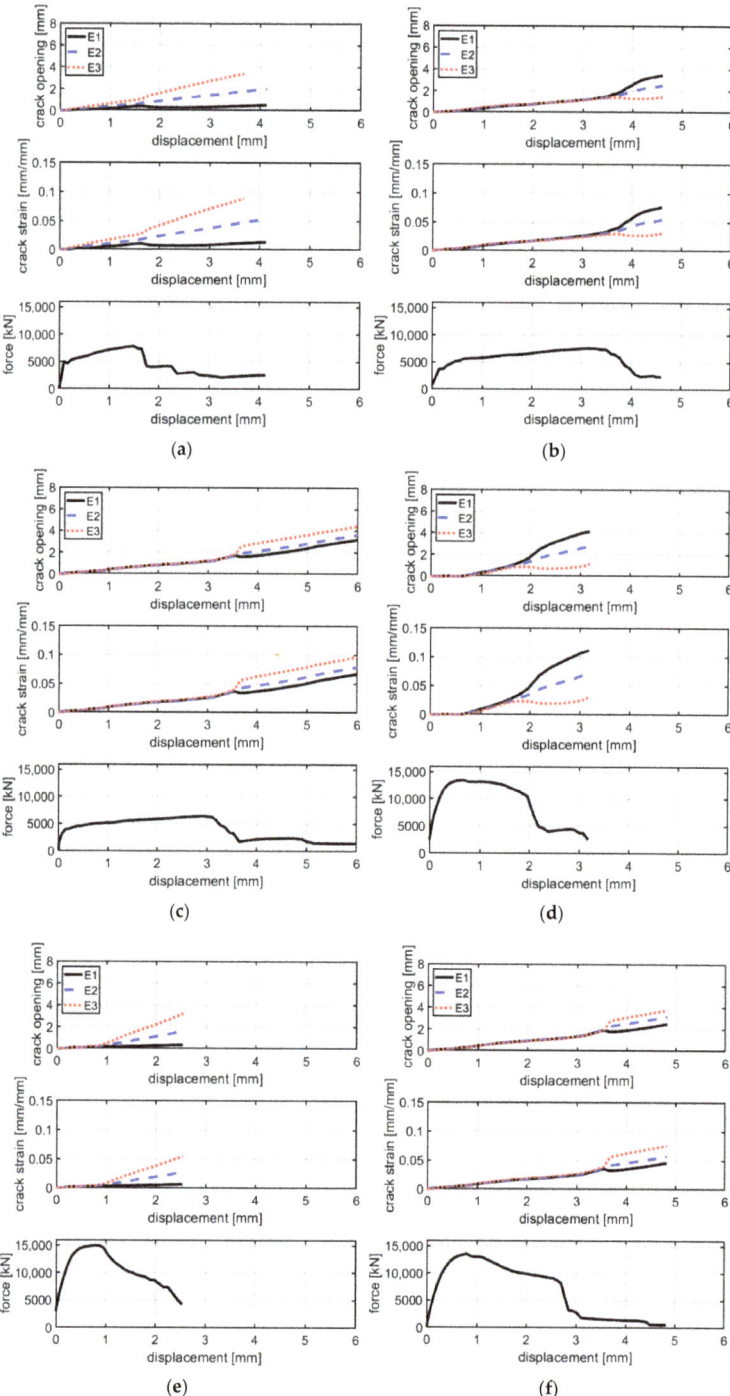

Figure 15. Graphs of crack opening/crack deformation/force-axial displacement of specimens BP-1-B (a), BP-2-B (b), BP-3-B (c), BPf-1-B (d), BPf-2-B (e) and BPf-3-b (f).

4.2.2. C-TRC vs. C-F/TRC

In terms of force–displacement and stress–strain curves, the behavior of the C-TRC specimens of series B, as shown in Figure 16a,b, is characterized by a first linear step, which lasts until approximately 4000–5000 N. Corresponding to the appearance of the first crack, the stiffness starts to decrease until the axial displacement is 2.5 mm and the load is approximately 8000 N. A drop in load precedes the last stage of the curves, corresponding to the tendency of the curves to have large displacements under a reduced increase in loads. The mode of failure of these specimens is similar to that of specimens of series A: the pull-out of textile fabric from the cementitious matrix. It can be seen from the results of these tests that the force–displacement behavior is slightly different compared with that of specimens of series A.

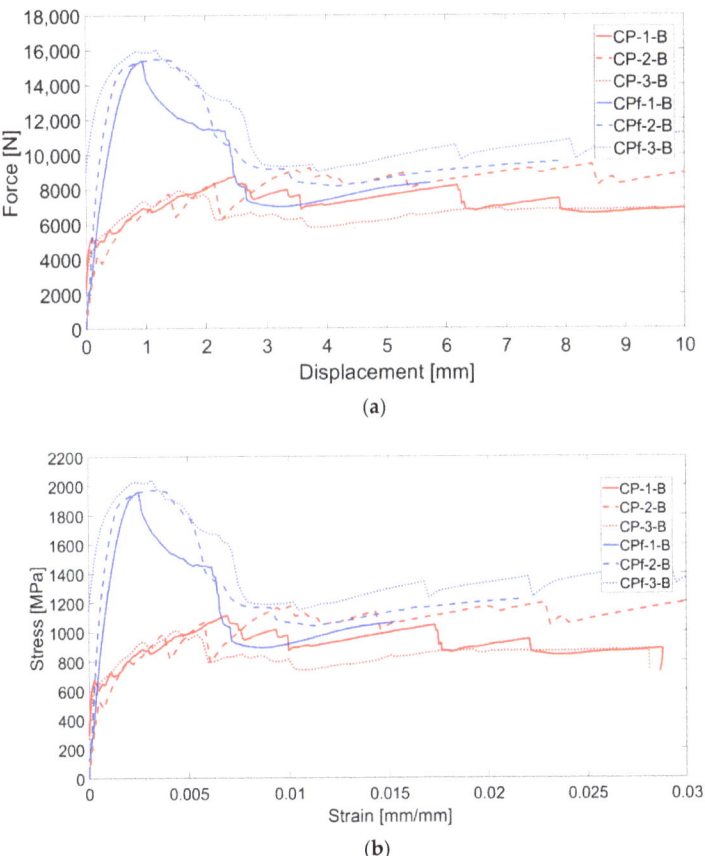

Figure 16. Force–displacement (**a**) and stress-strain (**b**) curves of C-TRC specimens (in red) and C-F/TRC specimens (in blue).

In Figure 17, a comparison between the force–displacement curves and the crack opening–axial displacement of the main crack is presented for each C-TRC specimen of series B. In this series, specimens are characterized by two main cracks. Therefore, several crack opening and crack deformation graphs are presented. Similarly to what happens in series A, the displacement measured by the virtual extensometer tends to increase with similar behavior, which is consistent with the pull-out of the textile fabric from the cementitious matrix. The two main cracks influence each other's behavior; when the rate of widening of the first crack decreases, that of the second one increases and vice versa.

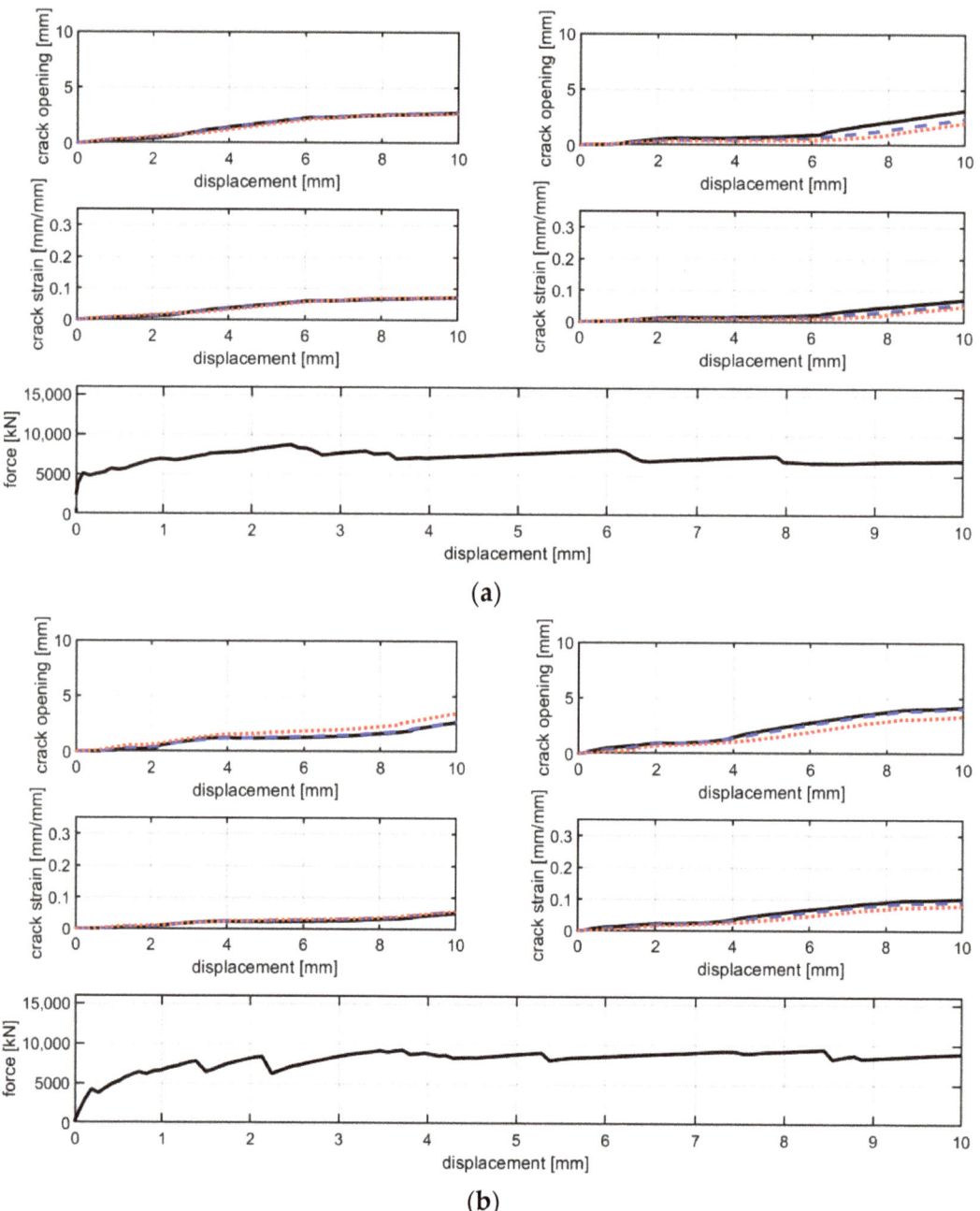

Figure 17. *Cont.*

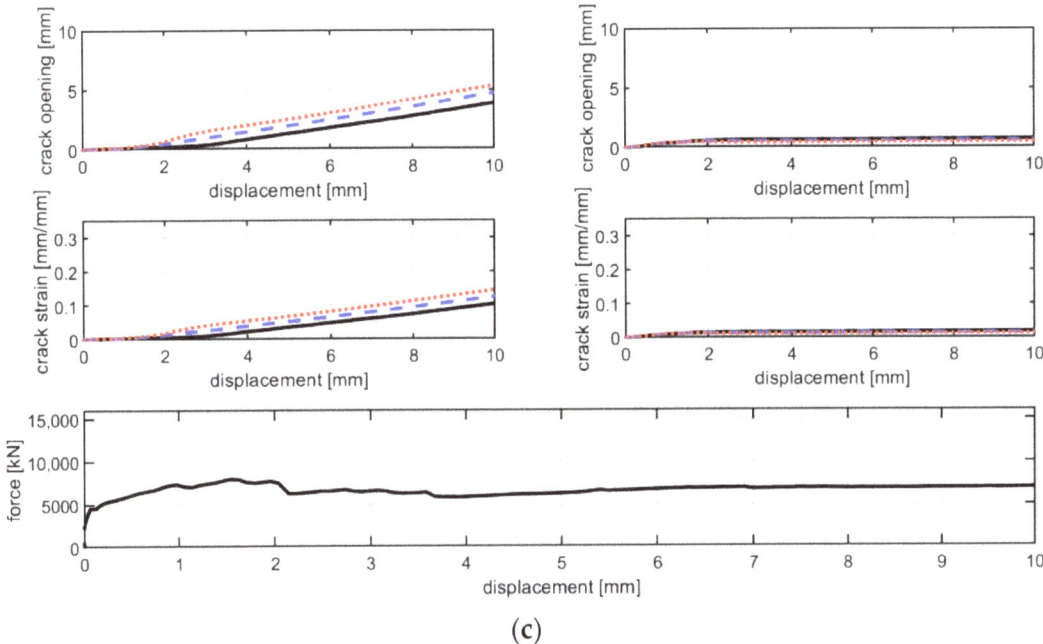

Figure 17. Graphs of crack opening/crack deformation/force-axial displacement of specimens CP-1-B (**a**), CP-2-B (**b**) and CP-3-B (**c**).

The force–displacement curves of C-F/TRC specimens of series B are characterized by behavior similar to that of series A. A first branch, which lasts until approximately 14,000 N, corresponds to the formation of the first cracks and the consequent decrease in stiffness. This leads to the maximum of the curve, in which the maximum load is reached. The maximum load is higher in these series of tests since specimens are slightly thicker compared with those of series A: approximately 18 mm instead of 15 mm (20% higher). Consequently, in the next stage, the load decreases until approximately 12,000–13,000 N. What follows is the loss of the bond between the textile fabric and the cementitious matrix, resulting in a drop in load of approximately 4000 N. The last stage of the curves is related to the pull-out of the rovings from the cementitious matrix. In Figure 18, during the second stage, at maximum load it can be observed that the principal cracks tends to widen. There, the crack proceeds by widening quite homogenously until the drop in load of the fourth stage. This is observable in the crack opening curves, where the displacement measured by the virtual extensometer increases similarly in all three curves.

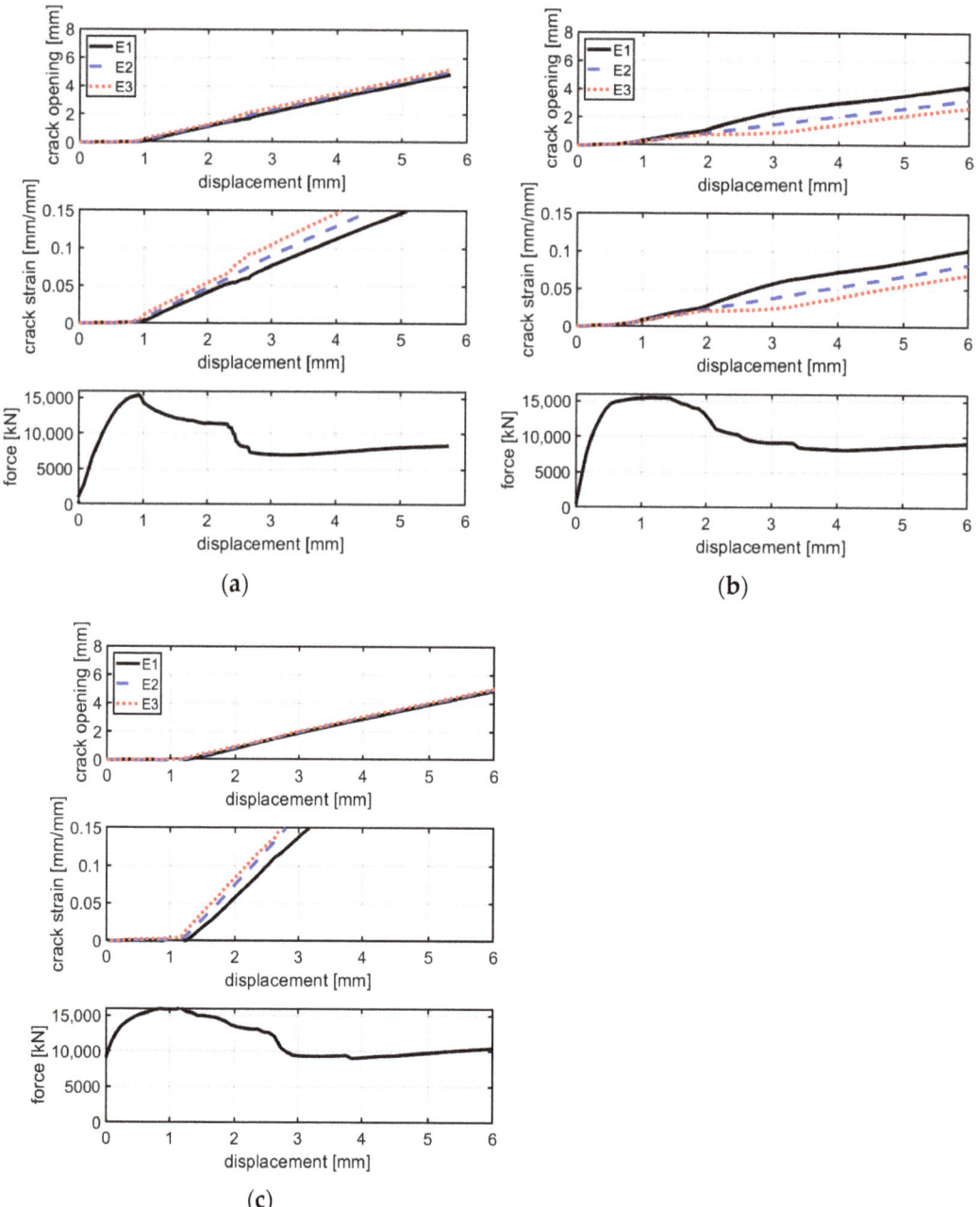

Figure 18. Graphs of crack opening/crack deformation/force-axial displacement of specimens CPf-1 (**a**), CPf-2 (**b**) and CPf-3 (**c**).

5. Discussion

5.1. B-TRC vs. B-F/TRC Series A

The behavior of the B-TRC specimens of series A is characterized by a first linear branch which is influenced by both the matrix and the textile fabric. After the first crack, the load is spread over the basalt textile fabric. During the failure, the principal crack tends to widen non-uniformly: one side of the crack tends to widen more, resulting in a "rotation" of the specimen during the test (Figure 6c). This suggests that the single rovings are not equally loaded at failure. This "rotation" is confirmed by the crack opening–axial displacement behavior (Figure 7a–e). The virtual extensometer applied on the crack shows that on one side the crack widens abruptly, while on the other side it tends to close. By assuming that the activation of the rovings is proportional to the displacement at the failure of the crack, it is possible to estimate the load at failure. For simplicity, a linear stress–strain distribution is considered along the principal crack (Figure 19).

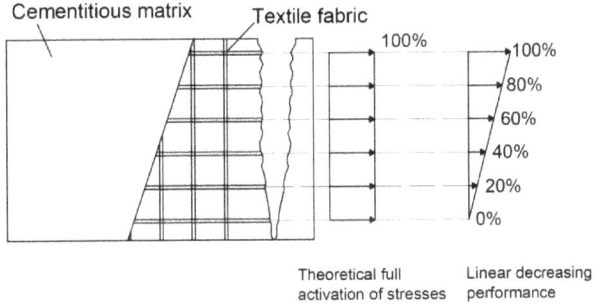

Figure 19. Theoretical and reduced distribution of stresses on basalt textile fabric.

This behavior is based on the assumption that the displacement along the crack develops linearly, where one side is fully activated and the other is not activated. The rovings in the middle are activated proportionally according to the linear behavior. This phenomenon, which leads to a non-uniform distribution of stresses through the textile fabric, may be due to the natural non-uniformity of the cementitious matrix, minimal geometric imperfections of the textile fabric and imprecision during the production process. Considering that the average tensile strength of the roving is approximately 1495 MPa [38] and the cross-section area of one roving is 1.3 mm^2, the theoretical maximum load of six rovings equally activated is approximately 11,661 N. Nevertheless, by assuming an activation of the textile fabric, with a linear activation of the rovings, the maximum load is approximately 5830 N, 5.6% lower compared with the average maximum load of the experimental results, 6178.3 N.

On B-F/TRC specimens of series A, the behavior of the distribution of stresses is affected by the presence of the short steel fibers in the cementitious matrix. Compared with B-TRC specimens, the higher load of the first branch is to be attributed to the short steel fibers admixed in the cementitious matrix, which improves the performance of the cementitious matrix in terms of tensile strength. With a reduction in stiffness the first crack appears, and a pattern of transversal cracks spreads along the specimens. This is a typical behavior related to the short fibers admixed in the matrix. During the propagation of cracks along the specimens, the principal crack, which widens abruptly during the failure of the specimen, seems to open abruptly with a crack width of 1.5–2.0 mm. During the widening of the principal crack, characterized by the pull-out of fibers, the load decreases, intercepting the curves of specimens B-TRC. In this phase, the load is increasingly transmitted to the basalt textile fabric. The load decreases progressively until approximately 6000 N; then, when the force–displacement curves of B-F/TRC specimens intercept the curves of B-TRC specimens, the textile fails due to the rupture of the basalt fibers. At this point, the mode of failure is similar to that of B-TRC specimens and the maximum load is mainly attributed to the presence of the short steel fibers in the cementitious matrix.

Concerning the average index of inelastic displacement μ_{av} in Table 4, it appears that the B-F/TRC specimens show a reduction of approximately 93.8% in μ_{av} compared with specimens of B-TRC. This might suggest that B-F/TRC is characterized by a poor postelastic behavior compared with B-TRC specimens. However, it seems that the force–displacement curves of B-F/TRC specimens tend to intercept those of B-TRC during the descendant branch. This leads us to believe that only considering the value of μ may not fully describe the post elastic capacity of the material. In fact, the failure of both B-TRC and B-F/TRC is related to the rupture of the textile fabric, with similar values of load and axial displacement. This leads to the observation that the actual axial displacement capacity before failure is comparable on both materials.

5.2. C-TRC vs. C-F/TRC Series A

The curves of tests on C-TRC specimens can be simplified by the multilinear curve in Figure 20a, which, apart from the first stage, is comparable to that obtained by Ortlepp et al. [41] from pull-out tests. The first stage of the multilinear curve lasts until the first crack appears and a consequent reduction in stiffness is observed (stage "1"). In the second stage, while reaching the maximum load, the adhesion between the textile fabric and the cementitious matrix is progressively activated and the first transversal crack initially formed is usually followed by the propagation of a second crack (stage "2"). In the third branch, the bond between the carbon textile fabric and the cementitious matrix is lost, which results in a reduction in the load in the force–displacement curve (stage "3"). Finally, in the fourth stage, the load slightly increases asymptotically to approximately 8000 N, resulting from the friction between the textile fabric and the cementitious matrix, leading to the pull-out of the textile (stage "4"). In the fourth stage, as shown in Figure 10, the curves are characterized by small drops in force, probably caused by the manner of widening of the crack: the widening is not uniform along the crack, but it is always more pronounced on one side, and after a small drop, it is more pronounced on the other side, as shown Figure 20b.

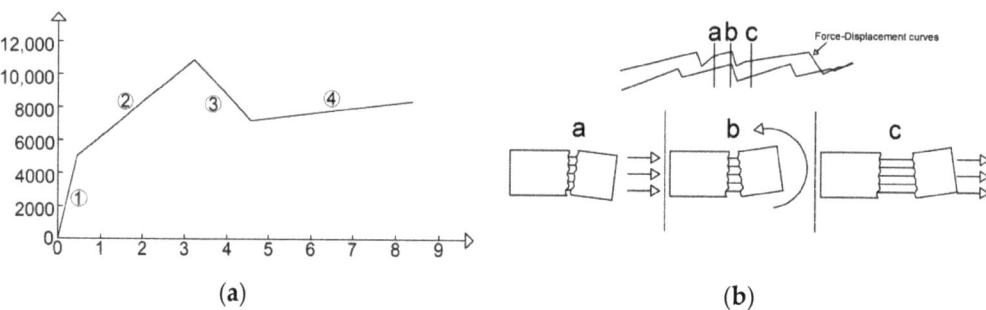

Figure 20. Simplified force–displacement curve of uniaxial tensile test on C-TRC specimens (a) and scheme of crack widening during the drop in load during stage 4 (b).

The force–displacement curves of the C-F/TRC specimens can be simplified and approximated with a multilinear behavior (Figure 21). The multilinear behavior seen in Figures 20 and 21 differs mainly in the first two stages. For C-F/TRC specimens, the maximum load is reached in the last part of the first almost linear stage. While approaching the last part of this branch, the first crack appears, leading to a reduction in stiffness (stage "1"). In the second stage of the curve, the load, differently from C-TRC curves, decreases until approximately 10,000 N. The load reached at the end of the second stage is close to that observed for C-TRC specimens. During this stage, it is assumed that, as for the C-TRC specimens, the textile fabric is progressively activated (stage "2"). Finally, the third and fourth stages are comparable to those of C-TRC specimens. The drop in load is attributed to the loss of bond between the rovings of the textile fabric and the cementitious matrix (stage "3"). The long final linear branch is caused by the friction between the rovings and the cementitious matrix during the pull-out of the textile fabric from the cementitious matrix.

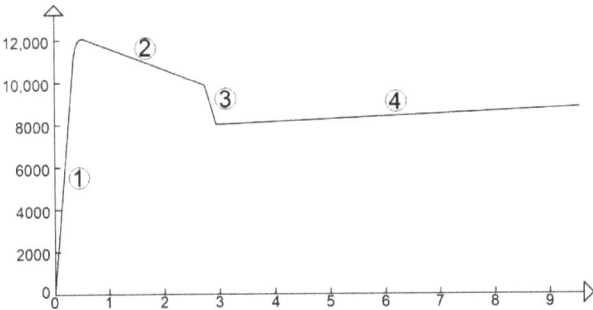

Figure 21. Simplified force–displacement curve of uniaxial tensile test on C-F/TRC specimens. The numbers 1–4 indicate the four stages of the curve.

Figure 12 shows the tendency of the principal cracks to open uniformly along the transversal direction. This information could imply that the contribution of each roving during the widening of the crack is similar.

Due to the definition of the index of inelastic displacement μ given here, the large increase in axial displacement at almost constant load characterized by the pull-out failure is not considered. However, μ expresses how C-TRC specimens realize larger axial displacement between the first crack and a drop of 20% in the maximum load.

Both C-TRC and C-F/TRC specimens failed with the pull-out of the textile fabrics from the matrix. This is a premature failure of the specimens, since the maximum tensile capacity of the carbon rovings has not been exploited. In fact, the maximum load that five rovings could carry is approximately 25,860 N, since the tensile transverse strength is 2841 MPa and the cross-section of each roving is 1.82 mm^2 [37]. The average maximum load reached by the C-TRC and C-F/TRC specimens is 10,296.7 and 12,159 N, respectively, 60% and 53% lower compared with the potential maximum load of the total of the five rovings of the specimens. This type of failure is caused by the insufficient bond between the textile fabric and the matrix, which prevents the rovings from reaching their maximum capacities. As previously mentioned, this eventuality could arise when the textile fabric and the cement-based matrix cannot impregnate the inner fibers of the roving. The degree of penetration of the matrix and the shape of the cross-section of the rovings are salient to the bond connection. Flat rovings, such as that of the basalt textile, have a better distribution of stress [26], while on the contrary, the carbon rovings adopted here are characterized by a circular-cross section, which possibly leads to a worse penetration of the matrix. However, it must be pointed out that the higher declared tensile strength of the carbon textile fabric, compared with that of the basalt textile fabric, may also have contributed to making pull-out failure the main mode of failure.

The considerations above could explain the different modes of failure between the specimens with carbon and basalt textile fabric.

5.3. B-TRC vs. B-F/TRC Series B

B-TRC and B-F/TRC specimens of both series A and B are characterized by similar force–displacement behavior. This may be due to the fact that the overlap length of specimens of series B is long enough to avoid premature failure and leads to the same mode of failure: rupture of the basalt textile. However, the average maximum load of specimens of series B is higher compared with that of series A.

On B-TRC specimens of series B, the difference between the experimental average maximum load and the hypothesized linear distribution is approximately 20%. This high difference may be attributed to the tendency of cracks to open more uniformly compared with those of series A. This would lead to a slightly better distribution of stresses between the rovings at failure.

5.4. C-TRC vs. C-F/TRC Series B

The different behavior between C-TRC specimens of series A and B could be explained by the formation of cracks closer to the extremity of the specimens. Specimens of series B are characterized by cracks that are situated at approximately 150 mm from the edge, while for series A this length is bigger, being 177 mm on average. This leads to a bond length that is approximately 20% shorter. This could be explained by the fact that due to the presence of the overlap, cracks are formed closer to the edges of the specimens. The average maximum load of specimens of series B is 8905.2 N and of series A 10,375.1 N, which is approximately 15% higher. This could be explained by the lower force transmitted between the cementitious matrix and the textile fabrics because of the lower bond length. Consequently, the drop in force in the third stage is lower for the same reason, related to the reduced bond length. Specimen CP-2-B is characterized by higher loads, probably due to the fact that the position of the crack leaves a bond length of 200 mm from the border of the specimens, similarly to specimens of series A.

As for specimens C-F/TRC of series B, observations similar to those for series A can be made: the mode of failure of specimens retrofitted with carbon textile fabric is still the pull-out of the textile fabric. The presence of overlap does not affect the mode of failure of the composite.

In this experimental test campaign, two different modes of failure have been observed: rupture of the textile fabrics and pull-out of the textile fabric from the cementitious matrix. The latter failure mode, observed here on C-TRC and C-F/TRC specimens and related to the insufficient bond between textile fabrics and cementitious matrix, has also been observed by Hojdys et al. [27]. By comparing his results and those of other articles [30,42–46], he observed that this mode of failure happens especially when a clamping system with bolted steel plates is used. He suggests that, to provide adequate length for the textile fabric, the test setup could be modified. The suggestions proposed in his article include the use of pneumatic or hydraulic gripping to increase the pressure of the clamps on the specimen, designing longer specimens and longer steel plates for the clamping system and lengthening the specimen outside the bolted steel plates. While the force–displacement behavior of B-TRC specimens is similar to the classic trilinear behavior of TRC specimens under tensile loads, the force–displacement (and stress–strain) behavior of C-TRC specimens, affected by the loss of bond between textile fabric and cementitious matrix, is similar to that observed by Ortlepp and Lorenz [41]. In her work, Ortlepp executed pull-out tests on TRM specimens characterized by a high-strength concrete matrix and a textile fabric composed by carbon filament yarns in warp directions and alkali-resistant glass fibers in weft direction.

The influence of admixing of short fibers in the cementitious matrix of TRM has been investigated by Zhou et al. [33]. In his study, 2% by volume of short steel fibers were admixed in the cementitious matrix. The tensile strength of the TRM plates increases in approximately 100% compared to the reference specimens. A similar result has been obtained in this work, in fact an increase of approximately 83% and 92% of the average maximum load has been observed on B-F/TRC of series A and B, respectively. On C-F/TRC specimens, the increase in tensile strength was lower, 18% and 76% on series A and B, respectively.

Finally, the considerations about the results reported in here are limited by the materials, the specimen's layouts and the test setup adopted. Therefore, because of the uniqueness of the combination of materials and the design of the specimens adopted here, results such as maximum tensile strength, and failure mode are specific to this work.

6. Conclusions

In this work, an extensive experimental test campaign was conducted in order to investigate the properties of textile-reinforced concrete (TRC) under tension. A total of 24 uniaxial tensile tests were conducted on specimens comprising one layer of carbon or basalt textile fabric immersed in a layer of inorganic matrix. Additionally, the influence of admixing short steel fibers in the cementitious matrix was investigated, which led to a new material called fiber/textile-reinforced concrete (F/TRC). Finally, the investigation involved the study of the behavior of TRC and F/TRC on the overlap of the textile. The results of the experimental test campaign are expressed in terms of force–displacement curves, stress–strain curves and crack opening–displacement curves. Due to the great variability of different textile fabrics, inorganic matrices and short fibers, each TRC and F/TRC system is unique and warrants further investigation. In this study, commercial high-performance concrete, short steel fibers and basalt and carbon textile fabrics are combined to realize unique TRC and F/TRC systems. Therefore, given the originality of the combinations of materials adopted here, an investigation on the mechanical properties of TRC and F/TRC is necessary. The results of this research will be used to better comprehend the behavior of existing reinforced concrete (RC) structures retrofitted with these systems. Furthermore, this is a fundamental step to refine the analytical formulations used to estimate the contribution of these systems to the increase in performance of the retrofitted elements.

From the analysis of the experimental test results, the following conclusions can be drawn:

- Carbon-retrofitted specimens perform better, in terms of maximum tensile load, compared with those retrofitted with basalt textile fabric. In this experimental test campaign, the average maximum loads are 66.7% and 22.2% higher for TRC specimens and 7.5% and 11.9% higher for F/TRC specimens of series A and B, respectively.
- Short steel fibers improve the performance of both basalt- and carbon-retrofitted specimens. The propagation of transversal cracks, which causes the reduction in stiffness of the first branch of the force–displacement curve, occurs at higher levels of load. The range of loads in which the stiffness of the force displacement curves decreases due to the propagation of transversal cracks increases from approximately 3000–4000 N to 9000–10,000 N in series A and from approximately 4000 N to 12,000–13,000 N in series B.
- Ductile failure characterizes carbon-retrofitted specimens. In this experimental test campaign, C-TRC and C-F/TRC specimens fail due to the pull-out of the textile fabric from the cementitious matrix. During the pull-out of C-TRC specimens, the average load is 85% and 84% of the average maximum load for series A and B, respectively. For C-F/TRC specimens, the average load during the pull-out of the textile fabric is approximately 72% and 60% of the average maximum load of series A and B, respectively.
- The assumed overlap length is long enough to avoid premature failures during the tests. The mode of failure of specimens with overlap of the textile fabric is similar to that of specimens without overlap of the textile fabric.
- A simple method to estimate the load capacity of basalt composites is proposed. The difference between the theoretical and experimental results is lower than 5.6% and 21% for series A and B, respectively.

Author Contributions: Conceptualization, G.M., N.R. and M.P.; methodology, G.M., N.R. and M.P.; validation, G.M. and N.R.; formal analysis, G.M.; investigation, G.M. and N.R.; resources, N.R.; data curation, G.M.; writing—original draft preparation, G.M.; writing—review and editing, N.R and M.P.; visualization, G.M.; supervision, N.R.; project administration, N.R.; funding acquisition, N.R. All authors have read and agreed to the published version of the manuscript.

Funding: This research was funded by the Austrian Research Promotion Agency (FFG), grant number 866881.

Institutional Review Board Statement: Not applicable.

Informed Consent Statement: Not applicable.

Data Availability Statement: The data presented in this study are available from the corresponding author upon reasonable request.

Acknowledgments: The authors would like to thank the companies PAGEL Spezial-Beton GmbH & Co. Kg, Essen, Germany and Hitexbau GmbH—Albani Group GmbH & Co. KG, Augsburg, Germany for providing the carbon textile fabrics and Kimia S.p.a, Perugia, Italy for providing the basalt textile fabrics. The scientific input provided by M. Pauletta was partly carried out within the Interconnected Nord-Est Innovation Ecosystem (iNEST) and partially funded by the European Union Next-Generation EU (PIANO NAZIONALE DI RIPRESA E RESILIENZA (PNRR)—MISSIONE 4 COMPONENTE 2, INVESTIMENTO 1.5—D.D. 1058 23/06/2022, ECS00000043). This manuscript reflects only the authors' views and opinion; neither the European Union nor the European Commission can be considered responsible for them.

Conflicts of Interest: The authors declare no conflict of interest. The funders had no role in the design of the study; in the collection, analyses, or interpretation of data; in the writing of the manuscript, or in the decision to publish the results.

References

1. Consiglio Nazionale Delle Ricerche (CNR). Istruzioni per la Progettazione, l'Esecuzione ed il Controllo di Interventi di Consolidamento Statico mediante l'utilizzo di Compositi Fibrorinforzati a Matrice Inorganica. 2020. Available online: https://www.cnr.it/it/node/9347 (accessed on 19 April 2022).
2. Mechtcherine, V.; Schneider, K.; Brameshuber, W. 2-Mineral-based matrices for textile-reinforced. In *Textile Fibre Composites in Civil Engineering*; Triantafillou, T., Ed.; Woodhead Publishing: Sawston, UK, 2016; pp. 25–43. [CrossRef]
3. Triantafillou, T.C.; Papanicolaou, C.; Zissimopoulos, P.; Laourdekis, T. Concrete Confinement with Textile-Reinforced Mortar Jackets. *ACI Struct. J.* **2006**, *103*, 28–37. [CrossRef]
4. Bournas, D.A.; Triantafillou, T.; Papanicolaou, C.G. Retrofit of Seismically Deficient RC Columns with Textile- Reinforced Mortar (TRM) Jackets. In Proceedings of the 4th Colloquium on Textile Reinforced Structures (CTRS4), Dresden, Germany, 3–5 June 2009; pp. 471–490.
5. Bournas, D.A.; Lontou, P.; Papanicolaou, C.; Triantafillou, T.C. Textile-Reinforced Mortar versus Fiber-Reinforced Polymer Confinement in Reinforced Concrete Columns. *ACI Struct. J.* **2007**, *104*, 740–748. [CrossRef]
6. De Domenico, D.; Maugeri, N.; Longo, P.; Ricciardi, G.; Gullì, G.; Calabrese, L. Clevis-Grip Tensile Tests on Basalt, Carbon and Steel FRCM Systems Realized with Customized Cement-Based Matrices. *J. Compos. Sci.* **2022**, *6*, 275. [CrossRef]
7. Rossi, E.; Randl, N.; Harsányi, P.; Mészöly, T. Experimental study of fibre-reinforced TRC shear strengthening applications on non-stirrup reinforced concrete T-beams. *Eng. Struct.* **2020**, *256*, 113923. [CrossRef]
8. Rossi, E.; Randl, N.; Mészöly, T.; Harsányi, P. Effect of TRC and F/TRC Strengthening on the Cracking Behaviour of RC Beams in Bending. *Materials* **2021**, *14*, 4863. [CrossRef]
9. Rossi, E.; Randl, N.; Harsányi, P.; Mészöly, T. Overlapped joints in Textile Reinforced Concrete with UHPC matrix: An experimental investigation. *Mater. Struct.* **2021**, *54*, 1–15. [CrossRef]
10. Mészöly, T.; Ofner, S.; Randl, N. Effect of Combining Fiber and Textile Reinforcement on the Flexural Behavior of UHPC Plates. *Adv. Mater. Sci. Eng.* **2020**, *2020*, 1–8. [CrossRef]
11. Rossi, E.; Randl, N.; Mészöly, T.; Harsányi, P. Flexural Strengthening with Fiber-/Textile-Reinforced Concrete. *ACI Struct. J.* **2021**, *118*, 4. [CrossRef]
12. Deng, M.; Dong, Z.; Zhang, C. Experimental investigation on tensile behavior of carbon textile reinforced mortar (TRM) added with short polyvinyl alcohol (PVA) fibers. *Constr. Build. Mater.* **2020**, *235*, 117801. [CrossRef]
13. Barhum, R.; Mechtcherine, V. Effect of short, dispersed glass and carbon fibres on the behaviour of textile-reinforced concrete under tensile loading. *Eng. Fract. Mech.* **2012**, *92*, 56–71. [CrossRef]
14. Rampini, M.C.; Zani, G.; Colombo, M.; di Prisco, M. Mechanical Behaviour of TRC Composites: Experimental and Analytical Approaches. *Appl. Sci.* **2019**, *9*, 1492. [CrossRef]
15. Ortlepp, R.; Ortlepp, S. Textile reinforced concrete for strengthening of RC columns: A contribution to resource conservation through the preservation of structures. *Constr. Build. Mater.* **2017**, *132*, 150–160. [CrossRef]
16. Wang, J.; Wan, C.; Zeng, Q.; Shen, L.; Malik, M.; Yan, D. Effect of eccentricity on retrofitting efficiency of basalt textile reinforced concrete on partially damaged masonry columns. *Compos. Struct.* **2020**, *232*, 111585. [CrossRef]
17. Faleschini, F.; Zanini, M.; Hofer, L.; Toska, K.; De Domenico, D.; Pellegrino, C. Confinement of reinforced concrete columns with glass fiber reinforced cementitious matrix jackets. *Eng. Struct.* **2020**, *218*, 110767. [CrossRef]
18. Faleschini, F.; Zanini, M.; Hofer, L.; Pellegrino, C. Experimental behavior of reinforced concrete columns confined with carbon-FRCM composites. *Constr. Build. Mater.* **2020**, *243*, 118296. [CrossRef]
19. Peled, A.; Bentur, A.; Mobasher, B. Textiles. In *Textile Reinforced Concrete*, 1st ed.; CRC Press: Boca Raton, FL, USA, 2017; pp. 1–473.

20. De Fazio, P. Basalt fiber: From earth an ancient material for innovative and modern application. *Energ. Ambient. Innov.* **2011**, *3*, 89–96.
21. Fiore, V.; Scalici, T.; Di Bella, G.; Valenza, A. A review on basalt fibre and its composites. *Compos. Part B Eng.* **2015**, *74*, 74–94. [CrossRef]
22. Di Ruocco, G. Basalt fibers: The green material of the XXI-century, for a sustainable restoration of historical buildings. *Vitr. Int. J. Archit. Technol. Sustain.* **2016**, *1*, 25–39. [CrossRef]
23. Wang, Q.; Ding, Y.; Randl, N. Investigation on the alkali resistance of basalt fiber and its textile in different alkaline environments. *Constr. Build. Mater.* **2021**, *272*, 121670. [CrossRef]
24. RILEM Technical Committee 232-TDT (W. Brameshuber). Recommendation of RILEM TC 232-TDT: Test methods and design of textile reinforced concrete. Uniaxial tensile test: Test method to determine the load bearing behavior of tensile specimens made of textile reinforced concrete. *Mater. Struct.* **2016**, *49*, 4923–4927. [CrossRef]
25. Hartig, J.; Jesse, F.; Schicktanz, K.; Häußler-Combe, U. Influence of experimental setups on the apparent uniaxial tensile load-bearing capacity of Textile Reinforced Concrete specimens. *Mater. Struct.* **2012**, *45*, 433–446. [CrossRef]
26. Hegger, J.; Will, N.; Bruckermann, O.; Voss, S. Load–bearing behaviour and simulation of textile reinforced concrete. *Mater. Struct.* **2006**, *39*, 765–776. [CrossRef]
27. Hojdys, Ł.; Krajewski, P. Tensile Behaviour of FRCM Composites for Strengthening of Masonry Structures—An Experimental Investigation. *Materials* **2021**, *14*, 3626. [CrossRef] [PubMed]
28. D'Antino, T.; Papanicolaou, C. Mechanical characterization of textile reinforced inorganic-matrix composites. *Compos. Part B Eng.* **2017**, *127*, 78–91. [CrossRef]
29. de Felice, G.; Garmendia, L.; Ghiassi, B.; Larrinaga, P.; Lourenço, P.; Oliveira, D.; Paolacci, F.; Papanicolaou, C.G. Mortar-based systems for externally bonded strengthening of masonry. *Mater. Struct.* **2014**, *47*, 2021–2037. [CrossRef]
30. Caggegi, G.P.L.C.; Ceroni, F.; De Santis, S.; Krajewski, P.; Lourenço, P.; Morganti, M.; Papanicolaou, C.; Pellegrino, C.; Pronta, A.; Zuccarino, L. Performance assessment of basalt FRCM for retrofit applications on masonry. *Compos. Part B Eng.* **2017**, *128*, 1–18. [CrossRef]
31. Larrinaga, P.; Chastre, C.; San-José, J.; Garmendia, L. Non-linear analytical model of composites based on basalt textile reinforced mortar under uniaxial tension. *Compos. Part B Eng.* **2013**, *55*, 518–527. [CrossRef]
32. Beßling, M.; Groh, M.; Koch, V.; Auras, M.; Orlowsky, J.; Middendorf, B. Repair and Protection of Existing Steel-Reinforced Concrete Structures with High-Strength, Textile-Reinforced Mortars. *Buildings* **2022**, *12*, 1615. [CrossRef]
33. Zhou, F.; Liu, H.; Du, Y.; Liu, L.; Zhu, D.; Pan, W. Uniaxial Tensile Behavior of Carbon Textile Reinforced Mortar. *Materials* **2019**, *12*, 374. [CrossRef]
34. Mattarollo, G.; Randl, N.; Pauletta, M.; Rossi, E. Experimental investigation on confinement of columns with TRC: A comparison between basalt and carbon textile fabrics. In Proceedings of the 14th fib International PhD Symposium in Civil Engineering, Rome, Italy, 5–7 September 2022; pp. 517–524.
35. Mattarollo, G.; Randl, N.; Pauletta, M.; Rossi, E. Confinement of Columns with Textile Reinforced Concrete: An Experimental Comparison between Basalt and Carbon Textile Reinforced Concrete. In Proceedings of the 6th fib International Congress, Oslo, Norway, 12–16 June 2022; Novus Press: Oslo, Norway, 2022; pp. 2286–2295.
36. Deutsches Institut für Bautechnik, "Verfahren zur Verstärkung von Stahlbeton mit TUDALIT (Textilbewehrter Beton)". 2016. Available online: https://www.dibt.de/de/service/zulassungsdownload/detail/z-3110-182 (accessed on 31 January 2022).
37. HITEXBAU, Albani Group. Art.279136 HTC 21/21-40 Technical Data Sheet. Available online: https://www.hitexbau.com/menu/products/ (accessed on 27 July 2021).
38. Kimia, S.P.A. Certificato di Valutazione Tecnica. 2022. Available online: https://www.kimia.it/sites/default/files/doc/CVT/cvt-kimia-compositi-frcm-n-207 (accessed on 13 October 2022).
39. Kimia, S.P.A. Kimitech BS ST 400 ST2-0319 Tessuti di Armatura in Fibra di Basalto per Rinforzi FRCM Techncal Data Sheet. Available online: https://www.kimia.it/it/prodotti/kimitech-bs-st-400 (accessed on 31 August 2020).
40. Park, R. Evaluation of ductility of structures and structural assemblages from laboratory testing. *Bull. N. Z. Soc. Earthq. Eng.* **1989**, *22*, 155–166. [CrossRef]
41. Ortlepp, R.; Lorenz, E. Bond Behavior of Textile Reinforcements—Development of a Pull-Out Test and Modeling of the Respective Bond versus Slip Relation. In *High Performance Fiber Reinforced Cement Composites 6*; RILEM Bookseries; Springer: Dordrecht, The Netherlands, 2012; pp. 1–8. [CrossRef]
42. De Santis, S.; de Felice, G. Tensile behaviour of mortar-based composites for externally bonded reinforcement systems. *Compos. Part B Eng.* **2015**, *68*, 401–413. [CrossRef]
43. Carozzi, F.G.; Bellini, A.; D'Antino, T.; de Felice, G.; Focacci, F.; Hojdys, Ł.; Laghi, L.; Lanoye, E.; Micelli, F.; Panizza, M.; et al. Experimental investigation of tensile and bond properties of Carbon-FRCM composites for strengthening masonry elements. *Compos. Part B Eng.* **2017**, *128*, 100–119. [CrossRef]
44. Leone, M.; Aiello, M.; Balsamo, A.; Carozzi, F.; Ceroni, F.; Corradi, M.; Gams, M.; Garbin, E.; Gattesco, N.; Krajewski, P.; et al. Glass fabric reinforced cementitious matrix: Tensile properties and bond performance on masonry substrate. *Compos. Part B Eng.* **2017**, *127*, 196–214. [CrossRef]

45. De Santis, S.; Ceroni, F.; de Felice, G.; Fagone, M.; Ghiassi, B.; Kwiecień, A.; Morganti, M.; Santandrea, M.; Valluzzi, M.; Viskovic, A. Round Robin Test on tensile and bond behaviour of Steel Reinforced Grout systems. *Compos. Part B Eng.* **2017**, *127*, 100–120. [CrossRef]
46. Caggegi, C.; Carozzi, F.; De Santis, S.; Fabbrocino, F.; Focacci, F.; Hojdys, Ł.; Lanoye, E.; Zuccarino, L. Experimental analysis on tensile and bond properties of PBO and aramid fabric reinforced cementitious matrix for strengthening masonry structures. *Compos. Part B Eng.* **2017**, *127*, 175–195. [CrossRef]

Disclaimer/Publisher's Note: The statements, opinions and data contained in all publications are solely those of the individual author(s) and contributor(s) and not of MDPI and/or the editor(s). MDPI and/or the editor(s) disclaim responsibility for any injury to people or property resulting from any ideas, methods, instructions or products referred to in the content.

Article

Impact of Polypropylene Fibers on the Mechanical and Durability Characteristics of Rubber Tire Fine Aggregate Concrete

Arash Karimi Pour [1], Zahra Mohajeri [2] and Ehsan Noroozinejad Farsangi [3,*]

[1] Department of Civil Engineering, University of Texas at El Paso (UTEP), El Paso, TX 79968, USA
[2] Centre for Transportation Infrastructure Systems (CTIS), Department of Civil Engineering, University of Texas at El Paso (UTEP), El Paso, TX 79968, USA
[3] Department of Civil Engineering, The University of British Columbia (UBC), Vancouver, BC V6T 1Z4, Canada
* Correspondence: ehsan.noroozinejad@ubc.ca

Abstract: In this research, the consequence of using rubber tire aggregates (RTA) on the durability and mechanical characteristics of polypropylene fibers (PF) reinforced concrete is evaluated. Fifteen concrete mixtures were produced and tested in the laboratory. RTA was utilized instead of fine natural aggregates (FNA) to the concrete at concentrations of 0%, 5%, 10%, 15%, and 20% by a volumetric fraction; also, the contents of PF in the concrete mixtures were 0%, 1%, and 2% by weight fraction. Finally, the following parameters were tested for all the mixtures: compressive and tensile resistances, fracture, changes in drying shrinkage, bulk electrical resistivity, elastic moduli, and resonance occurrences. The control sample was the one without RTA and PF. According to the results, by adding RTA to the mixtures, the shrinkage deformation amplified, but the PF addition caused a decrease in the shrinkage deformation. Furthermore, adding 0%, 5%, 10%, and 15% RTA, with 2% PF leads to an upsurge in the flexural resistance by 34%, 24%, 16%, and 6%, respectively, relative to the control sample without PF and RTA. Moreover, the fracture energy of mixtures increased by utilizing PF and RTA simultaneously.

Keywords: mechanical characteristics; polypropylene fiber; rubber tire aggregates; ultrasonic pulse rate; sustainable concrete

1. Introduction

Compiling RTA around the world can cause a huge environmental risk if we cannot get rid of them properly. One efficient method, from both environmental and economic points of view, is to recycle these waste materials by using them as an aggregate for producing cement-based concrete [1–4]. Therefore, utilizing this waste as an agent for concrete production has become popular all over the world [5–8]. To see the effect of different contents of RTA, as a partial substitution for cement, FNA and coarse natural aggregates (CNA), on the characteristics of concrete mixtures, much research was directed. The influence of RTA on the mechanical performance of high-strength concrete was evaluated by Abdelmonem et al. [9]. RTA was used as a substitution for aggregates at different contents by volumetric fractions- 0%, 10%, 20%, and 30%. The high-strength concrete impact performance, slump, water absorption, density, fracture energy, compressive, and tensile resistances were measured. Based on the result of raising the RTA content, the adsorption of water decreased. Additionally, by adding more RTA, the flexural, tensile, and compressive performance of mixtures was decreased. As a result, adopting improvement methods to increase the strength of concrete can make it possible to use RTA in producing green concrete. Zhu et al. [10] conducted research on the comportment of high-performance RTA-incorporated concrete on the microstructural scale. Two methods were applied for that purpose, the first one was the Mercury intrusion porosimeter and the second one was scanning electron microscopy. The result illustrated that adding 5% RTA to the concrete mixtures leads to a 3.5% increase in their porosity. In addition, by increasing the RTA content, the pores' size also increased, which leads to increasing the crack width which increases the need for solutions to improve the behavior of concrete.

In another study, Li et al. [11] assessed the characteristics of modified concrete with RTA on its surface. The findings revealed that the elastic moduli of the control mixture compared to the RTA-modified mixtures is lower. Consequently, the author applied a new correlation for electric modulus calculation for concrete mixtures according to RTA content. They recommended employing the additional materials for the RTA-incorporated concrete characteristics improvement. Based on the Strukar et al. research work [12], RTA as a partial substitution for FNA can be utilized, leading to deformation, ductility, and energy dissipation enhancement of the concrete mixtures. In another study, Li et al. [13] evaluated the concrete mechanical characteristics after adding RTA. According to the result, workability can be improved by adding RTA due to the particles' hydrophobic nature and a proper mixture. So, by applying 0–10% RTA to concrete mixtures, the flexural resistance is decreased about 8–20%. Contrarily, RTA utilization had a disadvantageous impact on the initial density of concrete mixtures. Therefore, the incorporation of new materials is necessary to compensate for the negative influence of RTA incorporation. In addition, RTA concrete outperformed concrete made with natural particles in terms of freeze-thaw, sulfuric and sulfate attacks, and electrical, and abrasion resistances. Guo et al. [14] assessed the damage statuses of RTA concrete mixtures with a correlation of quantitative cloud images. To reach this goal, RTA was incorporated at various rates, ranging from 10% to 30%. Alternatively, one mix without RTA served as the control. An experimental test-based numerical model on the randomness of concrete with RTA was developed and validated. The consequence of RTA on the thermal stability and mechanical characteristic of concrete mixtures were examined by Záleská et al. [15]. A correlation between thermal conductivity and moisture content was found, and concrete sample properties, including secant and dynamic elastic modulus, and compressive and flexural resistances, were also examined. The weight and thermal conductivity of the concrete sample were decreased by using RTA. Additionally, the water transport capabilities of concrete mixtures with RTA were not significantly impacted. Moreover, temperatures up to 300 °C did not significantly alter the features of concrete related to the integration of RTA, but temperatures above 400 °C caused noticeable alterations. A recent thorough evaluation of the mechanical characteristics of concrete mixtures with RTA was published by Roychand et al. [16]. Therefore, it was investigated how various mechanical characteristics of RTA were impacted by the size of aggregate, treatment method, and replacement fraction. The findings showed that adding RTA to concrete has a reasonable impact on its hardened characteristics, but that the fresh characteristics considerably deteriorate.

According to Feng et al. [17], strain rates have an impact on how well RTA concrete flexes dynamically. The three-point flexural test was used to evaluate the concrete mixtures. Five RTA contents of 0%, 10%, 20%, 30%, and 40% were used to substitute FNA inside the concrete mixture. When the RTA percentage reached up to 30%, the concrete mixture exhibited a sharper strain rate compared to the control mixture; nevertheless, larger substitution contents led to no change in the strain rate. Additionally, compared to the regular concrete mixture, the sample contains 30% RTA as a partial replacement and has higher deformability and less rate of cracking. External confinement's effect on the rubberized concrete's compressive performance was considered by Chan et al. [18], as an improvement method to mitigate the negative influence of RTA. Using polymers that were fiber-reinforced significantly increased the axial performance of rubber-modified concrete, based on the results. Furthermore, constrained RTA concrete behaved very differently relative to the control concrete. The performance of a modified RTA mixture with fiber-reinforced polymers under load was therefore forecasted using a novel model. Wang et al. [19] recently evaluated the durability and mechanical behavior of RTA concrete modified with PF. The utilized RTA content was 10% and 15%. Moreover, to decrease the detrimental impact of RTA on the concrete mixtures, 5% PF were added to the mixes. The outcomes showed that employing fibers and RTA together might greatly increase fracture energy. Fibers were added to RTA concrete, which significantly improved its flexural resistance, drying shrinkage, and deformation.

The subject of another research by Farhad and Ronny [20] was the investigation of the RTA concrete characteristic containing PF. RTA with sizes 2–5 mm were utilized as a substitution for 20% FNA. The content of fibers is determined considering the type of fibers and volumetric ratios of 0.1–0.25%. The fresh characteristics of the mixtures were examined by the workability and the J-ring tests, and the hardened characteristics were evaluated by the compressive stress–strain, compressive, and tensile resistance tests. According to the findings, by increasing the content of fibers, the detrimental impact on the rheological characteristics of the concrete that are modified by fibers increases. Regarding mechanical characteristics, as the content of PF was raised, the compressive resistance lessened. In another study, the mechanical characteristics of cement-based RTA mortars that were modified by fibers were evaluated by Nguyen et al. [21]. To prevent cracking at very early stages, a special type of steel fiber with high bond properties with 20, and 30% volume content was applied. To achieve tensile strength, strain capacity, and behavior after residual peak, direct tensile experiments were carried out. Moreover, for obtaining Young's modulus and compressive resistance, compressive experiments were conducted. According to the results, although adding RTA to concrete mixtures causes a reduction in compressive and tensile resistance and elastic modulus, it leads to an increase in strain capacity.

2. Research Significance

Literature reviews illustrate that a significant number of research studies have been performed on the evaluation of RTA-incorporated concrete. The application of RTA-based concrete has been limited to non-structural functions because of some of its significant weaknesses, such as early cracking, low strength, and stiffness, due to the inadequate bonding between RTA and concrete paste. Additionally, the previous finding showed that the incorporation of RTA to produce eco-friendly concrete led to reducing the mechanical characteristics of concrete. Therefore, using other improvement techniques is necessary to improve the performance of green concrete having RTA. In this regard, previous investigations showed that the incorporation of fibers significantly enhanced the mechanical characteristics of various types of concrete [22–30]. For this aim, PF was utilized to enhance the behaviors of concrete made with RTA. One noticeable point is that in most of the earlier investigations, the volume content of PF in the concrete mixtures was less than 1.5%, with high contents of RTA. Moreover, the RTA and PF effects on the mechanical experiments, freezing thaw, compressive and tensile strength, and slum were investigated. Furthermore, the modified concrete behavior after PF addition was assessed. Consequently, the concrete's mechanical and long-lasting behavior that is affected by PF and RTA addition was investigated. To achieve this goal, various PF and RTA were used to measure their effects on mechanical experiments, such as drying shrinkage, ultrasonic pulse velocity, fracture energy, bulk electrical performance, freeze-thaw damage, and length changing.

3. Materials and Methods

3.1. Polypropylene Fibers

To produce fiber-reinforced mixtures, 12 mm length PF were introduced at three volumetric fractions of 0%, 1% and 2%. The tensile resistance, density and melting point of used fibers are 360 MPa, 890 kg/m^3 and 160 °C, correspondingly.

3.2. Rubber Tire Aggregate

RTA comes from waste tires with particle sizes of 5–30 mm. RTA is used as a fractional substitution for FNA at five volumetric fractions- 0%, 5%, 10%, 15%, and 20%. Based on the Si et al. [31] study, RTA was submerged for 30 min in a NaOH solution with 40 g/L density and after that was washed and air-dried. Figure 1 shows a sample of utilized aggregates.

 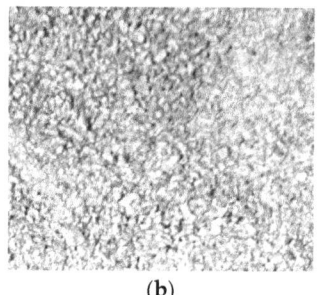

(a)   (b)

Figure 1. Utilized aggregates: (a) RTA and; (b) FNA.

3.3. Concrete

Ordinary Portland cement was utilized for concrete production in a drum-type mixer based on the ASTM C150/C150M [32]. At first, for one minute, all dry particles were mixed, after that PF was added, and mixed for some minutes. Then, for another minute, mixing continued with two-thirds of the water addition. Finally, the rest of the water and superplasticizer were blended. Table 1 illustrates the cement's chemical and physical characteristics. Moreover, the aggregates' mechanical characteristics are provided in Table 2, as per ASTM D2419 [33]. Figure 2 presents the aggregates' gradation graph. A total of five volumetric contents of RTA were utilized 0%, 5%, 10%, 15% and 20%; also, three weight contents of PF were incorporated: 0%, 1% and 2%. Accordingly, the mixture design can be seen in Table 3.

Table 1. Characteristics of the used cement.

Physical	
Feature	Value
Final set period	190 min
Initial set period	146 min
Exact surface	3215 cm^2/g
Precise gravity	3.20 g/cm^3
Autoclave extension	0.07%
Chemical	
SiO_2 (%)	22.34
Al_2O_3 (%)	5.12
Fe_2O_3 (%)	3.27
CaO (%)	58.29
MgO (%)	2.46
SO_3 (%)	1.97
Na_2O (%)	0.36
K_2O (%)	0.58
C_2S (%)	1.34
C_3A (%)	0.24
C_4AF (%)	2.34
Free CaO	1.69

Table 2. Characteristics of the used aggregates.

Fine Aggregates	
Feature	Value
Fineness moduli	2.65
Precise gravity	2.81 g/cm^3
Water absorption	1.69%
Extreme size	4.74 mm
Coarse Aggregates	
Precise gravity	2.37 g/cm^3
Water absorption	0.41%
Ultimate aggregate size	11.4 mm

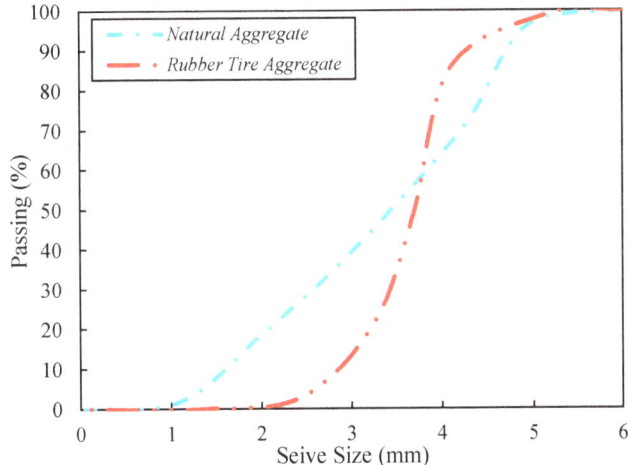

Figure 2. Used aggregates size distribution.

Table 3. Concrete mixes design (kg/m^3).

Mixes	Cement	PF	CAN	FNA	RTA	Superplasticizer	Water/Cement
0RTA-0PF	450	0	950	815	0	1.5	0.47
0RTA-1PF	450	89	950	815	0	2.0	0.47
0RTA-2PF	450	178	950	815	0	2.0	0.47
5RTA-0PF	450	0	950	797	18	2.0	0.47
5RTA-1PF	450	89	950	797	18	2.1	0.47
5RTA-2PF	450	178	950	797	18	2.2	0.47
10RTA-0PF	450	0	950	779	36	2.2	0.47
10RTA-1PF	450	89	950	779	36	2.2	0.47
10RTA-2PF	450	178	950	779	36	2.2	0.47
15RTA-0PF	450	0	950	760	54	2.2	0.47
15RTA-1PF	450	89	950	760	54	2.3	0.47
15RTA-2PF	450	178	950	760	54	2.3	0.47
20RTA-0PF	450	0	950	742	72	2.3	0.47
20RTA-1PF	450	89	950	742	72	2.3	0.47
20RTA-2PF	450	178	950	742	72	2.3	0.47

4. Results and Discussion

4.1. Slump

Slum flow is applied to evaluate the fresh concrete behavior. The goal of this is to test the sample's workability determination based on the ASTM C143 [34]. All the concrete mixture outcomes are shown in Figure 3. From the workability point of view, the size of all slums for fresh mixtures is in the range of 9.8, and 19.7 cm. Moreover, raising the RTA content, caused a decrease in the concrete slump, specifically at 20% rubber content and 2% PF, where the slump dropped roughly 49%. This phenomenon is due to the higher friction between the RTA and other concrete components. The noticeable point is that adding PF to mixtures did not affect the concrete air void significantly. The RTA naturally repel water, and as a result, this leads to an increase in the amount of air trapped in the mixture. Additionally, when friction increases at the contact surface of RTA, PF leads to PF gathering and air trapping. Alternatively, in the absence of PF, the reduction in the slump was lower. As a result, by adding 20% RTA, the slump reduction was about 18% in the absence of PF. Furthermore, adding 20% RTA with 1% PF together leads to a 31% decrease in a slump.

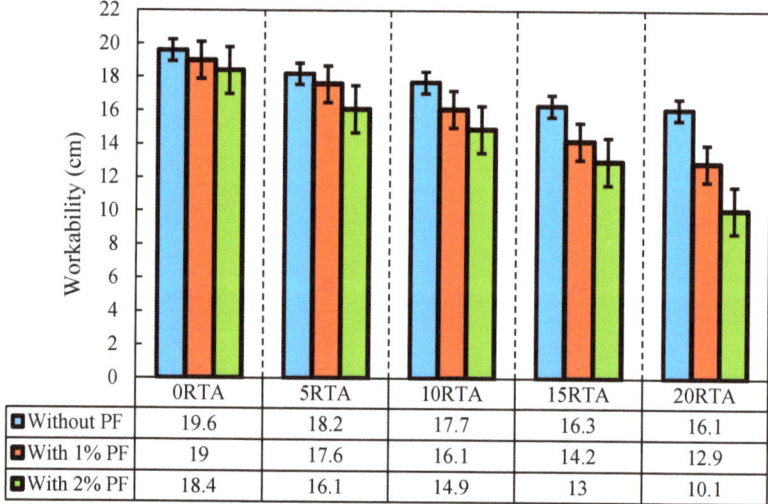

Figure 3. Effect of PF and RTA on the workability of samples.

4.2. Compressive and Tensile Resistances

After 28 days, tests were performed using the ASTM C496 [35] to determine the compressive and tensile resistances. The hydraulic jack maximum load was noted to calculate the tensile resistance, as illustrated in Figure 4. Three cylindrical samples with dimensions 300 mm by 150 mm were utilized for this experiment, and the load was applied to specimens. For this test, the load was applied at 1 mm/s. The average of the three specimens' outcomes was considered as these characteristics value that is calculated by Equation (1).

$$f_t = 2P/(\pi DL). \tag{1}$$

where f_t, P, D and L are the tensile resistance, the failure point, the samples' diameter, and the cylinder height, respectively.

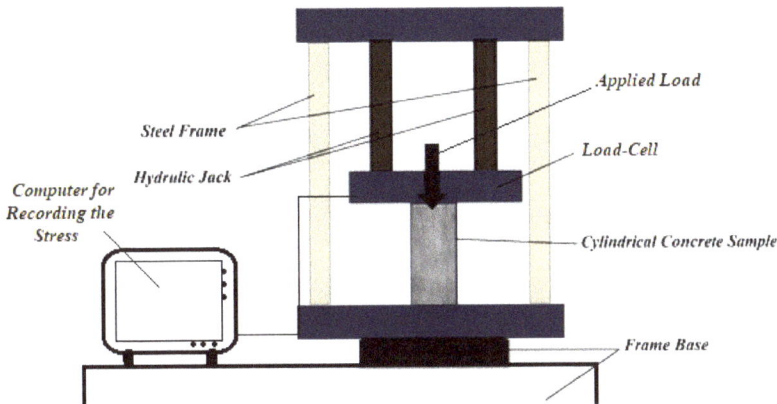

Figure 4. Used setup to determine compressive and tensile resistances.

The tensile resistance of RTA-based concretes that were modified with PF is illustrated in Figures 5 and 6. It should be noted that in Figure 6, the increase in the area enclosed by the graphs generally indicates the level of effectiveness. The tensile resistance is determined by the adhesion in the contact area of cement components and particles, and also mixture strength. PF enhances tensile resistance noticeably by reducing the maximum crack width. Consequently, by applying just 20% RTA, the value of tensile strength dropped 23%, whereas, utilizing 2% PF in the concrete mixture improved the tensile strength by about 38%, also, prohibited the tensile resistance decrease when RTA is employed as a fractional substitution for FNA. Furthermore, by applying 2% PF with the RTA contents 0%, 5%, 10% and 15%, the rubberized concrete enhancement was about 28%, 18%, 13% and 5%, respectively. The tensile resistance is significantly affected by utilizing PF because the fragile cement paste cannot bear tensile tension, and in that case, the cement matrix quality determines the crack's beginning. Therefore, the main cement paste crack typically appeared at the peak load. However, once the crack had already begun, the quick fracture spreading was significantly controlled by the bridging action of fibers. As a result, PF presence successfully prevented crack growth by enhancing the splitting tensile strength. Moreover, Figure 7 illustrates the PF's favorable effect on mixtures with various rubber content when it came to tensile resistance. Under axial tensile load, and when PF was applied, cracking just took place on the concrete surface, and the whole sample did not collapse. However, the samples were kept from completely disintegrating using RTA as NFA. Additionally, specimens lacking PF split into two pieces at failure; however, by adding PF, this behavior was averted. RTA usage reduces concrete's tensile resistance, although this can be made up for by PF addition. The use of fibers lengthens the concrete's failure time by making a bridge between any potential fractures that enhances the tensile resistance. As a result, as per ASTM C496 [35] to evaluate the specimens' function in the hardened state, the outcomes of measuring how PF and RTA affected the specimens' compressive resistance are depicted in Figure 8. Notably, the mixtures' age upon curing is also considered. Three cylindrical shape samples measuring 150 mm by 300 mm were subjected to a hydraulic jack for this goal.

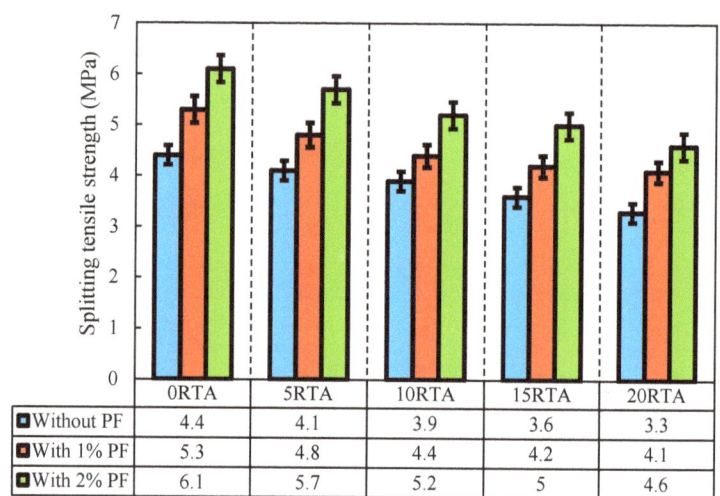

Figure 5. Impact of PF and RTA on the tensile resistance.

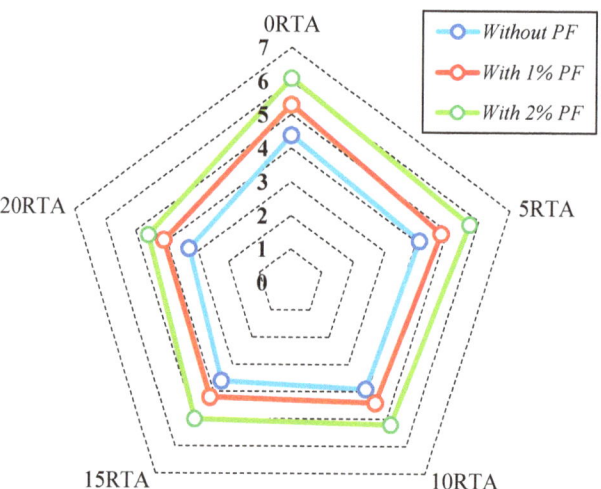

Figure 6. Influence of PF and RTA on the distribution of tensile resistance.

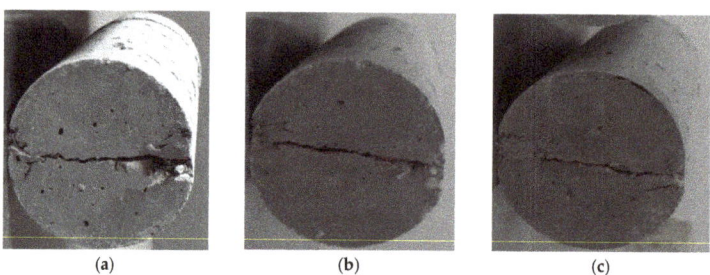

Figure 7. Failure in specimens due to tensile stress in the case of: (**a**) without RTA and PF; (**b**) without PF and with 20% RTA and; (**c**) with 2% PF and without RTA.

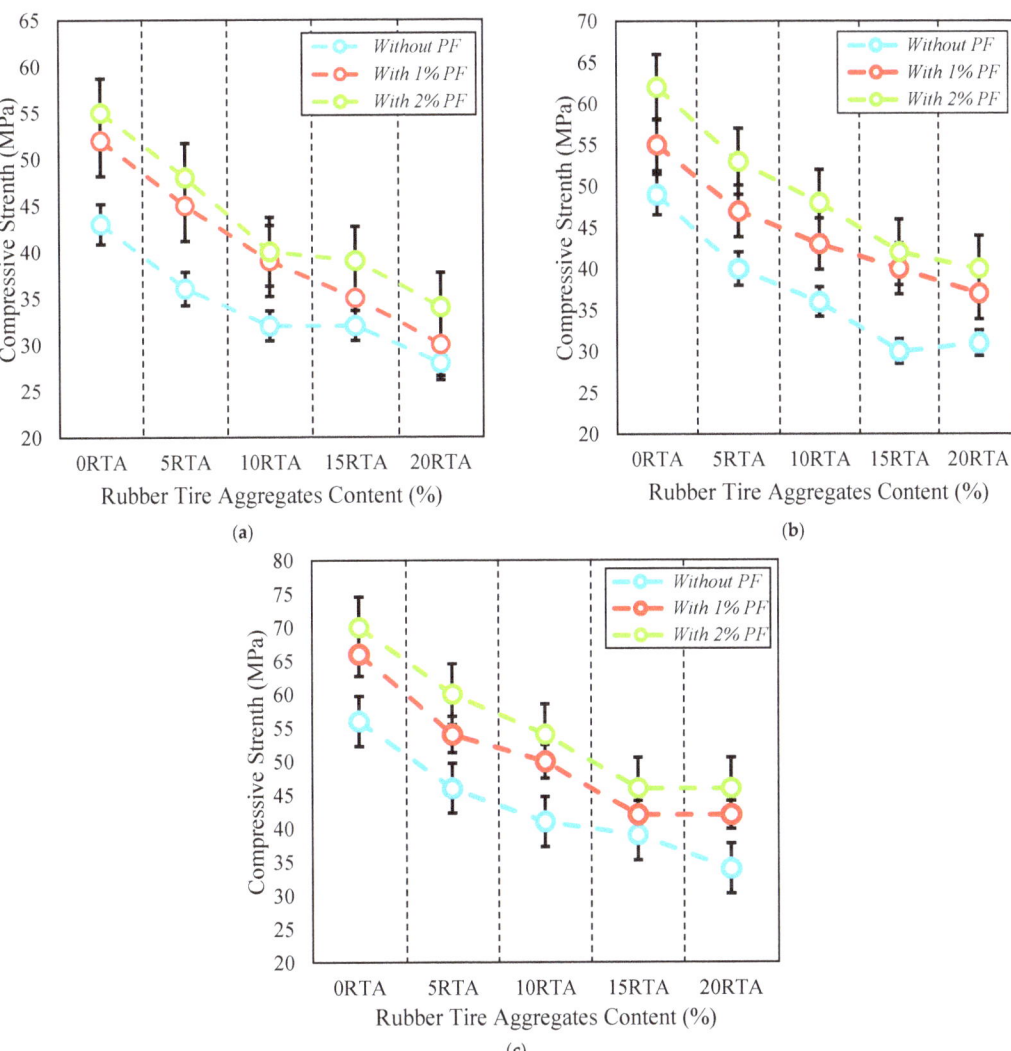

Figure 8. Compressive resistance of concrete having various PF and RTA fractions at: (**a**) 7 days; (**b**) 28 days and; (**c**) 56 days of curing.

As shown in Figure 8, employing RTA causes concrete's compressive resistance to decrease; nevertheless, this characteristic was significantly enhanced when PF was used. As compared to natural aggregate, RTA has a larger Poisson ratio and lower stiffness, which contributes significantly to the strength decrease. The deterioration of strength may also be caused by poor adherence to cement paste and RTA. The addition of RTA also had an impact on the stability of the concrete samples, causing the elastic modulus to drastically decrease and, therefore, the entire concrete mixture stiffness decrease. Consequently, adding 2% PF with 0%, 5%, 10%, 15% and 20% RTA improved the concrete's compressive resistance in 7 days by nearly 24%, 29%, 24%, 25% and 22%, correspondingly. As a result, comparable trends were seen at 14 and 28 days. In addition, as the curing time grew longer, the PF impact on the RTA-based concrete's compressive resistance increased. A similar decreasing pattern can be observed in tensile resistance reported by Wang et al. [36] for concrete having only RTA. In addition, by utilizing high content of RTA, Rubber aggregates' decrease impact will decline.

New relationships that considered the effects of both RTA and PF were developed in accordance with the findings of the compressive and tensile resistance tests, as shown in Figure 9. To forecast the tensile and compressive behaviors of RTA-based concrete, it is possible to utilize the suggested formulation with good values for fitting and R^2 greater than 92%, as provided in Equation (2). In this equation, f_t and f_c indicates the splitting tensile and compressive strengths, respectively. Regarding Figure 9, as anticipated, the compressive and tensile resistances both increased concurrently.

$$f_t = 0.25 f_c^{0.7} \tag{2}$$

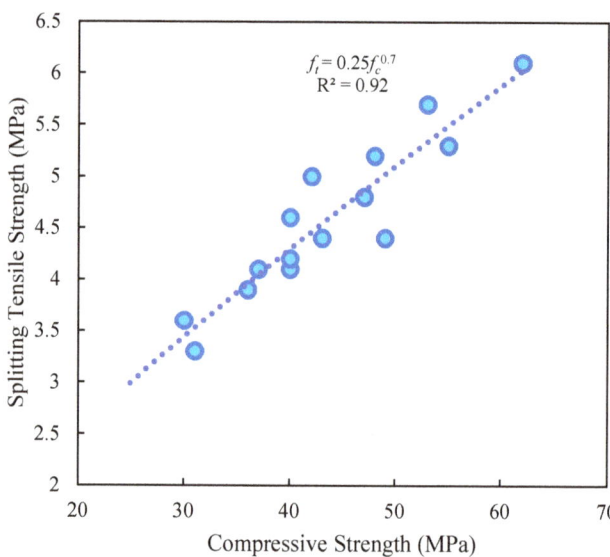

Figure 9. Relationship among compressive and tensile resistances of concrete containing different PF and RTA fractions.

4.3. Flexural Resistance and Fracture Energy

According to the JCI-002-2003 standard, the fracture energy of the sample was calculated [37]. Therefore, after 28 days, single-edge notched beams with the dimensions 102 × 102 × 381 mm were fabricated and evaluated using a three-point bending arrangement (Figure 10). The notch's width and depth are equal to 5 mm, and 30 mm, correspondingly. Loading continued until the samples' complete failure. Equation (3) was used to calculate the specimens' flexural resistance, where L, F, h, and b stand for the span width, maximum load, broken ligament height, and broken ligament width (area above the notch), respectively. The findings are shown in Figure 11.

$$\sigma = 3FL/\left(2bh^2\right). \tag{3}$$

Figure 10. Fracture energy test arrangement.

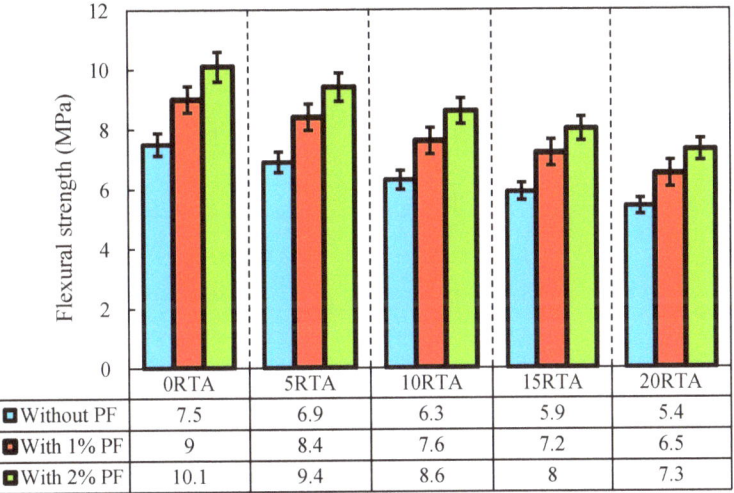

Figure 11. Impact of PF and RTA on the flexural resistance.

Therefore, the JCI-002-2003 standard's proposed formula [37] was used to calculate the fracture energy using Equation (4), where, where G_F, W_0, w_1, A_i, m_1, m_2, g and $CMOD_c$. indicates the fracture energy, the area below the load-displacement curve up to the failure of the specimen, work performed by deadweight of the specimen and loading, the area of the broken ligament, the mass of the samples, load point width, the whole length of the sample, the mass of the jig not devoted to the loading apparatus but located on the sample until failure, gravity and load-deformation at the time of rupture, correspondingly. The fracture energy is demonstrated in Figure 12.

$$\begin{cases} G_F = \frac{0.75W_0 + w_1}{A_{jig}} \\ w_1 = 0.75\left(\frac{S}{L}m_1 + 2m_2\right)g \times CMOD_c \end{cases} \quad (4)$$

As shown in Figure 11, introducing PF somewhat made up for the decrease in flexural resistance that occurred when RTA was substituted for the samples. As a result of PF's bridging function, increasing particle connection, and improvement of concrete's tensile and compressive resistances, particularly when RTA was not employed, PF significantly increased the flexural resistance of the specimens. The inelastic fracture load of rubber-based concretes in the absence of PF quickly decreased; however, once PF was added, everything changed completely, and the flexural resistance significantly increased. Therefore, by adding just 2% PF the flexural resistance, was improved by about 34%.

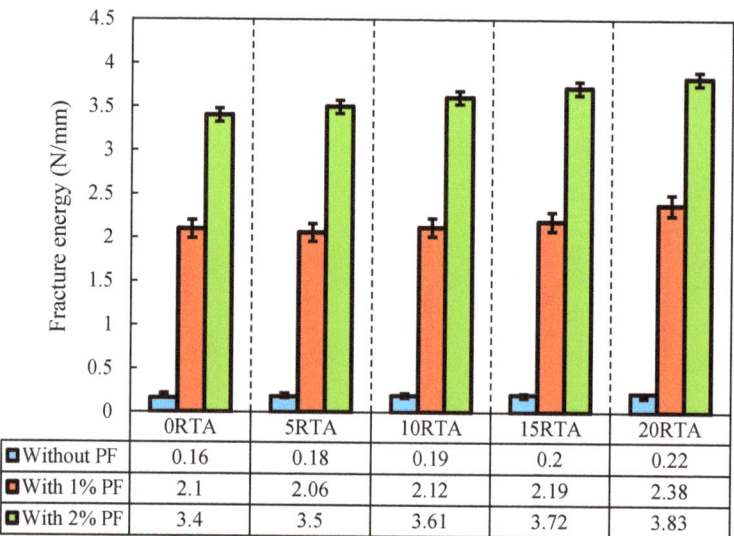

Figure 12. Effect of the PF and RTA on the fracture energy.

Furthermore, the addition of 2% PF together with 0%, 5%, 10%, and 15% RTA caused an improvement in the flexural resistance of the concrete mixtures by 34%, 24%, 16%, and 6%, correspondingly. In contrast, when RTA and PF were utilized in the mixture together, rubber particles increased the mixture's fracture energy (Figure 12). Following the test, the failure patterns of the various specimen kinds were also examined. Only a little single fracture was observed in the control sample, which lacked RTA particles and PF. However, when PF and RTA were combined, the failure mechanism and the crack formation scenario were clearly altered. Additionally, the considerable opening of the crack mouth in modified mixtures with PF demonstrates the fiber bridging effect; yet, the sharpness of the fracture did not alter in comparison to the control sample. RTA and PF were combined, and the stress-relieving properties of elastic rubbers changed the crack's formation course (Figure 12). As a result, ordinary concrete crumbled once the crack appeared and swiftly spread due to the low tensile resistance of the paste, demonstrating small fracture energy. Though, afterwards, the fracture started, the pull-out procedures of the arbitrarily positioned PF used a sizable amount of energy, greatly enhancing the fracture energy of the control mixture. The fracture energy was also raised with the inclusion of RTA, even while the flexural resistance was decreased with the addition of RTA and the reduction was elevated by increasing the RTA percentage in RTA-based concrete mixtures that were modified by PF. Figure 13 shows the flexural failure of specimens having RTA and PF.

The relations among flexural resistance and fracture energy, as well as among flexural and compressive resistances, are shown in Figures 14 and 15. As shown in Figure 14, employing PF reduced the slope of the figures. The very precise models given to estimate the link between fracture energy and flexural resistance with various RTA concentrations are shown in this image. In addition, Figure 15 projected formulas among specimens' flexural and compressive resistances have strong R-square values, indicating an upright fit, as provided in Equation (5). In this equation, f_r and f_c denote the flexural and compressive strengths of concrete samples, respectively. Additionally, as anticipated, the flexural resistance increased at the same time as the compressive resistance.

$$f_r = 0.31 f_c^{0.84} \tag{5}$$

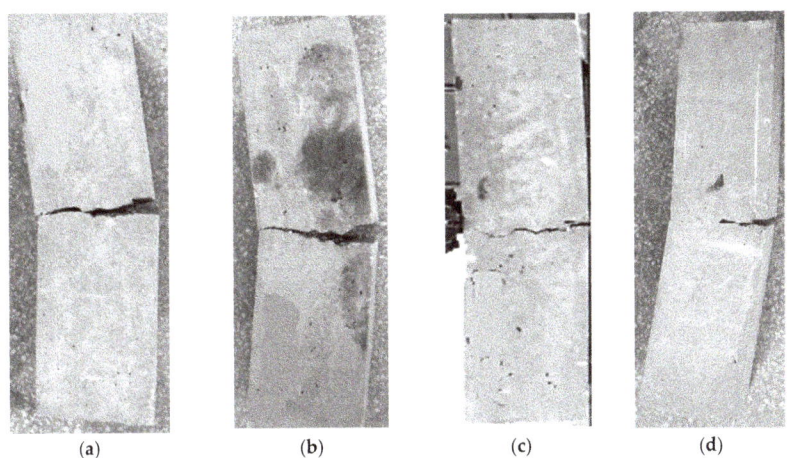

Figure 13. Failure in specimens due to flexure stress in case of: (**a**) without RTA and PF; (**b**) without PF and with 20% RTA; (**c**) with 2% PF and without RTA and; (**d**) with 2% PF and 20% RTA.

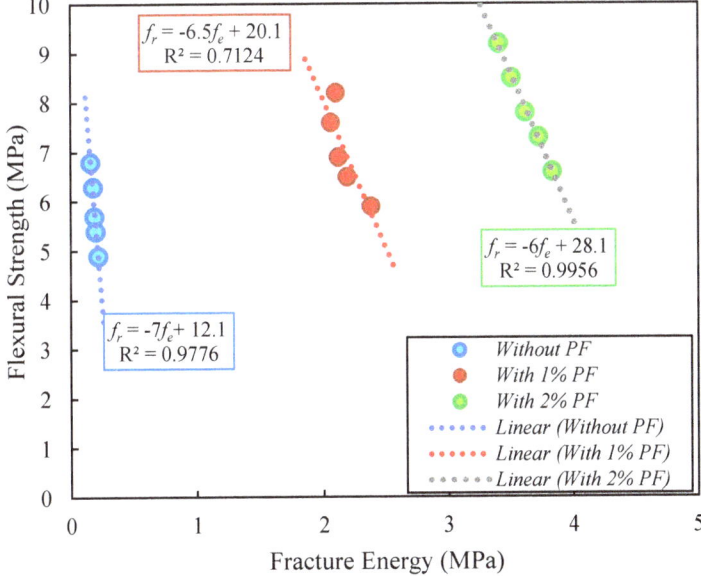

Figure 14. Relation among flexural resistance and fracture energy of sample containing different PF and RTA fractions

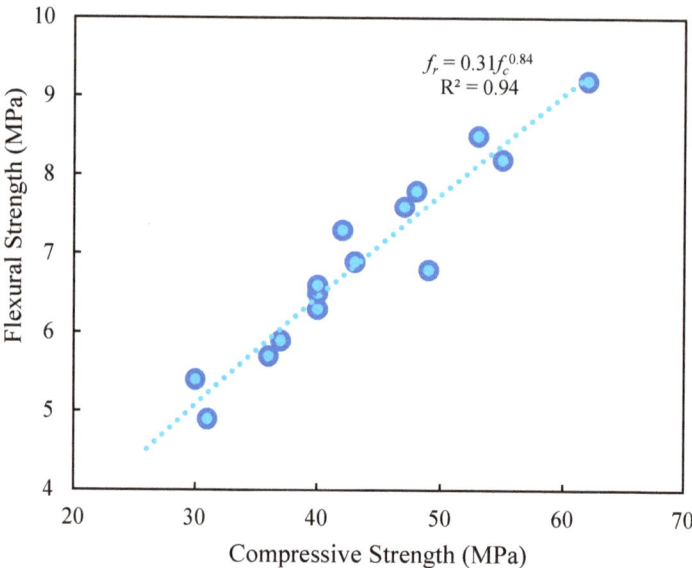

Figure 15. Relationship between compressive resistance and fracture energy of concrete containing different PF and RTA fractions.

4.4. Ultrasonic Pulse Velocity

As per the Guo et al. study [38], the ultrasonic pulse velocity measurement was conducted after 28 days. A sample with a thickness of 30 mm and a diameter of 100 mm was made for this test. The top and bottom outsides of samples were applied to two Olympus 5070 transducers at a frequency of 0.5 MHz, and the average of three samples was used to calculate the pulse speed. The findings are shown in Figure 16, which demonstrates that employing both PF and RTA has a significant negative impact on this attribute. Additionally, the value of the ultrasonic pulse velocity decreases in the underdeveloped area when 20% RTA is applied. So, the ultrasonic pulse velocity is decreased by roughly 21%, 28%, and 31%, respectively, when 20% RTA and 0%, 1%, and 2% PF are used. Since the RTA's elastic moduli are significantly lower than that of FNA and the cured cement matrix. Conversely, RTA increased the number of air spaces in the solid slides, which may have an impact on the pulse rate. Employing RTA causes concrete's ultrasonic pulse velocity to decrease; nevertheless, this characteristic was significantly enhanced when PF was used. As compared to natural aggregate, RTA has a larger Poisson ratio and lower stiffness, which contributes significantly to the ultrasonic pulse velocity decrease. The deterioration of strength may also be caused by poor adherence to cement paste and RTA. The addition of RTA also had an impact on the stability of the concrete samples, causing the ultrasonic pulse velocity to decrease. As a result, adding RTA to ordinary cement mix reduced its ultrasonic pulse velocity. Previous research has not employed this test to assess the behavior of PF-reinforced RTA, demonstrating its originality.

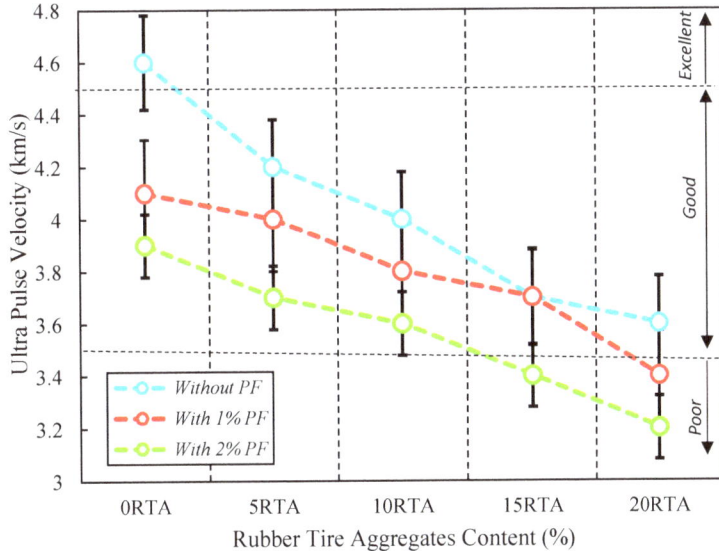

Figure 16. Effect of PF and RTA on the ultrasonic pulse velocity.

4.5. Drying Shrinkage

As per ASTM C157 [39], concrete's drying shrinkage performance was evaluated. For this test, 25 mm × 25 mm × 286 mm samples were tested at 1, 3, 5, 7, 11, 14, and 21 days of curing. After 24 h of curing, the samples were taken out of the molds and kept in the curing chamber at a temperature of 23 °C and relative humidity of 50%. Each sample's starting length was measured right after demolding, and Figure 17 shows how the length varies over time. The samples' lengths changed as the amount of RTA grew, as seen in Figure 17. Due to the lessened stiffness of RTA when compared with NFA, RTA could be easily deformed under the internal drying shrinkage stress; however, incorporating PF decreased the shrinkage length deviations and the amplified drying shrinkage caused by RTA could be properly limited with the application of PF. Furthermore, the length variations got worse at 14 days of cure and got worse till 21 days. As a result, utilizing RTA of 5%, 10%, 15%, and 20% increased the length variation at 28 days by 10%, 14%, 21%, and 27%, correspondingly. However, when 2% PF was added along with 0%, 5%, 10%, 15%, and 20% RTA, shrinkage decreased by roughly 37%, 28%, 21%, 20%, and 15% in comparison to the control specimen without PF and RTA. In addition, utilizing 2% PF has a greater impact than using 1% PF on minimizing the amount of shrinkage length changes.

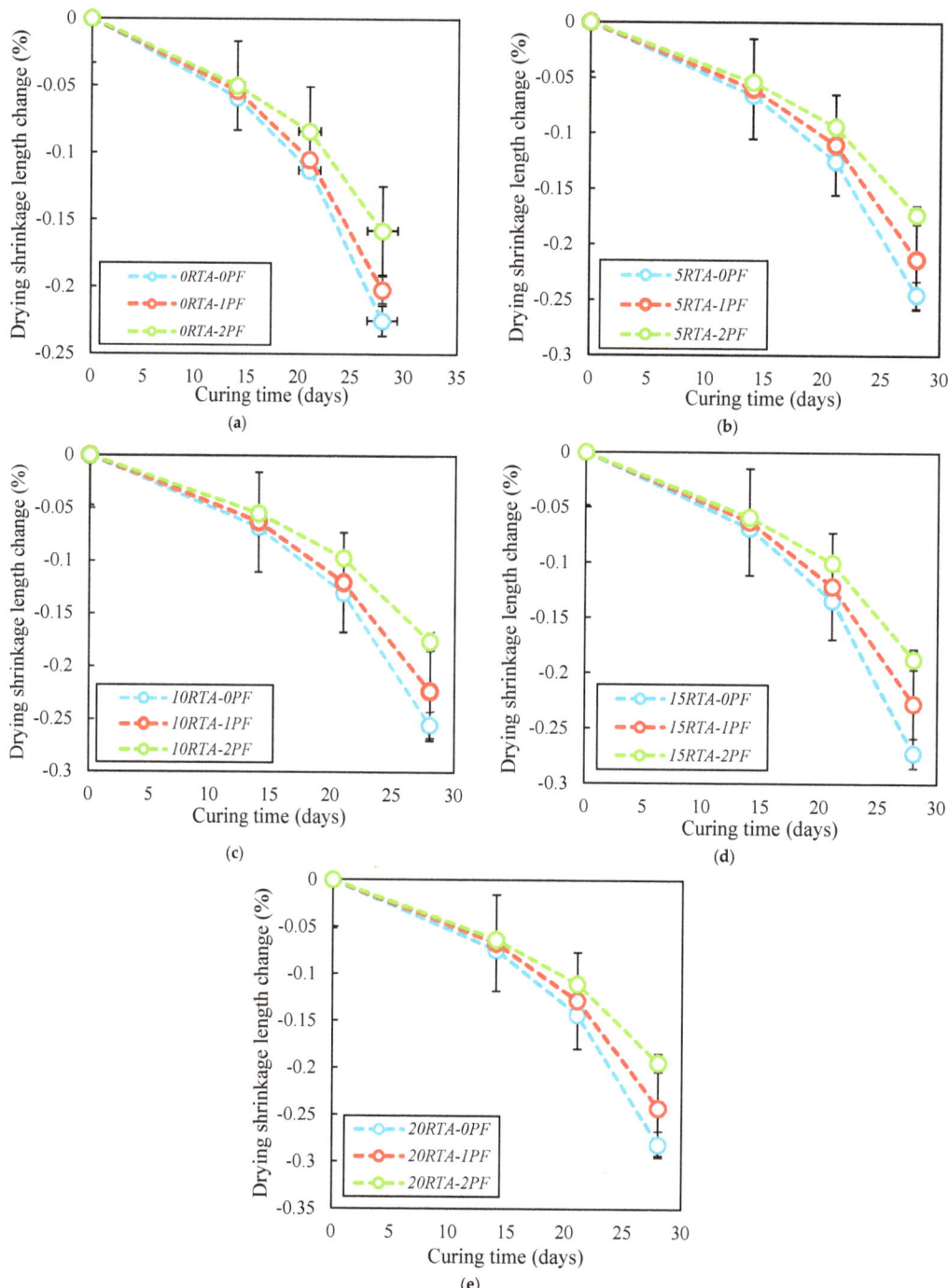

Figure 17. Influence of PF on the length variations due to shrinkage having: (**a**) 0% RTA; (**b**) 5% RTA; (**c**) 10% RTA; (**d**) 15% RTA and; (**e**) 20% RTA.

4.6. Bulk Electrical Resistivity

Hamed et al. [40] developed a novel approach for determining concrete's bulk electrical resistivity in 2015. For this aim, cylindrical samples with a diameter of 102 mm and a height of 204 mm were subjected to a frequency between 10 Hz and 10,000 Hz. The findings of three samples were averaged, and the results are shown in Figure 18. A total of 10% of RTA is the ideal amount to use to produce the highest possible bulk resistivity value. Therefore, by adding 2% PF and 10% RTA, this characteristic was enhanced by around 12%. The bulk resistivity is significantly improved by using PF, particularly when more than 10% RTA is used in place of natural aggregates. Furthermore, when more than 10% RTA was added, the resistance significantly decreased. The pathways of the pore solution can be obstructed by the rubber aggregates, Increasing the bulk electrical resistivity; nevertheless, employing extra RTA than 10% increased the absorbency within the mixture and increased slump. As a result, utilizing RTA of 5%, and 10% increased the bulk electrical resistivity of concrete by 8% and 10%, correspondingly. However, when 2% PF was added along with 0%, 5% and 10% RTA, the bulk electrical resistivity of concrete amplified by roughly 7%, 9.5% and 16%, in comparison to the control specimen without PF and RTA.

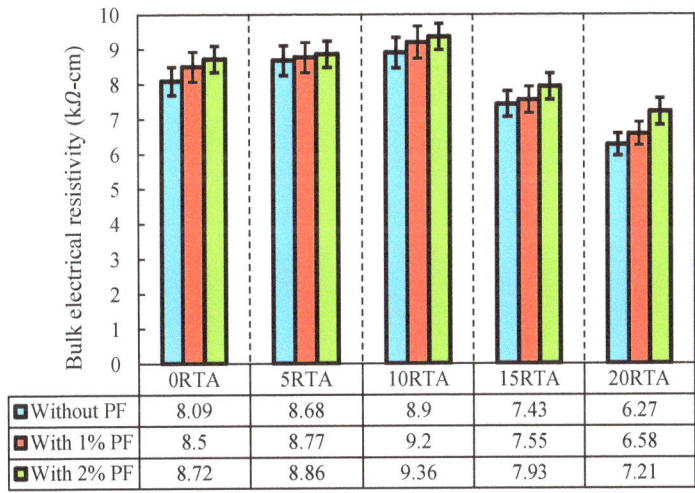

Figure 18. Impact of PF and RTA on the bulk electrical resistance.

4.7. Expansion

The quick alkali-silica response test to gauge concrete's length expansion is described in ASTM C1260 [41]. The specimens utilized in this test are the same size as those that were employed in the shrinkage test. To cure the samples, tap water was used in a container that was positioned in an oven set at 80 °C for 24 h before the samples were demolded. Following the early measurement, the samples were submerged in a second container containing a 40 g/L NaOH solution. Figure 19 shows the test marks as a function of the amount of PF and RTA used. After 7 days, the expansion rate significantly increased. Furthermore, the length expansion changes were significantly reduced because of the use of RTA. In addition, after 11 days, the length expansion changes were significantly slowed by 20% RTA (Figure 19e). Because of its bridging action, applying 2% PF delays the length expansion of mixtures having PF. The fibers may constrain the cement matrix since it tended to expand, which would prevent further growth of the expansion. Additionally, it should be highlighted that by boosting the amount of RTA, the impact of PF on the alkali-silica response growth is reduced. Consequently, when 20% RTA was employed, there was no discernible difference in length expansion between reinforcing concrete with 2% PF and 1% PF.

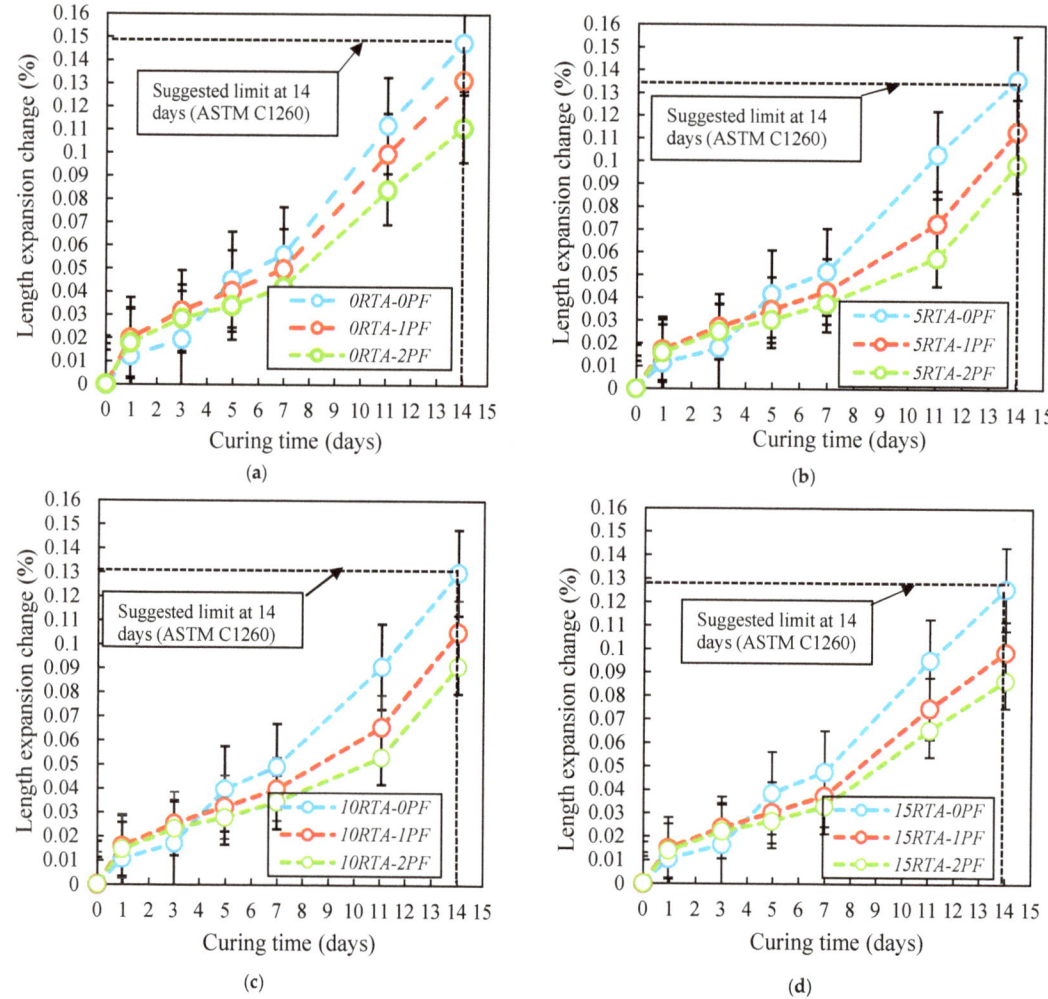

Figure 19. *Cont.*

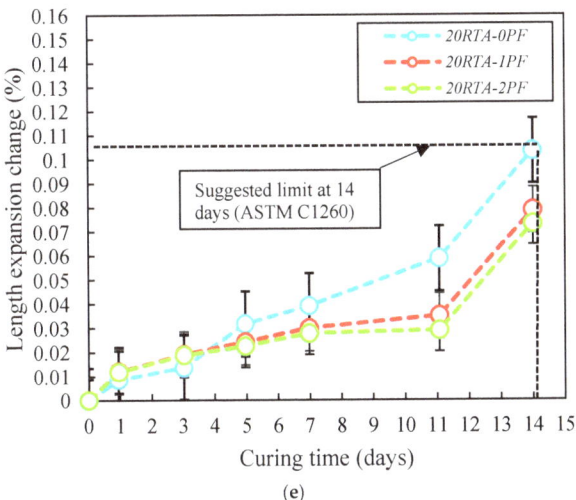

Figure 19. Influence of PF and RTA on the length growth. (**a**) without RTA, (**b**) with 5% RTA, (**c**) with 10% RTA, (**d**) with 15% RTA and (**e**) with 20% RTA

4.8. Freeze-Thaw Resistance

As per ASTM C666-15 [42], concrete specimens measuring 76 mm by 102 mm by 381 mm were created and kept in a freeze-thaw chamber. Therefore, after 28 days of underwater cure at 220 °C, the samples' early mass weight, length, and resonance frequency were noted. During the test, length variations and the dynamic elastic modulus were assessed under different freeze-thaw cycles (0, 36, 50, 100, 150, 200, 250, and 300). As a result, Figure 20 shows the relative modulus of dynamic elasticity. The increase in dynamic elastic modulus was a sign that the concrete specimens generated for this evaluation were of good quality and freeze-thaw resistant. As shown in Figure 20, utilizing RTA with up to 10% raises the relative dynamic elastic modulus, but using RTA with more than 10% causes a large decrease in this parameter. When compared to the original state, the hydration created a more solid structure inner of the specimens which results in improving the relative dynamic elastic modulus. For 10RTA-0PF specimens, the relative dynamic elastic modulus significantly increased after 50 cycles. Alternatively, when 15% or 20% RTA were utilized, adding PF decreased the dynamic elastic modulus (Figure 20c). It should be mentioned that the dynamic modulus changes significantly up to 35 cycles. The dynamic elastic moduli remained constant after 45 cycles, and a similar trend was seen for samples with various PF and RTA amounts. Therefore, using 10% RTA with 0%, 1%, and 2% PF, respectively, raised the property's maximum value by around 48%, 56%, and 51%. Almost the same results were reported by Alsaif et al. [43]. When RTA aggregates were used, the dynamic elastic modulus reduces by 10% in comparison with the control sample.

Moreover, the specimens' resonance frequency was evaluated. The resonant frequency of the samples during the freeze-thaw testing is demonstrated in Figure 21. The resonant frequency was dramatically lowered using RTA and PF. Furthermore, there were significant fluctuations in resonance frequency up to 50 cycles. However, when 15% and 20% RTA were used, this property did not stabilize until after 100 and 150 cycles, respectively. A concrete's resistance to freeze-thaw cycles is shown by an increase in the dynamic modulus of elasticity and a decrease in resonance frequency. Because RTA is considerably less rigid than traditional fine aggregate, the resonance frequency of concrete samples has been reduced. Additionally, the RTA may allow more air to enter the concrete samples, increasing the amount of air space in the cured concrete specimens.

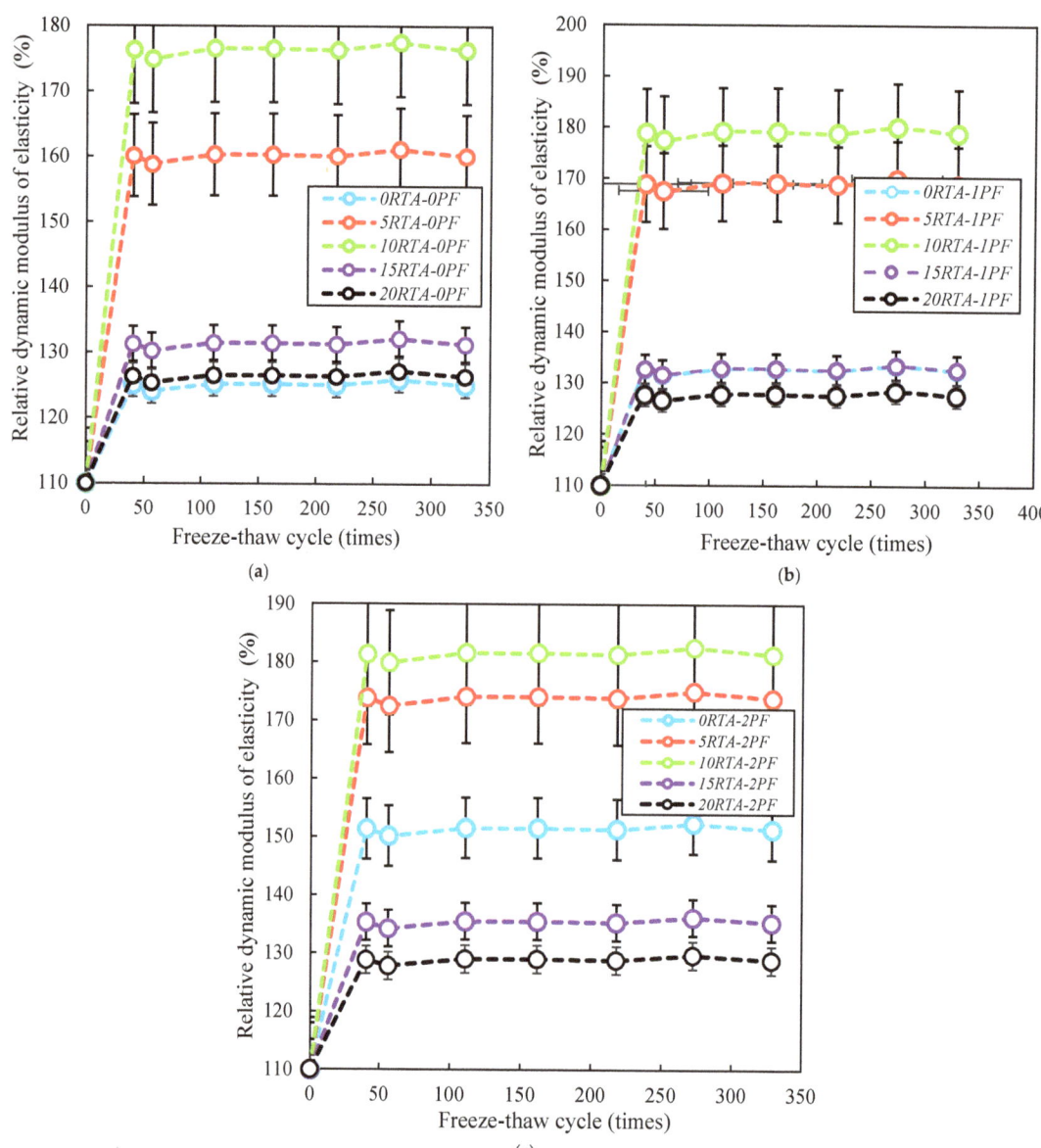

Figure 20. Dynamic elastic modulus of concrete having PF and RTA. (**a**) without PF, (**b**) with 1% PF and (**c**) with 2% PF

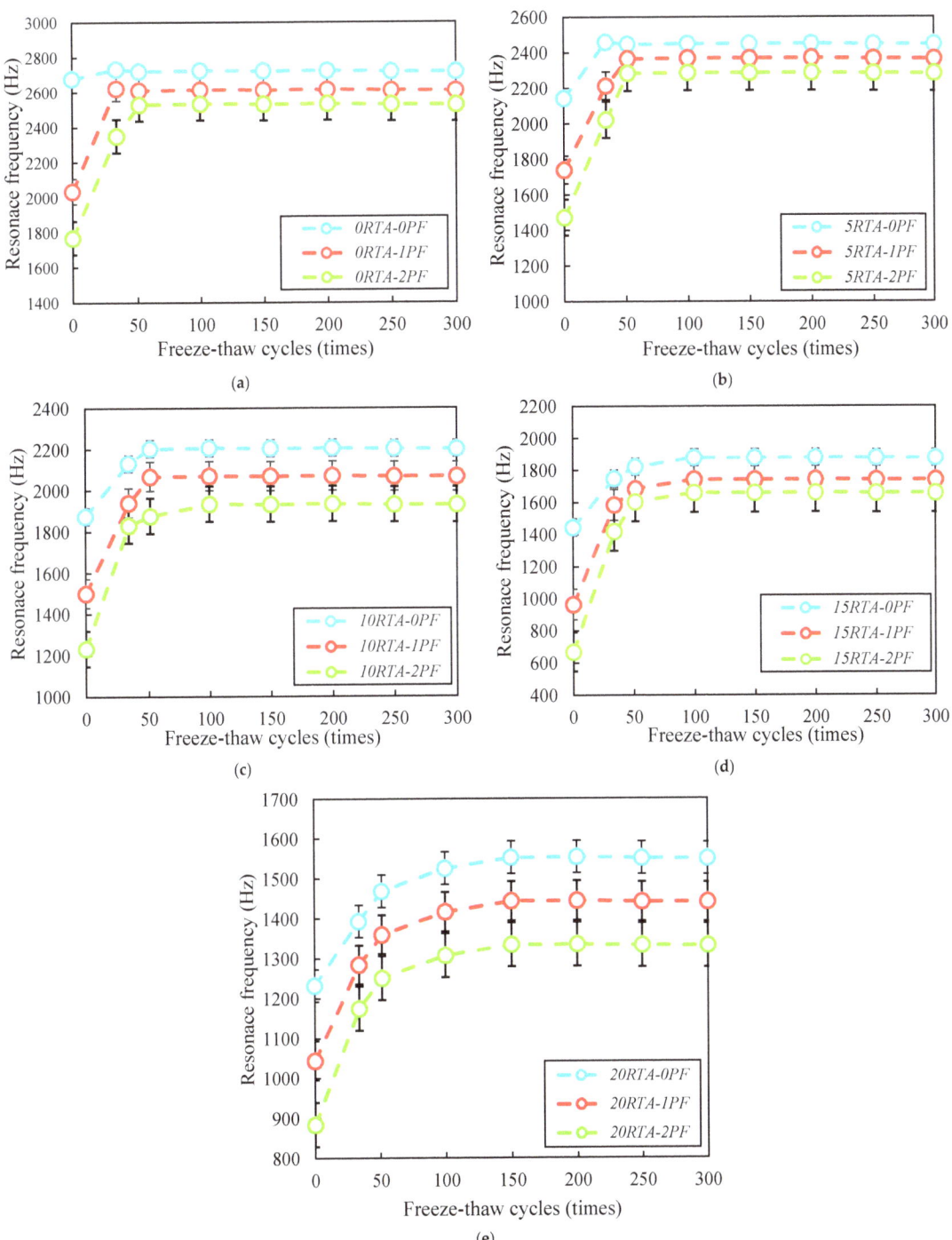

Figure 21. Influence of PF and RTA on the frequency: (**a**) Without RTA; (**b**) 5% RTA; (**c**) 10% RTA; (**d**) 15% RTA and; (**e**) 20% RTA.

4.9. Freeze-Thaw Damage Durability Aspect

As per the consequences of the dynamic elastic moduli utilizing the next equation, durability was taken into consideration:

$$DF = PN/M \qquad (6)$$

The durability feature of the samples, the relation dynamic elastic moduli at N cycles, the number of cycles until P ranges the detailed lowest rate for stopping the examination, and the quantified number of cycles until the experience is to be stopped is represented by the letters *DF, P, N,* and *M,* correspondingly. The durability of the samples is shown in Figure 22. As the freeze-thaw performance is improved, the durability factor increases. In relation to durability, 10% represents the RTA's ideal value because beyond that point, it significantly decreased. The inclusion of RTA has enhanced the air content, which might enhance freeze-thaw resistance. Additionally, as the bulk electrical resistivity result indicates that the transport property of the 10RTA-0PF sample is the lowest, samples containing 10% RTA may associate with improved durability. Additionally, incorporating PF increased durability, particularly when 2% PF was applied. Additionally, once 2% PF was added with 0%, 5%, and 10% RTA to create concrete, the durability factor increased by 13%, 31%, and 37%.

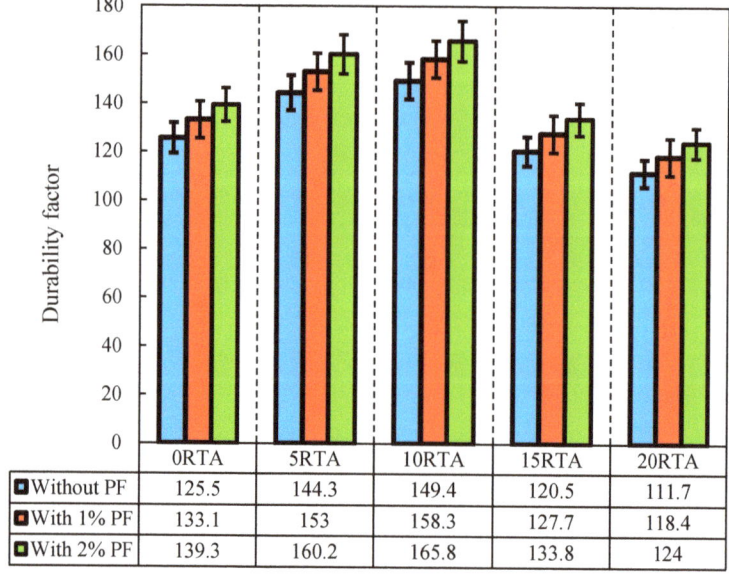

	0RTA	5RTA	10RTA	15RTA	20RTA
Without PF	125.5	144.3	149.4	120.5	111.7
With 1% PF	133.1	153	158.3	127.7	118.4
With 2% PF	139.3	160.2	165.8	133.8	124

Figure 22. Durability factor of the concrete having different PF and RTA fractions.

4.10. Length Change

At each exposure stage, the length variations of several samples were documented, as can be shown in Figure 23, where the length underwent a significant shift for up to 50 cycles. Due to continuous cement hydration, autogenous shrinkage occurred whereas samples were kept in the freeze-thaw chamber. The specimens' length gradually increased for the number of cycles less than 50, and then decreased by increasing the number of the freeze-thaw cycle. Additionally, by boosting the PF content, the specimens' length reduction became more pronounced. Additionally, by adding more RTA, the length of the samples was further reduced. Furthermore, by boosting the content of RTA, the variations in specimen length recovered less. Specimens made with RTA exhibit more significant shrinkage changes than those made with natural aggregates. The specimens' length roughly stabilized after 150 cycles.

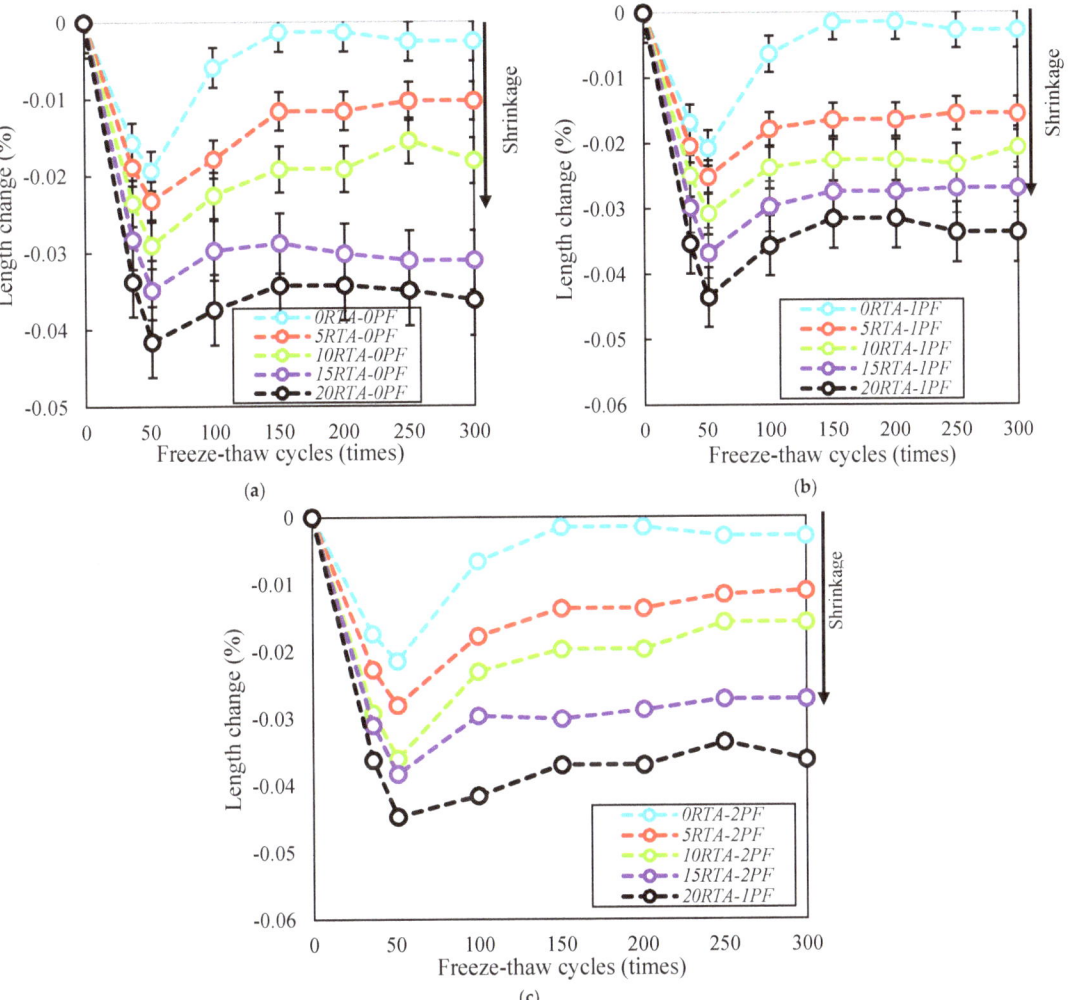

Figure 23. Change the length of the concrete mixtures having different PF and RTA fractions. (**a**) without PF, (**b**) with 1% PF and (**c**) 2% PF

5. Conclusions

The current investigation evaluated the characteristics of concrete having PF and RTA. Fifteen mixes were created with this objective in mind. Five ratios of RTA (0, 5, 10, 15, and 20%) were employed instead of FNA. The concrete mixes also included PF at three different volume contents: 0%, 1%, and 2%. Then, measurements were made of the specimens' workability, compressive and tensile and flexural resistances, ultrasonic pulse velocity, shrinkage, electrical resistance, dynamic elastic moduli, and resonance frequency. The subsequent conclusions could be made based on the achieved findings:

1. Workability was reduced by adding more RTA, especially when a combination of 20% RA and 2% PF was utilized, which resulted in a slump reduction of nearly 49%.

Conversely, when PF was not utilized in the concrete mixtures, workability was reduced less. So, this property was reduced by about 18% when only 20% RTA was used. Additionally, utilizing 1% PF and 20% rubber reduced slump by about 31%.

2. Utilizing RTA reduced the concrete's tensile resistance. While PF greatly enhanced this feature. Due to the minor value of RTA relative to FNA, the tensile strength decreased by 28% when 20% RTA was used. Though, incorporating 2% PF amplified the tensile resistance by 5% and prohibited the drop of tensile resistance because of the fiber-bridging character in preventing the quick cracks growth.
3. Compressive resistance of concrete was decreased when RTA was used; however, this characteristic was significantly increased when PF was incorporated. In comparison to concrete produced with merely 0%, 5%, 10%, 15%, and 20% RTA combining 2% PF increased the concrete's 7-day compressive resistance by about 24%, 29%, 24%, 25% and 22%, correspondingly.
4. By employing RTA, the specimens' flexural resistance was decreased while PF significantly enhances the flexural performance of the specimens, particularly when RTA was not utilized, due to its bridging function, which increases particle connection and elevates concrete's tensile and compressive resistances.
5. Incorporating SF and RTA significantly declined the ultrasonic pulse velocity due to the elastic modulus of RTA being lesser than that of the concrete paste and FNA, when 20% RTA was employed, the ultrasonic pulse velocity decreased in the weak zone.
6. Due to the RTA's lower stiffness than that of the FNA, and because of which RTA could easily deform under the internal drying shrinkage stress, the specimens' shrinkage length changes increased with the RTA content; however, the addition of PF decreased these length changes. Therefore, using 5%, 10%, 15%, and 20% RTA as a substitute for FNA increased shrinkage length at 28 days by 10%, 14%, 21%, and 27%, correspondingly.
7. A total of 10% of RTA was the optimal amount to maximize bulk resistivity. With the addition of 2% PF and 10% RTA, this characteristic was improved by roughly 12%. It is worth noting that the ideal rate to achieve the highest possible bulk resistivity was established based on the amounts of PF and RTA employed in this investigation. Additionally, SF significantly improved bulk resistivity, particularly when more than 10% RTA was utilized;
8. By adding RTA, the length growth under alkali-silica reaction assault was significantly reduced. Additionally, by adding more RTA, the impact of PF on the alkali-silica response development decreased, and there is no discernible alteration between the length growth of the sample with 2% PF reinforcement and concrete with 1% PF and 20% RTA.
9. According to the materials utilized in this investigation, 10% is the ideal amount of RTA to maximize the durability factor before it begins to significantly decline. Additionally, the durability factor increased by 13%, 31%, and 37%, respectively, when 2% PF was added together with 0%, 5%, and 10% RTA compared to the control mix.
10. Up to 50 freeze-thaw cycles resulted in a momentous change in the specimens' length. Due to ongoing cement hydration, autogenous shrinkage happened when the specimens were kept in the freeze-thaw chamber. At low cycle counts, the length changes of the specimens abruptly decreased and then gradually increased. The specimens' length reduction also got worse as PF content increased. Alternatively, it declined as the RTA content increased, and the length changes recovered less as the RTA increased.

Author Contributions: Conceptualization: A.K.P., Z.M. and E.N.F.; Formal analysis: A.K.P., Z.M.; Investigation: A.K.P., Z.M. and E.N.F.; Methodology: A.K.P., Z.M. and E.N.F.; Validation: A.K.P. and Z.M.; Visualization, A.K.P. and Z.M.; Writing—review & editing, A.K.P., Z.M. and E.N.F. All authors have read and agreed to the published version of the manuscript.

Funding: This research received no external funding.

Institutional Review Board Statement: Not applicable.

Informed Consent Statement: Not applicable.

Data Availability Statement: Some or all data, models, or codes that support the findings of this study are available from the corresponding author upon reasonable request.

Conflicts of Interest: On behalf of all authors, the corresponding author states that there is no conflict of interest.

References

1. Karimi Pour, A.; Jahangir, H.; Rezazadeh Eidgahee, D. A thorough study on the effect of red mud, granite, limestone and marble slurry powder on the strengths of steel fibres-reinforced self-consolidation concrete: Experimental and numerical prediction. *J. Build. Eng.* **2021**, *44*, 103398. [CrossRef]
2. Gencel, O.; Kazmi, S.M.S.; Munir, M.J.; Kaplan, G.; Bayraktar, O.Y.; Yarar, D.O.; Karimi Pour, A.; Ahma, M.R. Influence of bottom ash and polypropylene fbers on the physico-mechanical, durability and thermal performance of foam concrete: An experimental investigation. *Constr. Build. Mater.* **2021**, *306*, 124887. [CrossRef]
3. Karimi Pour, A.; Ghalehnovi, M.; Golmohammadi, M.; Brito, J. Experimental Investigation on the Shear Behaviour of Stud-Bolt Connectors of Steel-Concrete-Steel Fibre-Reinforced Recycled Aggregates Sandwich Panels. *Materials* **2021**, *14*, 5185. [CrossRef]
4. Sutcu, M.; Gencel, O.; Erdogmus, E.; Kizinievic, O.; Kizinievic, V.; Karimi Pour, A.; Muñoz Velasco, P. Low cost and eco-friendly building materials derived from wastes: Combined effects of bottom ash and water treatment sludge. *Constr. Build. Mater.* **2022**, *324*, 126669. [CrossRef]
5. Binti Katman, H.Y.; Khai, W.J.; Serkan Kırgız, M.; Nehdi, M.L.; Benjeddou, O.; Thomas, B.S.; Papatzani, S.; Rambhad, K.; Kumbhalkar, M.A.; Karimi Pour, A. Transforming Conventional Construction Binders and Grouts into High-Performance Nanocarbon Binders and Grouts for Today's Constructions. *Buildings* **2022**, *12*, 1041. [CrossRef]
6. Khitab, A.; Serkan Kırgız, M.; Nehdi, M.L.; Mirza, J.; de Sousa Galdino, A.G.; Karimi Pour, A. Mechanical, thermal, durability and microstructural behavior of hybrid waste-modified green reactive powder concrete. *Constr. Build. Mater.* **2022**, *344*, 128184. [CrossRef]
7. Fasihihour, N.; Mohebbi Najm Abad, J.; Karimi Pour, A.; Mohebbi, M.R. Experimental and numerical model for mechanical properties of concrete containing fly ash: Systematic Review. *Measurement* **2022**, *188*, 110547. [CrossRef]
8. Abdelmonem, A.; El-Feky, M.S.; El-Sayed, A.R.N.; Kohail, M. Performance of high strength concrete containing recycled rubber. *Constr. Build. Mater.* **2019**, *227*, 116660. [CrossRef]
9. Zhu, H.; Wang, Z.; Xu, J.; Han, Q. Microporous structures and compressive strength of high-performance rubber concrete with internal curing agent. *Constr. Build. Mater.* **2019**, *215*, 128–134. [CrossRef]
10. Li, Y.; Zhang, X.; Wang, R.; Lei, Y. Performance enhancement of rubberised concrete via surface modification of rubber: A review. *Constr. Build. Mater.* **2019**, *227*, 116691. [CrossRef]
11. Strukar, K.; Šipoš, T.K.; Miličević, I.; Bušić, R. Potential use of rubber as aggregate in structural reinforced concrete element—A review. *Eng. Struct.* **2019**, *188*, 452–468. [CrossRef]
12. Li, Y.; Zhang, S.; Wang, R.; Dang, F. Potential use of waste tire rubber as aggregate in cement concrete—A comprehensive review. *Constr. Build. Mater.* **2019**, *225*, 1183–1201. [CrossRef]
13. Guo, Q.; Zhang, R.; Luo, Q.; Wu, H.; Sun, H.; Ye, Y. Prediction on damage evolution of recycled crumb rubber concrete using quantitative cloud imagine correlation. *Constr. Build. Mater.* **2019**, *209*, 340–353. [CrossRef]
14. Záleská, M.; Pavlik, Z.; Čítek, D.; Jankovský, O.; Pavlíková, M. Eco-friendly concrete with scrap-tyre-rubber-based aggregate—Properties and thermal stability. *Constr. Build. Mater.* **2019**, *225*, 709–722. [CrossRef]
15. Roychand, R.; Gravina, R.J.; Zhuge, Y.; Ma, X.; Youssf, O.; Mills, J.E. A comprehensive review on the mechanical properties of waste tire rubber concrete. *Constr. Build. Mater.* **2020**, *237*, 117651. [CrossRef]
16. Feng, W.; Liu, F.; Yang, F.; Li, L.; Jin, L.; Chen, B.; Yuan, B. Experimental study on the effect of strain rates on the dynamic flexural properties of rubber concrete. *Constr. Build. Mater.* **2019**, *224*, 408–419. [CrossRef]
17. Chan, C.W.; Yu, T.; Zhang, S.S. and Xu, Q.F. Compressive behaviour of FRP-confined rubber concrete. *Constr. Build. Mater.* **2019**, *211*, 416–426. [CrossRef]
18. Wang, J.; Dai, Q.; Si, R.; Guo, S. Mechanical, durability, and microstructural properties of macro synthetic polypropylene (PP) fibre-reinforced rubber concrete. *J. Clean. Prod.* **2019**, *234*, 1351–1364. [CrossRef]
19. Farhad, A.; Ronny, G. Experimental investigation into the properties of self-compacting rubberised concrete incorporating polypropylene and steel fibers. *Int. Fed. Struct. Concr.* **2018**, *20*, 267–281.
20. Nguyen, T.H.; Toumi, A.; Turatsinze, A. Mechanical properties of steel fibre reinforced and rubberised cement-based mortars. *Mater. Des.* **2010**, *31*, 641–647. [CrossRef]
21. Karimi Pour, A. Experimental and numerical evaluation of steel fibres RC patterns influence on the seismic behaviour of the exterior concrete beam-column connections. *Eng. Struct.* **2022**, *263*, 114358. [CrossRef]
22. Karimi Pour, A.; Brito, J.; Ghalehnvoi, M.; Gencel, O. Torsional behaviour of rectangular high-performance fibre-reinforced concrete beams. *Structures* **2022**, *35*, 511–519. [CrossRef]
23. Karimi Pour, A.; Ghalehnovi, M.; Edalati, M.; Brito, J. Properties of Fibre-Reinforced High-Strength Concrete with Nano-Silica and Silica Fume. *Appl. Sci.* **2021**, *11*, 9696. [CrossRef]

24. Ghalehnovi, M.; Karimi Pour, A.; Anvari, A.; Brito, J. Flexural strength enhancement of recycled aggregate concrete beams with steel fbre-reinforced concrete jacket. *Eng. Struct.* **2021**, *240*, 112325. [CrossRef]
25. Ghalehnovi, M.; Karimi Pour, A.; Brito, J.; Chaboki, H.R. Crack Width and Propagation in Recycled Coarse Aggregate Concrete Beams Reinforced with Steel Fibres. *Appl. Sci.* **2020**, *10*, 7587. [CrossRef]
26. Karimi Pour, A. Effect of untreated coal waste as fine and coarse aggregates replacement on the properties of steel and polypropylene fibres reinforced concrete. *Mech. Mater.* **2020**, *150*, 103592. [CrossRef]
27. Rezaiee-Pajand, M.; Karimi Pour, A.; Mohebbi Najm Abad, J. Crack Spacing Prediction of Fibre-Reinforced Concrete Beams with Lap-Spliced Bars by Machine Learning Models. *Iran. J. Sci. Technol. Trans. Civ. Eng.* **2020**, *45*, 833–850. [CrossRef]
28. Anvari, A.; Ghalehnovi, M.; Brito, J.; Karimi Pour, A. Improved bending behaviour of steel fibres recycled aggregate concrete beams with a concrete jacket. *Mag. Concr. Res.* **2019**, *73*, 608–626. [CrossRef]
29. Farokhpour, M.; Ghalehnovi, M.; Karimi Pour, A.; Amanian, M. Effect of Polypropylene Fibers on the Behavior of Recycled Aggregate Concrete. In Proceedings of the 5th National Conference on Recent Achievements in Civil Engineering, Architecture and Urbanism, Tehran, Iran, 14 February 2019.
30. Ghalehnovi, M.; Farokhpour Tabrizi, M.; Karimi Pour, A. Investigation of the effect of steel fibers on failure extension of recycled aggregate concrete beams with lap-spliced bars. *Sharif J. Civ. Eng.* **2019**, *18*, 12.
31. Si, R.; Wang, J.; Guo, S.; Dai, Q.; Han, S. Evaluation of laboratory performance of self-consolidating concrete with recycled tire rubber. *J. Clean. Prod.* **2018**, *180*, 823–831. [CrossRef]
32. ASTM C150/C150M; Standard Specification for Portland Cement. Annual Book of ASTM Standards. ASTM International Standards Organization: West Conshohocken, PA, USA, 2012.
33. ASTM D2419; Standard Test Method for the Sand Equivalent Value of Soils and Fine Aggregate. ASTM International Standards Organization: West Conshohocken, PA, USA, 2002.
34. ASTM C143/C143M-15a; Standard Test Method for Slump of Hydraulic Cement Concrete. ASTM International: West Conshohocken, PA, USA, 2015.
35. ASTM C496/C496M-17; Standard Test Method for Splitting Tensile Strength of Cylindrical Concrete Specimens. ASTM International: West Conshohocken, PA, USA, 2017.
36. Wang, Y.; Chen, J.; Gao, D.; Huang, E. Mechanical properties of steel fibers and nano-silica modified crumb rubber concrete. *Adv. Civ. Eng.* **2018**, *10*, 124875.
37. JCI-S-002-2003; Method of Test for the Load-Displacement Curve of Fiber Reinforced Concrete by Use of Notched Beam. Japan Concrete Institute (JCI): Tokyo, Japan, 2003.
38. Guo, S.; Dai, Q.; Sun, X.; Sun, Y. Ultrasonic scattering measurement of air void size distribution in hardened concrete samples. *Constr. Build. Mater.* **2016**, *113*, 415–422. [CrossRef]
39. ASTM C157/C157M-17; Standard Test Method for Length Change of Hardened Hydraulic-Cement Mortar and Concrete. ASTM International: West Conshohocken, PA, USA, 2017.
40. Hamed, L.; Ghods, P.; Alizadeh, A.R.; Salehi, M. Electrical resistivity of concrete. *Concr. Int.* **2015**, *37*, 41–46.
41. ASTM C1260-14; Standard test method for potential alkali reactivity of aggregates (mortar-bar method). ASTM International: West Conshohocken, PA, USA, 2014.
42. ASTM C666/C666M-15; Standard test method for resistance of concrete to rapid freezing and thawing. ASTM International: West Conshohocken, PA, USA, 2015.
43. Alsaif, A.; Susan, A.; Bernal, S.A.; Guadagnini, M.; Pilakoutas, K. Freeze-thaw resistance of steel fibre reinforced rubberised concrete. *Constr. Build. Mater.* **2019**, *195*, 450–458. [CrossRef]

Article

Experimental Study on Secondary Anchorage Bond Performance of Residual Stress after Corrosion Fracture at Ends of Prestressed Steel Strands

Rihua Yang [1,2], Yiming Yang [1,*], Xuhui Zhang [3] and Xinzhong Wang [1]

1. School of Civil Engineering, Hunan City University, Yiyang 413000, China
2. Hunan Engineering Research Center of Development and Application of Ceramsite Concrete Technology, Hunan City University, Yiyang 413000, China
3. School of Civil Engineering and Mechanics, Xiangtan University, Xiangtan 411105, China
* Correspondence: yangyiming@hncu.edu.cn

Abstract: In order to explore the secondary bond anchorage performance between prestressed tendons and concrete after the fracture of steel strands in post-tensioned, prestressed concrete (PPC) beams, a total of seven post-tensioned, prestressed concrete specimens with a size of $3 \times 7\phi15.2$ mm were constructed firstly, and the steel strands at the anchorage end were subjected to corrosion fracture. Then, the pull-out test of the specimens was conducted to explore the secondary anchorage bond mechanism of the residual stress of prestressed tendons experiencing local fracture. Moreover, the influences of factors such as the embedded length, release-tensioning speed, concrete strength, and stirrup configuration on anchorage bond performance were analyzed. Finally, the test results were further verified via finite element analysis. The results show that the failure of pull-out specimens under different parameters can be divided into two types: bond anchorage failure induced by the entire pull-out of steel strands and material failure triggered by the rupture of steel strands. The bond anchorage failure mechanism between steel strands and the concrete was revealed by combining the failure characteristics and pull-out load–slippage relation curves. The bond strength between prestressed steel strands and concrete can be enhanced by increasing the embedded length of steel strands, elevating the concrete strength grade, and enlarging the diameter of stirrups so that the specimens are turned from bond anchorage failure into material failure.

Keywords: bond performance; bridge engineering; corrosion fracture; numerical simulation; secondary anchorage

1. Introduction

The concept of stress transfer and anchorage, which derives from pre-tensioned specimens, refers to the process of stress increases in prestressed tendons at the beam end from zero to the effective stress and ultimate stress [1]. The pre-tensioned specimens complete the prestress transfer and anchorage through the bonding between prestressed tendons and concrete, and the successful bonding between them is the basis for the normal service of pre-tensioned members. Therefore, it is necessary to study bond anchorage performance between steel strands and the concrete of pre-tensioned concrete members. The relevant research results can also provide a scientific basis for the safety assessment and maintenance decision-making processes of bridges in the future [2–4].

It is generally believed that the bonding force between steel strands and concrete is composed of adhesion, mechanical interlock, and friction [5]. The bond anchorage performance of steel strands is similar to that of plain steel bars in the early loading stage, i.e., the bonding force is mainly provided by adhesion. In the later loading stage, mechanical interlock plays a dominant role since the twisting process is adopted for steel strands, and slippage is accompanied by the rotation along the twisting direction [6–8]. The bond performance between steel strands and concrete is correlated with concrete strength, the thickness of the concrete cover, the stirrup ratio, and the shape of prestressed tendons [9–11]. In some studies, bond stress–slippage models have been proposed through experimentation and theoretical analysis. As a result, the nonuniform distribution characteristics of bond stress along the embedded length have been determined [12,13]. With an increase in concrete compressive strength, the average bond stress increases while the anchorage length decreases [14,15]. Because stirrups can confine the development of splitting cracks, the slip amount is greater when anchorage fails. In other words, with an increase in the stirrup ratio, both the bond strength and ultimate slip amount increase. To avoid the failure of the structure due to anchorage failure, many national codes have given calculation formulas for the anchorage length of flexural specimens [1,16,17]. However, the existing research results show that the calculation of the anchorage length given in the codes is partially conservative and safe [9,18,19].

For bonded post-tensioned, prestressed concrete (PPC) members, the prestress is transferred through the anchorage in the prestress application stage, and the anchorage bonding is completed through the bonding between prestressed tendons and concrete in the working stage [20]. Due to construction defects, however, PPC specimens may experience insufficient grouting, so that the structure develops from fully bonded prestressed concrete to partially bonded prestressed concrete or even unbonded prestressed concrete. The structure can still work normally without any bond failure thanks to the end anchorage. In practical construction, due to construction defects such as the untimely closure of the anchorage end or cavities in the concrete at the anchor sealing position, corrosion occurs in the end anchorage zone with the extension of the service time, which leads to anchorage failure [21–23]. It is considered that, after the anchorage failure of the PC specimens, which is similar to pre-tensioned specimens, the residual prestress can be transferred and anchored for the second time through the bond with concrete [24,25]. Its mechanism is similar to that of pre-tensioned tendons but there are also differences. On the one hand, the prestressed tendons of post-tensioned specimens are mostly placed in the specimens in the form of multiple steel strands and the prestressed tendons interact with each other [26]; on the other hand, the corrosion process of prestressed tendons is slow, which is different from the rapid release-tensioning of pre-tensioned tendons. Rapid release-tensioning increases the initial damage of concrete and reduces the bond stiffness, which affects the secondary anchorage bond performance of the residual stress.

The secondary transfer and anchorage of residual stress after the stress fracture of prestressed tendons have been preliminarily studied [25]. In this study, the flexural performance of the specimens after the local fracture of a single steel strand was explored. The results show that the flexural bearing capacity of the specimens experiencing end anchorage failure only declined by 7.2% [24]. The calculation method for the bonding length of a single steel strand after end fracture was obtained by setting ribbed pre-embedded pipelines. However, corrugated pipes are usually used in practical projects, which is obviously not consistent with a real-world situation [25]. At present, the bond performance of tendons and concrete after the anchorage failure at the end of post-tensioned specimens has not been systematically investigated.

Therefore, bond performance between the steel strands and concrete after the end anchorage failure of the 3 × 7ϕ15.2 mm post-tensioned, prestressed concrete specimens is studied, and the bond mechanism and bond failure mode are discussed through pull-out tests and simulation analysis. Then, the key parameters influencing bond performance are captured. Finally, a finite element modeling method is proposed, and its rationality is verified based on experimental results.

2. Test Introduction
2.1. Specimen Design

In this experiment, a total of seven post-tensioned PC specimens were designed and manufactured, numbered S1–S7 in turn. Considering that the web width at the end of a small box girder is usually about 300 mm, the specimen width was also set to this value in this experiment. The specimen height was determined by considering both the height–width ratio and experimental economy. The cross-sectional dimensions of all pull-out specimens were identical, being 300 mm × 500 mm. In addition, for the actual prestressed concrete box girder, the concrete strength grades are usually C40 and C50, and the concrete strength grade of some structures also reaches C60. As a result, the effect of these three strength levels on the test results is also considered. Similarly, three commonly used stirrup diameters of 8, 10, and 12 mm are used for the comparative analysis. Moreover, the anchorage length of prestressed tendons that were estimated according to the specification of GB50010-2020 [16] should be less than 1300 mm, and only two types of specimens with lengths of 1000 mm and 1250 mm are designed here. See Table 1 for the detailed dimensions of the specimens.

Table 1. Parameters of specimens S1–S7.

Specimen No.	Sectional Dimension (mm)	Length (mm)	Concrete Grade	Stirrup Diameter (mm)	Fracture Mode
S1	300 × 500	1000	C50	Φ8	Corrosion-induced fracture
S2	300 × 500	1250	C50	Φ8	Corrosion-induced fracture
S3	300 × 500	1000	C50	Φ8	Direct release-tensioning
S4	300 × 500	1000	C40	Φ8	Corrosion-induced fracture
S5	300 × 500	1000	C60	Φ8	Corrosion-induced fracture
S6	300 × 500	1000	C50	Φ10	Corrosion-induced fracture
S7	300 × 500	1000	C50	Φ12	Corrosion-induced fracture

A corrugated pipe with a diameter of 60 mm was reserved in the center of each specimen, seven-wire, twisted-steel strands (3 × 7Φ15.2 mm) were designed in the corrugated pipe, and eight Φ12 mm HRB400 deformed steel bars were longitudinally arranged outside the corrugated pipe; stirrups were arranged around the longitudinal reinforcement at a spacing of 100 mm. In order to study the influences of specimen length, concrete strength, and stirrup ratio on the stress transfer and anchorage performance of fractured prestressed tendons, two bond length values (1000 and 1250 mm), three concrete strength grades (C40, C50, and C60), and three stirrup configurations (Φ8@100 mm, Φ10@100 mm, and Φ12@100 mm) were designed for the test specimens. The stirrups were HPB300 plain steel bars. The reinforcement layout of each specimen is displayed in Figure 1.

The tension control stress of steel strands is 1395 MPa, and the measured yield strength is 1810 MPa. The material properties of steel strands and ordinary steel bars are listed in Table 2. The concrete was composed of 42.5# ordinary Portland cement, natural river sand, graded aggregate, and laboratory tap water mixed together. The tap water in the laboratory was used as mixing water. Commercial cement mortar with the same grade as concrete was used as grout. Standard cubic blocks were cast using the same batch of concrete and maintained together with the test specimens. The average compressive strength of the concrete of the C50, C40, and C60 specimens at 28 d was 53.5, 42.6, and 63.3 MPa, respectively.

Table 2. Material properties of steel strands and ordinary steel bars.

Diameter (mm)	Yield Strength (MPa)	Ultimate Strength (MPa)	Elastic Modulus (GPa)
15.2	1810	1915	195
12	476	612	200
8	263	366	210
10	285	357	210

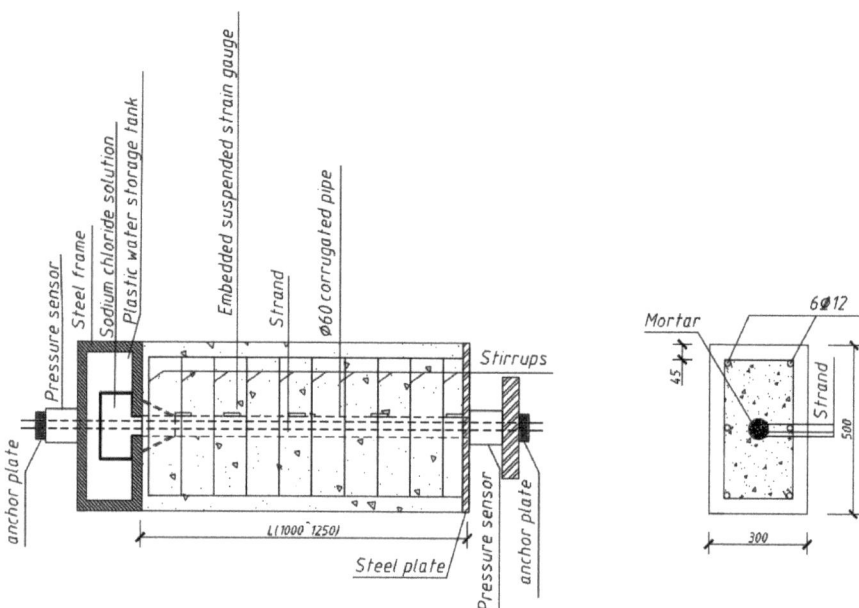

Figure 1. Dimensions and reinforcement of S1–S7 specimens (unit: mm).

2.2. Tensioning of Prestressed Tendons and Effective Prestress

During the tensioning process, the tensioning force was monitored with the pressure sensor, and, meanwhile, the elongation of the steel strands was recorded using a dial indicator. At 48 h after tensioning, the reading of the pressure sensor was obtained, and the effective prestress applied by the member was measured, as seen in Table 3.

Table 3. Effective prestress in each test stage.

Specimen No.	S1	S2	S3	S4	S5	S6	S7
Tensioning prestress (kN)	636.12	636.12	636.12	636.12	636.12	636.12	636.12
Effective prestress before the corrosion test (kN)	497.2	511.3	499.5	449.9	510.5	501.1	505.2
Effective prestress after the anchorage failure (kN)	497.2	511.3	426.3	449.9	510.5	501.1	505.2

2.3. Stress Release of Anchorage Parts at the End of Prestressed Tendons

After grouting and maintaining for 28 d, the anchorage zone at the ends of S1–S5 was locally corroded, and the corrosion mode was indoor electrochemical rapid corrosion. In this experiment, a local corrosion tank with a length of 20 cm was designed and fixed in the end area of concrete specimens using structural adhesive, and the tank was filled with 5% NaCl solution and installed with stainless-steel plates. During corrosion, the anode wire of the constant DC power source was turned on and connected to the steel strand, and the cathode wire of the power source was connected to the stainless-steel plate placed in the corrosion tank. At the same time, a current loop could be formed by pouring NaCl solution into the tank. Under the action of current, the anodic steel strand was corroded, as shown in Figure 2. During the test, only the corrosion at the end of the steel strand was considered. To avoid the influence of corrosion on ordinary steel bars, anti-corrosion treatment was performed by smearing epoxy resin when binding the ordinary reinforcement cage. The whole test was carried out in an environment with a temperature of 20 °C and humidity of

65%. It took about 5 d for the corrosion of the member. Through inspection, all prestressed tendons experienced corrosion fracture at the end, which was consistent with the test design, as shown in Figure 3.

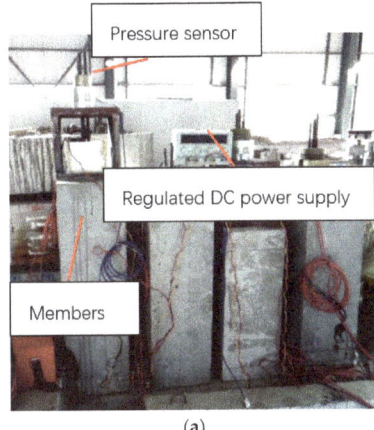 

(a) (b)

Figure 2. Corrosion position of test members: (**a**) corrosion process of specimen; (**b**) detailed structure of corrosion groove.

Figure 3. Corrosion fracture at end anchorage part of steel strands in test members.

When specimen S3 was tensioned and anchored, a metal block was inserted between the anchor cup at one end and the specimen. After 48 h from completing the tensioning, the metal block was removed to achieve direct release tensioning.

2.4. Pull-Out Test Design of Steel Strands and the Test Devices

To discuss the secondary anchorage test, the pull-out test was implemented on the uncorroded side after the corrosion fracture of the steel strands on one side of the member. For this purpose, connectors, pull rods, and anchorage devices were specially designed, as displayed in Figure 4. According to the rules of the concrete and steel strand grip test provided in the Testing Code of Concrete for Port and Waterway Engineering (JTS/T 236-2019) [27], the pull-out test device was self-designed. After completion, the test specimens were tensioned using a hydraulic jack with a range of 100 t, and the pull-out force was recorded with a pressure sensor.

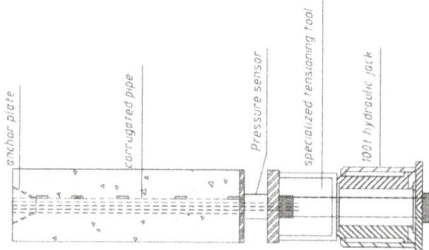 

Figure 4. Pull-out test device.

The steel strands were pre-tensioned before the formal pull-out test to eliminate the nonelastic deformation between test devices, and the pre-tensioning force was set to 5 kN. After pre-tensioning, the readings of the anchorage dynamometer and dial indicator were reset at the loading end, and the formal loading procedure was started. The load was controlled as per the load staging: initially, a 5 kN load was applied at each stage and 15 kN was applied at each stage after reaching 15 kN. Meanwhile, various test phenomena such as the displacement at the loading end and drawing end of the members, the readings of the anchorage dynamometer and suspension-type strain gauge, crack development, and the test sound under various loads were recorded. The pull-out test would be stopped immediately under any of the following circumstances during the test process: (1) the steel strand was pulled out or experienced fracture failure; (2) the slip amount at the loading end exceeded 30 mm, the load could not be continuously increased or the increment was very small, and the bond–slip failed.

3. Results and Discussion

3.1. Failure Mode

In this test, typical failure modes were observed in two categories: the first involved the steel strands, which were pulled out as a whole, resulting in bond–slip failure. However, it differed slightly from the individual strand slip process, in which three steel strands along with the concrete between them were entirely pulled out. The second category involved the fracture of steel strands resulting in material failure, as depicted in Figure 5. Specimens S1, S3, and S4 experienced bond slip failure, while specimens S5 to S7 exhibited material failure, as indicated in Table 4.

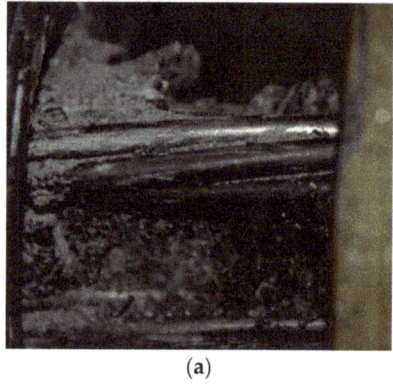 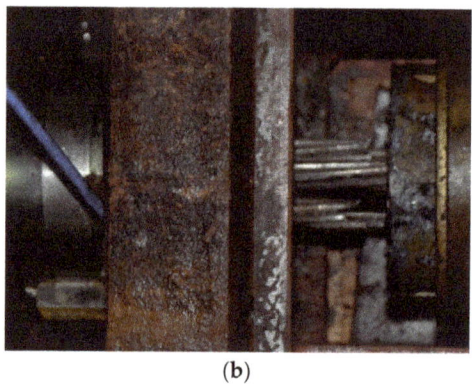

(a) (b)

Figure 5. Failure modes of specimens: (**a**) bond–slip failure; (**b**) material tensile failure.

Table 4. Ultimate bearing capacity in pull-out tests of specimens S1–S7.

Specimen No.	Failure Mode	Initial Tensile Force of Steel Strands (kN)	Ultimate Tensile Force of Steel Strands (KN)	Pull-Out Force (kN)	Ultimate Slippage at Pull-Out End (mm)	Ultimate Slippage at Tensioning End (mm)	Crack Type
S1	Pull-out as a whole	432	765	315	14.5	30.6	Type I
S2	Rupture	456	802	345	0.3	10.5	Type II
S3	Pull-out as a whole	425	670	245	15.1	32.5	Type I
S4	Pull-out as a whole	361	695	335	15.2	31.7	Type I
S5	Rupture	460	793	333	2.0	12.5	Type II
S6	Rupture	451	772	321	2.2	13.8	Type III
S7	Rupture	431	818	387	9.1	20.9	Type III

3.2. Crack Distribution

Different test specimens exhibit varying crack distributions, which can be classified into three types based on the pattern of cracks. For the first type, splitting cracks occurred in specimens S1, S3, and S4. As the pull-out load continued to increase, a new crack along the direction of the steel strands started appearing at the tensioning end. When the load further increased to about 50% of the ultimate load, the initial tensile crack width started increasing from the pull-out end to the free end, and several secondary splitting cracks along the length direction appeared beside the splitting cracks; moreover, the free end started slipping. As the load increased near the ultimate load, the main and secondary splitting cracks both ran through the whole specimen, accompanied by a very loud splitting sound. Subsequently, the steel strands were pulled out when the load reached the ultimate value, as shown in Figure 6.

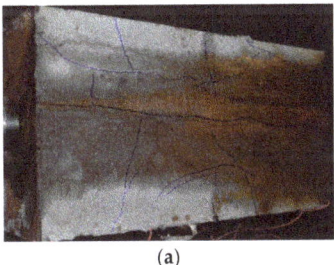

(a)   (b)

Figure 6. Crack diagram under ultimate state of type I specimens: (a) distribution of splitting cracksl (b) shape of the main splitting crack at the corrosion end.

For the second type, longitudinal initial cracks appeared after the initial tensioning of S5 was completed. During the test, the width of the initial tensile cracks increased subtly with the increase in the pull-out load; moreover, the pull-out force reached about 60% of the ultimate pull-out force, several longitudinal cracks were added (but only one splitting crack appeared when the ultimate load was reached), and all cracks did not completely run through the whole specimen. As shown in Figure 7, the load reached the limiting value, and a loud sound was heard; moreover, the steel strands were ruptured, leading to material failure that was accompanied by the rotation of the jack, which might be associated with the twists that formed in the steel strands.

The third type of cracks occurred in S6 and S7. No secondary splitting cracks were found in such specimens during the whole test process. Compared with type II specimens, there were more longitudinal cracks; however, the length was smaller. A loud sound was heard when the load reached the limiting value: the steel strands were ruptured, and material failure occurred, as shown in Figure 8.

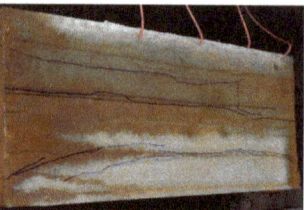

Figure 7. Crack diagram under ultimate state of type II specimens.

Figure 8. Crack diagram under ultimate state of type III specimens.

3.3. Load Bond–Slip Curves

The load bond–slip curves between the loading end and the free end of specimens S1–S7 under the pull-out force are exhibited in Figure 9. For the same specimen, the loading end slipped earlier than the free end, and the growth rate and limiting value of the slip amount were both greater than those at the free end. This was because, during the stress transfer of the pull-out force inside the specimen, stress loss would occur, and the pull-out force borne at the free end was much smaller than that at the pull-out end.

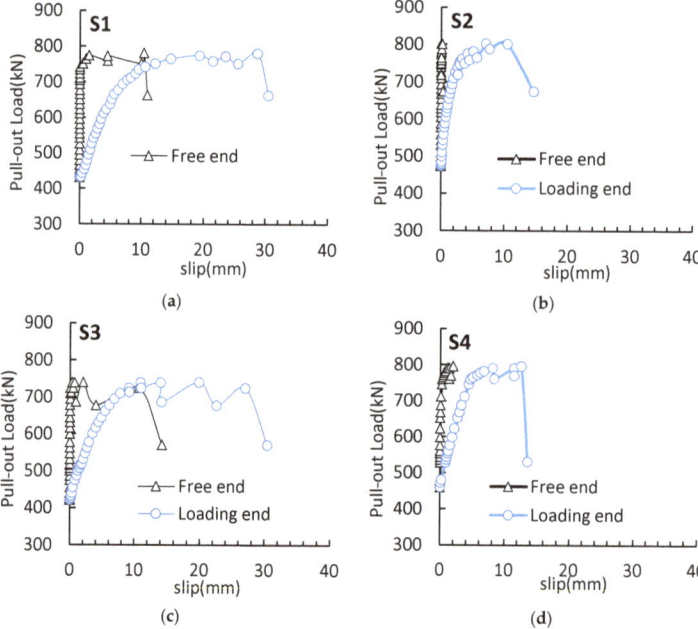

Figure 9. *Cont.*

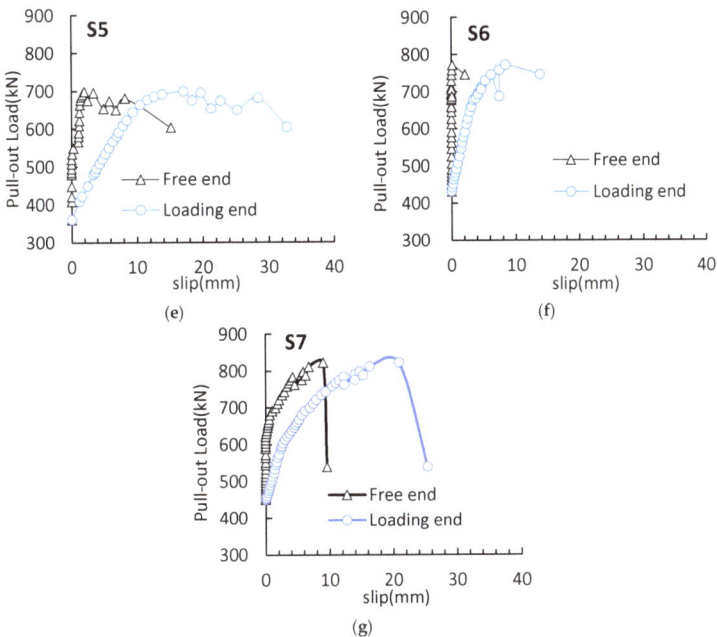

Figure 9. Load bond–slip curves at the loading end: (**a**) S1; (**b**) S2; (**c**) S3; (**d**) S4; (**e**) S5; (**f**) S6; (**g**) S7.

It could be known from the load bond–slip curves on the loading side that the secondary anchorage failure process of the steel strands could be divided into three stages. The first was the linear stage: when the tensile force was less than about 30% of the ultimate pull-out load, the slip amount at the loading end was basically linear with the pull-out load, and the sliding friction between steel strands and concrete played a major role; the second was the yield stage: when the pull-out load continued to increase to about 90% of the ultimate pull-out load, the slip growth of the pull-out end was obviously accelerated, and the radial component of the mechanical bite force between the steel strands and the grouting body continued to increase with the increase in the slip amount. Meanwhile, the bonding force between the steel strand and the concrete was gradually destroyed, the free end began slipping, and the stress of the steel strands gradually increased, entering the yield stage; the third stage was the failure stage: the pull-out force increased slightly, and the slip value increased rapidly. The steel strands were in the post-yield stage, the pull-out force suddenly decreased and then increased again, and the pull-out force fluctuated in a certain range until the steel strands were finally broken. Relative to the type II specimens in which steel strands were broken, for the type I specimens experiencing the pull-out failure of steel strands, the ultimate slip value at the pull-out end exceeded 30 mm upon failure, and the specimens failed to provide enough secondary anchorage strength.

3.4. Influencing Factors

In order to study the influence of different parameters on the bond anchorage performance of specimens, Figure 10 shows the load slip curves at the loading end under the influence of different specimen lengths, prestress release methods, concrete strength, and stirrup diameters. Figure 10a shows the pull-out test results of specimens with different concrete strengths. Comparing the data in Figure 10 and Table 4, the ultimate pull-out force of specimen S1 was 747 kN, which was 10.3% lower than that of specimen S4 and 5.8% higher than that of S5. Under the ultimate load, the relative displacement between the pull-out end and the free end was 16.1 mm for S1 and 16.5 mm for S4, which was 2.5% higher than that for S1. The relative displacement for S5 was 10.5 mm, which was 34.8%

less than that for S1. It could be seen that concrete strength had a significant influence on the pull-out performance of the specimens. The higher the concrete strength, the greater the ultimate pull-out force. Moreover, compared with the slippage and pull-out as a whole, the slip value at the loading end upon the tensile failure of the steel strands was significantly reduced, and the specimen stiffness was strengthened by high concrete tensile strength. This was because, the higher the concrete strength, the more obvious the radial squeezing effect on steel strands, increasing the circumferential tensile stress and delaying the appearance of both microcracks and splitting cracks in the specimen. Therefore, both the bond strength and stiffness of specimens significantly increased with the increase in concrete strength, and S1 and S4 developed from the first type of failure mode into the second type and experienced sufficient secondary anchorage.

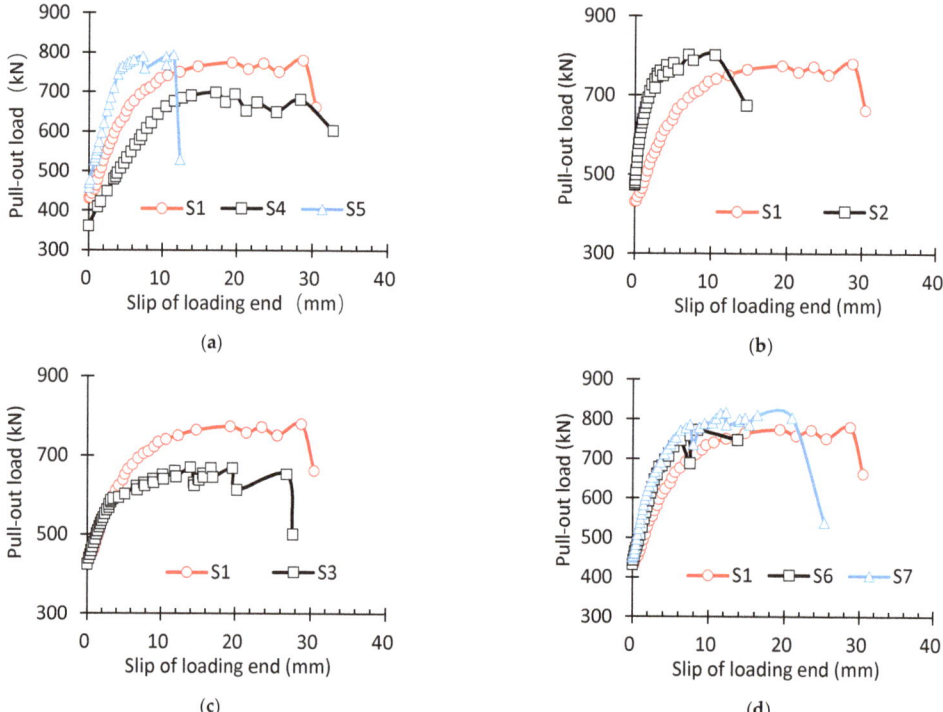

Figure 10. Comparison of load bond–slip curves under different test conditions: (**a**) influence of concrete strength on pull-out performance; (**b**) different specimen lengths; (**c**) different release-tensioning speeds and modes; (**d**) different stirrup configurations.

Figure 10b shows the pull-out test results of specimens with different lengths. Compared with specimen S1, the length of S2 only increased by 250 mm. Comparing the data in Figure 10b and Table 4, the ultimate pull-out load of S2 increased by 5.8% compared with that of S1. The maximum slip amount of S2 was only 48.1% of that of S1. This is because, when the embedded length of the specimen was greater than the secondary anchorage length, the effective bond length could be increased by increasing the specimen length, the actual bond strength was smaller than the ultimate bond strength, and the overlapping between the length influenced by the pull-out force and the transfer length of residual prestress at the corrosion end was delayed, the development of splitting cracks was also delayed, and, thus, specimen stiffness grew evidently. The ultimate bond force provided by the specimen was directly proportional to the embedded length of steel strands. Hence, when the embedded length of the steel strands was greater than the secondary anchorage length, the failure mode in the pull-out test was transformed from the first type—slippage failure as a whole—into the second type—material failure induced by the rupture of steel strands.

Figure 10c shows the test results of different release-tensioning methods. Both S1 and S3 experienced pull-out slippage failure as a whole. Compared with S1, the ultimate pull-out load of S3 was reduced by 10.3%, and both specimens were subjected to the slippage failure of steel strands as a whole. In the initial loading stage, almost no differences were found between S1 and S3. As the load increased, the stiffness of S3 gradually declined because the specimen needed a greater secondary transfer length due to sudden release tensioning, which led to a reduction in the effective bond length. Under the pull-out force, the pull-out bond length of the specimen overlapped with the secondary transfer length earlier; moreover, the concrete splitting cracks also developed earlier, and both the structural stiffness and ultimate bond strength decreased.

Figure 10d shows the pull-out test results of specimens with different stirrup configurations. As more stirrups were configured, both S6 and S7 experienced the tensile failure of steel strands, indicating that the increase in the stirrup ratio facilitated sufficient secondary anchorage. Comparing the data in Figure 10d and Table 4 shows that, compared with S1, the ultimate pull-out force of S6 increased by 3.3% while that of S7 grew by 9.5%. The pull-out performance of the specimens was significantly affected by concrete strength: the higher the concrete strength, the greater the ultimate pull-out force. In comparison with S1, S7 possessed a longer platform after the yielding of steel strands. This was because, with the increase in the diameter of the stirrups, the horizontal restraining effect provided by the stirrups was enhanced, which could effectively restrain and delay the development of splitting cracks, prevent the splitting failure of specimens, increase the structural bond strength, and enhance structural ductility. Comparing the load–slip curves of S1, S6, and S7, however, the difference among the three specimens in the slope of the load–slip curves was not evident in the initial loading stage, and the structural stiffness could not be significantly increased by increasing the stirrups. This was because the stirrup stress was relatively small in this case; moreover, the average stress of concrete within the wrapping scope of stirrups was small, and the horizontal restraining effect of stirrups was also weak, thus exerting a limited effect when increasing the structural initial stiffness.

4. Numerical Simulation Analysis

4.1. Modeling

To further explore the secondary anchorage performance of residual prestress after the end anchorage failure of the post-tensioned, prestressed members, the finite element calculation model of damaged concrete specimens was established using Abaqus (version of Abaqus CAE2016). The structure was established by using a discrete model, and the reduced integral hexahedral element (C3D8R) was used to simulate concrete. The size of the concrete unit is 20 mm × 18.75 mm × 20 mm.

In the modeling process, a constitutive concrete damage plasticity (CDP) model was adopted. Regarding the model parameter of the CDP model, the measured value of the cubic compressive strength of the concrete was used. The tensile strength, elastic modulus, and other parameters of concrete were calculated according to specification GB50010-2020 [16]. The values of other CDP model parameters are shown in Table 5. Additionally, both ordinary steel bars and prestressed steel bars were simulated using truss elements (T3D2), and a constitutive relation model of double-broken-line, equal-strength hardening was selected [28]. A total of 28,522 nodes and 25,723 elements were set in the model, and the overall structural model is displayed in Figure 11.

Table 5. CDP model parameters.

Poisson's Ratio	Expansion Angle	Eccentricity	Parameters Affecting the Yield Morphology of Concrete	Ratio of Ultimate Strength under Biaxial and Uniaxial Compression	Viscosity Coefficient
0.2	30	0.1	0.667	1.16	0.005

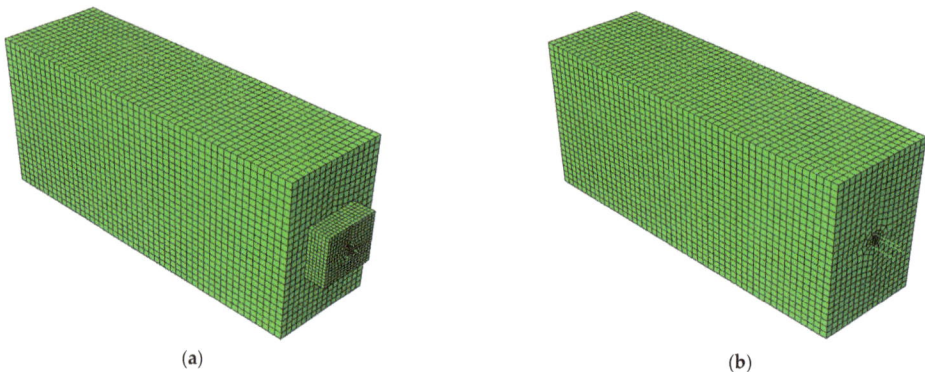

Figure 11. Overall structural model: (**a**) step 4 model; (**b**) step 6 model.

To consider the bond slip between the steel strand and concrete, two different nodes are set at the same positions as the steel strand and concrete, which are connected using nonlinear spring elements (SPRING2), and the bond–slip constitutive model in [29] is transformed into a spring constitutive model. In addition, the anchor pad is connected using the tie-in interaction module, and the connection between ordinary steel bars and concrete is embedded. To be consistent with the constraints in the actual working conditions, three boundary conditions, DX, DY, and DZ, are added to the constrained concrete element in the actual working conditions. By activating and passivating the boundary conditions, the constraint changes in the prestressed tension stage, corrosion stage, and drawing stage are simulated.

The external loads that need to be applied during the simulation process mainly include prestress and tensile force. The prestressing force is applied using the cooling method, while the pull-out load is applied to steel strand nodes through the direct loading method. The external loads that need to be applied during the simulation process mainly include prestress and tensile force. The prestressing force is applied using the cooling method, while the pull-out load is applied to the steel strand nodes through the direct loading method.

The static and general analysis steps were adopted as follows: In step 1, all structural elements were inactivated. In step 2, the concrete elements, anchor plates, steel strands, reinforcement cages, and boundary conditions beyond the grouting part were activated. In step 3, prestress was applied through the cooling method. In step 4, grouting elements were activated. In step 5, the steel strand elements and anchor plates at the fracture part were inactivated. In step 6, the anchor blocks on the pull-out load side were inactivated, and the pull-out load was applied to the steel strand nodes.

4.2. Verification of Calculation Results

Because static and general analysis steps were adopted, only S1–S2 and S3–S7 were subjected to the finite element method. The calculated and measured values of the ultimate pull-out load for the specimens are listed in Table 6. It could be seen that the relative error between the test value and the numerically calculated value of the ultimate pull-out load of each specimen was relatively small, with the maximum value only being 5.0%, and all calculated values relatively coincided with the test values.

Table 6. Comparison between test values and calculated values of ultimate bearing capacity in the pull-out test.

Ultimate Load (kN)	Specimen No.					
	S1	S2	S4	S5	S6	S7
Test value	771	802	695	793	772	818
Calculated value	753	791	730	809	783	797
Relative error (%)	2.3	1.4	5.0	2.1	1.4	2.5

See Figure 12 for the test value and numerically calculated value of the loading–end slip value of different specimens under different pull-out loads. It could be observed that the test value of the loading–end slip amount under different loads was relatively consistent with the calculated value, further verifying the effectiveness of the calculation model. Figure 12b needs to be adjusted to the following figure.

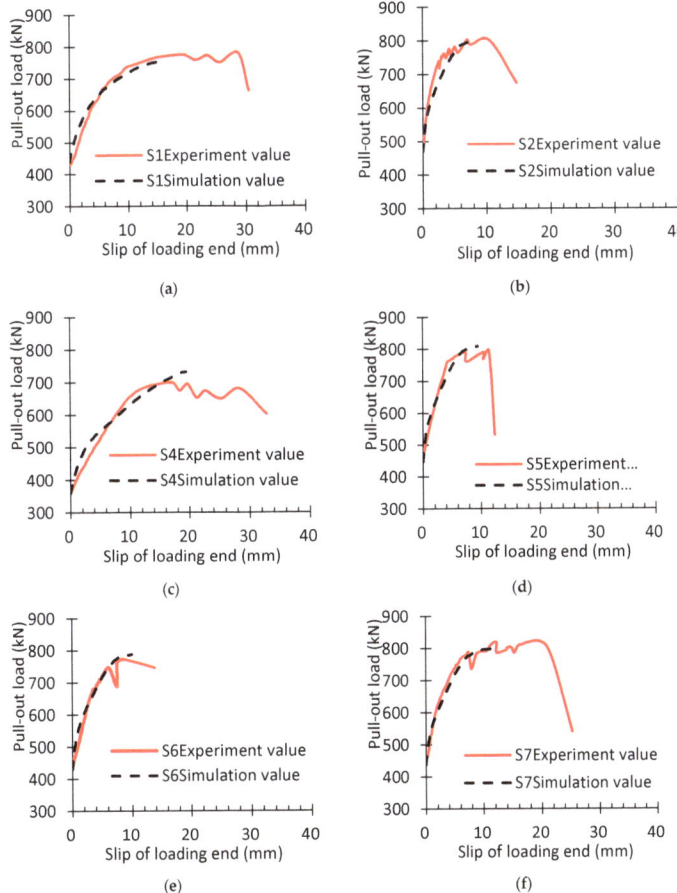

Figure 12. Comparison between test results and calculation results of load bond–slip curves: (a) S1; (b) S2; (c) S4; (d) S5; (e) S6; (f) S7.

In order to compare the damage status of concrete under the same level of the pull-out load and the secondary anchorage status of residual prestress in the specimens, the influence of the embedded length of steel strands, concrete strength, and stirrup configuration on the secondary anchorage of residual prestress was analyzed theoretically. Referring to the ultimate bearing capacity of each specimen, the pull-out load was 730 kN, and the concrete damage and cracking induced by tensioning and the stress state of concrete in each specimen are shown in Figure 13. It can be seen from Figure 13 that, when the stress of the steel strands reached 1730 MPa, splitting cracks appeared in all specimens, and one to four secondary splitting cracks were produced. Therein, the main splitting cracks already ran through the whole S4 specimen. The increasing embedded length of the steel strands could postpone the generation of the main splitting cracks; however, the effect was not evident in the free-end secondary splitting cracks. Increasing the concrete strength and the stirrup ratio could effectively constrain the generation of splitting cracks and enhance the splitting stress and ultimate bond strength of specimens, which coincided with the test results and further verified the effectiveness of the calculation model.

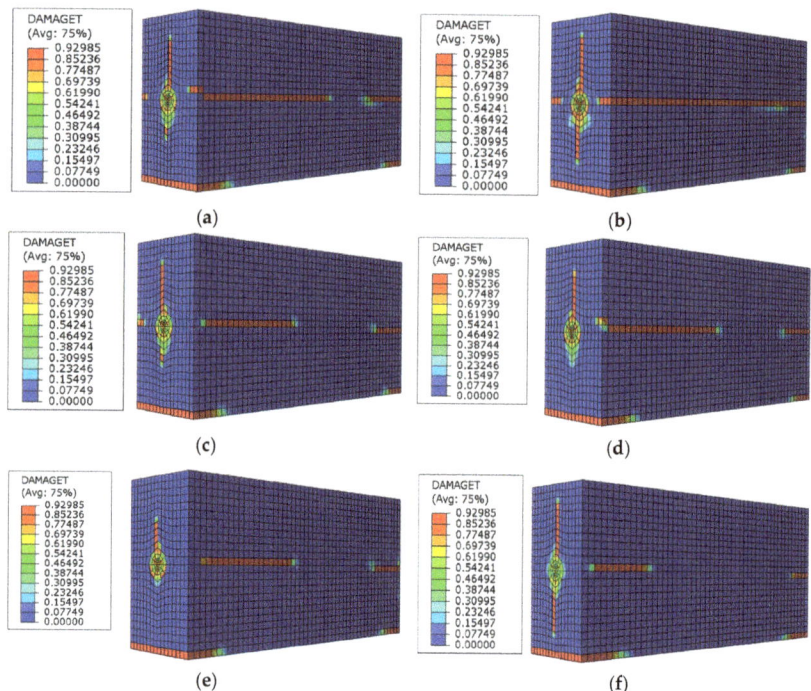

Figure 13. Comparison of tensioning-induced concrete damage and cracking of specimens under the pull-out load of F = 730 kN: (**a**) S1; (**b**) S2; (**c**) S4; (**d**) S5; (**e**) S6; (**f**) S7.

5. Conclusions

(1) After the end anchorage failure of post-tensioned, prestressed specimens, the secondary anchorage bond performance of residual prestress is affected by factors such as the embedded length of steel strands, release-tensioning speed, concrete strength, and stirrup ratio. The increase in the embedded length of steel strands and the strength grade of concrete is beneficial for improving the bonding performance of the specimens;

(2) The pull-out process of prestressed specimens can be divided into three stages: the linear stage, yield stage, and failure stage. After the maximum pull-out force is

reached, the specimen experiences bond failure or fracture failure in the prestressed steel strands;

(3) In this study, the construction condition considered is the corrosion fracture of end steel strands as a whole, which can be accompanied by corrosion damage along the full length of steel strands; however, this factor is not taken into account;

(4) The specimen sizes are evidently smaller than those in actual engineering structures, and the number of specimens is small. In addition, the secondary anchorage bond performance of residual stress after the end corrosion fracture of steel strands in practical engineering remains to be further investigated.

Author Contributions: Conceptualization, R.Y.; Data curation, X.Z.; Formal analysis, Y.Y.; Investigation, Y.Y. and X.Z.; Methodology, R.Y. and Y.Y.; Resources, R.Y.; Software, X.W.; Validation, X.W.; Writing—original draft, R.Y.; Writing—review and editing, R.Y., Y.Y., X.Z. and X.W. All authors have read and agreed to the published version of the manuscript.

Funding: This study was supported by the National Natural Science Foundation of China (grant no. 52208166), the Science and Technology Innovation Program of Hunan Province (2022RC1186), the Research Foundation of Education Bureau of Hunan Province (grant no. 22A0561), and the Aid program for the Science and Technology Innovative Research Team in the Higher Educational Institutions of Hunan Province.

Institutional Review Board Statement: Not applicable.

Informed Consent Statement: Not applicable.

Data Availability Statement: Data are contained within the article.

Conflicts of Interest: The authors declare no conflict of interest.

References

1. ACI Committee. *Building Code Requirements for Structural Concrete and Commentary*; American Concrete Institute: Farmington, MI, USA, 2014.
2. Zucca, M.; Reccia, E.; Longarini, N.; Eremeyev, V.; Crespi, P. On the structural behaviour of existing RC bridges subjected to corrosion effects: Numerical insight. *Eng. Fail. Anal.* **2023**, *152*, 107500. [CrossRef]
3. Yang, Y.; Liu, Z.; Tang, H.; Peng, J.J.C.; Materials, B. Deflection-based failure probability analysis of low shrinkage-creep concrete structures in presence of non-stationary evolution of shrinkage and creep uncertainties. *Constr. Build. Mater.* **2023**, *376*, 131077. [CrossRef]
4. Zhang, X.; Wang, H.; Zhang, Y.; Wang, L.J.C.; Materials, B. Corrosion of steel rebars across UHPC joint interface under chloride attack. *Constr. Build. Mater.* **2023**, *387*, 131591. [CrossRef]
5. Osborn, A.; Lanier, M.; Hawkins, N. Bond of prestressing strand to concrete. *PCI J.* **2021**, *66*, 28–48. [CrossRef]
6. Xu, Y.; Yu, B.; Zhu, L.; Zhao, H. Experimental Study on the Basic Performance of Strand and Anchorage Length. *Build. Struct.* **1996**, *3*, 34–38. [CrossRef]
7. Wang, L.; Yuan, P.; Xu, G.; Han, Y. Quantification of non-uniform mechanical interlock and rotation in modelling bond-slip between strand and concrete. *Structures* **2022**, *37*, 403–410. [CrossRef]
8. Gustavson, R.J.M. Structures, Experimental studies of the bond response of three-wire strands and some influencing parameters. *Mater. Struct.* **2004**, *37*, 96–106. [CrossRef]
9. Xie, X.; Zhou, W.; Yu, Y. Experimental analysis of bond property and anchorage length for strands. *J. Harbin Inst. Technol.* **2018**, *50*, 81–88. [CrossRef]
10. Zhang, F.; Zhao, G.-H.; Wu, Y.-F.; Zhang, Y. Effect of strand debonding on the shear strength of existing pretensioned PC hollow slab. *Eng. Struct.* **2023**, *291*, 116417. [CrossRef]
11. Lazić, Ž.; Marinković, S.; Koković, V.; Broćeta, G.; Latinović, M. Testing bond behaviour of an innovative triangular strand: Experimental setup challenges and preliminary results. *Gradevnski Mater. I Konstr.-Build. Mater. Struct.* **2022**, *65*, 93–103. [CrossRef]
12. Yi, J.; Wang, L.; Floyd, R.W. Bond strength model of strand in corrosion-induced cracking concrete. *Struct. J.* **2020**, *117*, 119–132. [CrossRef]
13. Mohandoss, P.; Pillai, R.G.; Gettu, R. Determining bond strength of seven-wire strands in prestressed concrete. *Structures* **2021**, *33*, 2413–2423. [CrossRef]
14. Martí-Vargas, J.R.; García-Taengua, E.; Serna, P. Influence of concrete composition on anchorage bond behavior of prestressing reinforcement. *Constr. Build. Mater.* **2013**, *48*, 1156–1164. [CrossRef]

15. Arezoumandi, M.; Looney, K.B.; Volz, J.S. Bond performance of prestressing strand in self-consolidating concrete. *Constr. Build. Mater.* **2020**, *232*, 117125. [CrossRef]
16. *GB50010-2020*; Code for Design of Concrete Structures. China Architecture and Building Press: Beijing, China, 2020.
17. American Association of State Highway Officials (AASHO). AASHTO (American Association of State Highway Transportation Officials). In *Standard Specification for Highway Bridges*; AASHTO: Washington, DC, USA, 2020.
18. Orr, J.J.; Darby, A.; Ibell, T.; Thoday, N.; Valerio, P. Anchorage and residual bond characteristics of 7-wire strand. *Eng. Struct.* **2017**, *138*, 1–16. [CrossRef]
19. Li, S.; Song, C. Experimental research on bond anchorage performance of 1860-grade high-strength steel strands and lightweight aggregate concrete. *Constr. Build. Mater.* **2020**, *235*, 117482. [CrossRef]
20. Dolan, C.W.; Hamilton, H.R. *Prestressed Concrete: Building, Design, and Construction*, 1st ed.; Springer: Berlin/Heidelberg, Germany, 2019.
21. Theryo, T.; Garcia, P. *Sunshine Skyway Bridge Post-Tensioned Tendons Investigation*; Parsons Brinckerhoff Quade and Douglas, Inc.: Tallahassee, FL, USA, 2001. [CrossRef]
22. Zhu, E.; Liu, C.; He, L.; Zhang, H.; Xie, N. Stress Performance Analysis on Corroded Pre-stressed Concrete Beam. *China Saf. Sci. J.* **2006**, *16*, 136–140. [CrossRef]
23. Anania, L.; Badalà, A.; D'Agata, G. Damage and collapse mode of existing post tensioned precast concrete bridge: The case of Petrulla viaduct. *Eng. Struct.* **2018**, *162*, 226–244. [CrossRef]
24. Yang, R.; Yang, Y.; Zhang, X.; Wang, X. An experimental study on secondary transfer performances of prestress after anchoring failure of steel wire strands. *Metals* **2023**, *13*, 1489. [CrossRef]
25. El Zghayar, E.; Mackie, K.R.; Haber, Z.B.; Potter, W. Secondary anchorage in post-tensioned bridge systems. *Struct. J.* **2013**, *110*, 629–638.
26. Asp, O.; Tulonen, J.; Laaksonen, A. Bond and re-anchoring tests of post-tensioned steel tendon in case of strand failure. In Proceedings of the Bond in Concrete 2022: Bond-Anchorage-Detailing, 5th International Conference, Stuttgart, Germany, 25–27 July 2022; pp. 981–991.
27. *JTS/T 236-2019*; Technical Specifications for Concrete Testing of Port and Waterway Engineering. China Communications Press Co., Ltd.: Beijing, China, 2019.
28. Wang, Y. *Abaqus Analysis User's Guide: Material*; China Machine Press: Beijing, China, 2021.
29. Yang, R.; Yang, Y.; Zhang, X.; Wang, X. Experimental Study on Performance of Local Bond-Slip Test of Steel Strand Tendons and Concrete. *Coatings* **2022**, *12*, 1494. [CrossRef]

Disclaimer/Publisher's Note: The statements, opinions and data contained in all publications are solely those of the individual author(s) and contributor(s) and not of MDPI and/or the editor(s). MDPI and/or the editor(s) disclaim responsibility for any injury to people or property resulting from any ideas, methods, instructions or products referred to in the content.

Article

Bond Modification of Carbon Rovings through Profiling

Paul Penzel [1,*], Maximilian May [2], Lars Hahn [1], Silke Scheerer [3], Harald Michler [3], Marko Butler [4], Martin Waldmann [1], Manfred Curbach [3], Chokri Cherif [1] and Viktor Mechtcherine [4]

[1] Institute of Textile Machinery and High Performance Material Technology (ITM), Technische Universität Dresden, 01062 Dresden, Germany
[2] CARBOCON GMBH, 01067 Dresden, Germany
[3] Institute of Concrete Structures (IMB), Technische Universität Dresden, 01062 Dresden, Germany
[4] Institute of Construction Materials (IfB), Technische Universität Dresden, 01062 Dresden, Germany
* Correspondence: paul.penzel@tu-dresden.de

Abstract: The load-bearing behavior and the performance of composites depends largely on the bond between the individual components. In reinforced concrete construction, the bond mechanisms are very well researched. In the case of carbon and textile reinforced concrete, however, there is still a need for research, especially since there is a greater number of influencing parameters. Depending on the type of fiber, yarn processing, impregnation, geometry, or concrete, the proportion of adhesive, frictional, and shear bond in the total bond resistance varies. In defined profiling of yarns, we see the possibility to increase the share of the shear bond (form fit) compared to yarns with a relatively smooth surface and, through this, to reliably control the bond resistance. In order to investigate the influence of profiling on the bond and tensile behavior, yarns with various profile characteristics as well as different impregnation and consolidation parameters are studied. A newly developed profiling technique is used for creating a defined tetrahedral profile. In the article, we present this approach and the first results from tensile and bond tests as well as micrographic analysis with profiled yarns. The study shows that bond properties of profiled yarns are superior to conventional yarns without profile, and a defined bond modification through variation of the profile geometry as well as the impregnation and consolidation parameters is possible.

Keywords: carbon reinforced concrete; bond behavior; bond mechanisms; profiling technology; tensile test; bond test

Citation: Penzel, P.; May, M.; Hahn, L.; Scheerer, S.; Michler, H.; Butler, M.; Waldmann, M.; Curbach, M.; Cherif, C.; Mechtcherine, V. Bond Modification of Carbon Rovings through Profiling. *Materials* **2022**, *15*, 5581. https://doi.org/10.3390/ma15165581

Academic Editor: Alessandro P. Fantilli

Received: 27 July 2022
Accepted: 11 August 2022
Published: 14 August 2022

Copyright: © 2022 by the authors. Licensee MDPI, Basel, Switzerland. This article is an open access article distributed under the terms and conditions of the Creative Commons Attribution (CC BY) license (https://creativecommons.org/licenses/by/4.0/).

1. Introduction

The use of fiber reinforcements in concrete has been established as an alternative to reinforcing steel in recent years. Research into the properties of TRC/CRC (textile/carbon reinforced concrete), TRM (textile reinforced mortar), or FRCM (fiber-reinforced cementitious matrices)—all synonyms for a composite made of continuous fiber-based reinforcement and a mineral-based matrix—is being conducted worldwide. Hereby, the studies focus primarily on strengthening concrete structures and fiber-reinforced composites as a high-performance material in civil engineering [1–6]. Impressions of the application potential in new constructions and building strengthening are, for example, filigree and precast concrete structures for bridges, façade panels, beams, shells, and pavilions, as well as strengthening of existing structures, especially bridges for shear and bending [7–13].

Continuous carbon fibers, also called rovings or Carbon Fiber Heavy Tows (CFHTs), are processed into grid-like structures or rods for use in concrete. Common carbon fiber rebars have diameters between 6 and 10 mm, thus in the size range of thin steel reinforcement rebars. They are preferably used in new building components. The carbon rovings that are processed into grid-like non-crimp fabrics (NCF) have significantly smaller diameters in the range of 1 to 3 mm. This makes them particularly suitable for use in filigree components such as façade panels or subsequent component reinforcement.

A capable bond between reinforcement and concrete is essential for load transfer and efficient utilization of the composite's components. This requires knowledge of the acting bond mechanisms for continuous-fiber-based reinforcements and their interaction with the surrounding concrete.

Rovings for concrete constructions consist of several thousand individual filaments, e.g., [3,14]. In combination with the solvent spinning of the fibers, the rovings are given a sizing to facilitate handling and reduce sensitivity to mechanical damage in the textile manufacturing process. After the textile processing of the rovings to grid-like structures, they are provided with impregnation in an online or offline process. At this moment, the impregnation and consolidation of the roving (a) create the bond between the filaments in the yarn (inner bond) and thus ensure that all filaments participate equally in the load transfer as far as possible [15,16], and (b) influence the bond between the roving and the surrounding concrete matrix (outer bond).

The bond mechanisms in TRC/CRC are the subject of extensive research, e.g., [17–20]. It was recognized relatively early on that an effective internal bond, which is achieved by an even impregnation and consolidation of the roving, is a prerequisite for effective yarn utilization, as otherwise, the edge filaments are subjected to significantly greater stress than the core filaments, resulting in premature failure, e.g., [17,21]. Today's common impregnations are generally based on styrene–butadiene rubber (SBR), epoxy resin (EP), or polyacrylate (PA). Each impregnation has a different effect on the internal and external bond but also the structural stability. SBR-impregnated reinforcements are usually more flexible and therefore suitable for curved shapes and reinforcements in existing structures (subsequent application), whereas stiff EP fabrics are preferred for precast slabs and plates. PA-impregnated grids with thermoplastic properties can be reshaped after textile processing and used for rectangular reinforcement structures.

The bond between rovings and concrete is similar to steel reinforced concrete, mainly based on three mechanisms: adhesion, mechanical interlock, and friction (e.g., [18,22–24]). Adhesion results from the ingrowth of hydration products into the impregnation layer and/or fiber strand [14], adhesion or chemical bond between impregnation and cement paste. A relative displacement between roving and concrete destroys adhesion and activates the frictional resistance, which depends primarily on the roughness at the interface fiber strand and concrete matrix. It is adjustable by material abrasion ([21,23]). The form fit (also or shear bond) is the most important bond component in ribbed reinforcing steels. However, so far, form fit is relatively low with TRC/CRC. Contributions to the mechanical interlocking can be a (periodic) widening of the yarns between the crossing points in a grid-like NCF, the yarn waviness due to the roving constriction with the knitting thread, or cross yarns firmly connected to the yarn in the main carrying direction (see especially [20,23,25]).

Due to the continuously ongoing development of raw materials, yarn processing techniques, impregnations, and concretes on the one hand and due to a large number of possible combinations of fiber reinforcements and concretes on the other, the bond mechanisms cannot be described in general terms to date, let alone allow a precise prediction and furthermore, a specific controlling of the bond behavior. However, important factors influencing the bond, such as the type and material characteristics of fiber and impregnation, the quantity of yarn or fabric geometry, textile binding and processing, concrete properties, and test conditions, such as temperature, are known, e.g., [20,22–24,26]. Accordingly, the proportion of the three relevant bond mechanisms in the overall bond resistance can vary. For textiles with low-modulus"soft"impregnation (e.g., SBR), the adhesive and frictional bond are decisive; for "stiff" textiles (high-modulus impregnation material such as EP or acrylate), the form-fit dominates [23,24]. In the case of conventional grid-like NCF, the form-fit effect is relatively small; therefore, the transmittable bond forces are comparatively low. Hence, the necessary reinforcement area is increased, which is further amplified by the imprecise predictable bond behavior resulting in additional over-dimensioning and high reduction factors for the design of the CRC. Therefore, the highly efficient use of the reinforcement structures is hindered.

Another difficulty in determining the bond resistance of CRC is, as yet, that one test setup has not been standardized. Single- or double-sided textile pull-out tests (e.g., [18,22,23]), overlap [22], and fiber strand pull-out tests [27] are used in various designs. From the tests, bond flow–crack opening, respectively, bond stress–slip relationships are derived (e.g., [18,25]). During all mentioned bond tests, different failures can occur. If the fiber strand breaks, the yarn tensile force is completely transferred into the matrix. In the case of yarn pull-out (interface failure), the bond length is too small—insufficient internal bond in the yarn results in telescopic yarn pull-out. In addition, the surrounding concrete can fail; in the case of newer generation textiles with higher yarn cross-sections and stiffer impregnation especially, there is a risk of splitting of the specimen (longitudinal cracking and/or splitting of the concrete cover), e.g., [20,22,23,25].

In summary, the bond behavior of continuous fiber-based reinforcement in concrete is not a trivial problem. For the application-specific design and construction of components, as well as strengthening layers made of CRC, a high-performance bond (high pull-out loads and bond stiffness) is a basic prerequisite. For increased material efficiency as well as suitability for use, a specific controllable composite would also be highly desirable in order to guarantee a long-lasting and predictable load-bearing capacity. According to our thesis, a defined profiling process of the yarns/rovings, including defined profile geometry, impregnation, and consolidation parameters, can specifically influence and control the bond resistance and, beyond that, the crack formation under load. Therefore, different profiles of carbon fiber rebars have already been studied in depth, and it has been shown that different geometries and varying rebar compositions determine the effectiveness of the TRC-composite [28,29]. This knowledge is now to be transferred to yarns because, in contrast to rods and rebars, yarns are flexible enough for further textile processing into windable, grid-like structures with better handling and higher productivity. In the paper, a promising approach for bond modification through targeted yarn profiling with adjustable profile, impregnation, and consolidation parameters, in addition to the first results of yarn analysis, as well as tensile and composite tests, are presented.

2. Yarn Profiling Technology

Different profiled rovings with varied impregnation and consolidation parameters were investigated for the development of CRC structures with controllable and predictable bond behavior on the basis of a defined form fit effect. In order to create profiled rovings with enhanced but also defined bond performance through a mechanical interlock with the concrete matrix yet maintaining high tensile properties and enough flexibility to be windable, a new profiling technology and innovative roving geometry were developed and patented at the Institute of Textile Machinery and High Performance Material Technology (ITM) [30]. The general process of the profiling method is shown in Figure 1. The characteristic of the roving geometry is the alternating, rectangular profile dents in the vertical and horizontal plane (Figure 2a,c), creating a so-called tetrahedral shape.

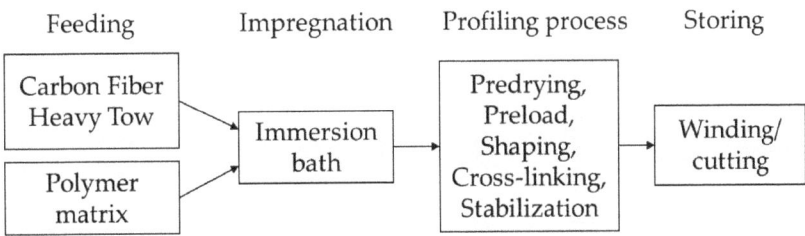

Figure 1. Schematic diagram of the profiling process.

To verify the qualities of the tetrahedral profiled rovings, the first tests were performed at ITM [31,32]. For the profiling, single Carbon Fiber Heavy-Tows (CFHTs) were impregnated in an aqueous polymer dispersion on an acrylate basis and placed in a developed profiling unit. The prototype unit consisted of profile bars that interlocked and formed a cavity when it was closed. The carbon roving acquired its profile as a negative form of the profiling unit. The profiling was permanently stabilized through the consolidation process of the polymer matrix under infrared (IR) radiation (see Figure 1).

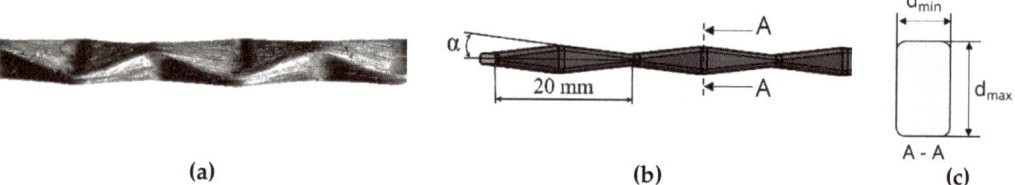

(a)   (b)   (c)

Figure 2. Profiled carbon roving with a tetrahedral geometry: (**a**) photography; (**b**) schematic illustration; (**c**) schematic cross-section.

During the profiling process, the CFHT acquired innovative and patented geometry in the form of a tetrahedral shape (Figure 2a). The profile is hereby defined by spacings between the dents in the same plane (horizontal or vertical) (here: 20 mm), the angular deviation from the linear orientation on the roving surface α (Figure 2b), and the difference between the minimum and maximum diameter in a dent (d_{min} and d_{max}, resp.; Figure 2c).

In contrast to other shaping processes (helix, spiral, twisted, braided, etc.), the tetrahedral geometry distinguishes itself through a uniform reorientation of all filaments. Due to the alternation of the rectangular profile dents in the horizontal and vertical plane, the filaments are reoriented in such a way that all have the same deviation and, therefore, the same length between neighboring profile dents (Figure 3). Hence, all filaments distribute the load under strain in the same way: evenly and maintain their high tensile properties.

Figure 3. Schematic filament orientation of tetrahedral profiled rovings.

Through the developed manufacturing process, a very good yarn impregnation with the impregnation agent and a very dense filament arrangement in comparison to conventional manufacturing methods, such as a combination of multiaxial warp knitting techniques with online or offline impregnation processes, is predicted (see also Section 4.2). The thesis is that this clearly increases material utilization of the rovings (see Section 4.3) because an improvement of the inner bond causes a more even roving activation.

In order to enable a continuous and productive profiling process with high reproducible quality, a laboratory unit was developed and built at ITM [32–34] (see Figure 4a).

In contrast to the prototype unit, it allowed continuous and endless production of the tetrahedral-shaped rovings. In principle, the laboratory unit worked according to the same process shown in Figure 1. The profiling was realized by an upper and lower circumferential chain with profiling tools that interlock when they meet. The yarn shape was limited and adjustable by the vertical distance of the two chains. During drying, cross-linking, and stabilization, the roving was clamped between the profiling tools of the chain (see Figure 4b).

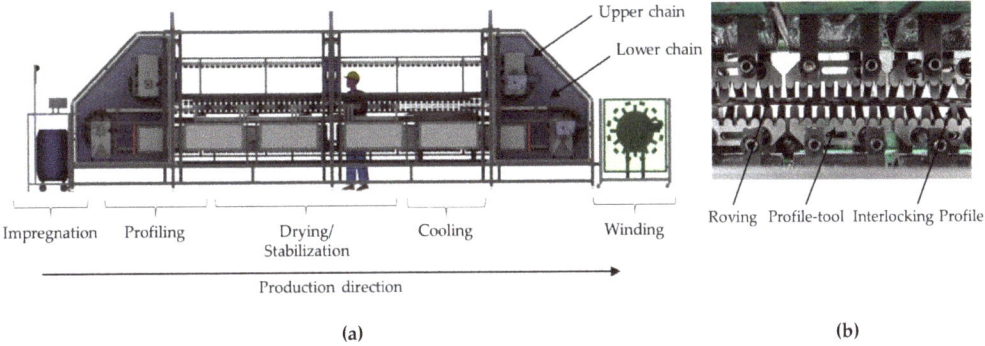

Figure 4. Function of the laboratory profiling unit: (**a**) schematic illustration; (**b**) interlocking profiling tools (**b**).

3. Materials, Test Program, and Testing Methods

3.1. Rovings with Different Configurations for Carbon Fiber Reinforcement

A carbon fiber heavy tow (CFHT) Teijin Tenax-E STS 40 F13 48K 3200 tex (Teijin Carbon Europe GmbH, Wuppertal, Germany) was selected to investigate the influence of profiling on the mechanical properties. All different rovings in this study were produced with this CFHT. Table 1 shows the properties of the dry yarn. The tensile strength was determined in single yarn tensile tests at ITM, according to ISO 3341 (see Sections 3.3 and 4.2, [35,36]).

Table 1. Properties of the dry yarn.

Property	Description/Values [a]
Fiber material	Teijin Tenax-E STS 40 F13 48K 3200 tex Carbon roving
Density in g/cm^3	1.77
Fineness in tex	3215
Tensile strength in MPa	1827 [b]
Elastic modulus in GPa	188
Ultimate strain in %	1.20

[a] Data according to the manufacturer's specifications, unless otherwise stated [37]. [b] Determined in single yarn tensile tests at the ITM acc. to ISO 3341 [36].

For the impregnation, profiling and consolidation of the rovings two different impregnation agents called TECOSIT CC 1000 (CHT Germany GmbH, Tübingen, Germany) and Lefasol BT 91001-1 (Lefatex Chemie GmbH, Brüggen, Germany), which are both polymeric dispersions with a solid content of ca. 50% were used. The exact product properties are listed in Table 2. The difference between the two impregnation agents could have a minor effect on the test results (variation of tensile properties) and will be validated in further studies; the test setup will remain the same.

Table 2. Properties of the impregnation agents, data according to manufacturer's specifications (Lefasol: [38], Tecosit: [39]).

Impregnation Agent				
Product Name	Characteristics	Base-Material	Solid Content in %	Linking Temperature in °C
TECOSIT CC 1000 (CHT Germany GmbH)	Aqueous polymer dispersion	Polyacrylate	47 ± 1	160
Lefasol BT 91001-1 (Lefatex Chemie GmbH)		Polystyrol	52 ± 1.5	150–160

In addition to the non-impregnated, straight rovings (dry yarn from a spool), impregnated, non-profiled rovings, and two profiled roving variants (medium and strong profile, see Table 3) were manufactured (impregnation and consolidation according to the general process in Figure 1 without profiling) and subsequently analyzed. Hereby, each variant was combined with the two different impregnation agents Tecosit and Lefasol. As a reference (short "Ref."), a single, straight roving extracted from the textile SITgrid 040 (Wilhelm Kneitz Solutions in Textile GmbH, Hof, Germany) with the same fiber material and impregnation agent Tecosit is used. The textile was tested during the project Carbon Concrete Composite (C^3) [40] and represents a reliable reference. Table 3 shows the basic properties of these rovings.

Table 3. Profile characteristics of the different rovings.

Configuration	Geometry	Dimension (~)	Cross-Section	Illustration
Roving				
Without defined profile				
Dry yarn	Band-shaped	Variable (no internal bond)	—	
Impregnated roving	Circular	d = 2 mm		
Roving from textile	Elliptical	d_1 = 3.3 mm d_2 = 1.3 mm		
With defined profile				
Tetrahedral profiled roving	Medium profile	d_{diff} = 0.6 mm α = 3°		
Tetrahedral profiled roving	Strong profile	d_{diff} = 1.0 mm α = 5°		

For the investigation of the influence of the profiling process on the roving properties, tetrahedral profiled rovings were produced on the discontinuously working prototype unit (Series index "P") as well as on the continuously working laboratory unit (see Table 4). The function of the prototype unit is described in Section 2 and is distinguished from the laboratory unit through a static shaping process and constantly applied pressure during the consolidation. The laboratory unit produced profiled rovings with different profile configurations (see Table 3).

The profile characteristics of the different rovings are shown in Table 3. Hereby, the profile of the tetrahedral-shaped rovings was characterized by the difference between the minimum and maximum diameter in a profile dent (smallest cross-section) and the angle of the filament orientation. According to Figure 2, the angle α was hereby determined as the tangent between the distance between two neighboring profile dents in the vertical and horizontal plane (10 mm) and the difference between the minimum and maximum diameter. The impregnated roving with no profile showed a circular shape with a diameter of about 2 mm, and the single roving extracted from the textile showed an elliptical cross-section due to the warp knitting process and the fixation with the knitting thread.

Table 4. Properties of different rovings; Index "P" for prototype unit.

Roving Configuration	Sample	Parameter				
		Roving Geometry	Profile Unit	Impreg-Nation Material	Solid Content in %	Consoli-Dation Time in Min
Rovings without profile						
Dry yarn	Series 0	-	-	-	-	-
Impregnated roving	Series 0L	Circular		Lefasol	50	4
	Series 0T			Tecosit		
Roving from textile (Ref.)	Series R	Elliptical				unknown
Profiled rovings from prototype unit						
Profiled roving	Series 2_P	Tetrahedral Strong	Prototype unit	Lefasol	50	4
	Series 4_P			Tecosit		
Profiled rovings from laboratory unit with different profiles and impregnation agents						
Profiled roving	Series 1	Tetrahedral Medium	Laboratory unit	Lefasol	50	4
	Series 2	Tetrahedral Strong				
	Series 3	Tetrahedral Medium		Tecosit		
	Series 4	Tetrahedral Strong				
Profiled rovings from laboratory unit with different solid content and consolidation						
Profiled roving	Series 4_30%	Tetrahedral Strong	Laboratory unit	Tecosit	30	4
	Series 4_40%				40	
	Series 4_10 min				50	10

In order to investigate the influence of the solid content of the impregnation as well as the consolidation parameters on the bond behavior (specific pull-out load and bond stiffness) of tetrahedral profiled rovings, test specimens with the same profile characteristics (strong profile—Series 4) were produced on the laboratory unit with different solid contents (30%, 40%, 50%) (see Table 4). To investigate if an intensified consolidation of the profiled rovings has an influence on the bond behavior, the series with the highest solid content (50%) and, therefore, presumably the highest bond performance was consolidated for 4 min as well as for 10 min (see Table 4). Hereby, the consolidation time varied through different production speeds of the continuously working Profiling unit. For the consolidation (drying and stabilization), the impregnated roving was positioned between several opposites positioned IR-modules from OPTRON GmbH (Germany) Typ IRDS750 SM 3kW (400 V) fast middle wave with 90% power (2.7 kW) and a distance of 50 mm to the roving. The solid content of the impregnation varied by adding water to the polymeric dispersion.

3.2. Concrete Matrix

Fiber-based reinforcements are very often embedded in cementitious matrices with small maximum grain sizes (e.g., [14]). For such fine concrete matrices, the compressive strength and the flexural tensile strength are usually determined according to DIN EN 196-1 [41] after 28 days. Three standard prisms with a cross-section of 40 × 40 mm and a length of 160 mm were concreted per batch. First, the bending tensile strength was evaluated in a three-point bending test [41]. The compressive strength was then determined in a uniaxial compression test on the resulting two prism halves.

In the course of the initial trials on profiled yarns presented here, two different fine concretes were used. One was the TF 10 CARBOrefit® (PAGEL Spezial-Beton GmbH & Co. KG, Essen, Germany) fine concrete. This cement-based, fine concrete dry mix has been established for the subsequent strengthening of structures in Germany over the last years [42,43]. The maximum grain size of the mixture is 1 mm. Only water needs to be added to the ready-mix. The soft plastic consistency is suitable for laminating in layers and for spraying. The factory-guaranteed properties are summarized in Table 5. These minimum values were met in all test series. The second mixture—a fine concrete-dry-mix (called BMK 45-220-2, consisting of binder material BMK-D5-1 from Dyckerhoff, Germany; KSM Compact III by KSM-Babst GmbH, Germany; fine sand BCS 412 from Strobel, Germany; sand 0/2 from Ottendorf, Germany; superplasticizer PCE SP VP-16-0205-02 from MC-Bauchemie, Germany and water)—was used for the pull-out tests at the Institute of Construction Materials (IfB) of TU, Dresden. The concrete properties were determined on 40 × 40 × 160 mm prisms according to DIN EN 196-1 [41]; the mean values are also listed in Table 5.

Table 5. Concrete properties, minimum values after 28 days.

Concrete Property	TF 10 CARBOrefit® Fine Concrete [42]	BMK-45-220-2
Compressive strength in MPa	≥80	≥105
Bending tensile strength in MPa	≥6	≥11.5
Maximum grain size in mm	1	2

3.3. Test Program and Test Setups

In order to visualize the impregnation quality of the yarns and to analyze the influence of the shaping process on the filaments, micrographic analyses were carried out on different roving sections (cross- and longitudinal sections). Hereby, EP-resinated roving samples were examined with a reflected-light microscope (Zeiss AxioImager.M1m from the Carl Zeiss AG, Jena, Germany) with a bright field and magnification factor of 200.

Tensile tests on single impregnated rovings are less time-consuming and a fast method to obtain a statement on the change in load-bearing capacity as a result of further processing, e.g., profiling. They are suitable for production control and characterization of the influence of various production parameters. The tests were conducted on the basis of DIN EN ISO 10618 [44] (see also [35,45]. The free yarn length was 200 mm. The ends of the profiled yarns were clamped with metal clamps with a steel file cut. For this purpose, the single impregnated rovings (without profile, with profile, and from textile) were resinated in the clamping area (Section 3.4) and clamped between two pneumatic pressured steel clamps (50 × 60 mm) with a file cut surface at 35 bar. Figure 5 shows the principle of the clamping on the left and the test facility on the right. All tests were performed with the testing machine Zwick 100 from ZwickRoell GmbH & Co. KG (Germany). The test speed was 3 mm/min. The entered force was measured with a 100 kN force tensor, and the elongation of the roving was determined with an optical laser system consisting of two length variation sensors and reflex markers, which were fixed on the roving prior to the test. The modulus of elasticity was calculated from the applied force during a roving elongation from 0.15 to 0.9%.

All tests on the CRC specimens took place at 20 °C, 28 days after casting. For the tensile test on textile reinforced concrete, usually, fabric sections are embedded in fine-grained concrete. In addition to the tensile strength, the cracking behavior of the composite can be analyzed. The specimen dimensions essentially depend on the grid-like non-crimp fabric geometry (specimen width), the yarn thicknesses (specimen thickness), and the fabric's load-bearing capacity. Detailed recommendations can be found in [23,46]. The testing was conducted according to these suggestions. However, individual yarns which were embedded parallel and as stretched as possible in the concrete were tested. The test setup is shown in Figure 6. The test specimen was clamped into the testing machine at both ends. The specimen length included sufficiently long anchorage areas and a centric measuring section of at least 200 mm, which was not influenced by the lateral pressure in the load introduction area. The displacement transducer (DD1) was highly visible, clamped to the specimen in the middle area, and used to record the strain in the free measuring length.

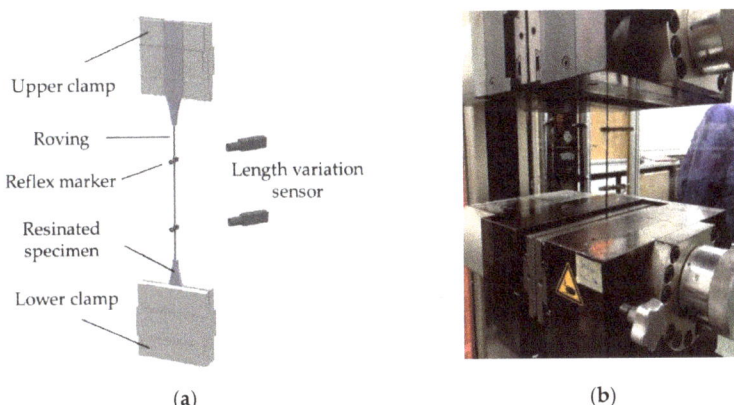

Figure 5. Test setup for yarn tensile test on basis of DIN EN ISO 10618 at ITM: (**a**) clamping principle; (**b**) test stand. [35] .

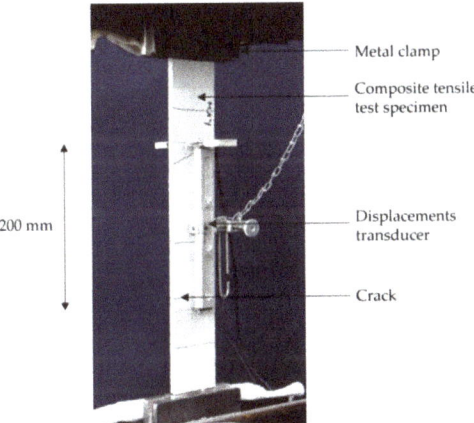

Figure 6. Characterization of carbon-reinforced concrete samples; test setup and measurement equipment for tensile tests on the composite.

There are various possibilities for the characterization of the bond between textile reinforcement and concrete; however, there is no standardized test method yet.

Therefore, a test method suitable for single yarn pull-out tests was examined, mainly to understand whether this test method is suitable for the investigation of profiled yarns.

Single yarn pull-out (YPO) tests were conducted at the IfB in order to analyze the characteristic bond–slip behavior of single rovings with different profile properties (e.g., [27,47,48]). In this type of experiment, individual rovings were embedded in cubic concrete blocks. The upper block provided an embedment length of 50 mm at the top roving section. The lower block possessed an increased embedment length of 90 mm at the bottom roving section for a defined roving fixation. The concrete cover was 40 mm. The specimens were fixed in an upper and lower specimen holder, and the pull-out force—slip–deformation curve was measured by a single-sided pull-out in the upper concrete block with a controlled quasi-static load (Figure 7). The pull-out (slip) deformation was measured by an optical system consisting of laser sensors and aluminum clips, which were fixed to the yarn.

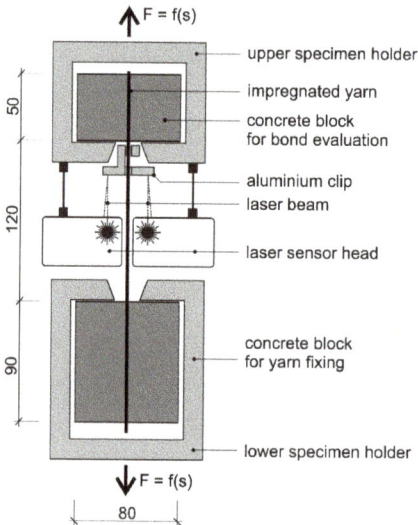

Figure 7. Schematic test-setup for single yarn pull-out (YPO) test (dimensions in mm) [27,47,48].

3.4. Specimens Manufacturing

Short roving sections of about 10 mm were placed in cylinders of 20 mm diameter and fully resinated for the microscopic examinations. After one day of drying, the front side was ground with sandpaper and finally polished.

450 mm long roving sections were cut to size for the yarn tensile tests. Then, the rovings were stretched and clamped in a frame. With the help of metal molds, the ends were cast in epoxy resin. Figure 8 shows a sample ready for testing.

Figure 8. Tensile test specimen of a profiled roving with EP-resinated ends.

To determine the tensile strength of rovings embedded in TF10 CARBOrefit© fine concrete, six tensile specimens with three profiled yarns each and a concrete cover of 5 mm were produced for each yarn series by laminating. This was done in a formwork in which the individual yarns were fixed and aligned with a yarn spacing of 13 mm. Then, the fine-grained concrete was filled in; first, a bottom layer (Figure 9) was subsequently slightly compressed. The top concrete layer was filled in and smoothed in a second step. The 120 cm long, 1 cm thick, and 33 cm wide plate was then covered with damp cloths. The plate was stored in water from the 2nd to the 7th day. From day 8 to day 28, in a climate chamber. Before the tensile tests, the plate was sawn in 5.2 cm wide stripes containing three yarns each.

Specimens for the YPO tests were made by embedding single profiled carbon rovings as well as rovings with no profile and warp knitted rovings (reference) in the self-compacting fine-grained concrete BMK 45-220-2 in a cube formwork (Figure 10). One specimen consisted of two centered concrete blocks at the yarn ends and a free yarn segment of 120 mm in between the blocks. This was a clearly defined area in which composite failure could occur. The specimens were stored for seven days underwater and stored for additional 21 days in a climate chamber (20 °C and 65% relative humidity).

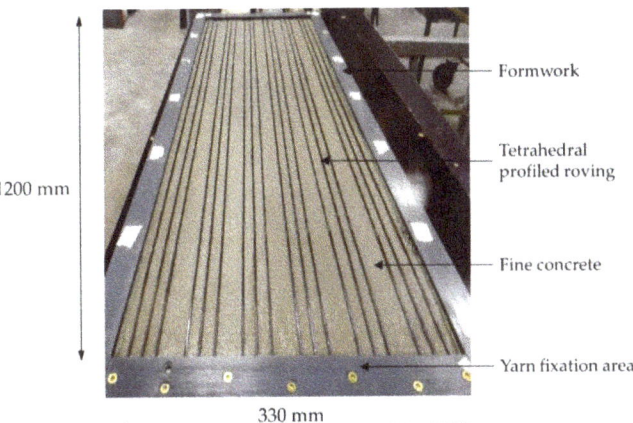

Figure 9. Laminating of tensile specimen: first concrete layer and straight yarns (three per sample).

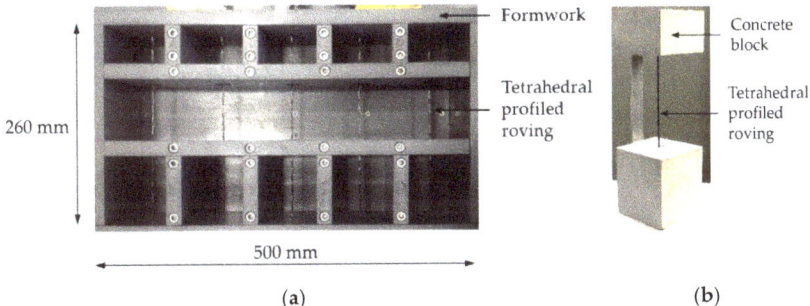

Figure 10. YPO specimen: (**a**) formwork with single yarns; (**b**) test specimen ready for pull-out.

4. Results and Discussion

4.1. Processing Quality of Profiled Rovings

In order to evaluate the processing quality, especially the filament arrangement of the profiled rovings, microsection analyses were conducted on different roving sections (Figure 11). The microscopic tests were performed at ITM's textile–physical testing laboratory. At least five cross- and longitudinal sections of profiled rovings were analyzed. The microsections showed almost no air gaps between the filaments or polymer accumulations. From this, it is deduced that a very dense filament arrangement was achieved (Figure 11a). The longitudinal section (Figure 11b) visualized the filament course along the roving axis, where the filaments showed no apparent damage or deviation from the linear orientation between the profile sections. In the profile sections, the filaments showed a dense arrangement, which resembled the cross-section analysis. The dark spots in the longitudinal section (Figure 11b) are surface irregularities (filament detachment) due to preparation (cut of the roving) which caused the light to reflect away.

In conclusion of the microscopic test series, the dense filament arrangement increased the inner bond of the impregnated roving and, according to HAHN et al., resulted in higher material utilization because almost no air gaps or polymer accumulations disturbed the load transmission between the single filaments [15]. The mechanical characterization of the tensile properties of the profiled rovings themselves, embedded rovings in concrete, as well as their bond behavior is discussed in Sections 4.2–4.4.

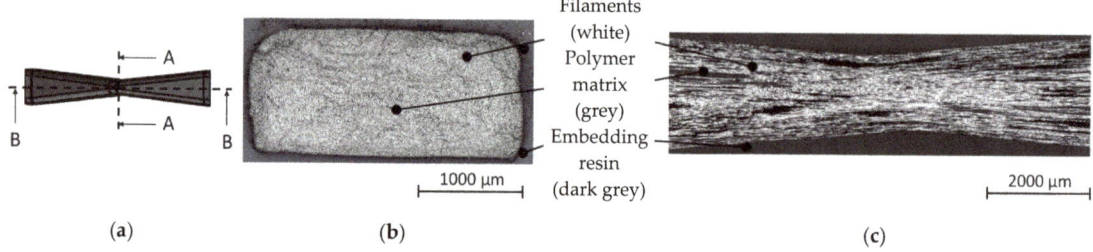

Figure 11. Exemplary microsection of profiled rovings: (**a**) schematic illustration of the microsection directions; (**b**) cross-section (A–A); (**c**) longitudinal section along the roving axis (B–B).

4.2. Tensile Strength of Single Rovings

The following diagrams illustrate the mean values of the tensile tests of the different series of fiber strand configurations with their standard deviation. For each series, at least seven to ten single specimens were tested according to DIN EN ISO 10618 [44]. An important part of the study was a comparison between tensile properties of dry rovings (Series 0, compare Table 4), impregnated rovings from a reference textile (Series R), impregnated rovings with no profile (Series 0T/0L), and impregnated rovings with a defined tetrahedral profile. This was to conclude the influence of the profile on tensile properties of the roving (Series 3/4). Hereby, the determined tensile strength (N/mm^2) refers in all tests (dry and consolidated rovings) to the measured force (in N) divided by the dry and compact filament area of 1.81 mm^2. The composite dimensions of the impregnated rovings were neglected for the calculation of the tensile strength and Young's Modulus because only the filaments transmit the load. The Young's Modulus is the quotient of the absolute tensile strength difference and total elongation in the range of 0.15% and 0.9% elongation.

The diagrams in Figure 12 show the determined tensile strength and Young's Modulus of dry CF-rovings (Series 0) in comparison to impregnated rovings with two different impregnation agents (Lefasol—Series 0L, Tecosit—Series 0T). The single standard deviation is specified with error bars.

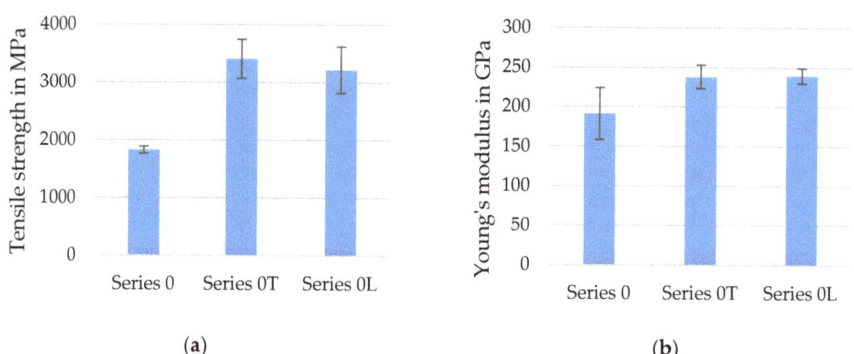

Figure 12. Tensile strength (**a**) and Young's Modulus (**b**) of dry and impregnated CF-rovings.

As expected, tensile strength significantly increased by a factor of two through the impregnation of the rovings in contrast to dry rovings. The high tensile strength indicated a good and even distributed polymeric impregnation of the rovings resulting in an improved internal bond and load transmission between the filaments across the roving cross-section. A further influence was the different test setups. The dry rovings were tested according to ISO 3341 [36] with wrap specimen holders, as described in [35], whereas the impregnated rovings were tested with resinated clamping areas in metal clamps (see Section 3.4), allowing a better load introduction into the roving. The theoretical tensile strength of the carbon roving with 4300 MPa according to the datasheet [37] (density of 1.77 g/cm^3) could not be achieved due to the inhomogeneous load introduction of the dry roving. The slightly different composition of the impregnation agents caused a little difference in achievable tensile properties. The tensile strength of impregnated rovings with Tecosit (Series 0T) was about 5% higher compared to impregnated rovings with Lefasol (Series 0L). Young's Modulus of both impregnations was almost identical at around 235 GPa. Because of the better performance of impregnated, unprofiled rovings with Tecosit impregnation in contrast to Lefasol impregnated rovings, the following diagrams for the evaluation of the tensile properties of profiled rovings will only show results from profiled rovings with the impregnation agent Tecosit.

The averaged tensile properties of at least seven to ten tetrahedral profiled rovings with different profile configurations (medium profile—Series 3, strong profile—Series 4) are shown in the diagrams in Figure 13 and compared to impregnated rovings with no profile (Series 0T) and single rovings from the reference textile SITgrid 040 (Series R). All tested specimens were impregnated with Tecosit.

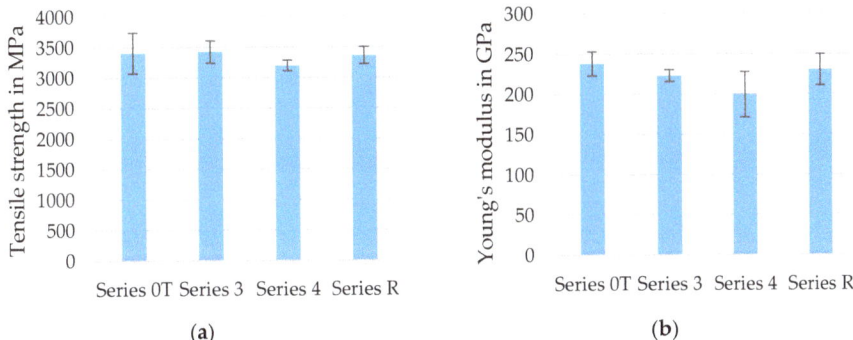

Figure 13. Tensile strength (a) and Young's Modulus (b) of CF-rovings with different profiles.

The tensile strength of the impregnated rovings without profile (Series 0T), the extracted rovings from the SITgrid 040 textile (Series R), and the middle profiled rovings (Series 3) were ~3400 MPa similarly high. The rovings with a stronger profile (Series 4) showed ~3200 MPa, a slight decrease in their tensile strength compared to the other variants.

These results also mainly applied to the E-Modulus of the different rovings. The rovings with no profile achieved ~240 GPa, the highest due to the straight-oriented filaments. The extracted rovings from the grid-like non-crimp fabric SITgrid 040 showed ~230 GPa, a slightly reduced E-Modulus. The thesis is that the roving with no profile (Series 0T) was more uniformly impregnated, which resulted in a more uniform utilization of the single filaments. In addition, the roving was impregnated in a state of almost no tension, whereas the roving from the reference textile (Series R) was warp knitted before impregnation. The constriction of the roving with the knitting thread impeded a uniform impregnation [15]. The profiled rovings (Series 3/4) showed a slight reduction in Young's Modulus to around 200 GPa. This was a logical consequence of profiling because the filaments were slightly deviated from an absolute straight orientation along the yarn axis, and therefore the applied load was not induced straight into the filaments, resulting in a slightly incomplete utilization of the anisotropic fiber properties. The strongly profiled rovings had the lowest tensile stiffness compared to the other profile configurations. Nonetheless, the tensile properties of the profiled rovings were very high due to the almost uniform reorientation of all the filaments during the shaping process (see Section 4.1).

In conclusion, the profiled rovings achieved a high tensile strength of over 3000 MPa in addition to a high tensile stiffness of around 200 GPa.

4.3. Tensile Strength of Concrete Embedded Rovings

Results from tensile tests with embedded carbon fiber heavy tows (CFHT) (impregnation: Lefasol, compare Table 4) are shown. These yarns were profiled in the prototype unit (Series 2_P). Of the total of six samples produced, only five could be included in the evaluation, as the DD1 had slipped in the first attempt. Figure 14 shows the single yarn tension–elongation curves of individual test specimens (each specimen numbered 1 to 5) with a very low scatter and the averaged mean curve (red) of all tested specimens.

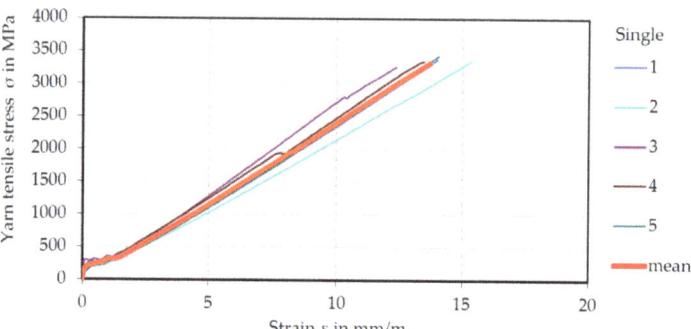

Figure 14. Stress–strain curves of CF-rovings profiled in the prototype unit. 1: specimen 1; 2: specimen 2; 3: specimen 3; 4: specimen 4; 5: specimen 5, mean: average of all tested specimens.

A mean yarn tensile stress of 3382 MPa (refers to a dry and compacted filament area of 1.81 mm^2) was determined, which is right in the range of the failure strength in the yarn tensile tests (see Section 4.2). This was further confirmation that the profiling does not cause any process-related damage to the carbon fibers. Furthermore, it can be stated that the two experimental methods (tensile test of single yarns and tensile test of embedded yarns) are well suitable to determine the tensile strength of carbon rovings.

Figure 15 shows fractured test specimens of concrete embedded rovings after testing.

Figure 15. Tensile test specimen of concrete embedded rovings after failure.

It is visible, that yarn rupture occurred in all tests, indicating that the bond was sufficient enough to transfer the full load from the concrete matrix into the rovings. The profiling of the yarns was still clearly visible after the tensile test. There was no pull-out of the yarns from the areas of load application. At the moment of failure, a complete spalling of the concrete occurred in the measuring area. No splitting or delamination cracks could be observed during the tests.

Yarns profiled in the laboratory unit were also tested. Figure 16 shows the averaged mean value yarn tensile stress–strain curves of six samples from Series 2_P (prototype unit, Lefasol), Series 1 (laboratory unit, Lefasol), and Series 3 (laboratory unit, Tecosit). All values were in the order of magnitude of the single yarn tension test and referred to the compact filament area of 1.81 mm^2; the results in the laboratory and the prototype system were very similar. The failure in each case was yarn breakage. The scatter was again very moderate, which means that the tensile properties are reproducible. The failure strength of the Lefasol impregnated yarns (Series 1) was slightly lower than that of Tecosit yarns (Series 3), which agrees with the results in Section 4.2. The varying slope of the curves indicates differences in yarn stiffness. This and the formation of cracks were not systematically investigated during these initial tests and therefore not discussed in detail here. Such considerations are the subject of ongoing and planned research.

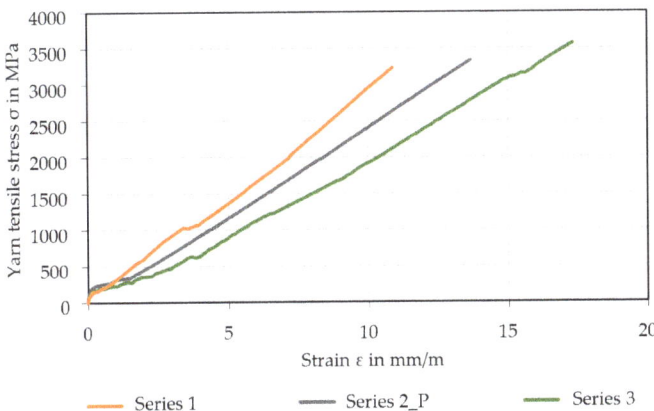

Figure 16. Averaged stress–strain curves of different profiled CF-rovings.

4.4. Bond Behavior of Concrete Embedded Rovings

In order to compare the bond behavior of the rovings with different profiles as well as varied impregnation and consolidation parameters, pull-out tests were carried out at the Institute of Construction Materials (IfB) of the TU, Dresden, according to the described test setup (Section 3.3). On average, four test specimens per configuration were tested.

The following diagrams show the averaged specific pull-out load– slip-deformation curves and the bond strength of the rovings, which was equal to the maximum pull-out load. Hereby, the specific pull-out load, and thus bond strength (in N/mm), refers to the measured bond force (in N) divided by the bond length. Because no tested specimen showed shear cone failure or partial separation of concrete from the sample surfaced, the initial bond length was a constant 50 mm.

The diagrams in Figure 17 illustrate the bond strength (b) and the measured curves (a) of continuously produced profiled rovings with long consolidation (Series 4_10 min) as well as of profiled rovings from the discontinuously working prototype unit (Series 4_P). Both were compared to impregnated rovings without profile (Series 0T) and fiber strands extracted from the reference textile SITgrid 040 (Series R).

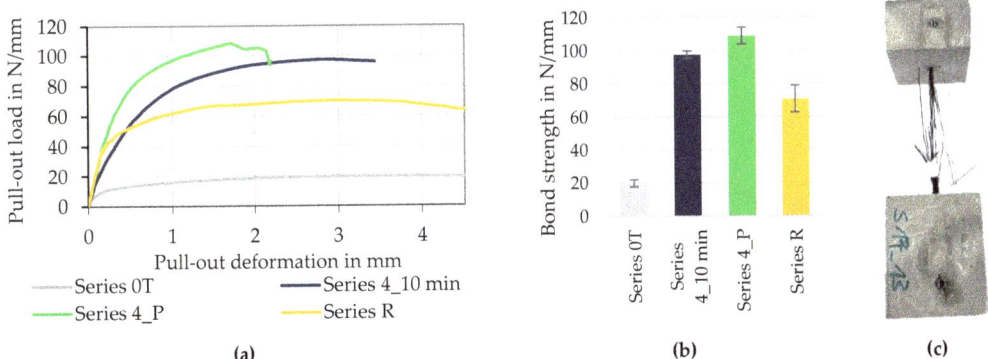

Figure 17. Specific pull-out load–slip deformation curves (a) and bond strength (b) of CF-rovings with different profile configurations and a specimen for the bond tests after failure (c).

Hereby, the profiled rovings achieved their highest bond strength at ~100 N/mm, which was up to 40% higher compared to the reference (~70 N/mm) and almost five times the bond strength of the straight roving (~20 N/mm). The pull-out curves proved that by the defined profiling of the roving, the transmittable bond force can be increased and controlled, depending on the profile configuration.

As can be seen in diagram (a), all processed rovings initially showed approximately the same level of bond stiffness. However, the reference rovings (Series R) did not have a distinct profile geometry except for the oscillating geometry of the oval cross-section due to the warp knitting process, which caused a lower form-fit effect compared to the profiled rovings (Series 4). Therefore, the bond force was lower than for profiled rovings. As expected, the straight rovings without profile (Series 0T) showed the lowest bond performance (~20 N/mm). Here, the bond mechanism was mainly of a chemical nature due to adhesion between the roving surface and the surrounding concrete matrix.

During tests, all profiled rovings showed yarn rupture; thus, only incomplete pull-out curves were recorded. It indicated that the bond length of 50 mm is sufficient for a complete transfer of the tensile load inside the yarn to the concrete matrix via bond forces. Straight rovings and yarns from the reference textile (Ref.), were pulled out completely. Thus, the bond strength could be measured. Therefore, the strongly profiled rovings showed a much better bond performance compared to non-profiled reinforcement yarns.

The profiled rovings from the prototype unit (Series 4_P) achieved ~110 N/mm, the highest bond strength, as well as a higher bond stiffness compared to the strongly profiled rovings from the laboratory unit (Series 4_10 min). One reason is seen in the different profiling processes. Profiling in the prototype unit was a static process. Constant high pressure was applied—shaping and compaction of the roving were very uniform (see Section 2). It is believed that the profile produced in this way offers greater resistance to deformation under load than the yarns shaped in the laboratory plant. Here, the profiling tools were moved, and at the current stage of development, a truly constant pressure could not yet be guaranteed. The targeted control of the process parameters is subject to development to date.

Figure 18 illustrates bond behavior (a) and bond strength (b) of continuously produced profiled rovings with different profile configurations (no profile—Series 0T, medium profile—Series 3, and strong profile—Series 4). In comparison to rovings with no profile (~20 N/mm), medium-profiled rovings reached ~40 N/mm, about twice the bond strength. The stronger profiled rovings achieved ca. 80 N/mm, four times the bond force compared to rovings with no profile. Furthermore, profiled rovings showed a much higher bond stiffness indicated by the steeper increase at the beginning of the pull-out load–slip deformation curve. Hereby, especially the profiled rovings showed a continuous increase in the bond force and a distinct plateau at the maximum bond force, which indicated a stable mechanical interlock of the roving and the surrounding concrete matrix.

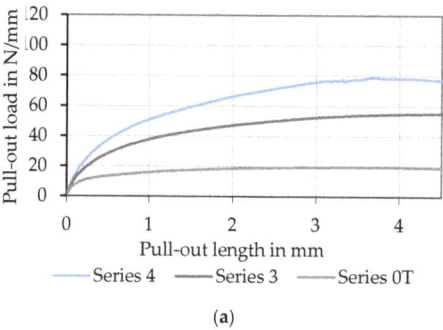
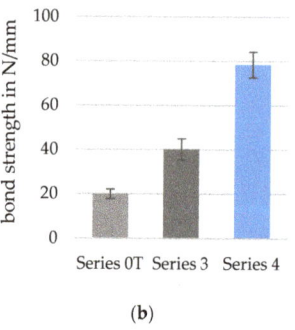

(a) (b)

Figure 18. Specific pull-out load–slip deformation curves (a) and bond strength (b) of tetrahedral profiled CF-rovings with different profile configurations.

Figure 19a shows the bond behavior of tetrahedral profiled rovings with strong profile (Series 4) in dependence on the solid contents of the impregnation Tecosit. It is evident that a reduced solid content of the polymeric dispersion led to a decrease in bond strength and stiffness. Furthermore, the reduction in solid content from 50% to 30% resulted in a decrease in transmittable bond loads of about 40% (from ~80 N/mm to ~50 N/mm). A possible reason is a smaller resistance against deformation of the profile during pull-out due to reduced impregnation content. The hypothesis is that the profile of the roving deformed differently under stress (depending on impregnation and consolidation parameters), resulting in a less corrugated profile and hence, reduced bond properties when the solid content of the impregnation is reduced.

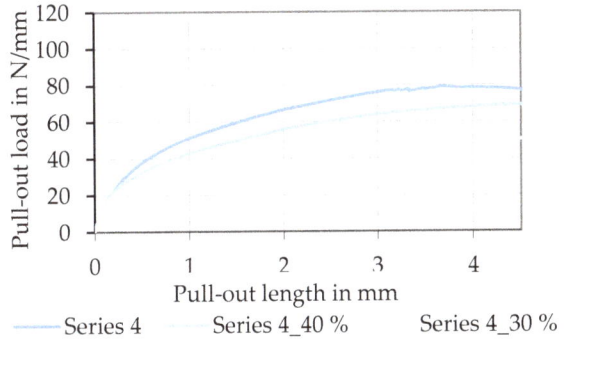
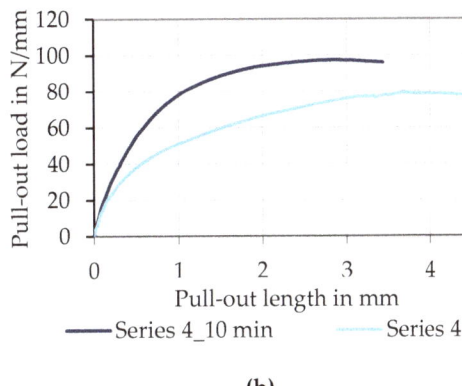

(a) (b)

Figure 19. Specific pull-out load–slip deformation curves of tetrahedral profiled CF-rovings (strong profile) with different solid contents of Tecosit (**a**) and different consolidation times (**b**).

The influence of consolidation time on rovings with the same profile and solid content is illustrated in Figure 19b. Herby, a longer-lasting consolidation by a heat input with IR radiation for 10 min (IR-specification see Section 3.1) resulted in a significantly higher maximum bond strength as well as a higher bond stiffness (steeper increase in the curve). According to the previously formulated thesis, an intensified consolidation of the roving resulted in higher resistance against deformation of the profile. It therefore allowed the transmission of higher bond forces before failure or deformation of the rovings. The thesis will be tested in further research studies via an optical analysis of the roving geometry during tensile tests with a high-speed camera, which allows detecting small deformations or changes in the profile during the applied tensile stress.

5. Conclusions and Outlook

In summary, the profiled carbon rovings transmitted higher pull-out loads compared to unprofiled rovings and rovings extracted from warp knitted textiles. Due to the gentle shaping process and the good penetration of the impregnating agent, the rovings with a tetrahedral shape had high tensile strengths.

It can be stated that:
- Better dense filament arrangement and better material utilization can be achieved by good penetration of impregnation agent and immediately following shaping of rovings (see Section 4.1)
- The developed shaping process created profiled rovings with a defined tetrahedral geometry that showed almost no decrease in their tensile properties ($\leq 10\%$) compared to impregnated rovings with no profile (see Figure 13).

- Tetrahedral-shaped rovings showed up to 500% the concrete bond strength compared to rovings with no profile (see Figure 17) and 140% of warp knitted rovings (that showed a slight waviness and roving constriction).
- Bond strength and bond stiffness depended on the profile geometry, as well as impregnation and consolidation parameters (see Figures 18 and 19); a defined variation of the stated parameters enabled a modification of the bond behavior.
- A strong profile in combination with an intensive (long) consolidation and a high solid content of the impregnation agent (50%) resulted in a higher bond performance (see Figure 19) with a maximum bond strength of about 100 N/mm.

Based on the results, the tetrahedral shaping process of impregnated rovings shows a high potential to create high-performance textile reinforcement structures with concrete bond-optimized behavior and high tensile properties. Primarily, the complete form-fit-based anchoring between profiled roving and surrounding concrete matrix resulted in an increase in the maximum transmittable bond strength and high bond stiffness with up to five times the values of consolidated straight rovings. The reduction in the required bond length enabled a better material efficiency, especially considering the energy-intensive production of carbon fibers and reinforcements.

In summary, it can be stated that we see further potential in the profiling process to maximize bond strength and bond stiffness by optimization of the impregnation (yarn spreading for even impregnation distribution) and consolidation process (focused energy input for intensified consolidation). Additionally, the investigation of influential parameters on bond behavior and their targeted adjustment could enable a predictable design of CRC structures with specific and application-oriented bond behavior. Therefore, extensive basic research is planned to find a method for the targeted adjustment of strength and composite properties through the defined and variable profiling of carbon yarns in addition to a numerical description of the bond behavior.

Carbon fiber reinforcements with specifically adjustable properties will clearly increase the material efficiency of carbon-reinforced concrete in the future in the areas of new construction and strengthening. In the context of new construction, we are thinking of currently researched, material-efficient structural elements [49]. In the case of component strengthening, for example, shortened end anchorage and overlap lengths will improve handling. Additionally, the lower material consumption reduces costs, which increases competitiveness compared to other reinforcement methods.

Author Contributions: Conceptualization, P.P., S.S. and L.H.; methodology, P.P.; validation, P.P., M.M. and S.S.; formal analysis, P.P. and S.S.; investigation, P.P. and M.M.; data curation, M.B. and M.M.; writing—original draft preparation, P.P. and S.S.; writing—review and editing, S.S.; visualization, P.P., M.M. and M.B.; supervision, L.H., H.M., M.W.; M.C., C.C. and V.M.; project administration, L.H., H.M., M.C., C.C. and V.M.; funding acquisition, L.H., H.M., M.C., C.C. and V.M. All authors have read and agreed to the published version of the manuscript.

Funding: The results presented here were obtained over a longer period. Mention should be made of the following:

- Financial support for setting up the profiling units as part of a Twenty20 investment project/C^3 project (funding code 03ZZ03X03, within C^3–Carbon Concrete Composite, funding body: BMBF).
- The IGF research project 21375 BR of the Forschungsvereinigung Forschungskuratorium Textil e. V. is funded through the AiF within the program for supporting the "Industriellen Gemeinschaftsforschung (IGF)" from funds of the Federal Ministry for Economic Affairs and Climate Action on the basis of a decision by the German Bundestag.

Institutional Review Board Statement: Not applicable.

Informed Consent Statement: Not applicable.

Data Availability Statement: Data is contained within the article.

Acknowledgments: In addition to the funding agencies, the authors would like to thank our colleagues in the laboratories, Martin Waldmann, Johannes Wendler (both ITM, support with the tensile tests), and Philipp Kunze (help with the YPO test). We would also like to thank all involved companies for their technical support and the provision of test material, as well as all other partners who supported us in the research on this topic.

Conflicts of Interest: The authors declare no conflict of interest. The funders had no role in the design of the study; in the collection, analyses, or interpretation of data; in the writing of the manuscript, or in the decision to publish the results.

References

1. Hegger, J.; Will, N. Textile-reinforced concrete: Design models. In *Textile Fibre Composites in Civil Engineering*, 1st ed.; Triantafillou, T.C., Ed.; Woodhead Publ.; Elsevier: Cambridge, UK, 2016. Available online: https://www.sciencedirect.com/book/9781782424468/textile-fibre-composites-in-civil-engineering (accessed on 10 August 2022).
2. Peled, A.; Bentur, A.; Mobasher, B. *Textile Reinforced Concrete*; Series: Modern Concrete Technology No 19; CRC Press: Boca Raton, FL, USA, 2017. Available online: https://www.taylorfrancis.com/books/mono/10.1201/9781315119151/textile-reinforced-concrete-alva-peled-barzin-mobasher-arnon-bentur (accessed on 10 August 2022).
3. Jesse, F.; Curbach, M. Verstärken mit Textilbeton. In *BetonKalender 2010*; Bergmeister, K., Fingerloos, F., Wörner, H.D., Eds.; Ernst & Sohn: Berlin, Germany, 2010; Chapter VII; pp. 457–565. [CrossRef]
4. Koutas, L.N.; Tetta, Z.; Bournas, D.A.; Triantafillou, T.C. Strengthening of Concrete Structures with Textile Reinforced Mortars: State-of-the-Art Review. *J. Compos. Constr.* **2019**, *23*, 03118001-1–03118001-20. [CrossRef]
5. Carloni, C.; Bournas, D.A.; Carozzi, F.G.; D'Antino, T.; Fava, G.; Focacci, F.; Giacomin, G.; Mantegazza, G.; Pellegrino, C.; Perinelli, C.; et al. Fiber Reinforced Composites with Cementitious (Inorganic) Matrix. In *Design Procedures for the Use of Composites in Strengthening of Reinforced Concrete Structures–State-of-the-Art Report of the RILEM Technical Committee 234-DUC 2016*; Pellegrino, C., Sena-Cruz, J., Eds.; Springer: Dordrecht, The Netherlands, 2016; Chapter 9; pp. 349–392.
6. Scheerer, S.; Schladitz, F.; Curbach, M. Textile reinforced Concrete—From the idea to a high performance material. In *Proceedings of the FERRO-11 & 3rd ICTRC (PRO 98), Aachen, Germany, 7–10 June 2015*; Brameshuber, W., Ed.; S.A.R.L. Rilem Publ.: Bagneux, France, 2015; pp. 15–33. Available online: https://www.semanticscholar.org/paper/TEXTILE-REINFORCED-CONCRETE-%E2%80%93-FROM-THE-IDEA-TO-A-Scheerer-Schladitz/e290c2ae3140887d0dccecc97d7de66498066638 (accessed on 10 August 2022).
7. Rempel, S.; Will, N.; Hegger, J.; Beul, P. Filigrane Bauwerke aus Textilbeton–Leistungsfähigkeit und Anwendungspotenzial des innovativen Verbundwerkstoffs. *Beton-und Stahlbetonbau* **2015**, *110*, 83–93. [CrossRef]
8. Papanicolaou, C.G. Applications of textile-reinforced concrete in the precast industry (Chapter 10). In *Textile Fibre Composites in Civil Engineering*; Triantafillou, T.C., Ed.; Woodhead Publishing/Elsevier: Amsterdam, The Netherlands, 2016; pp. 227–244. [CrossRef]
9. Naaman, A.E. TRC products: Status, outlook, and future directions (Chapter 18). In *Textile Fibre Composites in Civil Engineering*; Triantafillou, T.C., Ed.; Woodhead Publishing/Elsevier: Amsterdam, The Netherlands, 2016; pp. 413–439. [CrossRef]
10. Scheerer, S.; Chudoba, R.; Garibaldi, M.P.; Curbach, M. Shells made of Textile Reinforced Concrete—Applications in Germany. *J. Int. Assoc. Shell Spat. Struct. J. IASS* **2017**, *58*, 79–93. [CrossRef]
11. Bournas, D. Strengthening of existing structures: Selected case studies (Chapter 17). In *Textile Fibre Composites in Civil Engineering*; Triantafillou, T.C., Ed.; Woodhead Publishing/Elsevier: Amsterdam, The Netherlands, 2016; pp. 389–411. [CrossRef]
12. Erhard, E.; Weiland, S.; Lorenz, E.; Schladitz, F.; Beckmann, B.; Curbach, M. Anwendungsbeispiele für Textilbetonverstärkung–Instandsetzung und Verstärkung bestehender Tragwerke mit Textilbeton. *Beton-und Stahlbetonbau* **2015**, *110*, 74–82. [CrossRef]
13. Herbrand, M.; Adam, V.; Classen, M.; Kueres, D.; Hegger, J. Strengthening of Existing Bridge Structures for Shear and Bending with Carbon Textile-Reinforced Mortar. *Materials* **2017**, *10*, 1099. [CrossRef] [PubMed]
14. Lieboldt, M. Feinbetonmatrix für Textilbeton; Anforderungen–baupraktische Adaption–Eigenschaften. *Beton-und Stahlbetonbau* **2015**, *110*, 22–28. [CrossRef]
15. Hahn, L.; Rittner, S.; Nuss, D.; Ashir, M.; Cherif, C. Development of Methods to Improve the Mechanical Performance of Coated Grid-Like Non-Crimp Fabrics for Construction Applications. *Fibres Text. East. Eur.* **2019**, *27*, 51–58. [CrossRef]
16. Hahn, L. Entwicklung einer In-situ-Beschichtungs- und Trocknungstechnologie für multiaxiale Gelegestrukturen mit hohem Leistungsvermögen. Ph.D. Thesis, Technische Universität Dresden, Dresden, Germany, 15 July 2020.
17. Jesse, F. Tragverhalten von Filamentgarnen in zementgebundener Matrix. Ph.D. Thesis, TU Dresden, Dresden, Germany, 2 July 2005. Available online: https://nbn-resolving.org/urn:nbn:de:swb:14-1122970324369-39398 (accessed on 10 August 2022).
18. Lorenz, E. Endverankerung und Übergreifung textiler Bewehrung in Betonmatrices. Ph.D. Thesis, TU Dresden, Dresden, Germany, 11 June 2015. Available online: https://nbn-resolving.org/urn:nbn:de:bsz:14-qucosa-170583 (accessed on 10 August 2022).
19. Kulas, C. Zum Tragverhalten getränkter textiler Bewehrungselemente für den Betonbau. Ph.D. Thesis, RWTH Aachen University, Aachen, Germany, 2013. (online 2014). Available online: http://nbn-resolving.org/urn:nbn:de:hbz:82-opus-49432 (accessed on 10 August 2022).

20. Preinstorfer, P. Zur Spaltrissbildung von textilbewehrtem Beton. Ph.D. Thesis, TU Wien, Wien, Austria, 2019. Available online: https://resolver.obvsg.at/urn:nbn:at:at-ubtuw:1-127879 (accessed on 10 August 2022).
21. Banholzer, B. Bond Behaviour of a Multi-Filament Yarn Embedded in a Cementitious Matrix. Ph.D. Thesis, RWTH Aachen, Aachen, Germany, 19 August 2004. (online 2005). Available online: https://publications.rwth-aachen.de/record/59781/files/Banholzer_Bjoern.pdf (accessed on 10 August 2022).
22. Schütze, E.; Curbach, M. Zur experimentellen Charakterisierung des Verbundverhaltens von Carbonbeton mit Spalten als maßgeblichem Versagensmechanismus. *Bauingenieur* **2019**, *94*, 133–141. [CrossRef]
23. Bielak, J.; Spelter, A.; Will, N.; Classen, M. Verankerungsverhalten textiler Bewehrungen in dünnen Betonbauteilen. *Beton-und Stahlbetonbau* **2018**, *113*, 543–550. [CrossRef]
24. Preinstorfer, P.; Kromoser, B.; Kollegger, J. Kategorisierung des Verbundverhaltens von Textilbeton. *Bauingenieur* **2019**, *94*, 416–424. [CrossRef]
25. Preinstorfer, P.; Kromoser, B.; Kollegger, J. Einflussparameter auf die Spaltrissbildung in Textilbeton. *Beton-und Stahlbetonbau* **2018**, *113*, 784–794. [CrossRef]
26. Donnini, J.; Corinaldesi, V.; Nanni, A. Mechanical properties of f_{rcm} using carbon fabrics with different coating treatments. *Compos. Part. B Eng* **2016**, *88*, 220–228. [CrossRef]
27. Schneider, K.; Michel, A.; Liebscher, M.; Mechtcherine, V. Verbundverhalten mineralisch gebundener Bewehrungsstrukturen aus Carbonfasern bis 500 °C. *Beton-und Stahlbetonbau* **2018**, *113*, 886–894. [CrossRef]
28. Pritschow, A. Zum Verbundverhalten von CFK-Bewehrungsstäben in Bauteilen aus ultrahochfestem Beton. Ph.D. Thesis, Universität Stuttgart, Stuttgart, Germany, 2016. Available online: http://elib.uni-stuttgart.de/handle/11682/8817 (accessed on 10 August 2022).
29. Schumann, A. Experimentelle Untersuchungen des Verbundverhaltens von Carbonstäben in Betonmatrices. Ph.D. Thesis, TU Dresden, Dresden, Germany, 30 November 2021. Available online: https://nbn-resolving.org/urn:nbn:de:bsz:14-qucosa2-732979 (accessed on 10 August 2022).
30. Waldmann, M.; Rittner, S.; Cherif, C. Bewehrungsstab zum Einbringen in eine Betonmatrix sowie dessen Herstellungsverfahren, ein Bewehrungssystem aus mehreren Bewehrungsstäben sowie ein Betonbauteil. Technische Universität Dresden. DE 10 2017 107 948 A1. 12.04.17. Available online: https://patentscope.wipo.int/search/de/detail.jsf?docId=WO2018189345 (accessed on 10 August 2022).
31. Cherif, C. Neuartige Profil-Carbonrovings für die Betonbewehrung. Proc. of 11. Carbon- und Textilbetontage, 2019, 18/19. Available online: https://www.carbon-textilbetontage.de/wp-content/uploads/2019/10/2019_C3_Tagungsband_final_web.pdf (accessed on 10 August 2022).
32. Penzel, P.; May, M.; Hahn, L.; Cherif, C.; Curbach, M.; Mechtcherine, V. Tetrahedral Profiled Carbon Rovings for Concrete Reinforcements. *Solid State Phenom.* **2022**, *333*, 173–182. [CrossRef]
33. Cherif, C.; Hahn, L. Fortschritte bei Fertigung von profilierten Carbonpolymergarnen mit höchsten Verbundeigenschaften. *TUDALIT-Magazin* **2020**, 1. Available online: http://tudalit.de/wp-content/uploads/2020/02/TUDALIT-Heft22.pdf (accessed on 10 August 2022).
34. Freudenberg, C.; Friese, D. *Carbonbetontechnikum Deutschland–Laboranlage zur Fertigung von profilierten Carbonpolymergarnen mit höchsten Verbundeigenschaften*; Final report of project "Zwanzig20 Carbon Concrete Composite–C^3: Investitionsvorhaben"; ITM TU Dresden: Dresden, Germany, 2020. [CrossRef]
35. Wendler, J.; Hahn, L.; Farwig, K.; Nocke, A.; Scheerer, S.; Curbach, M.; Cherif, C. Entwicklung eines neuartigen Prüfverfahrens zur Untersuchung zugmechanischer Kennwerte von Fasersträngen für textile Bewehrungsstrukturen. *Bauingenieur* **2020**, *95*, 325–334. [CrossRef]
36. Textilglas–Garne-Bestimmung der Reißkraft und Bruchdehnung; ISO 3341. Beuth: Berlin, Germany, May 2005. Available online: https://www.beuth.de/de/norm/iso-3341/33736152 (accessed on 10 August 2022).
37. Teijin Carbon Europe GmbH (Ed.) Tenax Filament Yarn–Produktdatenblatt (EU). Datasheet, Version 1.1. Available online: https://www.teijincarbon.com/fileadmin/PDF/Datenbl%C3%A4tter_dt/Product_Data_Sheet__EU_Filament___DE_.pdf (accessed on 10 August 2022).
38. Lefatex Chemie GmbH (Ed.) *Lefasol BT 91001-1**; Datasheet, 05.04.2017, REV: Stand 01; Lefatex Chemie GmbH: Brüggen, Germany, 2017.
39. CHT Germany GmbH (Ed.) *TECOSIT CC 1000**; Datasheet, 14.02.2022, REV: Stand 01; 40.39.; CHT Germany GmbH: Tübingen, Germany, 2022.
40. C^3–Carbon Concrete Composite e. V.: Homepage. Available online: https://www.bauen-neu-denken.de/c3-vorhaben/ (accessed on 10 August 2022).
41. Prüfverfahren für Zement-Teil 1: Bestimmung der Festigkeit; DIN EN 196-1:2016-11; Beuth: Berlin, Germany, November 2016. Available online: https://www.beuth.de/de/norm/din-en-196-1/252980793 (accessed on 10 August 2022).
42. CARBOCON GMBH: CARBOrefit®-Verfahren zur Verstärkung von Stahlbeton mit Carbonbeton. *Allgemein bauaufstitliche Zulassung/Allgemeine Bauartgenehmigung*; Dresden, Germany, 2021. Available online: https://cloud.carborefit.de/index.php/s/PgMiAwyYMTz8Gse?dir=undefined&path=%2FCARBOrefit%C2%AE%20Zulassung&openfile=27808 (accessed on 10 August 2022).

43. Scheerer, S.; Schütze, E.; Curbach, M. Strengthening and Repair with Carbon Concrete Composites—The First General Building Approval in Germany. In *Proceedings of the SHCC4—Int. Conf. on Strain-Hardening Cement-Based Composites, Dresden, Germany, 18–20 September 2017*; Mechtcherine, V., Slowik, V., Kabele, P., Eds.; Springer: Dordrecht, The Netherlands, 2018; pp. 743–751.
44. *Kohlenstofffasern–Bestimmung des Zugverhaltens von harzimprägnierten Garnen; DIN EN ISO 10618:2004-11*; Beuth: Berlin, Germany, November 2004; Available online: https://www.beuth.de/de/norm/din-en-iso-10618/73061591 (accessed on 10 August 2022).
45. Hinzen, M. Prüfmethode zur Ermittlung des Zugtragverhaltens von textile Bewehrung in Beton. *Bauingenieur* **2017**, *92*, 289–291. [CrossRef]
46. Schütze, E.; Bielak, J.; Scheerer, S.; Hegger, J.; Curbach, M. Einaxialer Zugversuch für Carbonbeton mit textiler Bewehrung | Uniaxial tensile test for carbon reinforced concrete with textile reinforcement. *Beton-und Stahlbetonbau* **2018**, *113*, 3–47. [CrossRef]
47. Schneider, K.; Michel, A.; Liebscher, M.; Terreri, L.; Hempel, S.; Mechtcherine, V. Mineral-impregnated carbon fibre reinforcement for high temperature resistance of thin-walled concrete structures. *Cem. Concr. Compos.* **2019**, *97*, 68–77. [CrossRef]
48. Kruppke, I.; Butler, M.; Schneider, K.; Hund, R.-D.; Mechtcherine, V.; Cherif, C. Carbon Fibre Reinforced Concrete: Dependency of Bond Strength on Tg of Yarn Impregnating Polymer. *Mater. Sci. Appl.* **2019**, *10*, 328–348.
49. Beckmann, B.; Bielak, J.; Bosbach, S.; Scheerer, S.; Schmidt, C.; Hegger, J.; Curbach, M. Collaborative research on carbon reinforced concrete structures in the CRC/TRR 280 project. *Civil. Eng. Des.* **2021**, *3*, 99–109. [CrossRef]

Article

Prediction of Bonding Strength of Externally Bonded SRP Composites Using Artificial Neural Networks

Sofija Kekez and Rafał Krzywoń *

Department of Structural Engineering, Silesian University of Technology, 44100 Gliwice, Poland; sofija.kekez@polsl.pl
* Correspondence: rafal.krzywon@polsl.pl; Tel.: +48-32-237-22-62

Abstract: External bonding of fiber reinforced composites is currently the most popular method of strengthening building structures. Debonding performance is critical to the effectiveness of such strengthening. Many models of bond prediction can be found in the literature. Most of them were developed based on laboratory research, therefore, their accuracy with less popular strengthening systems is limited. This manuscript presents the possibility of using a model based on neural networks to analyze and predict the debonding strength of steel-reinforced polymer (SRP) and steel-reinforced grout (SRG) composites to concrete. The model is built on the basis of laboratory testing of 328 samples obtained from the literature. The results are compared with a dozen of the most popular analytical methods for predicting the load capacity. The prediction accuracy in the neural network model is by far the best. The total correlation coefficient reaches a value of 0.913 while, for the best analytical method (Swiss standard SIA 166 model), it is 0.756. The sensitivity analysis confirmed the importance of the modulus of elasticity and the concrete strength for debonding. It is also interesting that the width of the element proved to be very important, which is probably related to the low variability of this parameter in the laboratory tests.

Keywords: steel-reinforced polymer; strengthening of concrete; bonding strength; artificial neural networks

Citation: Kekez, S.; Krzywoń, R. Prediction of Bonding Strength of Externally Bonded SRP Composites Using Artificial Neural Networks. Materials 2022, 15, 1314. https://doi.org/10.3390/ma15041314

Academic Editor: Andreas Lampropoulos

Received: 23 December 2021
Accepted: 7 February 2022
Published: 10 February 2022

Copyright: © 2022 by the authors. Licensee MDPI, Basel, Switzerland. This article is an open access article distributed under the terms and conditions of the Creative Commons Attribution (CC BY) license (https://creativecommons.org/licenses/by/4.0/).

1. Introduction

High-strength fiber-reinforced composites have been used to strengthen concrete structures since the 1980s. Over the past four decades, external bonding of composites has become the most popular method of retrofitting structures, not only concrete, but also masonry, wood, and even steel. During this time, technology has improved and new types of fiber have been introduced. Fiber-reinforced polymer (FRP) composites are distinguished by an excellent strength-to-weight ratio, several times better than that of traditional steel. The currently perceived disadvantage of fiber composites is their difficult disposal and reuse. Steel-reinforced polymer (SRP) composites do not have this drawback. Although they are characterized by a slightly higher weight than carbon fiber composites, other parameters, especially tensile strength and modulus of elasticity, are comparable.

Present studies demonstrate the effectiveness of SRP composites in strengthening residential structures. After replacing the polymer matrix with cement, or even lime grout, they are perfect for repairing historic buildings. The lack of standards and design handbooks is an obstacle to the popularization of SRP composites. However, some of the procedures can be directly derived from the manuals developed for FRP composites with slightly modified mechanical properties. The issue of the bond of SRP composites may raise the greatest doubt here. SRP composites are reinforced with wire strands with a diameter of 0.25–0.35 mm. They are fixed in the wet lay-up process, which is why the fiber content in the laminate is much lower than the carbon fiber strips produced in the pultrusion process. Therefore, the relationship of the strength of SRP tapes to the cross-sectional area can even be reduced by up to three times [1], leading to a greater width and thickness of the plate. It can be expected that increasing the width would have a positive effect on the bond, but increasing the thickness would have a negative effect. The study by Mitoldis [2] showed a clear difference in adhesion and slip compared to carbon-fiber-reinforced polymer (CFRP) composites. On the other hand, Papakonstantinou [3] has shown that existing design standards can be successfully used in the design of SRP reinforced beams, but with a slightly lower safety margin. Furthermore, research [4] shows that most analytical models and design standards guarantee acceptable accuracy in predicting the bond strength of SRP composites. The results of debonding tests collected during previous analyses were used to build a model based on artificial neural networks. The created model and the results are discussed in this paper.

An artificial neural network is a machine learning technique within the wider family of artificial intelligence. It is based on the theory of connectionism, which was first proposed during the 1940s to simulate the processing of the human brain. However, the concept was not widely used until the development of information technology, which allowed its reopening and further deployment [5]. Currently, artificial neural networks (ANNs) serve for classification, i.e., the prediction of a categorical value, or regression, i.e., the prediction of a numerical value [6]. The basic concept of ANNs is grounded in the learning of patterns from the presented examples in a supervised or unsupervised manner, in other words, with or without the target values, respectively. The most used learning algorithm is the feed-forward backpropagation (BP) algorithm because of its simplicity and applicability. The BP algorithm is based on the "backpropagation learning rule" which was established in 1985 as a solution for issues occurring in single-layer or bilayer networks [6]. It represents a generalization of the delta rule and functions as a gradient descent technique of error minimization by incremental adjustment of the connection weight between the layers of a multilayer perceptron (MLP) [7]. Solving civil engineering problems conventionally involves time-consuming empirical methods or the proposal of highly complex analytical expressions. Even then, the solutions imply some type of simplification due to the limitations of the method or the complex nature of the problem. Soft computing methods such as MLPs present a time- and cost-friendly alternative to solve any problem at hand including structural health monitoring [8], structural engineering [9], gas flow [10], seismic engineering [11], etc. Furthermore, neural network-based parameter sensitivity analysis is gaining more traction in civil engineering systems due to its remarkable ability to explain the nonlinear relationships between the explicative and the response variables of a certain problem [12].

Artificial neural networks and other machine learning methods have been used to predict the behavior of elements reinforced with various FRP composite materials. Notable works include a study by Koroglu [13], which deals with the prediction of the bonding strength of FRP rebars in concrete using an ANN. The study shows the efficiency of machine learning compared to analytical methods. Furthermore, the research by Mansouri and Kisi [14] brings forward the application of ANNs and adaptive neuro-fuzzy inference systems for the prediction of debonding strength for masonry elements retrofitted with FRP composites; the works of Mashrei et al. [15], Cascardi and Micelli [16], and Jahangir and Eidgahee [17] investigated whether the application of ANNs for the prediction of bonding strength between FRP strips and concrete can give better results than the existing analytical models. It has been shown that the ANN approach may present a more efficient option. Similarly, this paper uses ANNs to predict the bonding strength of SRP composites externally bonded to concrete structures. It compares the results of ANNs with the analytical models that are traditionally used and investigates the dependencies of the ANNs variables, postulated by the learning process of the ANN, which is obtained by the sensitivity analysis of the working ANN model. The relevant information for building a comprehensive dataset was obtained from the literature [18–24].

2. Bond-Slip Models of Externally Bonded Composites

Debonding can be simply defined as the loss of bond of the composite overlay to the substrate. In practice, several ways of failure are distinguished under this phenomenon. Debonding may occur along the length of the reinforcement in the area of the highest tensile force or the anchoring zone. It can refer to the adhesive layer, but often to the contact layer in the substrate as well, or it can appear as an interlayer in the composite itself. According to Teng [25], debonding failure models can be classified as plate end interfacial debonding, concrete cover separation, intermediate flexural crack-induced debonding, and critical diagonal shear crack-induced debonding. This ordination may be complemented by debonding initiated as interlaminar failure in the matrix-fiber interface [26]. Debonding is typically simulated in a simple laboratory shear test. Depending on how the sample is clamped and how the force is applied, this may be a double or a single, push or pull-shear test. These tests directly show the bond in the end anchorage zone; however, current research proves that this testing method can also be used in the simulation of intermediate debonding [27]. An alternative and less commonly used bond test method is the notched beam test.

Bond performance is a result of the cooperation of several components with different geometries and mechanical properties. In addition, it is also influenced by the application conditions and the environmental conditions of use. Due to the number of factors and the complexity of the bond process, most mathematical bond models are based directly on the results of laboratory tests. Usually, they are classified into three categories, as follows:

- fully empirical models, based on the regression of test data, such as Tanaka [28], Hiroyuki and Wu [29], and Maeda [30],
- fracture mechanics-based models including Taljsten [31], Niedermeier [32], Yuan and Wu [33], and Lu et al. [34],
- design models, usually based on simple assumptions, such as Dai et al. [35], Brosens and van Germet [36], Khalifa et al. [37], Yang et al. [38], Adhikary and Mutsuyoshi [39], Sato et al. [40], Chen and Teng [41], DeLorenzis et al. [42], and Seracino et al. [43].

A detailed description of these models can be found inter alia in Table 1 [4].

Table 1. Predicted experimental bond strength ratios for SRP composites bonded to concrete [4].

Model	Mean	R
Tanaka [28]	0.699	0.220
Hiroyuki and Wu [29]	0.746	0.503
Maeda et al. [30]	1.14	0.751
Taljsten [31]	0.761	0.748
Nidermeier [32]	0.763	0.629
Yuan and Wu [33]	0.763	0.747
Lu et al. [34]	0.873	0.676
Dai et al. [35]	1.55	0.734
Brosens and van Germet [36]	0.937	0.408
Khalifa et al. [37]	0.754	0.729
Yang et al. [38]	0.528	0.657
Adhikary and Mutsuyoshi [39]	2.00	0.476
Sato et al. [40]	1.81	0.573
Chen and Teng [41]	0.882	0.726
DeLorenzis et al. [42]	1.63	0.728
Seracino et al. [43]	0.752	0.716
JCI 2003 [44]	0.941	0.738
SIA 166/2004 [45]	0.911	0.756
CNR-DT200R1/13 [46]	0.928	0.718
Fib Bulletin 90/2019 [47]	0.962	0.755

Evaluation of the above-mentioned bond models for SRP strengthening (carried out by this author) showed that most of them are in relatively good agreement with the test results [4]. As can be seen in Table 1, the best prediction accuracy is given by Lu et al. [34], Chen and Teng [41], and most of the design standards.

When analyzing the predicted experimental ratio, most models can be considered rather conservative. Some models significantly increase the load capacity (including Dai et al. [35], Brosens and Germet [36], and Adhikary and Mitsuyoshi [39]). The largest scatter in the results concerns the proposals by Dai et al. [35], Adhikary and Mitsuyoshi [39], and DeLorenzis et al. [42]. The common feature of these models is the independence of the effective anchorage length.

3. Artificial Neural Networks

Artificial neural network (ANN) models have been developed to predict the bonding strength of the externally bonded SRP composite to the concrete element. All models are feed-forward backpropagation networks with a sigmoid activation function and a linear transfer function. Data were collected from the works of Figeys et al. [18], Matana et al. [19], Mitoldis et al. [20], Napoli et al. [21], and Ascione et al. [22–24]. Table 2 briefly summarizes the most relevant information from these studies, which are also incorporated in the dataset. The comprehensive dataset was halved, so that the first half could be used for training and the other half is used for testing the ANN model. The dividing of the dataset represented the realistic behavior of the established model, i.e., the level of fitting with the unseen data by the network. Finally, the working model was established and trained using the entire dataset.

Table 2. List of geometry and material properties collected from the available experimental data.

Reference	Number of Specimens	b [mm]	f_c [MPa]	b_f [mm]	t_f [mm]	L [mm]	E_f [GPa]
Figeys [18] [1]	7	100	35	95	0.601	150–200	177.6
Mantana [19] [1]	12	191	14.8	51	0.483	102–305	179.1
Mitoldis [20] [1]	8	100	22.4	50–80	0.562	150–300	221.4
Napoli [21] [1]	19	200	15.2–39.7	100	0.084–0.381	150–300	206.6
Ascione [22] [1]	129	200	13–45	20–100	0.084–0.381	100–350	190
Ascione [23] [1]	62	200	19.3–25.6	100	0.084–0.381	100–350	182.1–183.4
Ascione [24] [2]	83	200	13–40	50–100	0.084–0.254	100–350	182.1–183.4

[1] epoxy adhesive; [2] grout adhesive; b—sample width; b_f—width of SRP tape; f_c—concrete strength; t_f—effective SRP thickness; L—bond length; E_f—modulus of elasticity of SRP.

The entire dataset was randomly divided while ensuring that each half included data from every referenced source. Both the training and the testing sets consisted of 171 data tuples. Input data included information on the following: sample width, sample thickness, concrete compressive strength, concrete tensile strength, concrete modulus of elasticity, tape width, tape thickness, anchorage length of the tape, modulus of elasticity of SRP, the tensile strength of SRP, and the type of adhesive between the concrete element and the SRP tape. The target was the debonding force of the SRP tape from the concrete element. All input data, except the type of adhesive, were numerical. The numerical values of the input and target data were processed with min/max normalization within the [0, 1] range and, as such, were presented to the network. The types of adhesives between the SRP tape and the concrete element were epoxy and grout, presented to the network as the values 0 and 1, respectively. Figure 1 shows the distribution of values of the numerical input parameters throughout the entire dataset.

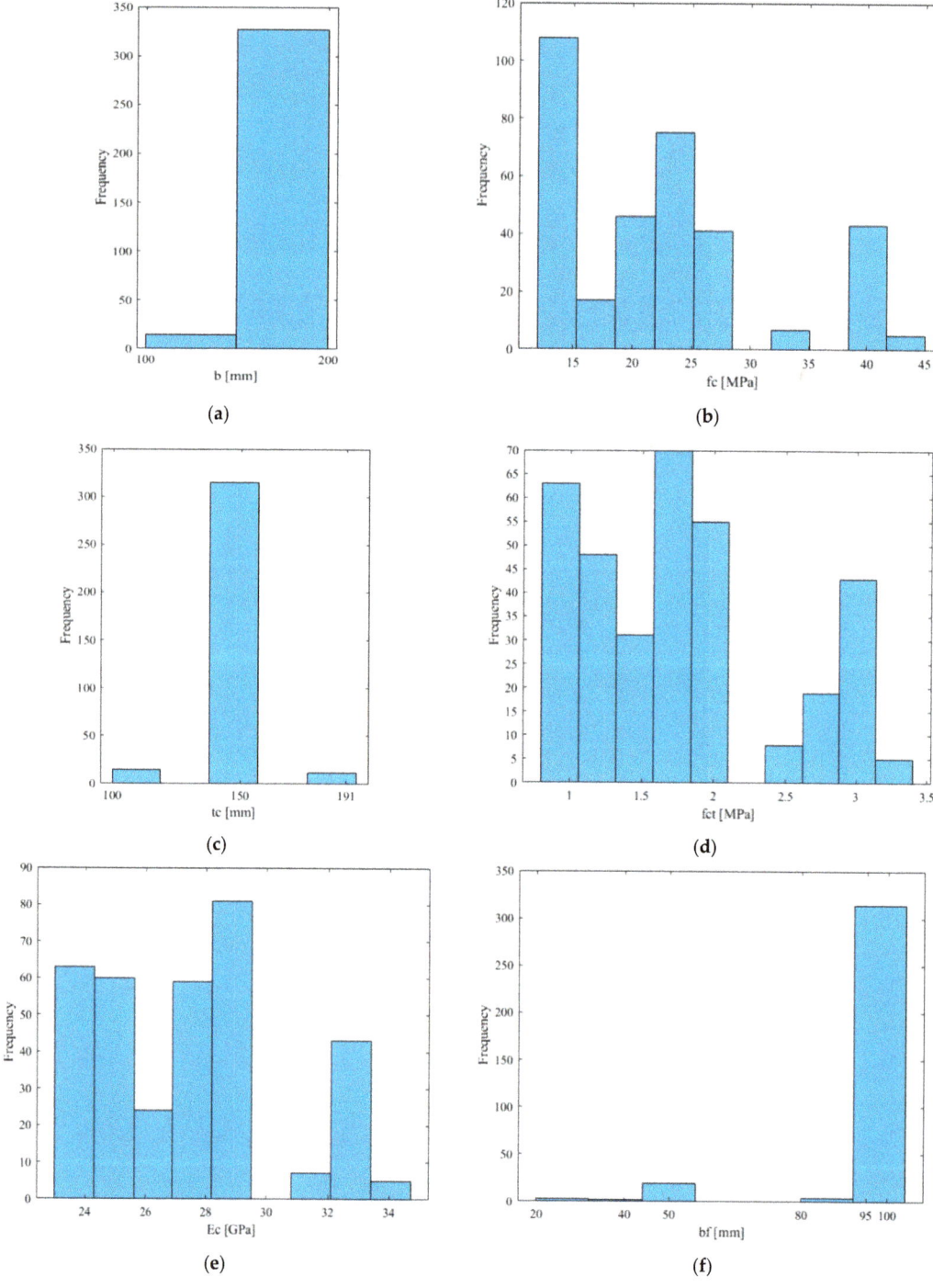

Figure 1. *Cont.*

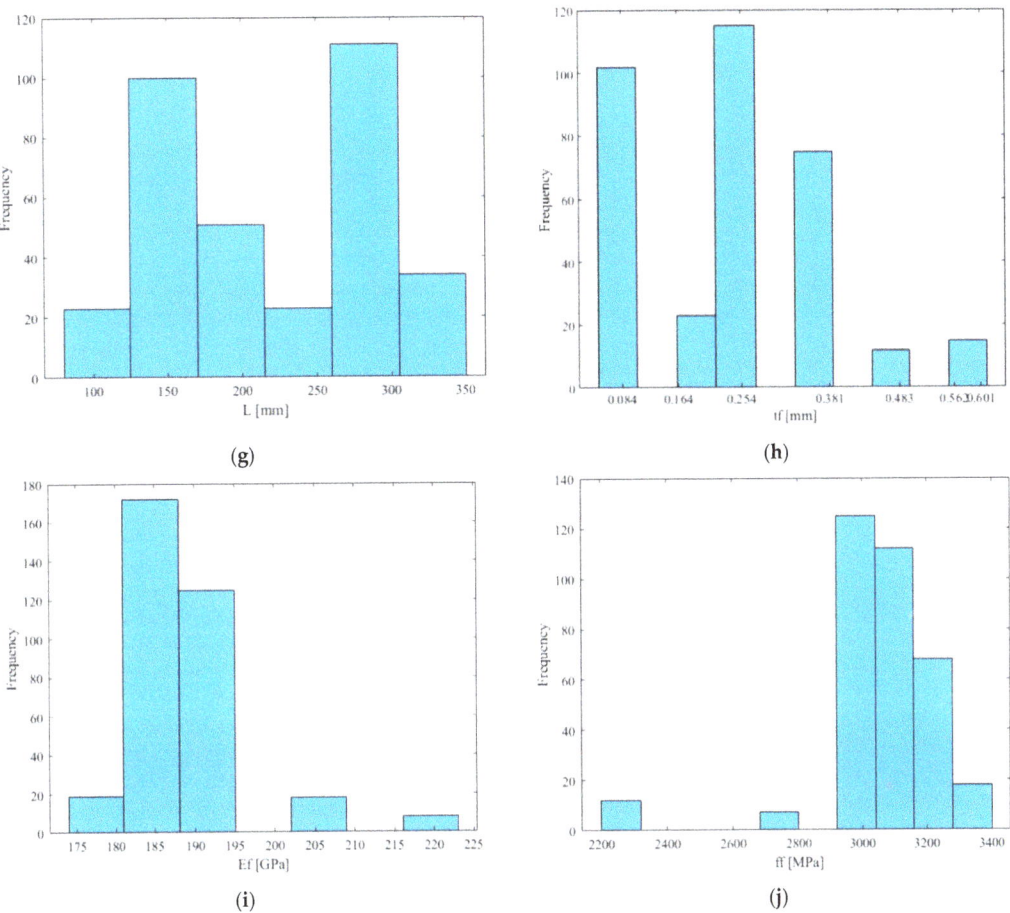

Figure 1. Input parameters: (**a**) sample width—b; (**b**) concrete compressive strength—f_c; (**c**) sample thickness—t_c; (**d**) concrete tensile strength—f_{ct}; (**e**) concrete modulus of elasticity—E_c; (**f**) SRP tape width—b_f; (**g**) anchorage length—L; (**h**) SRP tape thickness—t_f; (**i**) SRP modulus of elasticity—E_f; (**j**) SRP strength—f_f.

3.1. Establishing the Working ANN Model

Neural network models were developed using Matlab R2020b. The supervised training of the models was carried out with the following learning parameters: unipolar sigmoid activation function, linear transfer function, Levenberg–Marquardt algorithm, one hidden layer, 1000 epoch limit, 10^9 momentum, 10^{-6} learning rate, and six-fold cross-validation. Training was set to stop when the network did not improve after six consecutive validation checks. The procedure for establishing a working ANN model for the prediction of bonding strength was set as follows. Firstly, the training was performed on the initial model. The optimization of the initial model was then provided and the training repeated for the optimized model. The testing was performed on the optimized model to establish the quality of generalization. Since the optimized model was tested on new unseen data, it may be assumed that the results would show the realistic behavior of the network. Finally, the final working ANN model was established and trained on the full dataset.

3.1.1. Initial Model

The initial model included the simplest network architecture, having the input layer with eleven neurons, one hidden layer with an equal number of neurons as the input layer, and the output layer with one neuron. The training set was pre-processed using min/max normalization with the [0, 1] range for the numerical values, and zero or one values for the descriptive input. During the training process, 15% of the set was used for validation to ensure that the training process stopped when the six consecutive validation checkpoints showed no improvement in the training.

3.1.2. Optimization of the Initial Model

Optimization of the initial model served to establish the optimal number of neurons in the hidden layer. An improved topology of the network implied a better generalization and contributed to the stability of the network. Optimization was indicated by the level and change of the mean squared error for the varying number of neurons in the hidden layer. To this end, consecutive training was performed for models with an iterated number of neurons from one to fifty in the hidden layer.

The literature gives general recommendations on the number of neurons in the hidden layer. It is considered that for the network of this size in terms of the input neurons and the batch size, a very high number of neurons in the hidden layer would surely cause an overfitting of the network. The overfitting leaves the network unable to generalize and perform the prediction when presented with the new data. Hence, it is considered that the highest number of neurons in the hidden layer should not exceed fifty. After the iterations finished, the change in the mean squared error (MSE) was observed for all iterations and compared to establish which number of neurons in the hidden layer gave the satisfactory results and exhibited stability. On the other hand, it is important to secure the network from the occurrence of underfitting as well. Practically, the lowest number of neurons in the hidden layer may not be less than the number of the input neurons. Furthermore, the literature often gave the recommendation that the number of neurons in the hidden layer should not be under N_i+2, where N_i is the number of input neurons. Underfitting may be spotted as the occurrence of extremely low MSE, giving an overly positive result of the network's behavior.

3.1.3. Training and Testing of the Optimized Model

When the optimization process was concluded and the optimal network topology set, the new model was then independently trained with the previously described training set. The training was performed with the same learning parameters as the initial ANN model. The testing of the optimized model followed the training process. It was performed using the second half of the dataset, which presents entirely new data for the neural network. Hence, the testing process can show the real behavior of the network, i.e., the capability of generalization when the network is presented with unknown data. It is expected that proper generalization implies good prediction and similar results, after training and testing, may confirm the stability of the network.

3.1.4. Working ANN Model

The final ANN model presents the working model which can be further used for predictions of the bonding strength of the externally bonded SRP composite to the concrete element. This model used the previously established and tested topology of the optimized ANN model; however, it was trained using the entire dataset. After the successful training process, the weights and bias were fixed, and the network presented a ready-to-use model.

3.2. Sensitivity Analysis

The sensitivity analysis served to show the absolute or relative contribution of each input parameter to the output value. It was necessary to understand the relationship and influence of the input parameters on the problem that the ANN learns to solve [7].

Except for showing the contribution of each input parameter, the sensitivity analysis may also influence the topology of the final working model, because it may show that some parameters could hinder or slow down the learning process. On the other hand, it shows which parameters are crucial for the learning process, as well as the dependence of the output value on each input parameter. This analysis was provided using the weights method, otherwise known as the Garson's algorithm [48]. The algorithm was created for supervised neural networks with a single output, to describe the relative importance of the input parameters by deconstructing the model weights. The mathematical description of the algorithm for a network with a single hidden layer is as follows:

$$D_{ij} = \frac{|W_{ij}|}{\sum_{i=1}^{n_i}|W_{ij}|} \quad (1)$$

$$RC_i = \frac{\sum_{j=1}^{n_j} D_{ij}}{\sum_{j=1}^{n_j}\sum_{i=1}^{n_i} D_{ij}} \quad (2)$$

where n_i and n_j are the numbers of input and hidden neurons, respectively, W_{ij} is the weight corresponding to the i-th input and the j-th hidden neuron, and RC_i is the relative importance of the i-th input.

4. Results

The performance of the ANN model is described by the mean squared error MSE, root mean squared error RMSE, and the regression coefficient R. The MSE and RMSE represent the average squared and average root squared difference between the output and the target value, respectively, which tends to zero as the prediction becomes more accurate. The R value is the primary parameter that shows the correlation of the output compared to the target value, which tends to a value of one, as the prediction becomes more accurate. The regression coefficient is usually expressed as the total of R values for training, testing, and validation. Additionally, the error distribution shows the general behavior of the network and the relationship between the error during the training, testing, and validation.

4.1. Training and Optimization of the Initial ANN Model

As mentioned previously, the initial model was trained with the architecture including eleven neurons in the input and the hidden layer and a single neuron in the output layer. The results of the initial training show that there is room for improvement. Table 3 shows that, although the RMSE value is quite low, the regression coefficient of 0.85 for training implies that the learning process should be improved. Figure 2 shows the relationship between the target and the output values after training the initial model.

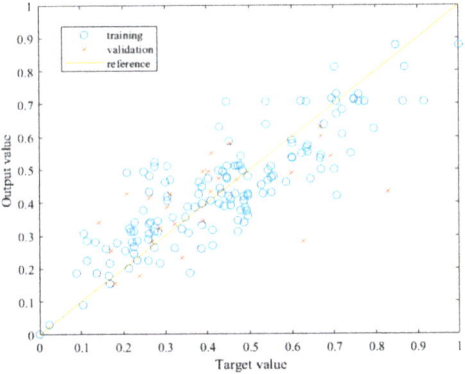

Figure 2. Relationship between the target and output values of the initial model.

Table 3. Comparison between the initial and the optimized ANN model.

Model	R Training	R Validation	R Total	MSE	RMSE
Initial	0.84959	0.60249	0.82339	0.0106	0.00991
Optimized	0.88602	0.75031	0.8675	0.0099	0.0076

Optimization of the initial model has been performed by simply iterating the number of neurons in the hidden layer and observing the MSE for training and validation. To establish the optimal number of neurons in the hidden layer, fifty iterations were carried out. The results of the optimization process are presented in Figure 3. It shows that overfitting occurs with 49 neurons in the hidden layer. Vis-à-vis, underfitting may have occurred with 5, 8, 9, 12, and 13 neurons in the hidden layer. The literature recommends the highest number to be equal to $2Ni + 1$, where Ni is the number of input neurons. Thus, the preferred number of hidden neurons should be between fifteen and thirty, which is supported by the results of the optimization of the network. A sudden drop or increase in the error for consecutive iterations may imply instability; hence, the error should show a relatively minimal change for several consecutive iterations. The behavior is somewhat steady within the range of 18 to 23 neurons, especially when the training MSE values are observed. Furthermore, within this range, the closest result between the training and the validation MSE occurs when the number of hidden neurons is 21, and thus it is considered to be an optimal value.

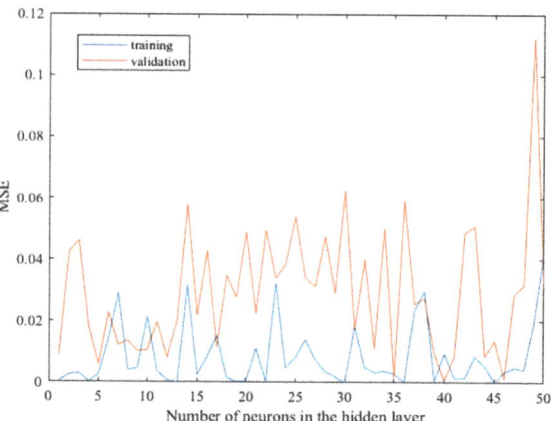

Figure 3. MSE values after optimization of the initial model.

4.2. Training the Optimized ANN Model

The optimized ANN model has been trained in exactly the same manner as the initial model, so that a realistic comparison between the two models can be achieved. The only difference is the number of neurons in the hidden layer, which is equal to 21 for the optimized model. Table 3 shows the comparison of the regression coefficients and the mean squared error between the initial and optimized models. The improvement is visible; the error of the optimized model shows a value closer to zero and regression coefficients show an improvement in the learning process and the prediction accuracy. Figure 4 compares the error distributions for both models. It may be observed that the error distribution of the optimized model exhibits a more uniform decrease in the error with less outliers and more symmetrical distribution around zero error.

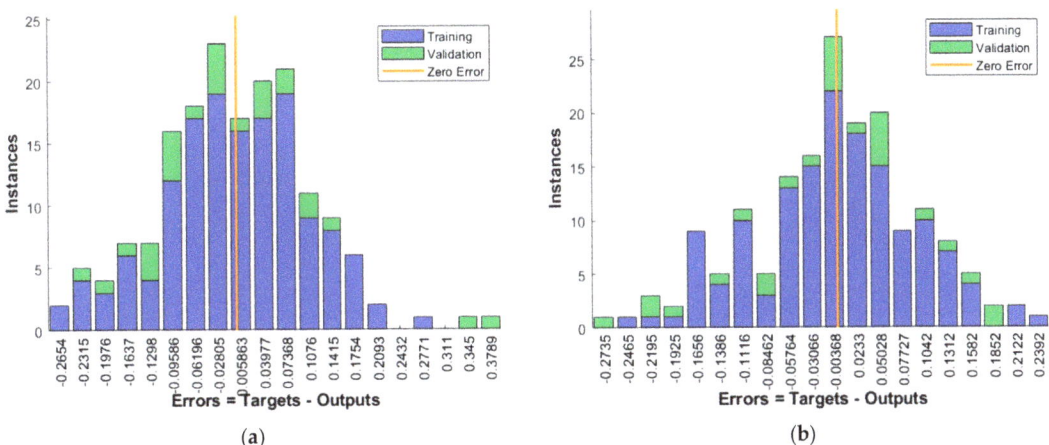

Figure 4. Error histogram for: (**a**) Initial model; (**b**) Optimized model.

Figure 5 compares the relationship between the target and the output values for the initial and the optimized model after training. The improvement is reflected in the significant increase in the training regression coefficient after optimization. Moreover, the validation R value shows a much higher increase which, in turn, gives the overall R value of the optimized model equal to almost 0.87. This implies a better prediction accuracy of the optimized model in comparison to the initial prediction. It may be assumed that the final model will show even better results because the training will be provided with twice as many data tuples.

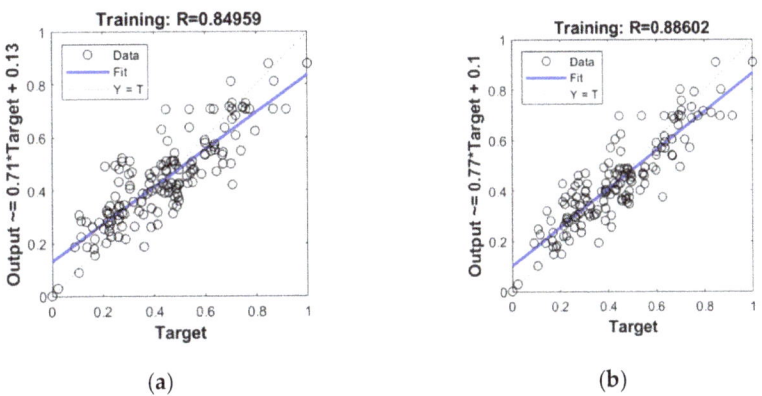

Figure 5. Regression after training for: (**a**) Initial model; (**b**) Optimized model.

4.3. Testing the Optimized ANN Model

After it had been concluded that the optimization of the initial model was successful, testing by introducing the data which the network had never seen presented the final verification for the optimized ANN model. This consists of introducing the second half of the dataset, data that were randomly selected during the division of the set, to the network. The testing of the trained network was performed by introducing the new set and calling a simulation function to the trained network. The results are presented by relating the target and the output values, as shown in Figure 6. The total R value after testing exceeds 0.91, which implies great success in the network generalization capability. Figure 7 shows the comparison of the results after training and after testing the optimized network. The network exhibits very similar behavior after training and testing, indicating that the architecture is suitable and that the network is stable with good generalization. The prediction accuracy is at a satisfactory level, given that the data were obtained through different sources, and the input data were somewhat repetitive.

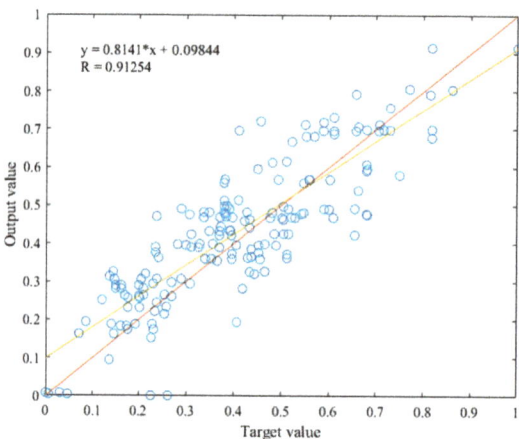

Figure 6. Relationship between target and output value after testing.

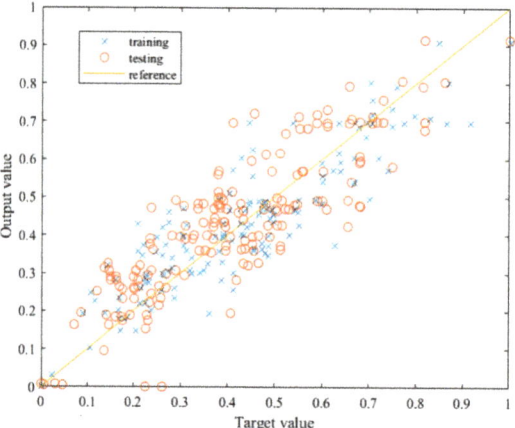

Figure 7. Relationship between target and output values after training and testing.

4.4. Training of the Working ANN Model

The final ANN model was established, tested, and confirmed to be successful in predicting the bond strength of the externally bonded SRP composite to the concrete element. In order to obtain the working model, the neural network was trained once more, using the entire dataset. Then, the weights and bias were set and fixed, and the neural network used for prediction. The learning parameters were kept from the previous models, the number of neurons in the hidden layer was 21, and the subset ratio was 80/15/5 for training/testing/validation. Figure 8 shows the error distribution and the relationship between the target and the output values, while Table 4 shows the values of the regression coefficients and the mean squared error for the working ANN model. The improvement is best seen when observing the error distribution which takes a Gaussian zero-centered shape. The lack of outliers is visible in both Figure 8a,b and, lastly, the regression coefficients experience a significant increase. It may be observed that the prediction accuracy of the working model exceeds 90%, which indicates a very reliable neural network.

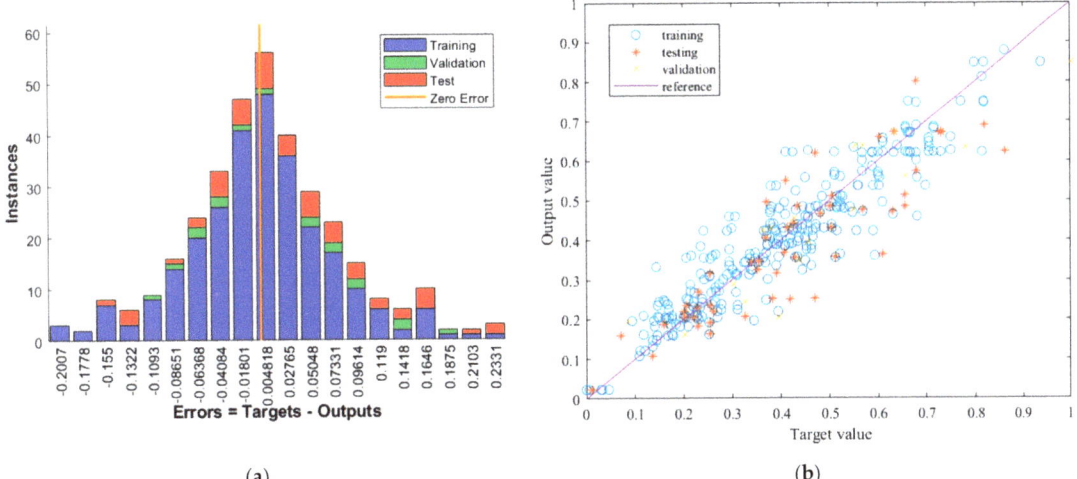

Figure 8. (**a**) Error distribution for the working model; (**b**) Relationship between target and output values for the working model.

Table 4. Results of the working ANN model.

Model	R Training	R Validation	R Testing	R Total	MSE	RMSE
Working	0.92367	0.90783	0.87864	0.91338	0.0073	0.00023

4.5. Sensitivity Analysis

The sensitivity analysis of the working ANN model has been provided using Garson's algorithm, i.e., the weights method, according to Equations (1) and (2). Table 5 shows the values of the weights connecting the neurons in the input and the hidden layer. These values are used to calculate and determine the relative contribution of each input parameter regarding the output. Figures 9 and 10 show the contribution of the input parameters. The results of the analysis show that all input parameters have a very close level of importance to the output. Relatively speaking, the modulus of elasticity of the SRP composite and the width of the concrete element have the highest importance to the output. Analysis shows that the anchorage length of the SRP composite is the least important to the bonding strength. Additionally, the thickness of the concrete element and the compressive strength of the concrete show low importance regarding the bonding strength for externally bonded SRP composites to the concrete element. However, none of the input parameters show less than a 50% contribution to the output value, meaning that none of them may be excluded from the dataset.

Table 5. Connection weights between the neurons in the input and the hidden layer.

Hidden/Input	1	2	3	4	5	6	7	8	9	10	11
1	−0.636	0.707	−0.534	−0.392	0.487	−0.315	−0.667	−0.350	0.735	−0.294	0.151
2	0.220	−0.179	−0.942	0.458	−0.140	0.648	−0.123	0.243	1.164	0.768	−0.725
3	0.869	−0.322	0.604	0.374	−0.508	−0.695	0.225	−0.248	0.683	0.792	0.612
4	0.808	−0.866	−0.555	−0.363	−0.298	−0.214	0.564	0.361	−0.586	0.244	−0.321
5	0.937	0.329	0.305	0.125	0.641	0.328	−0.399	−1.250	0.453	0.501	−0.474
6	−0.612	−0.658	0.017	0.398	0.309	−0.303	−0.459	−0.706	−0.617	0.799	0.622
7	0.026	−0.190	−0.592	0.073	−0.202	−0.638	−0.662	−0.539	−0.247	−0.402	−0.796
8	−0.745	0.763	0.816	0.029	−0.570	−0.670	0.096	0.240	−0.762	−0.548	0.257
9	0.374	0.309	0.272	−0.685	0.694	−0.344	−0.631	0.571	0.739	−0.310	0.234
10	−0.789	−0.136	0.101	0.353	−0.799	0.067	−0.632	−0.826	−0.024	0.026	−0.464
11	0.297	−0.317	−0.279	0.447	−0.384	0.813	−0.200	−0.520	0.846	−0.641	0.443
12	−0.195	−0.258	−0.439	0.809	0.615	0.047	0.045	−0.139	1.389	−1.057	1.407
13	−0.969	−0.416	0.948	−0.283	0.186	−0.755	0.097	−0.434	0.055	−0.269	−0.401
14	0.738	0.347	0.001	−0.245	−0.101	0.154	0.046	−0.110	−0.466	1.051	0.266
15	0.584	−0.654	−0.688	−0.625	0.424	0.229	0.231	0.091	0.573	0.209	−0.645
16	−0.701	0.048	0.123	0.455	−0.953	0.754	−0.150	0.071	−0.883	0.602	−0.331
17	0.909	−0.300	−0.557	−0.173	−0.024	−0.199	1.039	0.753	0.426	−0.065	0.451
18	−0.500	−0.339	−1.039	−0.644	−0.327	0.003	−0.076	−1.405	0.507	−0.472	0.472
19	0.321	−0.838	−0.861	0.110	0.399	0.605	−0.564	0.524	0.567	−0.955	0.532
20	0.206	0.258	−0.391	−0.450	0.749	−0.818	0.283	0.408	0.222	0.225	−0.370
21	0.901	0.192	0.369	−0.427	0.743	0.360	0.161	0.392	−1.031	−0.399	−0.945

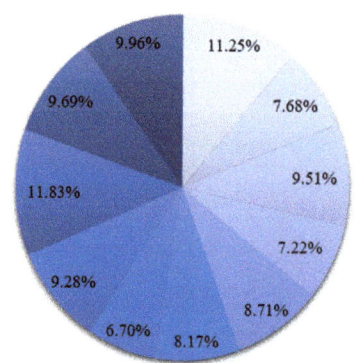

Figure 9. Contribution of the input parameters.

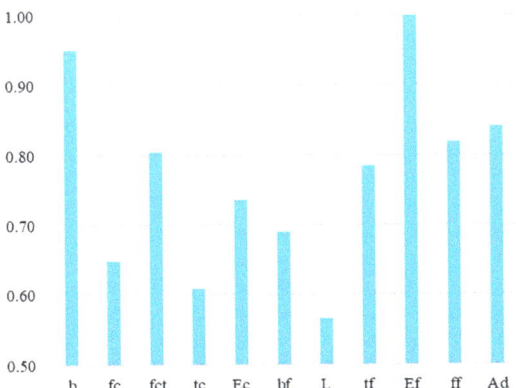

Figure 10. Relative importance of the input parameters.

5. Discussion of Results

The comparison of Tables 1 and 4 clearly shows that the ANN method guarantees a much better quality of results than any of the analytical methods evaluated. The total correlation coefficient R_{total} was equal to 0.91338, which is more than satisfactory and only slightly lower than the one obtained by the authors of a similar analysis of the adhesion of CFRP composites [15,17]. However, it should be noted that, in the case of CFRP composites, due to their popularity, much larger datasets of research results are disposable, which directly affects the behavior of the ANN model.

Mukhopadhyaya and Swamy [49] pointed out that increasing the elastic modulus of the bonded fiber composite results in a higher value of interfacial stresses. Teng [25], who also noted that the elastic modulus does not affect the location of the peak value, found similar conclusions. The authors cited also indicate similar findings regarding the thickness of the composite. Similarly, an additional layer of laminates increase stress (Shahawy et al. [50]). It is proof of the key role of laminate stiffness in interfacial stress values and, thus, the probability of premature debonding. The described effect has been proven by the sensitivity analysis in this work. The modulus of elasticity E_f and the thickness of the laminate t_f are the most important input parameters (Figure 10).

In most cases, delamination occurs in the contact layer between the adhesive and the concrete or the cover layer; therefore, it is commonly considered that the bond is significantly affected by the strength of the concrete and the preparation of the concrete surface [51]. The concrete tensile strength f_{ct} is the fourth most important parameter indicated by the sensitivity analysis. A similar meaning can be assigned to the Ad parameter, which represents the type of adhesive. Admittedly, the mechanical properties of the adhesive were not entered into the model. Only the division into epoxy resin and grout was parameterized. Epoxy adhesives are characterized by much better strength and bond properties, which were rightly indicated in the analysis.

Surprisingly, the lowest impact of the bond length L may be caused by the specificity of the test data. Most analytical methods define the effective bond length along which most of the interfacial load is transferred. For the bond length, which exceeds the L_e, the bond strength does not increase significantly. The effective bond length depends mainly on the stiffness of the composite plate. There is no consensus as to what this length is; for example, Sato [40] gives values of around 45 mm, while Brosens and Germet [36] suggest over 275 mm. The effective bond length for a single layer of SRP, calculated according to the above-mentioned analytical methods [30,32,41,45,47], is in the range of 50–160 mm and for most cases does not exceed the anchorage length provided in the test samples (Table 2 and Figure 1). Therefore, the slight influence of the anchorage length observed confirms a limited transfer of the interfacial force over the effective bond length.

The second result that requires comment is the significant effect of the sample width b. This problem may be related to the ratio of the composite width b_f to the sample width b, as this influences the stress distribution in the concrete. For the tested models, this proportion is usually equal to 0.5; therefore, slight changes in the width b could significantly affect the distribution of stresses and, thus, the debonding strength.

6. Conclusions

This paper describes an innovative approach for estimating the bond strength of SRP to concrete, based on the artificial neural network model. The developed model is trained on the basis of the experimental data gathered from published literature. The model is used to predict the bonding strength and further compared to some of the analytical bond-slip models from the literature. The results obtained show good agreement with the laboratory data collected. The working ANN model performs significantly better than other models in estimating the bonding strength. The sensitivity analysis concludes that the architecture of the working model is also optimal in terms of the number of input neurons. None of the input parameters can be excluded from the network, as all of them carry a high level of importance to the output value.

Undoubtedly, the study shows the potential of neural networks as a supporting tool for structural engineers; however, the main disadvantage of this method is that it is a 'black box' that cannot derive any universal equation and cannot function without a training base. On the other hand, it is only a matter of time before such datasets will be automatically created by the internet bots.

Author Contributions: Conceptualization, R.K.; methodology, R.K.; software, S.K.; formal analysis, S.K.; investigation, R.K.; resources, R.K.; data curation, S.K.; writing—original draft preparation, R.K. and S.K.; writing—review and editing, R.K. and S.K. All authors have read and agreed to the published version of the manuscript.

Funding: This research was funded by the Silesian University of Technology (Grant no BK-225/RB6/2022 03/060/BK_22/1026).

Institutional Review Board Statement: Not applicable.

Informed Consent Statement: Not applicable.

Data Availability Statement: The data presented in this study are available on request from the corresponding author.

Conflicts of Interest: The authors declare no conflict of interest.

References

1. Rizkalla, S.; Rosenboom, O.; Miller, A.; Walter, C. *Value Engineering and Cost Effectiveness of Various Fiber Reinforced Polymer (FRP) Repair Systems*; Technical Report; Department of Civil Engineering, North Carolina State University Raleigh: Raleigh, NC, USA, June 2007.
2. Mitolidis, G.J.; Salonikios, T.N.; Kappos, A.J. Mechanical and Bond Characteristics of SRP and CFRP Reinforcement—A Comparative Research. *Open Constr. Build. Technol. J.* **2008**, *2*, 207–216. [CrossRef]
3. Papakonstantinou, C.G.; Kakae, C.; Gryllakis, N. Can Existing Design Codes Be Used to Design Flexural Reinforced Concrete Elements Strengthened with Externally Bonded Novel Materials? *IOP Conf. Ser. Mater. Sci. Eng.* **2018**, *371*, 250–257. [CrossRef]
4. Krzywoń, R. Assessment of existing bond models for externally bonded SRP composites. *Appl. Sci.* **2020**, *10*, 8593. [CrossRef]
5. Kekez, S.; Kubica, J. Application of Artificial Neural Networks for Prediction of Mechanical Properties of CNT/CNF Reinforced Concrete. *Materials* **2021**, *14*, 5637. [CrossRef]
6. Cihan, M.T. Prediction of Concrete Compressive Strength and Slump by Machine Learning Methods. *Adv. Civ. Eng.* **2019**, *1*, 3069046. [CrossRef]
7. Kröse, B.; van der Smagt, P. *An Introduction to Neural Networks*; University of Amsterdam: Amsterdam, The Netherlands, 1996.
8. Matos, M.A.S.; Pinho, S.T.; Tagarielli, V.L. Application of machine learning to predict the multiaxial strain sensing response of CNT polymer composites. *Carbon* **2019**, *146*, 265–275. [CrossRef]
9. Fahmy, A.S.; EL Madawy, M.E.; Cobran, Y.A. Using artificial neural networks in the design of orthotropic bridge decks. *Alex. Eng. J.* **2016**, *55*, 3195–3203. [CrossRef]

10. Sattari, M.A.; Roshani, G.H.; Hanus, R.; Nazemi, E. Applicability of time-domain feature extraction methods and artificial intelligence in two-phase flow meters based on gamma-ray absorption technique. *Measurement* **2021**, *168*, 108474. [CrossRef]
11. Shin, J.; Scott, D.W.; Stewart, L.K.; Jeon, J.S. Multi-hazard assessment and mitigation for seismically deficient RC building frames using artificial neural network models. *Eng. Str.* **2020**, *207*, e110204. [CrossRef]
12. Cao, M.; Alkayem, N.F.; Pan, L.; Novak, D. Advanced Methods in Neural Networks-Based Sensitivity Analysis with their Applications in Civil Engineering. In *Artificial Neural Networks–Models and Applications*; Chapter 13; IntechOpen Book Series; IntechOpen: London, UK, 2016; pp. 335–353. [CrossRef]
13. Köroğlu, M.A. Artificial neural network for predicting the flexural bond strength of FRP bars in concrete. *Sci. Eng. Compos. Mater.* **2019**, *26*, 12–29. [CrossRef]
14. Mansouri, I.; Kisi, O. Prediction of debonding strength for masonry elements retrofitted with FRP composites using neuro fuzzy and neural network approaches. *Compos. Part B* **2015**, *70*, 247–255. [CrossRef]
15. Mashrei, M.A.; Seracino, R.; Rahman, M.S. Application of artificial neural networks to predict the bond strength of FRP-to-concrete joints. *Constr. Build. Mater.* **2013**, *40*, 812–821. [CrossRef]
16. Cascardi, A.; Micelli, F. ANN-Based Model for the Prediction of the Bond Strength between FRP and Concrete. *Fibers* **2021**, *9*, 46. [CrossRef]
17. Jahangir, H.; Eidgahee, D.R. A new and robust hybrid artificial bee colony algorithm–ANN model for FRP-concrete bond strength evaluation. *Compos. Struct.* **2021**, *257*, 113160. [CrossRef]
18. Figeys, W.; Schueremans, L.; Van Gemert, D.; Brosens, K. A new composite for external reinforcement: Steel cord reinforced polymer. *Constr. Build. Mater.* **2008**, *22*, 1929–1938. [CrossRef]
19. Matana, M.; Nanni, A.; Dharani, L.; Silva, P.; Tunis, G. Bond Performance of Steel Reinforced Polymer and Steel Reinforced Grout. In Proceedings of the International Symposium on Bond Behaviour of FRP in Structures 381 (BBFS 2005), IFRPC, Hong Kong, China, 7–9 December 2005.
20. Mitolidis, G.I.; Kappos, A.J.; Salonikios, T.N. Bond tests 382 of SRP and CFRP–strengthened concrete prisms. In Proceedings of the Fourth International Conference on FRP Composites in Civil Engineering (CICE2008), Zurich, Switzerland, 22–24 July 2008.
21. Napoli, A.; de Felice, G.; De Santis, S.; Realfonzo, R. Bond behaviour of Steel Reinforced Polymer strengthening systems. *Compos. Struct.* **2016**, *152*, 499–515. [CrossRef]
22. Ascione, F.; Napoli, A.; Realfonzo, R. Experimental and analytical investigation on the bond of SRP systems to concrete. *Compos. Struct.* **2020**, *242*, 112090. [CrossRef]
23. Ascione, F.; Lamberti, M.; Napoli, A.; Razaqpur, G.; Realfonzo, R. An experimental investigation on the bond behavior of steel reinforced polymers on concrete substrate. *Compos. Struct.* **2017**, *181*, 58–72. [CrossRef]
24. Ascione, F.; Lamberti, M.; Napoli, A.; Realfonzo, R. Experimental bond behavior of Steel Reinforced Grout 400 systems for strengthening concrete elements. *Constr. Build. Mater.* **2020**, *232*, 117105. [CrossRef]
25. Teng, J.G.; Chen, J.F.; Simth, S.T.; Lam, L. *FRP-Strengthened RC Structures*; John Wiley & Sons Ltd.: Chichester, UK, 2002.
26. Santandrea, M.; Focacci, F.; Mazzotti, C.; Ubertini, F.; Carloni, C. Determination of the interfacial cohesive 310 material law for SRG composites bonded to a masonry substrate. *Eng. Fail. Anal.* **2020**, *111*, 104322. [CrossRef]
27. Seracino, R. Axial intermediate crack debonding of plates glued to concrete surfaces. In Proceedings of the FRP Composites in Civil Engineering, Hong Kong, China, 12–15 December 2001; pp. 365–372.
28. Tanaka, T. Shear Resisting Mechanism of Reinforced Concrete Beams with CFS as Shear Reinforcement. Ph.D. Thesis, Hokkaido University, Kitaku, Japan, 1996.
29. Hiroyuki, Y.; Wu, Z. Analysis of debonding fracture properties of CFS strengthened member subject to tension. In Proceedings of the 3rd International Symposium on Non-Metallic (FRP) Reinforcement for Concrete Structures, Japan Concrete Institute, Sapporo, Japan, 14–16 October 1997; pp. 287–294.
30. Maeda, T.; Asano, Y.; Sato, Y.; Ueda, T.; Kakuta, Y. A study on bond mechanism of carbon fiber sheet. In Proceedings of the 3rd International Symposium Non-Metallic (FRP) Reinforcement for Concrete Structures, Japan Concrete Institute, Sapporo, Japan, 14–16 October 1997; pp. 279–285.
31. Taljsten, B. Strengthening of concrete prisms using the plate bonding technique. *Int. J. Fract.* **1996**, *82*, 253–266. [CrossRef]
32. Niedermeier, R. Stellungnahme zur Richtlinie für das Verkleben von Beton-bauteilen durch Ankleben von Stahllaschen. In *Schreiben 1390 Vom 30.10.1996 Des Lehrstuhls Für Massivbau*; Technische Universität München: Munich, Germany, 1996.
33. Yuan, H.; Wu, Z. Interfacial fracture theory in structures strengthened with composite of continuous fiber. In Proceedings of the Symposium of China and Japan: Science and Technology of the 21st Century, Tokyo, Japan, 3–4 December 1998; pp. 142–155.
34. Lu, X.Z.; Teng, J.G.; Ye, L.P.; Jiang, J.J. Bond-slip models for FRP sheets/plates bonded to concrete. *Eng. Struct.* **2005**, *27*, 920–937. [CrossRef]
35. Dai, J.; Ueda, T.; Sato, Y. Development of the nonlinear bond stress-slip model of fiber reinforced plastics sheet-concrete interfaces with a simple method. *J. Compos. Constr.* **2005**, *9*, 52–62. [CrossRef]
36. Brosens, K.; van Gemert, D. Anchoring stresses between concrete and carbon fiber reinforced laminates. In Proceedings of the 3rd International Symposium Non-Metallic (FRP) Reinforcement for Concrete Structures, Japan Concrete Institute, Sapporo, Japan, 14–16 October 1997; pp. 271–278.
37. Khalifa, A.; Gold, W.J.; Nanni, A.; Aziz, A. Contribution of externally bonded FRP to shear capacity of RC flexural members. *J. Compos. Constr.* **1998**, *2*, 195–202. [CrossRef]

38. Yang, Y.X.; Yue, Q.R.; Hu, Y.C. Experimental study on bond performance between carbon fibre sheets and concrete. *J. Build. Struct.* **2001**, *22*, 36–42.
39. Adhikary, B.B.; Mutsuyoshi, H. Study on the bond between concrete and externally bonded CFRP sheet. In Proceedings of the 6th International Symposium on Fiber Reinforced Polymer Reinforcement for Concrete Structures (FRPRCS-5), Cambridge, UK, 16–18 July 2001; pp. 371–378.
40. Sato, Y.; Asano, Y.; Ueda, T. Fundamental study on bond mechanism of carbon fiber sheet. *Concr. Libr. Int. JSCE* **2001**, *37*, 97–115. [CrossRef]
41. Chen, J.F.; Teng, J.G. Anchorage strength models for FRP and steel plates bonded to concrete. *J. Struct. Eng.* **2001**, *127*, 784–791. [CrossRef]
42. De Lorenzis, L.; Miller, B.; Nanni, A. Bond of fiber-reinforced polymer laminates to concrete. *ACI Mater. J.* **2001**, *98*, 256–264.
43. Seracino, R.; Saifulnaz, M.R.R.; Ohlers, D.J. Generic debonding resistance of EB and NSM plate-to-concrete joints. *J. Compos. Constr.* **2007**, *11*, 62–70. [CrossRef]
44. Japan Concrete Institute. Technical report of technical committee on retrofit technology. In Proceedings of the International Symposium on Latest Achievement of Technology and Research on Retrofitting Concrete Structures, JCI, Kyoto, Japan, 14–15 July 2003.
45. SIA Norm 166. *Klebebewehrung*; Schweizerischer Ingenieur-und Architekten-Verein: Zurich, Switzerland, 2004; p. 44.
46. CNR-DT 200 R1. *Guide for the Design and Construction of Externally Bonded FRP Systems for Strengthening Existing Structures*; CNR Advisory Committee on Technical Recommendations for Construction: Rome, Italy, 2013; p. 152.
47. Fib Bulletin 90. *Externally Applied FRP Reinforcement for Concrete Structures*; Fédération Internationale Du Béton (Fib): Lausanne, Switzerland, 2019; p. 229.
48. Garson, G.D. Interpreting neural network connection weights. *AI Expert* **1991**, *6*, 46–51.
49. Mukhopadhyaya, P.; Swamy, N. Interface shear stress: A new design criterion for plate debonding. *J. Compos. Constr.* **2001**, *5*, 35–43. [CrossRef]
50. Shahawy, M.A.; Arockiasamy, M.; Beitelman, T.; Sowrirajan, R. Reinforced concrete rectangular beams strengthened with CFRP laminates. *Compos. Part B* **1996**, *27*, 225–233. [CrossRef]
51. Bizindavyi, L.; Neale, K.W. Transfer Lengths and Bond Strengths for Composites Bonded to Concrete. *J. Compos. Constr.* **1999**, *3*, 153–160. [CrossRef]

Article

Salt Spray Resistance of Roller-Compacted Concrete with Surface Coatings

Huigui Zhang, Wuman Zhang * and Yanfei Meng

School of Transportation Science and Engineering, Beihang Univerisity, Beijing 100191, China; sy1913202@buaa.edu.cn (H.Z.); sy2013201@buaa.edu.cn (Y.M.)
* Correspondence: wmzhang@buaa.edu.cn

Abstract: In order to evaluate the feasibility of surface coatings in improving the performance of RCC under salt spray conditions, sodium silicate (SS), isooctyl triethoxy silane (IOTS), and polyurea (PUA) were used as surface coatings to prepare four types of roller-compacted concrete (RCC): reference RCC, RCC-SS, RCC-IOTS, and RCC-PUA. A 5% sodium sulfate solution was used to simulate a corrosive marine environment with high temperatures, high humidity, and high concentrations of salt spray. This study focuses on investigating various properties, including water absorption, abrasion loss, compressive strength, dynamic elastic modulus, and impact resistance. Compared to the reference RCC, the 24 h water absorption of RCC-SS, RCC-IOTS, and RCC-PUA without salt spray exposure decreased by 22.8%, 77.2%, and 89.8%, respectively. After 300 cycles of salt spray, the abrasion loss of RCC-SS, RCC-IOTS, and RCC-PUA reduced by 0.3%, 4.4%, and 34.3%, respectively. Additionally, their compressive strengths increased by 3.8%, 0.89%, and 0.22%, and the total absorbed energy at fracture increased by 64.8%, 53.2%, and 50.1%, respectively. The results of the study may provide a reference for the selection of coating materials under conditions similar to those in this study.

Keywords: salt spray; wear resistance; impact resistance; surface coating; roller-compacted concrete (RCC)

Citation: Zhang, H.; Zhang, W.; Meng, Y. Salt Spray Resistance of Roller-Compacted Concrete with Surface Coatings. *Materials* **2023**, *16*, 7134. https://doi.org/10.3390/ma16227134

Academic Editors: Yuri Ribakov and Andreas Lampropoulos

Received: 18 October 2023
Revised: 6 November 2023
Accepted: 10 November 2023
Published: 12 November 2023

Copyright: © 2023 by the authors. Licensee MDPI, Basel, Switzerland. This article is an open access article distributed under the terms and conditions of the Creative Commons Attribution (CC BY) license (https://creativecommons.org/licenses/by/4.0/).

1. Introduction

The service life of marine concrete structures is contingent upon concrete durability, often impacted by environmental factors, particularly in harsh conditions [1,2]. The salinity-induced corrosion prevalent in marine environments poses a significant threat to critical infrastructure [3–5]. To address durability issues arising from scouring and the multi-factor coupling in marine concrete, researchers focus on two aspects: enhancing permeability resistance through admixture additions [6] and bolstering durability via the application of surface coatings [7]. These measures aim to extend the service life of marine concrete structures.

Concrete surface coatings, essential for protecting structures in corrosive environments [7–9], fall into three main categories: organic coatings, inorganic coatings, and organic–inorganic composite coatings, depending on their chemical composition [10,11]. Organic coatings primarily comprise polymers like polyacrylate, epoxy resin, polyurethane, and fluorine resin [12,13]. Inorganic coatings include water-soluble silicates, silica sol, phosphates, and more. The growing attention to organic–inorganic composite coatings stems from their flexible composition and synergistic performance benefits. Polymer cement-based coatings, representing this category, demonstrate commendable mechanical properties, corrosion resistance, and weather resistance.

Mehdi et al. [14]. identified epoxy polyurethane and aliphatic acrylic as the most effective coatings for reducing chloride ion penetration and extending the service life of concrete structures. Almusallam et al. [15] compared the durability of epoxy- and polyurethane-coated concrete to that of acrylic, polymer, and chlorinated rubber coatings, finding the former to be superior. Elnaggar et al. [16] developed asphaltic polyurethane coatings with varying NCO/OH ratios, demonstrating high performance in aggressive environments. The Center for Innovative Grouting Materials and Technology devised tests and analytical models to evaluate epoxy- and polyurethane-coated concrete performance [17].

Additionally, Shi et al. [18] observed that polymer coatings enhanced the resistance of surface layer concrete to chloride ion diffusion. Maj and Ubysz [19] investigated the factors contributing to the loss of adhesion of polyurea coatings to concrete substrates in chemically aggressive water tanks. Santos et al. [20] proposed polyurea coatings as a retrofit option for non-load-bearing concrete masonry walls. Arabzadeh et al. [21] assessed superhydrophobic nanomaterial-based coatings on concrete surfaces for water repellency. Yin et al. [22] developed superhydrophobic coatings based on bionic mineralization to enhance marine concrete durability. Moon et al. [23] reported that calcium–silicate compound coatings improved resistance to chloride penetration, freezing–thawing, and carbonation in concrete specimens. Luo et al. [10] integrated kaolinite nanosheets into permeable epoxy resin, resulting in a high-adhesion, barrier-performance organic–inorganic composite coating. Li et al. [24] enhanced the waterproofing and chloride resistance of concrete by designing a nano-polymer-modified cementitious coating, incorporating nano-SiO_2 or nano-TiO_2 suspensions into an acrylic emulsion.

The cost-effectiveness and environmentally friendly nature of surface coatings have led to widespread use in various protective engineering applications [25,26]. However, the variety of surface coatings complicates the selection process. Even with similar generic chemical compositions, these coatings offer varying levels of protection, making the right choice challenging [27,28]. In addition, the environmental conditions in the island salt spray zone are different from those in the tidal and submerged zones [29]. In the South China Sea, the average annual temperature is as high as 28.6 °C, and the road surface temperature is as high as 60 °C in summer. The salinity of surface seawater ranges from 33.0 to 33.5; thus, the islands are characterized by high temperatures, high humidity, and high concentrations of salt spray.

In this study, sodium silicate (SS), isooctyl triethoxy silane (IOTS), and polyurea (PUA) were used as the surface coatings. Roller-compacted concrete (RCC), commonly used for airport runways, was prepared with and without a surface coating. Since the abrasion resistance and impact resistance requirements of the airport runways are higher than those of other ordinary building structures, these two properties of RCC, exposed to salt spray, were tested in this study. The microstructures and pore size distribution were also measured. The objective of this study was to evaluate the feasibility of surface coatings in improving the performance of RCC used in island airport runways under salt spray conditions.

2. Experimental Produce

2.1. Raw Materials

The chemical composition of 42.5-grade Portland cement was detailed in Table 1, with river sand's fineness modulus specified as 2.43 and coarse aggregate exhibiting a particle size range of 5–15 mm. Workability enhancement utilized a water-reduction agent. RCC mix proportions, computed following GJB 1578-1992 [30], are presented in Table 2. Surface coatings encompassed sodium silicate (SS), isooctyl triethoxy silane (IOTS), and polyurea (PUA) (see Table 3).

Table 1. Chemical compositions of cement (%).

SiO_2	Al_2O_3	Fe_2O_3	CaO	MgO	SO_3	Na_2O	K_2O	TiO_2
23.1	7.1	3.67	57.59	2.18	2.65	0.18	0.72	0.34

Table 2. Mix proportions of RCC (kg/m³).

Cement	Water	Fine Aggregate	Coarse Aggregate	Water-Reducing Agent
315	109	895	1207	8.7

Table 3. Surface-coating materials.

	Solution Concentration	Surface Treatment Age
Sodium silicate (SS)	20%	7 day
Isooctyl triethoxy silane (IOTS)	99%	28 day
Polyurea (PUA)	80%	28 day

2.2. Samples Preparation

The size of the prismatic specimen was 100 mm × 100 mm × 400 mm, and the side length of the cubic specimen was 150 mm. Concrete mixture was poured into the test molds in three layers, and each layer was compacted with a vibrating hammer for 30 s. After 24 h, the specimens were demolded and cured under the standard conditions.

The treatment process of the coating materials was carried out in accordance with the manufacturer's recommendations for use.

In the case of treating RCC with sodium silicate (RCC-SS), the process occurs at 7 days of curing. After removing the specimens from the curing room, their surfaces were brushed with a wire brush. Subsequently, sodium silicate, dissolved in warm water, was evenly sprayed onto the specimen surfaces. This treatment was repeated every two hours for a total of four applications. Finally, the treated specimens were returned to the standard curing chamber and allowed to cure until reaching 28 days.

For RCC treated with isooctyl triethoxy silane (RCC-IOTS), the process was initiated at 28 days of curing. After removing the specimens from the curing room, their surfaces were brushed with a wire brush. Isooctyl triethoxy silane was sprayed onto the specimen surfaces and left for 6 h before a second round of spraying was conducted.

In the case of RCC treated with polyurea materials (RCC-PUA), components A and B were mixed in a ratio of 1:0.45, and a specified amount of butyl acetate was added as a diluent. The treatment was also carried out at 28 days of curing. After removing the specimens from the curing room and brushing their surfaces with a wire brush, the polyurea mixture was uniformly sprayed onto the specimen surfaces. The specimens were then left at room temperature until the polyurea mixture hardened.

In this study, there was one group of reference specimens (without surface treatment) and three groups of surface-treated specimens, which were RCC-SS, RCC-IOTS, and RCC-PUA. The above four groups of specimens were subjected to the performance tests mentioned below, and each group of specimens contained three specimens. The average of the test results of the three specimens was used for comparative analysis.

Cubic specimens are used in the water-absorption tests, mass change, and abrasion tests. Prismatic specimens are used in the dynamic elastic modulus and impact tests. The compressive strength of the specimen is determined by using the prisms that break in the impact test, and a 100 mm × 100 mm × 10 mm steel plate is placed on the upper and lower surfaces, respectively, along the length of the specimen, so that the area of the compression surface of the specimen is 100 mm × 100 mm.

2.3. Salt Spray Cycles

Salt spray conditions in the South China Sea were simulated via indoor salt spray tests. RCC underwent corrosion tests in an automatic machine (see Figure 1) following GB10125-1997 [31] for cyclic exposure to salt spray conditions. A 5% Na_2SO_4 (w/w) was used for salt spray, and the deposition rate was 1.2 mL/(80 $cm^2 \cdot h$). After the specimen reached the age of maintenance, the specimen was uniformly arranged in the specimen holder to ensure that the upper surface of the specimen was horizontal and the interval between the specimens was more than 100 mm. Each cycle started with 4 h of salt spray at 26.5 °C with 96% humidity, followed by 2 h of drying conditions without salt spray at 50 °C. In order to minimize the effect of salt spray inhomogeneity on the results, the left, right, front, and rear specimens were repositioned every 5 salt spray cycles.

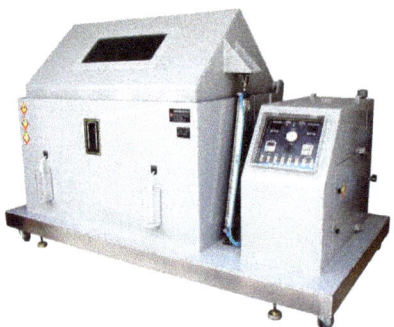

Figure 1. Automatic machine for cyclic exposure to salt spray conditions.

2.4. Abrasion Resistance

Abrasion resistance, assessed using JTG E30-2020 [32], was measured with a 200 N load on the pressure head. The horizontal pallet rotated at a speed of 17.5 r/min, with a transmission ratio of 35:1 between the pallet and the spindle (see Figure 2). Abrasion resistance was evaluated based on mass loss per unit area.

Figure 2. Abrasion test machine.

2.5. Impact Resistance

For impact resistance testing, the study employed a drop hammer impact tester (INSTRON 9350 HV, Norwood, MA, USA; see Figure 3). The tester, featuring a square shape with a 75 mm diameter circular area at the center and a double-layer pneumatic clamp, was used. During the impact test, the crosshead, holding the drop hammer, was released, allowing it to fall vertically along two guide frames and impact the specimen within the circular area. A sensor automatically recorded load, displacement, and time data to monitor changes. Simultaneously, a computer data acquisition system integrated the force–displacement curve to determine variations in impact energy absorbed by the specimen. The test utilized a hemispherical indenter measuring 12.6 mm in diameter and weighing 12.250 kg, adjusting the indenter's height to achieve initial impact energies.

Figure 3. Drop hammer impact tester.

3. Results and Discussion

3.1. Water Absorption of RCC without Salt Spray Cycles

Water absorption stands as a crucial transport property of concrete, as it serves as the primary avenue for the infiltration of aggressive ions. This penetration through water absorption is a key contributor to durability-related damage and the subsequent degradation of performance. While diffusion does contribute to ionic transport, studies indicate that individual diffusion is a notably slow process. Consequently, water absorption emerges as the dominant mechanism. Theoretically, considering its prevalence, water absorption can be viewed as a representative descriptor that effectively mirrors the durability of concrete [33].

The water absorption of RCC at 24 h is presented in Figure 4. The water absorption of RCC-SS, RCC-IOTS, and RCC-PUA surface coatings decreased by 22.8%, 77.2%, and 89.8%, respectively. RCC with surface coatings had significantly reduced water absorption compared to RCC without surface coatings. This indicates that the three coating materials effectively act as barriers, thereby significantly reducing the water permeability of RCC. In particular, the PUA surface coating forms a complete, smooth, and dense isolation layer on the surface of RCC (see Figure 5). Therefore, among the three surface coating materials, RCC with PUA surface coating exhibits the lowest water-absorption rate.

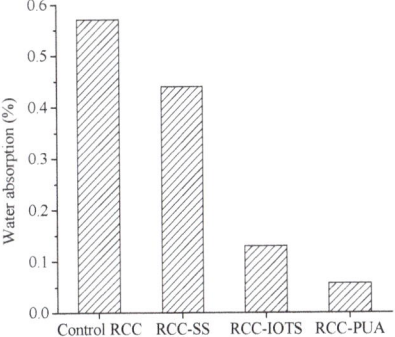

Figure 4. Water absorption of RCC.

Franzoni et al. [34] noted that concrete treated with SS demonstrated approximately half the 7-day water absorption compared to untreated samples. Almusallam et al. [15] and Zhu et al. [35] also highlighted that surface treatment enhanced resistance to capillary water absorption. The findings of this study are consistent with the above results. Moreover, the water absorption of concrete is directly linked to its impermeability, with lower water absorption typically indicating excellent impermeability. These outcomes are consistent with Mehdi et al.'s [14] discovery that PUA surface coating reduced chloride ion diffusion.

Figure 5. Surface of RCC-PUA.

3.2. Mass Change of Specimen

In a sodium sulfate environment, excessive crystal formation during subsequent salt spray cycles imposes expansive stress on the cement matrix. This stress, in turn, induces the development of micro-cracks and a decline in compressive strength [36]. Beyond osmotic and crystallization pressures, Na_2SO_4 exhibits a reversible transformation between its dehydrated and anhydrous states. Studies indicate that Na_2SO_4 can generate pore pressures ranging from 400 to 5000 psi, whereas $Na_2SO_4 \cdot 10H_2O$ can induce pore pressures of 1000–1200 psi [37].

Furthermore, sulfate ions react with the hydration products of cement to produce new products that expand in volume, such as gypsum and ettringite (see Equations (1) and (2)). The volume expansion of ettringite leads to the expansion of existing cracks and the creation of new cracks in the concrete, ultimately reducing the strength of the concrete under sulfate attack [36,38].

$$Na_2SO_4 + Ca(OH)_2 + 2H_2O \rightarrow CaSO_4 \cdot 2H_2O + 2NaOH \quad (1)$$

$$C\text{-}A\text{-}H + 3CaSO_4 \cdot 2H_2O + 2H_2O \rightarrow C_3A \cdot 3CaSO_4 \cdot 32H_2O \quad (2)$$

The initial physical sulfate attacks tend to augment the weight of concrete. However, subsequent physical and chemical assaults typically lead to concrete cracking and mass loss.

The mass change rate of RCC under salt spray cycles is given in Figure 6. For both the reference RCC and RCC-SS, the mass change rate initially increases and then decreases as the number of salt spray cycles increases. Zhang et al. [2] also found a similar pattern of mass change. The maximum rate of mass increase occurs after 200 cycles of salt spray. The initial increase in the change rate is primarily attributed to increased salt permeation and crystallization. The crystals fill and cover the pores, resulting in an increase in concrete mass [2,39]. As the cycles progress, the decrease in the change rate is mainly caused by the intensified corrosive effect of salt in the salt spray, leading to localized delamination of the surface layer [2]. However, the extent of delamination at this stage is still smaller than the mass increase resulting from salt filling and crystallization in the concrete pores. Consequently, the mass change rate remains positive but exhibits a decreasing trend.

For RCC treated with IOTS, the mass loss rate of the specimens subjected to 50 cycles of salt spray is 0.03%. After 100 cycles of salt spray, the rate of mass change of the specimens first approaches zero and then increases with the number of salt spray cycles. At 300 cycles of salt spray, the mass increase rate is 0.15%, which is only about 50% of the control RCC subjected to the same number of salt spray cycles. For RCC-PUA, the rate of mass loss of the specimens remains at approximately 0.04% as the number of salt spray cycles increases.

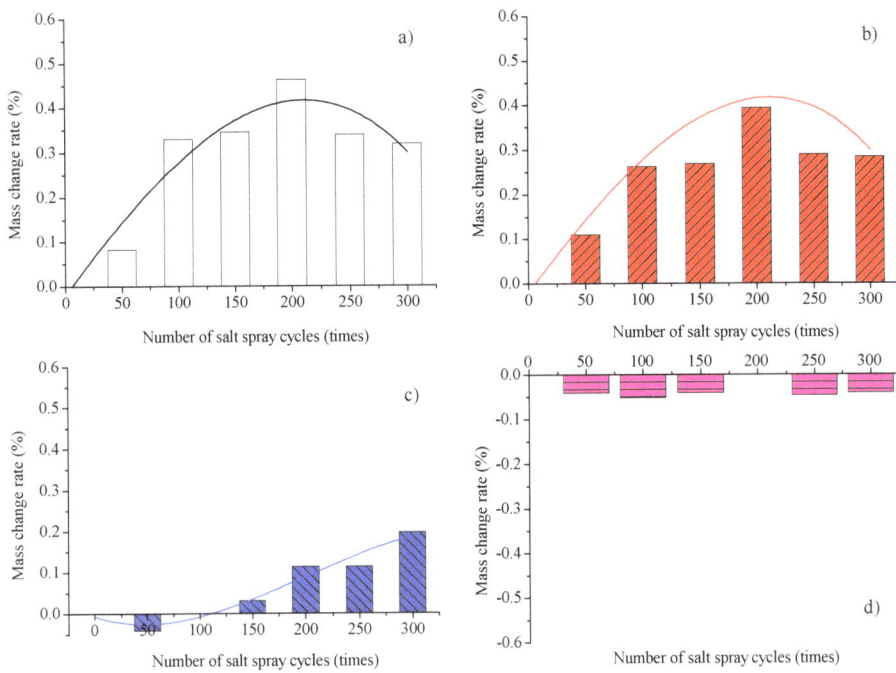

Figure 6. Change rate of mass; (**a**) Control RCC; (**b**) RCC-SS; (**c**) RCC-IOTS; (**d**) RCC-PUA.

Therefore, both RCC-IOTS and RCC-PUA surface coatings exhibit favorable resistance to salt spray corrosion when considering the criterion of mass change in a specimen under salt spray cycles.

3.3. Abrasion Resistance

The use of concrete in runway construction exposes it to rubbing, scraping, skidding, and sliding due to the impact loads from surface movement. These actions contribute to the deterioration of concrete surfaces. Surface fractures lead to a reduction in concrete thickness, resulting in a smoother surface and an increase in dust accumulation. These factors collectively weaken the concrete, posing a threat to flight safety. Therefore, it is imperative for concrete runways to possess adequate abrasion resistance—a property that shields the hardened concrete surface from wear caused by abrasive forces. Ensuring the abrasion resistance of concrete runways and pavements is crucial in preventing surfaces from becoming overly polished, thus maintaining optimal skid resistance.

The surface changes of the reference RCC with 300 cycles of salt spray are shown in Figure 7. It can be observed that with an increasing number of rotations of the grinding head, the surface wear of the hardened cement paste becomes more evident, and the area of exposed coarse aggregates in the specimen increases. The surface changes of the other RCC with surface coatings are found to be similar to those of the control RCC with 300 cycles of salt spray. Exposed aggregates are visible on the sample surfaces, signaling a decline in the efficacy of the surface treatment layers. Eventually, these layers are fully removed, indicating a complete loss of their protective effect [40].

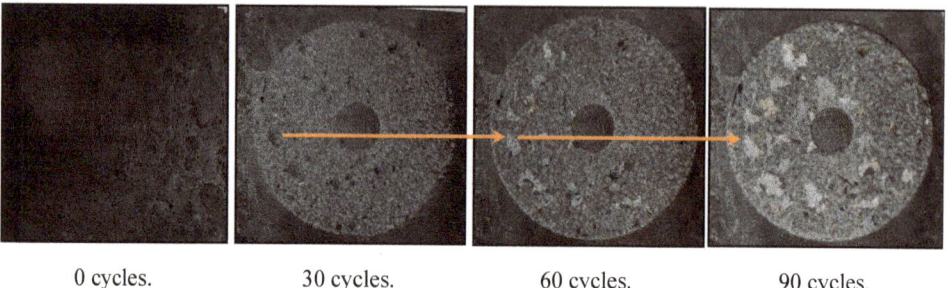

Figure 7. Surface change of control RCC.

The abrasion loss per unit area of RCC with 30 and 90 rotations of the grinding head are presented in Figure 8a,b, respectively. It is clear that the abrasion loss per unit area of reference RCC and RCC-SS is very close. However, Franzoni et al. [34] discovered that treatment with sodium silicate (SS) yielded the most effective performance in enhancing concrete's surface abrasion resistance, attributed to the substantial thickness of the resulting external layer. The difference between the results of this study and those of Franzoni et al. [34] is mainly due to the thickness of the SS coating; a larger coating thickness tends to increase the wear resistance of the specimen, whereas in, this study, only 20% SS solution was sprayed on the specimen four times at 2 h intervals, which produced a thinner SS coating; therefore, the determination of the coating thickness is as critical as the selection of the coating material.

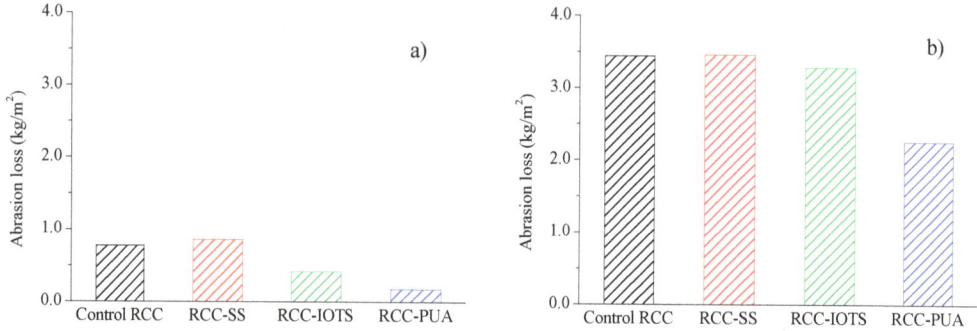

Figure 8. Abrasion loss. (**a**) 30 cycles; (**b**) 90 cycles.

Both RCC-IOTS and RCC-PUA exhibit significantly lower abrasion losses per unit area. After 90 cycles of grinding head rotation, RCC-IOTS and RCC-PUA experienced abrasion losses of 3.29 kg/m^2 and 2.26 kg/m^2, respectively—indicating reductions of 4.4% and 34.3% compared to the control RCC. Wu et al. [40] also noted that PUA surface treatments led to a notable decrease in concrete mass loss, signifying enhanced resistance to debris flow abrasion. Baltazar et al. [41] observed that as long as the PUA protective layer remained intact on the concrete surface, it imparted excellent abrasion resistance. The findings in this study underscore that PUA treatment excels in both abrasion resistance and salt spray resistance.

3.4. Dynamic Elastic Modulus and Compressive Strength

The percentage change in dynamic elastic modulus of RCC with 300 cycles of salt spray is shown in Figure 9a. It can be observed that all RCC samples experience varying degrees of increase in dynamic elastic modulus after the salt spray cycles. Zhang et al. [2]

also reported similar findings when concrete specimens were subjected to salt spray cycles. RCC-IOTS showed the highest increase, reaching 9.1%, while RCC-PUA showed the lowest increase, at 0.6%. There are two possible reasons for the increase in the dynamic elastic modulus. Firstly, during the 300 salt spray cycles, the specimens undergo a 60-day period, enabling continued cement hydration and subsequent increase in the dynamic elastic modulus. Secondly, a large amount of salt penetrates into RCC specimens during the salt spray cycles. The salt crystallizes within the pores and fills some of them during the drying process, thereby increasing the compactness and reducing the porosity of RCC to some extent. Both factors contribute to the increase in the dynamic elastic modulus of RCC. The relatively minimal increase in the elastic modulus of RCC-PUA is attributed to the formation of a complete, smooth, and dense sealing layer on the specimen's surface. Under such conditions, only the first mechanism mentioned above, related to cement hydration, significantly impacts the increase, while the second mechanism has minimal influence.

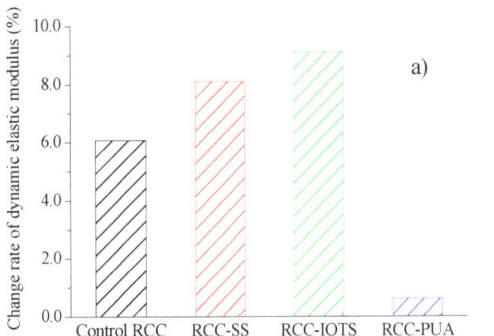 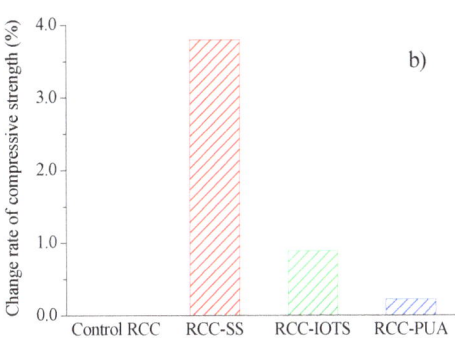

Figure 9. Change rate of dynamic elastic modulus and compressive strength. (**a**) dynamic elastic modulus, (**b**) compressive strength.

The relative change in the compressive strength of RCC with surface coatings after 300 cycles of salt spray is depicted in Figure 9b. It clearly shows that RCC-SS exhibits the highest compressive strength, followed by RCC-IOTS, while RCC-PUA shows the least improvement. The compressive strength of RCC-SS, RCC-IOTS, and RCC-PUA is 3.8%, 0.89%, and 0.22% higher than that of control RCC after 300 cycles of salt spray. These results demonstrate that all three surface coatings have enhanced the resistance of RCC to salt spray corrosion to varying degrees.

3.5. Impact Resistance

In the drop hammer impact test, each impact's energy is controlled at 13 J, and continuous impact loads are applied until the fracture of the specimen occurs. Figure 10 illustrates the relationship between the impact force and time during the first impact after the specimens are subject to 300 cycles of salt spray. It is evident that different groups of RCC exhibit varying peak impact forces under approximately the same impact energy. RCC-PUA displays the highest peak force at 21.9 kN, indicating the highest surface hardness after the salt spray cycles. Wu et al. [40] observed that concrete treated with PUA exhibited a harder surface compared to non-coated concrete.

The reference RCC followed with a peak impact force of 20.7 kN, while RCC-SS and RCC-IOTS show similar peak impact forces at around 19.2 kN, with two peaks observed. The second peak is likely attributed to the loosening of the surface layer of the specimens due to corrosion. When the impact force is applied, the loose surface concrete becomes compacted and comes into contact with the non-corroded and harder concrete in the interior. In addition, some of the following factors may also cause a second wave peak [42,43]: (1) the presence of eccentricity or an overly sharp head of the hammer may result in multiple peaks during impact; (2) non-homogeneous concrete may experience multiple peaks of strain under impact, resulting in corresponding peaks of impact force (which is the case with the results of the present study); (3) the flatness of the impact surface affects the absorption and transfer of energy, which results in multiple peaks of impact force; (4) multiple peaks of impact force may also occur when only a portion of the hammer head contacts the concrete.

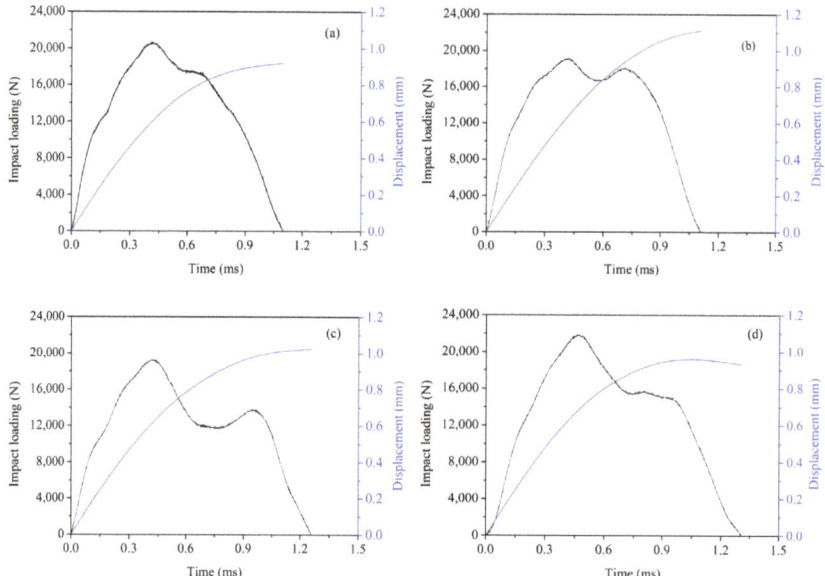

Figure 10. Impact test. (**a**) Control RCC; (**b**) RCC-SS; (**c**) RCC-IOTS; (**d**) RCC-PUA.

Table 4 presents the impact energy per impact, loss rate of the dynamic elastic modulus, and number of impacts until specimen fracture. It can be observed that the reference RCC fractures at the second impact, while other RCCs with different surface coatings fracture at the third impact. The impact energy before fracture is approximately 13 J for all specimens but decreases at the point of fracture. The total absorbed energy at fracture increases by 64.8%, 53.2%, and 50.1% for the different surface-coated RCC specimens. During the first impact, the control RCC exhibits the highest loss rate of the dynamic elastic modulus at 10.3%, followed by RCC-PUA at 2.6%. RCC-SS and RCC-IOTS show very low loss rates of dynamic elastic modulus at 0.4% and 0.2%, respectively. Overall, the surface coatings significantly improve the impact resistance of RCC under salt spray conditions. Wu et al. [40] used a drop-weight impact test to assess the impact resistance of concrete, noting no apparent failure and only changes in color on the PUA surface. The ductile behavior of PUA material contributed significantly to the impact resistance of concrete treated with PUA.

Table 4. Impact resistance.

	RCC		RCC-SS			RCC-IOTS			RCC-PUA		
Impact number (times)	1	2	1	2	3	1	2	3	1	2	3
Impact energy (J)	12.9	9.1	12.9	13.0	10.4	13.0	12.9	7.8	13.0	13.1	6.9
Loss rate DEM * (%)	10.3	/	0.4	3.2	/	0.2	61.5	/	2.6	51.3	/

* DEM is dynamic modulus of elasticity.

3.6. Microstructure and Pore Structures

Figure 11 presents the microstructures of RCC subjected to 300 cycles of salt spray. After the salt spray cycles, crystalline fillers are present in the surface pores of the control RCC, as well as in RCC-SS and RCC-IOTS. Although a thin protective film may form only on the pore wall, salt can still enter the pores of RCC-SS and RCC-IOTS, similar to the uncoated RCC, and crystallize during the drying process. Nonetheless, due to the presence of the protective coating, the direct contact between the salt and concrete layer is prevented until the protective coating is damaged. However, the PUA surface coating forms a complete and dense protective layer on the surface, effectively isolating the internal pores from salt penetration.

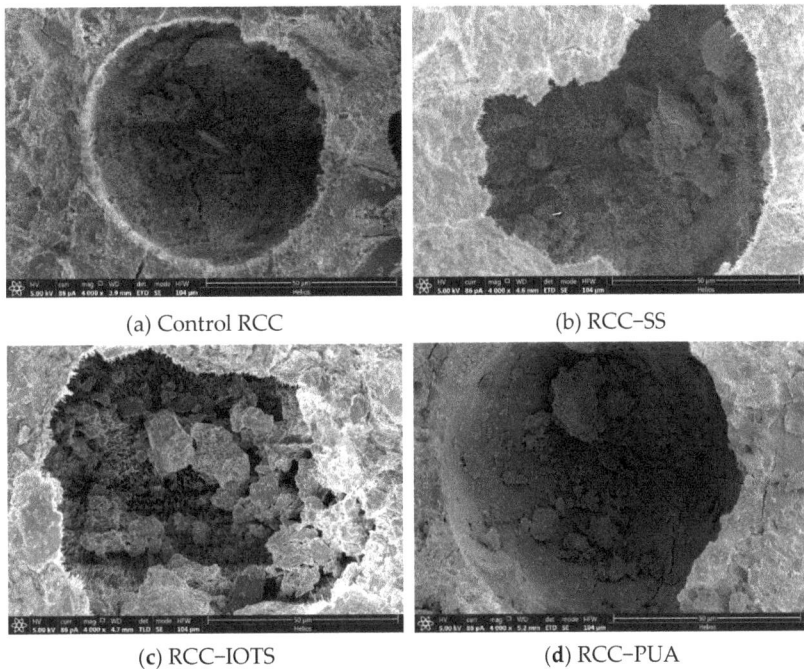

(a) Control RCC (b) RCC–SS

(c) RCC–IOTS (d) RCC–PUA

Figure 11. Microstructure of RCC.

Elemental analysis of the fillers in the internal pores shows sodium (Na) contents of 4.6%, 3.4%, 2.8%, and 0.5% for the control RCC, RCC-SS, RCC-IOTS, and RCC-PUA, respectively. Furthermore, this study revealed that surface coatings, particularly PUA, can effectively deter the infiltration of sodium sulfate into concrete. After 300 salt spray cycles, concrete treated with PUA exhibited the lowest sodium (Na) content.

In addition, clear microcracks are observed in the internal and edge regions of the pores in the uncoated RCC, which indicates that the internal pores have suffered damage after 300 cycles of salt spray. Although the pores of RCC-SS and RCC-IOTS are filled with crystals and no significant microcracks are found, the changing trend of the pore size distribution (see Figure 12) indicates a movement towards larger pore sizes, suggesting some degree of damage after 300 cycles of salt spray for all three surface-coated RCCs.

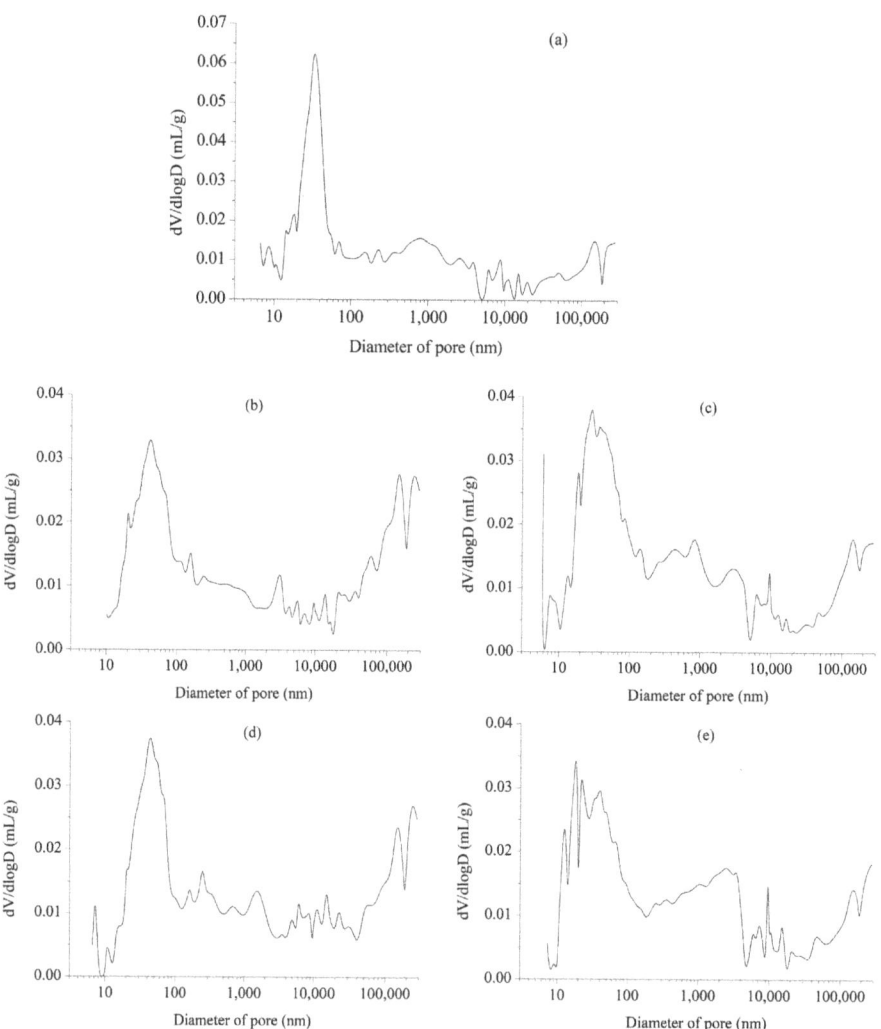

Figure 12. Pore size distribution of RCC. (**a**) RCC under standard conditions; (**b**) Control RCC; (**c**) RCC-SS; (**d**) RCC-IOTS; (**e**) RCC-PUA.

3.7. Discussion

There are three main types of hydrophobic surface treatments: (1) surface coatings, which form a continuous film of varying thickness on the surface; (2) pore filling, which acts as a localized pore barrier; and (3) impregnation or pore lining, which involves lining the pores along the entire surface of the concrete.

The SS coating undergoes hydrolysis at room temperature, forming an interconnected network structure, as shown in the following reaction:

$$Na_2O \cdot nSiO_2 + (2n+1)H_2O \rightarrow 2NaOH + nSi(OH)_4 \quad (3)$$

$$nSi(OH)_4 \rightarrow [Si(OH)_4]_n \xrightarrow{-2nH_2O} [-Si-O-Si-]_n \quad (4)$$

The resulting thin film material adheres to the concrete surface, creating a separation between the concrete and its surrounding environment. However, due to the solubility of Na+ in water, the water glass-formed film generally exhibits moderate impermeability. Nevertheless, when the surface water glass penetrates into the interior of the concrete, it reacts with the cement hydration product $Ca(OH)_2$, generating hydrated calcium silicate gel that fills the concrete pores, making it more compact and enhancing the durability of concrete [44,45]. Therefore, several properties of RCC-SS in this study were also improved, with a 22.8% reduction in water absorption, a 3.8% increase in compressive strength, and a 64.8% increase in impact energy absorption.

The chemical molecular structure of IOTS is shown in Figure 13. IOTS can impart excellent hydrophobicity to concrete surfaces without altering their surface microstructures.

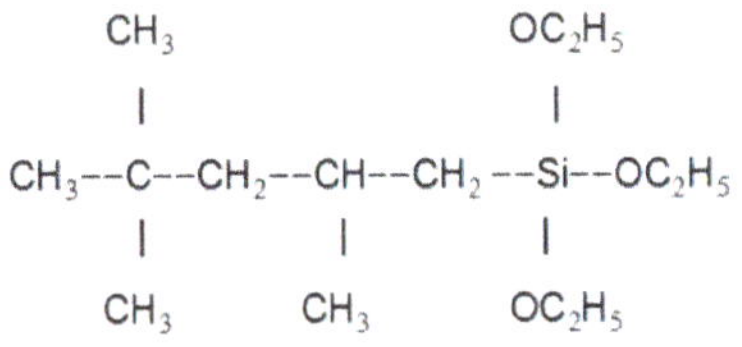

$$CH_3-\underset{\underset{CH_3}{|}}{\overset{\overset{CH_3}{|}}{C}}-CH_2-\underset{\underset{CH_3}{|}}{CH}-CH_2-\underset{\underset{OC_2H_5}{|}}{\overset{\overset{OC_2H_5}{|}}{Si}}-OC_2H_5$$

Figure 13. Chemical molecular structure of IOTS.

This phenomenon is consistent with Wenzel's theory, which suggests that rough mortar samples can be changed to hydrophobic surfaces after modification with low surface energy materials. This process involves the hydration of IOTS to form silanols (Si-OH), followed by the reaction of silanol with C-S-H, $Ca(OH)_2$, ettringite, and quartz sand through -OH group reactions. Subsequently, the two -OH groups of IOTS form Si-O-Si bonds via condensation and release water in the process. As a result, a continuous self-assembled molecular film of IOTS is formed on the surface of the hydrated products. The presence of -CH_3 and -CH_2 groups in IOTS effectively reduces the surface energy of the cement matrix and significantly improves its hydrophobicity [46]. Similarly, several properties of RCC-IOTS in this study were significantly improved, with a 77.2% reduction in water absorption, a 4.4% reduction in the abrasion loss per unit area, and a 53.2% increase in impact energy absorption.

PUA is a block polymer material, as given in Figure 14 [47]. It consists of hard segments and soft segments. The hard segments are uniformly distributed in the soft segment matrix at room temperature to form an interconnected network microstructure.

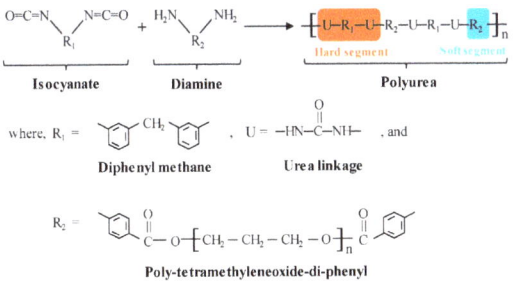

Figure 14. Chemical molecular structure of PUA [47].

The strength of PUA primarily relies on its hard segments, while the elongation is determined by the soft segments. Due to its unique properties, PUA forms a seamless, leak-free membrane, making it highly suitable for enduring continuous ponding water conditions. Moreover, PUA coatings are renowned for their exceptional durability, offering remarkable elongation and tensile strength, making them an excellent choice for various surface coating applications [48,49]. Several properties of RCC-PUA in this study were significantly improved, with an 89.8% reduction in water absorption, a 34.3% reduction in the abrasion loss per unit area, and a 50.1% increase in impact energy absorption.

4. Conclusions

This study presented the deterioration of RCC both with and without surface coatings during salt spray cycles. The following conclusions can be drawn from the test results:

1. Prior to salt spray exposure, RCC-SS, RCC-IOTS, and RCC-PUA exhibited 24 h water absorption rates 22.8%, 77.2%, and 89.8% lower than those of the control RCC, respectively.
2. After 300 cycles of salt spray, the abrasion loss per unit area of RCC-SS, RCC-IOTS, and RCC-PUA is reduced by 0.3%, 4.4%, and 34.3%, respectively, compared to the control RCC.
3. The compressive strength of RCC-SS, RCC-IOTS, and RCC-PUA is higher by 3.8%, 0.89%, and 0.22%, and the total absorbed energy at fracture is 64.8%, 53.2%, and 50.1% higher than that of control RCC, respectively.
4. Crystalline fillers are found in the pores of control RCC, RCC-SS, and RCC-IOTS, excluding RCC-PUA. However, the volume percentage of small pores in all RCCs decreases, while the volume percentage of large pores increases.

Author Contributions: H.Z.: conceptualization, methodology, resources, visualization, and writing—original draft. W.Z.: supervision, writing—review and editing. Y.M.: investigation and data. All authors have read and agreed to the published version of the manuscript.

Funding: This research is funded by the National Natural Science Foundation of China, grant numbers 51978026 and 51678022.

Data Availability Statement: Data are contained within the article.

Conflicts of Interest: The authors declare that they have no known competing financial interest or personal relationships that could have appeared to influence the work reported in this paper.

References

1. Yi, Y.; Zhu, D.; Guo, S.; Zhang, Z.; Shi, C. A review on the deterioration and approaches to enhance the durability of concrete in the marine environment. *Cem. Concr. Compos.* **2020**, *113*, 103695. [CrossRef]
2. Zhang, H.G.; Zhang, W.M.; Meng, Y.F.; Li, H.H. Deterioration of sea sand roller compacted concrete used in island reef airport runway under salt spray. *Constr. Build. Mater.* **2022**, *322*, 126523. [CrossRef]
3. Bader, M.A. Performance of concrete in a coastal environment. *Cem. Concr. Compos.* **2003**, *25*, 539–548. [CrossRef]
4. Seleem, H.; Rashad, A.M.; El-Sabbagh, B.A. Durability and strength evaluation of high-performance concrete in marine structures. *Constr. Build. Mater.* **2010**, *24*, 878–884. [CrossRef]
5. Ma, D.; Zhang, M.; Cui, J. A review on the deterioration of mechanical and durability performance of marine-concrete under the scouring action. *J. Build. Eng.* **2023**, *66*, 105924. [CrossRef]
6. Duan, A.; Dai, J.G.; Jin, W.L. Probabilistic approach for durability design of concrete structures in marine environments. *J. Mater. Civ. Eng.* **2015**, *27*, A4014007. [CrossRef]
7. Zhang, M.; Xu, H.; Liu, A.P.Z.X.; Tao, M. Coating performance, durability and anti-corrosion mechanism of organic modified geopolymer composite for marine concrete protection. *Cem. Concr. Compos.* **2022**, *129*, 104495. [CrossRef]
8. James, A.; Bazarchi, E.; Chiniforush, A.A.; Aghdam, P.; Hosseini, M. Rebar corrosion detection, protection, and rehabilitation of reinforced concrete structures in coastal environments: A review. *Construct. Build. Mater.* **2019**, *224*, 1026–1039. [CrossRef]
9. Yin, B.; Wu, C.; Hou, D.; Li, S.; Jin, Z.; Wang, M.; Wang, X. Research and application progress of nano-modified coating in improving the durability of cement-based materials. *Prog. Org. Coat.* **2021**, *161*, 106529. [CrossRef]
10. Luo, S.; Wei, J.X.; Xu, W.; Chen, Y.; Huang, H.L.; Hu, J.; Yu, Q.J. Design, preparation, and performance of a novel organic–inorganic composite coating with high adhesion and protection for concrete. *Compos. Part. B-Eng.* **2022**, *234*, 109695. [CrossRef]
11. Lu, S.; Zhao, P.; Liang, C.; Liu, L.; Qin, Z.; Wang, S.; Hou, P.; Lu, L. Utilization of Polydimethylsiloxane (PDMS) in polymer cement-based coating to improve marine environment service performance. *Constr. Build. Mater.* **2023**, *367*, 130359. [CrossRef]
12. Guo, S.; Zhang, X.; Chen, J.; Mou, B.; Shang, H.; Wang, P.; Zhang, L.; Ren, J. Mechanical and interface bonding properties of epoxy resin reinforced Portland cement repairing mortar. *Constr. Build. Mater.* **2020**, *264*, 120715. [CrossRef]
13. Huang, H.; Fang, S.; Luo, S.; Hu, J.; Yin, S.; Wei, J.; Yu, Q. Multiscale modification on acrylic resin coating for concrete with silicon/fluorine and graphene oxide (GO) nanosheets. *Constr. Build. Mater.* **2021**, *305*, 124297. [CrossRef]

14. Mehdi, K.M.; Shekarchi, M.; Hoseini, M. Time-dependent performance of concrete surface coatings in tidal zone of marine environment. *Constr. Build. Mater.* **2012**, *30*, 198–205.
15. Almusallam, A.A.; Khan, F.M.; Dulaijan, S.U.; Al-Amoudi, O.S.B. Effectiveness of surface coatings in improving concrete durability. *Cem. Concr. Compos.* **2003**, *25*, 473–481. [CrossRef]
16. Elnaggar, E.M.; Elsokkary, T.M.; Shohide, M.A.; El-Sabbagh, B.A.; Abdel-Gawwad, H.A. Surface protection of concrete by new protective coating. *Constr. Build. Mater.* **2019**, *220*, 245–252. [CrossRef]
17. Vipulanandan, C.; Liu, J. Polymer Coatings for Concrete Surfaces: Testing and Modeling. In *Handbook of Environmental Degradation of Materials*, 3rd ed.; Kutz, M., Ed.; William Andrew Publishing: Norwich, NY, USA, 2018; pp. 69–94.
18. Shi, L.; Liu, J. Effect of polymer coating on the properties of surface layer concrete. *Procedia Eng.* **2012**, *27*, 291–300. [CrossRef]
19. Maj, M.A.; Ubysz, A. The reasons for the loss of polyurea coatings adhesion to the concrete substrate in chemically aggressive water tanks. *Eng. Fail. Anal.* **2022**, *142*, 106774. [CrossRef]
20. Santos, A.P.; Chiquito, M.; Castedo, R.; López, L.M.; Gomes, G.; Mota, C.; Fangueiro, R.; Mingote, J.L. Experimental and numerical study of polyurea coating systems for blast mitigation of concrete masonry walls. *Eng. Struct.* **2023**, *284*, 116006. [CrossRef]
21. Arabzadeh, A.; Ceylan, H.; Kim, S.; Gopalakrishnan, K.; Sassani, A.; Sundararajan, S.; Taylor, P.C. Superhydrophobic coatings on Portland cement concrete surfaces. *Constr. Build. Mater.* **2017**, *141*, 393–401. [CrossRef]
22. Yin, B.; Xu, H.F.; Fan, F.Y.; Qi, D.M.; Hua, X.J.; Xu, T.Y.; Liu, C.H.; Hou, D.S. Superhydrophobic coatings based on bionic mineralization for improving the durability of marine concrete. *Constr. Build. Mater.* **2023**, *362*, 129705. [CrossRef]
23. Moon, H.Y.; Shin, D.G.; Choi, D.S. Evaluation of the durability of mortar and concrete applied with inorganic coating material and surface treatment system. *Constr. Build. Mater.* **2007**, *21*, 362–369. [CrossRef]
24. Li, G.; Ding, Y.G.; Gao, T.Y.; Qin, Y.M.; Lv, Y.J.; Wang, K.J. Chloride resistance of concrete containing nanoparticle-modified polymer cementitious coatings. *Constr. Build. Mater.* **2021**, *299*, 123736. [CrossRef]
25. Ilango, N.K.; Gujar, P.; Nagesh, A.K.; Alex, A.; Ghosh, P. Interfacial adhesion mechanism between organic polymer coating and hydrating cement paste. *Cem. Concr. Compos.* **2020**, *115*, 103856. [CrossRef]
26. Shu, X.; Zhao, Y.; Liu, Z.; Zhao, C. A Study on the mix proportion of fiber-polymer composite reinforced cement-based grouting material. *Constr. Build. Mater.* **2022**, *328*, 127025. [CrossRef]
27. Basheer, P.A.M. *Surface Treatments for Concrete*; Keynote Lecture, Session on Preventive Measures COST 509 Workshop; European Commission: Seville, Spain, 1995.
28. American Concrete Institute. *Guide to Durable Concrete*; ACI Committee Report ACI 201.2R-01; American Concrete Institute: Farmington Hills, MI, USA, 1997.
29. Yu, H.; Da, B.; Ma, H.; Dou, X.; Wu, Z. Service life prediction of coral aggregate concrete structure under island reef environment. *Constr. Build. Mater.* **2020**, *246*, 118390. [CrossRef]
30. *GJB 1578-92*; Technical Standard for Airport Pavement Cement Concrete Mix Design. General Logistics Department of the Chinese People's Liberation Army: Beijing, China, 2018.
31. *GB10125-1997*; Corrosion Tests in Artificial Atmospheres-Salt Spray Test. The State Bureau of Quality and Technical Supervision: Beijing, China, 1997.
32. *JTG E30-2020*; Testing Methods of Cement and Concrete for Highway Engineering. Ministry of Transport of the People's Republic of China: Beijing, China, 2020.
33. Zhuang, S.Y.; Wang, Q.; Zhang, M.Z. Water absorption behaviour of concrete: Novel experimental findings and model characterization. *J. Build. Eng.* **2022**, *53*, 104602. [CrossRef]
34. Franzoni, E.; Pigino, B.; Pistolesi, C. Ethyl silicate for surface protection of concrete: Performance in comparison with other inorganic surface treatments. *Cem. Concr. Compos.* **2013**, *44*, 69–76. [CrossRef]
35. Zhu, Y.G.; Kou, S.C.; Poon, C.S.; Dai, J.G.; Li, Q.Y. Influence of silane-based water repellent on thedurability properties of recycled aggregate concrete. *Cem. Concr. Compos.* **2013**, *35*, 32–38. [CrossRef]
36. Zhang, W.M.; Gong, S.; Kang, B. Surface Corrosion and Microstructure Degradation of Calcium Sulfoaluminate Cement Subjected to Wet-Dry Cycles in Sulfate Solution. *Adv. Mater. Sci. Eng.* **2017**, *2017*, 1464619. [CrossRef]
37. Brown, P.W. Thaumasite formation and other forms of sulfate attack. *Cem. Concr. Compos.* **2022**, *24*, 301–303. [CrossRef]
38. Bonakdar, A.; Mobasher, B. Multi-parameter study of external sulfate attack in blended cement materials. *Constr. Build. Mater.* **2010**, *24*, 61–70. [CrossRef]
39. Ozaki, S.; Sugata, N. Long-term durability of reinforced concrete submerged in the sea. In *Concrete under Severe Conditions 2: Environment and Loading, Proceedings of the Second International Conference on Concrete under Severe Conditions, Tromsø, Norway, 21–24 June 1998*; Odd, K.S., Gjørv, E., Banthia, N., Eds.; CRC Press: Tromsø, Norway, 1998; pp. 448–457.
40. Wu, F.; Chen, X.Q.; Chen, J.A. Abrasion resistance enhancement of concrete using surface treatment methods. *Tribol. Int.* **2023**, *179*, 108180. [CrossRef]
41. Baltazar, L.; Santana, J.; Lopes, B.; Rodrigues, M.P.; Correia, J.R. Surface skin protection of concrete with silicate-based impregnations: Influence of the substrate roughness and moisture. *Constr. Build. Mater.* **2014**, *70*, 191–200. [CrossRef]
42. Money, M.W. *Instrumental Falling Weight Impact Testing of Polymethacrylate and High Density Polyethylene*; University of London: London, UK, 1988.
43. *ASTM D3763-10*; High Speed Puncture Properties of Plastics Using Load and Displacement Sensors. ASTM: West Conshohocken, PA, USA, 2010.

44. Song, Z.; Xue, X.; Li, Y.; Yang, J.; He, Z.; Shen, S. Experimental exploration of the waterproofing mechanism of inorganic sodium silicate-based concrete sealers. *Constr. Build. Mater.* **2016**, *104*, 276–283. [CrossRef]
45. Pan, X.; Shi, Z.; Shi, C.; Hu, X.; Wu, L. Interactions between inorganic surface treatment agents and matrix of Portland cement-based materials. *Constr. Build. Mater.* **2016**, *113*, 721–731. [CrossRef]
46. Song, Z.; Lu, Z.; Lai, Z. Influence of hydrophobic coating on freeze-thaw cycle resistance of cement mortar. *Adv. Mater. Sci. Eng.* **2019**, *2019*, 8979864. [CrossRef]
47. Li, Y.; Chen, C.; Hou, H.; Cheng, Y.; Gao, H.; Zhang, P.; Liu, T. The influence of spraying strategy on the dynamic response of polyurea-coated metal plates to localized air blast loading: Experimental investigations. *Polymers* **2019**, *11*, 1888. [CrossRef]
48. Bahei-El-Din, Y.A.; Dvorak, G.J. Behavior of sandwich plates reinforced with polyurethane/polyurea interlayers under blast loads. *J. Sandw. Struct. Mater.* **2007**, *9*, 261–281. [CrossRef]
49. Gardner, N.; Wang, E.; Kumar, P.; Shukla, A. Blast mitigation in a sandwich composite using graded core and polyurea interlayer. *Exp. Mech.* **2012**, *52*, 119–133. [CrossRef]

Disclaimer/Publisher's Note: The statements, opinions and data contained in all publications are solely those of the individual author(s) and contributor(s) and not of MDPI and/or the editor(s). MDPI and/or the editor(s) disclaim responsibility for any injury to people or property resulting from any ideas, methods, instructions or products referred to in the content.

Article

Evaluating the Impact of Concrete Design on the Effectiveness of the Electrochemical Chloride Extraction Process

Zofia Szweda

Department of Building Structures, Faculty of Civil Engineering, Silesian University of Technology, 44-100 Gliwice, Poland; zofia.szweda@polsl.pl

Abstract: This paper presents a simple comparative method for evaluating the impact of concrete design on the effectiveness of repair with the electrochemical chloride extraction (ECE) process of reinforced concrete structures. This comparison covered two concretes with different types of used cement. Penetration of chloride ions to induce corrosion processes was accelerated with the electric field. However, the corrosion process itself occurred naturally. When the corrosion process was found to pose a risk to the reinforcement, the profile of chloride ion concentration was determined at the depth of concrete cover. Corrosion current intensity during migration and extraction processes of chloride ions was measured with the LPR method. Then, this serious condition for the structure was repaired with electrochemical chloride extraction. Rates of chloride extraction were determined from the derived concentration profiles. It should be noted that the critical concentration $C_{crit} = 0.4\%$ at the rebar surface was reached after 21 days of the migration process. Moreover, after the same time of extraction, the concentration was reduced by 95% at the rebar surface, which could suggest that extraction rate was slower than chloride ion migration to concrete within the electric field. Using the migration coefficient for predicting the extraction time, as well as ignoring the variability of the extraction coefficient and the initial concentration over time, may result in too short or unnecessarily long extraction times.

Keywords: chloride migration; corrosion risk; chloride extraction time; electrochemical chloride extraction; reinforced concrete structures; repair; rehabilitation

Citation: Szweda, Z. Evaluating the Impact of Concrete Design on the Effectiveness of the Electrochemical Chloride Extraction Process. *Materials* 2023, 16, 666. https://doi.org/10.3390/ma16020666

Academic Editor: Andreas Lampropoulos

Received: 31 October 2022
Revised: 31 December 2022
Accepted: 4 January 2023
Published: 10 January 2023

Copyright: © 2023 by the author. Licensee MDPI, Basel, Switzerland. This article is an open access article distributed under the terms and conditions of the Creative Commons Attribution (CC BY) license (https:// creativecommons.org/licenses/by/ 4.0/).

1. Introduction

The presence of chloride ions in de-icing agents and the coastal environment, as well as groundwater and municipal wastewater, is defined as the main cause of the risk of corrosion of reinforced concrete structures. Exceeding the critical concentration of chlorides at the depth of reinforcement leads to the initiation and development of a very dangerous type of steel corrosion, the so-called pitting corrosion. Chloride ions contribute to the destruction of the protective layer of passive reinforcing steel in concrete. Then, the mechanism of destruction consists of reducing the cross-section of the reinforcing bars and the accumulation of corrosion products at the surface of steel. The reduced cross-section leads to structural failure. On the other hand, the accumulated products of corrosion cause the bursting of the concrete cover. As the reinforcement is exposed to weather conditions, the corrosion processes accelerate [1].

Unfortunately, this critical content of chloride ions at the surface of the reinforcement is not clearly defined. According to the European [2] and British Standards [3] $C_{crit} = 0.4\%$ of cement by weight in reinforced concrete structures and $C_{crit} = 0.1\%$ of cement by weight in prestressed structures are considered as the critical chloride content, at which corrosion process can develop. The American Standards [4,5] allow for $C_{crit} = 0.1\%$ of cement by weight in reinforced concrete structures and $C_{crit} = 0.06\%$ of cement by weight in prestressed structures. Moreover, the standard [6] allows for $C_{crit} = 0.2\%$ of cement by weight in reinforced concrete structures and $C_{crit} = 0.08\%$ of cement by weight in prestressed structures.

The content of hydroxide ions in the concrete pore liquid is an additional factor which affects the initiation rate of corrosion processes. Haussman [7] developed the relationship between chloride and hydroxide ions by determining critical value, at which corrosion processes can be observed, as the ratio of chloride to hydroxide ions equal to 0.6. In numerous papers [8–10], the relationship between the content of free chloride ions and concentration of hydroxyl ions in concrete was used to express critical content of chlorides, but its value varied within a wide range from 0.3 to 40.

Concrete additives, such as pulverised fly ash and ground granulated blast furnace slag, affected the threshold values of chloride ion concentration. Differences were also observed in determining the threshold value of chloride ion content depending on whether they were added to concrete mix or originated from the external environment [11].

Another difficulty is the fact that corrosion processes of the reinforcement are not visible to the naked eye because they run under the surface of concrete cover. Qualitative evaluation of reinforcement corrosion can be performed with such electrochemical methods such as measuring potential and resistivity of concrete cover [12–14].

The assessed corrosion risk to the structure requires its immediate rehabilitation [15]. Since 1985, the ECE (Electrochemical Chloride Extraction) method has been used in many countries for reinforced concrete structures contaminated with chlorides [16–18].

It is also very important to precisely determine the development of corrosion in reinforced concrete structures before and after the application of electrochemical repair method. The repair time should be predicted more precisely and its effectiveness should be evaluated while analysing corrosion measurements using the LPR and EIS techniques [19].

Duration of the extraction process has been so far based on the migration coefficient determined from the migration process of chlorides in concrete [20]. Apart from poor effectiveness and repeatability of the standard methods of determining the migration coefficient, the obtained values of this coefficient could not be used to precisely determine the duration of the extraction process [21]. In the majority of papers describing the extraction method, this process was performed on laboratory specimens of cement pastes and grouts. Some papers presented the works conducted on the specimens of concrete with the ribbed reinforcement [22], however the extraction process is usually performed on the specimens of concrete mix, into which chloride ions are added directly after being dissolved in batched water [23]. There are still very few examples of tests performed under the laboratory conditions similar to the original ones, that is, with concrete and ribbed reinforcement and associated with chloride ions migrating from the external environment. [24].

This paper presents a simple comparative method for evaluating the impact of concrete design on the effectiveness of electrochemical extraction chlorides from concrete. This method was introduced in the paper [19] for one concrete. In this paper, the method was applied to a different type of concrete and used to compare both concretes at the same time. The tests described in that paper had two aims: the first was to test the method on the second concrete (it will be necessary to test this method on many different concretes); the second aim was to compare the effectiveness of Electrochemical Chloride Extraction (ECE) in both concretes using this method.

In this method, chloride ions were at first added from the external environment. The process of ion penetration was accelerated with the action of electric field. The specimens consisted of tested concrete types and contained reinforcement made of ribbed steel. When the corrosion process was found to pose a risk to the reinforcement, the profile of chloride ion concentration was determined at the depth of concrete cover. Then, this serious condition for the structure was repaired with electrochemical chloride extraction (ECE). Corrosion current intensity during migration and extraction processes of chloride ions were measured with the LPR method. After obtaining the satisfactory values of corrosion current, distribution of chloride ion concentration was checked at the depth of concrete cover. Then, the extraction coefficient was calculated from the distributed concentration of chloride ions at the depth of concrete cover after the relevant duration of this process. Linear variability of such values as boundary concentration of chloride ions and extraction coefficient over time were also included in this method in a simplified way. The extraction process was numerically modelled. This new approach is used to more precisely determine values of extraction coefficients for existing structures and to assess the effective duration of chloride extractions both for the existing structure and the process of designing and verifying properties of new concrete types.

2. Materials

The tests were performed on two types of concrete mix. Concretes C1 and C2 were made of ordinary concrete with different type of used cement. Specimens made of the same preparations were used in the work [25] to determine values of the diffusion coefficient of chloride ions. Concrete C1 contained CEM I 42.5 R cement. While concrete C2 contained blast-furnace cement with lowered content of alkalis—CEM III/A 42.5 N-LH/HSR/NA. The detailed compositions of mixes are presented in Table 1.

Table 1. Composition, Properties and compressive strength of concrete mixtures.

Constituent	C1	C2
	CEM I 42.5 R	CEM III/A 42.5 N-LH/HSR/NA
Cement	324	
Sand (0–2 mm)	722	
Gravel (2–8 mm)	512	
Gravel (8–16 mm)	681	
Water	162	
w/c	0.5	
Compressive strength f_{cm} MP	54.2	49.5
Volume weight γ_b kg/m^3	2271	2269

3. Test Methods

All of the tests were conducted and the specimens were prepared at the Laboratory of Civil Engineering of the Silesian University of Technology. Six cylindrical-shaped test specimens 1 with a diameter of 100 mm and a height of 60 mm were prepared from each concrete type. Ribbed rebars 2 with ø12 mm, made of steel B500SP, were placed in these specimens in the direction perpendicular to the cylinder axis. The most common diameter used for the main reinforcement was 12 mm and the concrete cover of 20 mm was adjusted to this diameter. The specimens were prepared as described in the paper [19]. Figure 1 shows the specimens prepared for testing during curing, before attaching plastic tanks made of PVC pipes to the upper surface of these elements.

Figure 1. Specimens of both types of concrete prepared for testing during curing.

3.1. Migration of Chloride Ions Accelerated with the LPR Method with Simultaneous Control over Corrosion

Prior to chloride diffusion to concrete accelerated by the electric field, the polarization tests were performed on all of the specimens with the LPR method (Figure 2) to determine corrosion potential of the specimen reinforcement in the passive state. Electrochemical measurements are greatly influenced by humidity and temperature. Therefore, all electrochemical studies were made under the same conditions. Before the tests, the specimens were immersed in water for 72 h in order to stabilize the half-cell potential and avoid overload in the potentiostat. In this case, the corrosion rate was not controlled by oxygen diffusion to the steel surface [26]. The measurements were performed in a three-electrode arrangement, where steel rebar was used as the working electrode 2. The counter electrode 4 was made of stainless-steel sheet, whose shape was adjusted to the test specimens. The reference electrode 5 (Cl^-/AgCl,Ag) was placed on the cylinder surface. It rested against walls of the plastic tank tightly fixed to the specimen. To provide satisfactory conductivity, a felt separator soaked with distilled water was placed both on the top element of the tank and in the bottom tank. The specimens were soaked with water by immersion for ca. 72 h. Then, the LPR tests were performed using the potentiostat 6 Gamry Reference 600 by Gamry Instruments, Warminster, Pennsylvania, United States of America in the potentiostatic mode within a range of frequencies 10 mHz–100 kHz at an amplitude of 10 mV over the corrosion potential of the reinforcement.

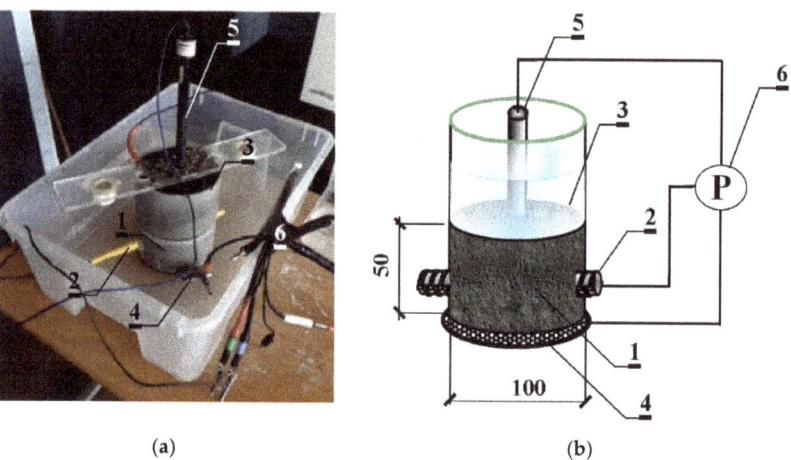

Figure 2. The applied test stand for polarization tests with the LPR method (**a**) view; (**b**) scheme: 1—concrete test specimen, 2—ribbed rebar ø12 mm made of steel B500S (working electrode), 3—plastic tank, 4—auxiliary electrode, 5—(Cl^-/AgCl,Ag) electrode as the reference electrode, 6—Gamry Reference 600 potentiostat with a computer unit and Gamry software.

After taking from water, the specimens were connected to the potentiostat 6 and changes in gradually stabilizing potential were observed with the reference electrode 5 for 60–120 min. When potential changes were at the level of 0.1 mV/s, LPR methods were performed on the steel reinforcement in concrete. The reinforcement was polarized at a rate of 1 mV/s within the range of potential changes from −150 mV to +50 mV regarding the corrosion potential.

The very long duration of chloride diffusion in concrete was shortened with the accelerated electromigration of chloride ions, whereas corrosion processes induced by the critical content of chloride ions in concrete took place naturally. The test specimens 1 were subjected to accelerated migration of chlorides using the electric field. Eighteen specimens grouped into six elements connected to the power source were simultaneously subjected to testing. The tests were performed in two independent test sets. Each test set was supplied with 18 V direct current 2. The specimens were placed on a big rectangular electrode (anode) made of titanium mesh 3 (coated with a thin layer of platinum) immersed in tap water at the bottom of a shallow tank 4. Plastic tanks 5 placed on the top were filled with 3% NaCl to a height of 7 cm. A round stainless-steel electrode (cathode) 6 with a diameter adjusted to the tank hole was placed on the top each specimen inside each tank. The process of chloride electrodiffusion was interrupted every 7 days to monitor the development of corrosion processes of the reinforcement by measuring corrosion potential. Electrochemical measurements were taken each time after 3 days from switching off the electricity supply to avoid polarization of the tested reinforcement (Figure 3).

Figure 3. The test stand for migration of chloride ions to concrete accelerated with the electric field: (**a**) 1—concrete test specimen, 2—electric circuit of 18 V, 3—titanic anode coated with platinum, 4—tank with distilled water, 5—small plastic tanks with 3% NaCl, 6—stainless steel cathode; (**b**) stabilised laboratory feeder KP 16,103 used as source of 18 V direct current.

The results from polarization tests on the selected specimens C1.1; C1.2; C2.1 and C2.2 are presented in Figures 4–6.

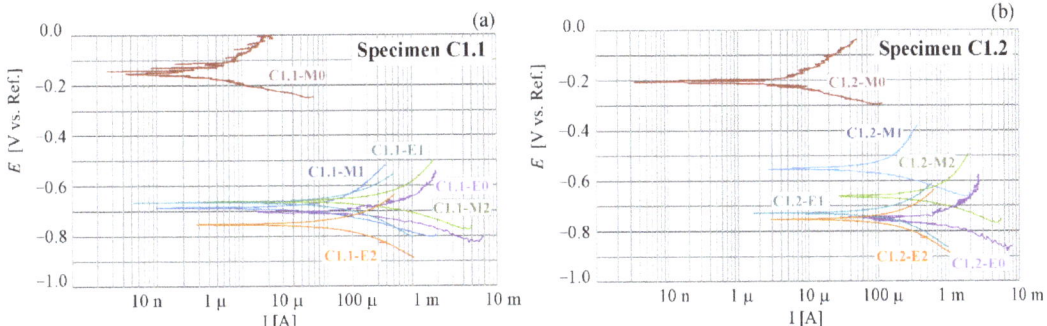

Figure 4. Potentiodynamic polarization curves for steel reinforcement in concrete C1 obtained for selected specimens: (**a**) C1.1 and (**b**) C1.2; M0 before chloride migration, M1 after 7 days, M2 after 14 days of chloride migration and E0 after 21 days of chloride migration and E1 after 10 days, E2 after 21 days of chloride extraction.

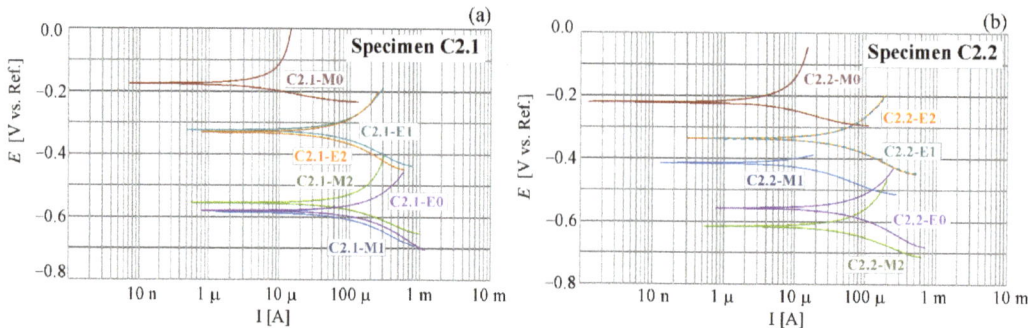

Figure 5. Potentiodynamic polarization curves for steel reinforcement in concrete C2 obtained for selected specimens: (**a**) C2.1 and (**b**) C2.2; M0 before chloride migration, M1 after 7 days, M2 after 14 days of chloride migration and E0 after 21 days of chloride migration and E1 after 10 days, E2 after 21 days of chloride extraction.

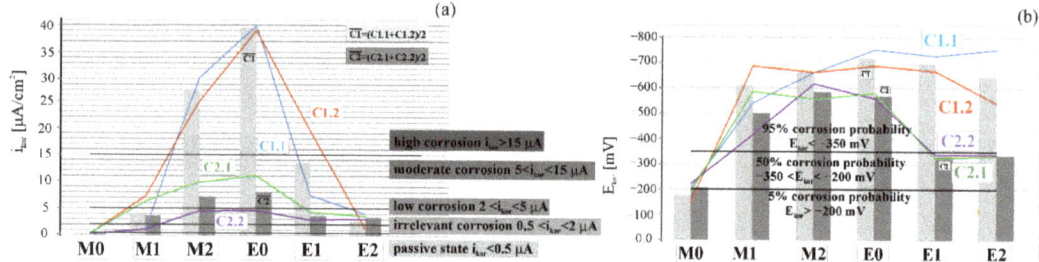

Figure 6. Distribution: (**a**) of corrosion current densities and (**b**) corrosion potential obtained selected specimen C1.1, C1.2, C2.1 and C2.2: M0—before chloride migration), M1—after 7 days, M2—after 14 days of migration, E0—after 21 days of migration, E1—after 10 days of chloride extraction and E2—after 21 days of extraction.

3.2. Material Tests—Determination of Distribution of Chloride and Hydroxide Ion Concentration at the Depth of Concrete Cover

When initiation of corrosion processes was found on the basis of interpretation of polarization measures with the LPR method, electrodiffusion was not continued and chloride profile at the depth of reinforcement cover was determined. For that purpose, "Profile Grinding Kit" by German Instruments was used to collect layers of concrete from the cover of two specimens made of each type of concrete. Crushed concrete was collected from 10 levels by 2 mm-thick layers. Later, the material from two specimens collected from the same level was mixed to average results of chloride ion concentrations in concrete (Figure 7).

(a) (b)

Figure 7. Profile Grinding Kit for concrete: (**a**) crushed concrete, (**b**) reference specimens from which concrete was collected by layers with this kit.

By mixing crushed concrete with distilled water (1:2 ratio), ten solutions were prepared from each of the three specimens, which roughly modelled pore solution and represented averaged chemical properties of concrete tested. Concentrations of chloride ions in these solutions were measured with the multi-functional multimeter CX-701 by Elmetron with an ion-selective electrode to determine concentrations of chloride ions. Concentrations of chloride ions obtained from the chemical analysis of the tested solutions were then converted into the mean volumetric concentrations in concrete 2 mm sampling sections of total chlorides ions ($\overline{C}_{t\ exp}$ (mole/m³)) and their concentrations (C (%)) expressed for reference as a percentage of the weight of cement in concrete according to the following expressions:

$$\overline{C}_{t\ exp} = \frac{m_{t\ exp}}{V_c} = \frac{2\ c_{sol} \cdot \rho_{cb}}{\rho_w}, \quad C = \frac{\overline{C}_{t\ exp}}{\rho_{cem}}, \quad (1)$$

where c_{sol} is chloride ion concentration (kg/m³) determined in the tested solution, $m_{t\ exp}$—mass (kg) of chlorides in the powdered concrete drilled from a sampling section, V_c—volume (m³) of the drilled concrete from sampling section in an intact state, ρ_{cb}—bulk density (kg/m³) of concrete, $\rho_w = 1000$ kg/m³—density of water, ρ_{cem}—density (kg/m³) of cement per a concrete unit volume. Precise details of the used experimental and calculation procedure for the determination of $\overline{C}_{t\ exp}$ can be found in the article [27]. The stationery pH meter was used for simultaneous measurements conducted for all ten solutions of pore water. Calculated concentrations and pH values are shown as diagrams in Figure 4a,b.

The determined pH values were used to verify the Hausmann criterion, whose simplified version is expressed by the following expression:

$$\left.\frac{[Cl^-]}{[OH^-]}\right|_{crit.} \leq 0.6 \quad (2)$$

Molar concentration of chloride ions $[Cl^-]$ was calculated by taking $M_{Cl} = 0.035453$ kg-mole as molar mass of chloride, and molar concentration of hydroxide ions $[OH^-]$ was defined as pH function from the relationship $[OH^-] = 10^{pH-14}$. Calculated concentrations and pH values are shown as diagrams in Figure 8c,d.

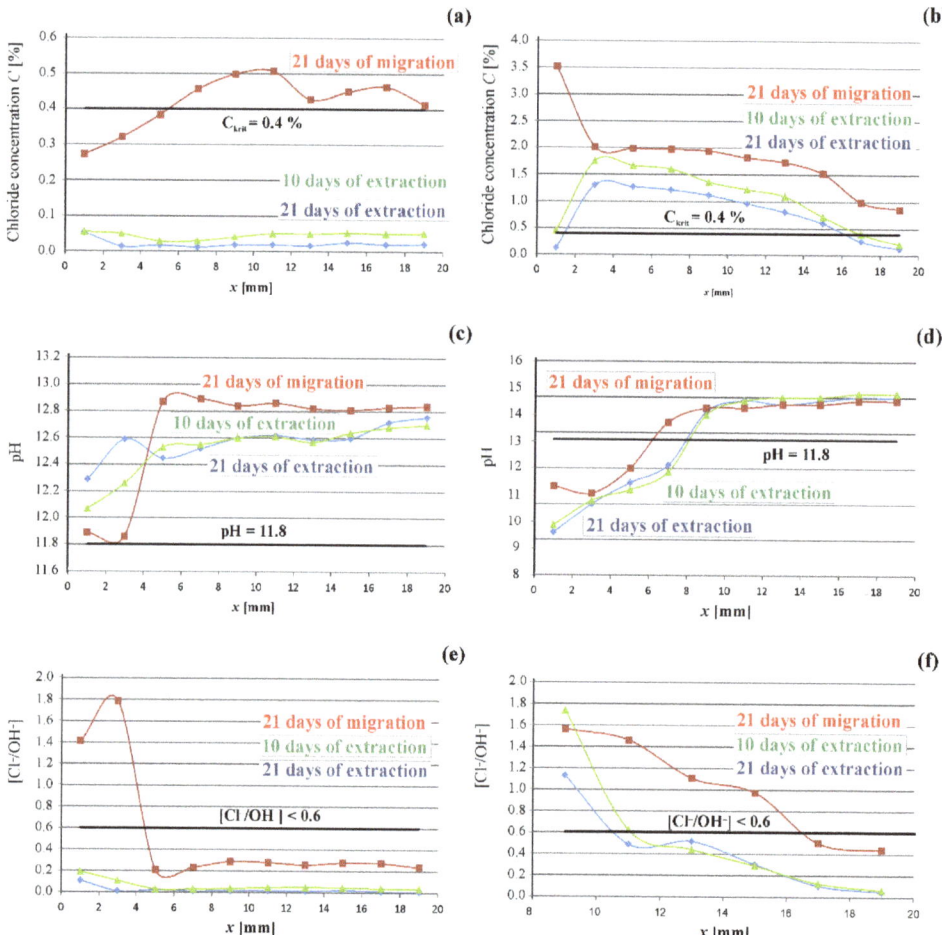

Figure 8. Test results obtained for reinforcement concrete C1 and C2 after 21 days of migration chloride ions, after 10 days and 21 days of extraction processes: (**a**,**b**) profiles of chloride concentrations; (**c**,**d**) distribution of pH values; and (**e**,**f**) distributions in the direction x of concrete cover thickness, values of concentration ratios of chloride and hydroxide ions—the Hausman criterion.

3.3. Electrochemical Chloride Extraction (ECE) with Simultaneous Control over Corrosion Processes with the Electrochemical Method

Assuming that chloride profile in relation to concrete cover depth in all of the tested specimens was similar to profiles from two tested specimens of each concrete type, the electrochemical extraction began. Such an assumption is necessary to continue testing the specimens because they are destroyed while concentration is determined. However, the averaged properties of the tested concrete were obtained by combining material from two specimens. The extraction process was performed simultaneously in six specimens 1 prepared from one concrete type and placed at a test stand similar to the one used in the electromigration process. This time, however, rebars 2 used as the cathode were connected to the negative pole 6 of 18 V DC, and the anode mesh 7 was connected to the positive pole. Tap water, into which the mesh was immersed, was used as the electrolyte. The specimens were protected with foil against drying, and water evaporating during the tests was refilled. For the group of three specimens, chlorides were extracted from each concrete type for

10 days. A week after completing the extraction process, polarization of the reinforcement was measured to evaluate effects of this process. Then, chloride profiles and pH distribution in relation to the concrete cover depth were determined within the group of two tested specimens in a way similar as described above. Extraction for the group of other two specimens of each concrete type was longer, it lasted for 11 days. Again, a week after completing the extraction process, by analogy to the above activities, polarization of the reinforcement was measured and the pore solution was tested. Figures 4 and 5 illustrate results from the polarization measurements, whereas chloride profiles and distribution of pH values are shown in Figure 8.

3.4. Determination of Coefficients of Migration and Extraction in Concrete

Value (D_e (m^2/s))—coefficient of chloride extraction was determined, similarly to in the paper [21], on the basis of matching the diagram of chloride concentration obtained from the calculated distribution of chloride ions according to the solution of diffusion equation (where according to [21] migration coefficient was introduced into the diffusion equation) to the concentration of these ion determined during the tests and expressed with reference to the cement weight:

$$C_{cal} = C_{0,cal}\left(1 - \text{erf}\frac{x}{2\sqrt{D_{m,e}\,t}}\right) \qquad (3)$$

($C_{0,cal}$ (%)) is calculated concentration of chloride ions at the element edge with reference to the weight of cement, erf—the Gauss error function, (t(s))—duration of chloride ion migration or extraction from concrete.

To determine the most convergent computational and experimental results, the lowest s—value the mean square error, was calculated from the following expression:

$$s = \sqrt{\frac{\sum_{i=1}^{n}[C_{cal}-C]^2}{n-1}} \qquad (4)$$

(C(%)) is measured while measuring the concentration of chloride ions within a distance x from the element edge, C_{cal}—chloride ion concentration within a distance x from the edge element calculated from the Equation (1), (%) chloride weight to cement weight, n—number of concrete layers, from which chloride concentration is determined. The calculated values of extraction coefficients are presented in Table 2.

Table 2. Numerical modeling of the extraction process.

		24 (1)	96 (4)	120 (5)	144 (6)	192 (8)	216 (9)	240 (10)	288 (12)	384 (16)	504 (21)	650 (27)
Time of Extraction t (Hour (Days))												
initial concentration (%)	C1	0.6	0.42	0.36	0.3	0.18	0.12	0.06	0.06	0.04	0.04	-
	C2	3.3	3.2	3.17	3.13	3.07	3.03	3.00	2.91	2.73	2.50	2.27
coefficient of extraction (10^{-10} m^2/s)	C1	5	5	4	4	3	2	2	2	1	1	-
	C2	0.9	0.83	0.81	0.79	0.74	0.72	0.70	0.63	0.48	0.3	0.12

4. Discussion of the Results Obtained

4.1. Results of Polarization Measurements for Reinforcement

By registering the potential changes as a function of the system response expressed as current density, a polarization curve is obtained (Figures 4 and 5). The semi-logarithmic polarization curve is the basis for the graphical determination of the corrosion current density. The results of such tests are the corrosive current densities, clearly defining the corrosion rate of the reinforcement. The corrosion current (i_{corr} (µA)) can be calculated via polarization resistance (R_p (kΩ)) obtained by LPR measurement according to the Stern-Geary equations [28]

$$R_p = \left.\frac{dE}{di}\right|_{i\to 0,\ E\to E_{corr}}, i_{corr} = \frac{b_a b_c}{2.303 R_p (b_a + b_c)},\qquad(5)$$

where b_a and b_c are constants of anodic and cathodic reactions, respectively, coefficients of rectilinear slope for segments of polarization curves—anodic b_a and cathodic b_c.

The corrosion current density clearly determines the corrosion intensity of steel because, according to Faraday's law, the mass of losses (Δm (mg)) is proportional to the flowing current (I_{corr} (μA/cm^2))

$$\Delta m = k I_{corr} t,\ I_{corr} = \frac{i_{corr}}{A},\qquad(6)$$

where k is electrochemical equivalent, t—time. The above relationship shows the correlation of the corrosion current density with the linear corrosion rate (V_r (mm/year) expressed as follows:

$$V_r = 0.011\ i_{corr}\qquad(7)$$

Corrosion rate (V_r (mm/year)) is determined from the average cross-section loss around the bar circumference, in mm, per 1 operational year of the structure. The detailed results from the analysis with the calculated densities for corrosion current are shown in Tables A1 and A2 of the Appendix A. For an easier comparative evaluation of the obtained test results, corrosion current densities i_{corr} and E_{corr} based on values from Tables A1 and A2, are presented in Figure 5.

The LPR tests were conducted on each test element from each measuring series. The whole period of testing produced a total of 56 polarization curves, in which exemplary shapes for four selected measuring elements are illustrated in Figures 4 and 5. The specimens, for which a number of measurements were taken during the whole research process, were analysed. They were C1.1, C1.2 and C2.1, C2.2 specimens (M0: reference measurements prior to chloride migration to concrete, M1: measurement taken after 7 days of chloride migration t concrete, M2: measurement taken after another 7 days of chloride migration to concrete, E0: measurement taken after another 7 days of chloride migration to concrete and after observation of corrosion in the specimens directly prior to extraction, E1: measurement taken after 10 days of chloride extraction, E2: measurement taken after another 11 days of chloride extraction).

A similar change in distribution of polarization curves over time was observed for the specimens C1.1 and C1.2 (Figure 4). After the first reference measurement of corrosion potential, taken prior to migration ($E_{corr} = 205(C1.1\_M0); 151(C1.2_{M0})$ mV), other measurements taken after migration, and also after chloride extraction from concrete produced the results close to the mean value of corrosion potential $\overline{E}_{corr} = 682(C1.1)$ mV for the specimen C1.1 and $\overline{E}_{corr} = 648(C1.2)$ mV for the specimen C1.2. On the other hand, values of the measured corrosion current ranged from the reference value measured prior to migration ($i_{corr} = 0.21(C1.1\_M0); 0.18(C1.2\_M0)$ μA) to the maximum value obtained prior to extraction ($i_{corr,max} = 38.94(C1.1\_E0); 37.98(C1.2\_E0)$ μA). Based on the paper [29], reference values of corrosion current in the specimens C1 could be regarded as the values which signalled the unexpected corrosion, and the maximum values $i_{corr,max}$ could indicate high corrosion activity.

Some similarity was also found in the behaviour of the specimens C2.1 and C2.2 made of concrete C2, but the observed trend was not the same as in the specimens made of concrete C1 (Figure 5). After the first reference measurement of corrosion potential, taken prior to migration ($E_{corr} = 221(C2.1\_M0); 191(C2.2\_M0)$ mV), other measurements taken after migration produced the results close to the mean value of corrosion potential $\overline{E}_{corr} = 531(C2.1)$ mV for the specimen C2.1 and $\overline{E}_{corr} = 574(C2.2)$ mV for the specimen C2.2. Then, during extraction these values fluctuated around

$\overline{E}_{corr} = 338 (C2.1)$ mV for the specimen C2.1 and $\overline{E}_{corr} = 328 (C2.2)$ mV for the specimen C2.2. On the other hand, values of the measured corrosion current ranged from the reference value ($i_{corr} = 0.23 (C2.1\_M0); 0.28 (C2.2\_M0)$ µA) to the maximum value ($i_{corr} = 11.03$ (C2.2_E0) µA). Based on the paper [29], reference values of corrosion current could be regarded as the values which signalled the unexpected corrosion, and the maximum values $i_{corr,max}$ measured in the specimens made of concrete C2 indicated the moderate corrosion activity.

Figure 6a presents a comparison of results from six measurements of corrosion current density i_{corr} of the steel reinforcement in concrete from two chosen test elements made of tested concretes. Figure 6b shows a comparison of results from six measurements of corrosion potential E_{corr} of the steel reinforcement in concrete from two chosen test elements made of tested concretes. Taking into account the assumptions described in the papers [29,30], the first reference measurement taken prior to migration indicated that both corrosion potential (E_{corr} ($\overline{E}_{corr} = 178(C1); 206(C2)$) < 350 mV) and corrosion current intensity ($i_{corr} < 0.3$ µA) suggested the passive state of all test elements. Another measurement taken after 7 days of chloride ions migration under the accelerated action of the electric field and 3 days after switching off the system indicated the onset of corrosion in all four specimens on the basis of corrosion potential values and intensity of corrosion current. However, a similar increase in average intensity of corrosion current (C1($\overline{\Delta i}_{corr}$ = 3.91; 3.41 (C2) µA) was observed for both types of concrete. However, in the case of the first type of concrete values of corrosion, the potential increased by almost 1.5 times greater than in the second type ($\overline{\Delta E}_{corr}$ = 435 (C1); 294(C2) mV). After another 7-day charging with chloride ions, a massive increase was found for concrete C1(C1: $\overline{\Delta i}_{corr}$ = 22.93 µA), and for concrete C2 an increase was similar as after the first week of charging and was (C2: $\overline{\Delta i}_{corr}$ = 3.44 µA). Corrosion potential, however, increased nearly twice in concrete C2: $\overline{\Delta E}_{corr}$ = 86 mV than in concrete C1: $\overline{\Delta E}_{corr}$ = 48.5 mV). Then, after another 7 days of charging (and 3 days of waiting for restraining rebars) and prior to extraction, a control measurement of polarization was taken. Values obtained from this measurement suggested the developed corrosion in all of the specimens. A particularly high intensity of corrosion current i_{corr} = 38.94 µA was observed in the specimens C1.1 and C1.2 i_{corr} = 37.98 µA. Much lower values of corrosion current (C2.1(i_{corr} = 11.03 µA); C2.2(i_{corr} = 4.59 µA)) were obtained for the specimens made of concrete C2. After the first 10-day extraction, a significant drop in corrosion current intensity was observed in all four specimens. The most significant drop was found in the specimen C1.1 (Δi_{corr} = 31.53 µA) at the simultaneous drop in corrosion potential (ΔE_{corr} = 24 mV), and the smallest drop was in the specimen C2.1 (Δi_{corr} = 1.68 µA) at significantly reduced potential (ΔE_{corr} = 221 mV). After another extraction, average values of corrosion current in concrete C1 ($\overline{\Delta i}_{corr}$ = 10.87 µA) and corrosion potential ($\overline{\Delta E}_{corr}$ = 50 mV) dropped. In concrete C2, however, the average value of corrosion current slightly dropped ($\overline{\Delta i}_{corr}$ = 0.28 µA) and the average value of corrosion potential ($\overline{\Delta E}_{corr}$ = 3.5 mV) slightly increased. As it could be observed, despite of high values of corrosion current in the specimens made from concrete C1, this intensity could be, however, the average value of corrosion current slightly dropped ($\overline{\Delta i}_{corr}$ = 0.28 µA) and the average value of corrosion potential ($\overline{\Delta E}_{corr}$ = 3.5 mV) slightly increased. As it could be observed, despite the high values of corrosion current in the specimens made from concrete C1, this intensity could be "reduced" after extraction, and consequently corrosion could be inhibited due to a 10-day extraction and concentration of chloride ions on steel surface lower than the value recommended by the standards: [2,3] amounting to C_{crit} = 0.4% of cement by weight. Another week of extraction did not bring such spectacular changes in corrosion current intensity in the specimens tested, particularly in concrete C2. Although final values of corrosion potential in concrete C1(C1.1(E_{corr} = 540 mV); C1.2(E_{corr} = 752 mV) and C2 (C2.1(E_{corr} = 332 mV); C2.2(E_{corr} = 580 mV) generally showed a downward trend, when compared to maximum values measured for C1.1 (an increase by 0.1%) C1.2 (a drop by 22%), C2 (a drop by

22%); C2.2 (a drop by 40%) the majority of these results did not reach the value taken for passivated steel according to [29,30].

Although final values of corrosion current intensity in concrete C1(C1.1: i_{corr} = 3.96 µA; C1.2: i_{corr} = 1.11 µA) and C2(C2.1(i_{corr} = 3.07 µA); C2.2(i_{corr} = 3.56 µA) showed a downward trend, when compared to maximum values measured for C1.1 (a drop by 90%);) C1.2 (a drop by 23%);), C2 (a drop by 68%);); C2.2 (a drop by 33%) the majority of these results did not reach the value taken for passivated steel according to [29,30]. Similar trends were found in the papers [26,31–33].

4.2. Results from Material Tests on Concentration of Chloride and Hydroxide Ions in Concrete

Chloride concentration and pH were determined in pore solutions representing ten 2-mm layers. The test results obtained for two concretes are presented in Figure 8 as the distribution of chloride concentrations and pH values in the direction of the depth of reinforcement cover. As shown in Figure 9a, chloride concentration after 21 days since their electrodiffusion was at the level of ca. 0.4% [2] of cement by weight in concrete C1 near the rebar. Concentration of chloride ions along the total depth of concrete cover in concrete C2 exceeded the critical value. According to the standard criterion [2], the risk of reinforcement corrosion was probable in both cases. Measurements of reinforcement polarization in fact confirmed this assumption and indicated rather high values of corrosion current after 21 days of chloride migration to concrete, both in the specimens made of concrete C1 and C2. Howevrer, after a week of charging concrete with chloride ions corrosion current values were increasing in both concretes, which could suggest potential corrosion at concentrations of chloride ions lower than the standard concentration [2] C_{crit} = 0.4%. This deduction was confirmed by final measurements when concentration at the whole depth of concrete cover in C1 was lower than a stricter value C_{crit} = 0.1 recommended by the American standards [4,5]. The corrosion current, despite a significant drop by 90%, compared to the maximum value (the specimen C1.1) did not reach density of corrosion current assumed in the papers [29,30] as the value for steel with no risk of corrosion. When concentration of chloride ions in concrete C2 reached C_{crit} = 0.1% at the surface of reinforcing steel, corrosion current densities (C2.1(i_{corr} = 3.56 µA); C2.2(i_{corr} = 3.07 µA) were characteristic for steel with moderate corrosion activity according to the criterion described in the papers [29,30]. Very similar values were obtained for concrete C1 (C1.1(i_{corr} = 3.96 µA); C1.2(i_{corr} = 1.11 µA) at five times lower concentration of chloride ions C = 0.02% at the steel surface.

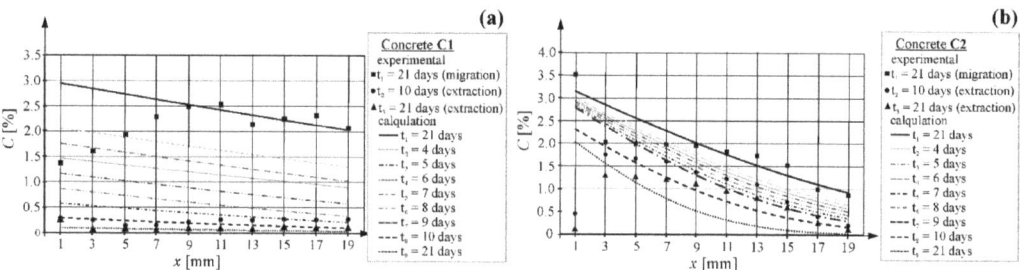

Figure 9. Distribution of chloride concentration in concrete obtained in the extraction process—calculated and obtained from tests: (**a**) concrete C1; (**b**) concrete C2.

Taking into account the additional presence of hydroxide ions, it can be observed that using the Hausmann criterion and evaluating the results from measuring corrosion current in concrete C1, the whole process of migration and extraction of chloride ions is safe and should not reach the state of corrosion risk according to this criterion. However, the measurements of corrosion did not confirm this assumption. A totally different interpretation could be made for the results obtained for the limit value of corrosion risk for the ratio $[Cl^-]/[OH^-] \leq 0.1$.

4.3. Results of Chloride Extraction from Concretes

A diagram in Figure 8a presenting the distribution of chloride ion concentration after 21 days of migration, and then after 10 days and 21 days of extraction in concrete C1 leads to conclusions that a 10-day electrochemical extraction clearly reduced concentration of chloride ions at the reinforcement surface. Another 11 days of desalination only slightly reduced the concentration of chloride ions.

Concrete C2 demonstrated a similar trend. A more considerable drop in chloride ion concentration was noticed during the first phase of extraction than in the second one (Figure 8b). These observations were consistent with the observations made by other researchers [18,19], who claimed a slowdown in chlorides migration over time.

Therefore, the extraction coefficient was expected not to be the constant value during electrochemical extraction of chlorides from concrete. Moreover, the concentration at the edge of the element was observed not to be constant, but it was decreasing as the extraction progressed. Moreover, the calculated values of chloride concentrations at the element edge did not agree with the values obtained from the chemical analysis. A drop in concentration at the element edge was caused by wetting the specimen with felt, which caused rediffusion of chloride ions at the edge of the test element. The equivalent shape of concentration curves is often based on tests on in situ diffusion of chlorides when chloride ions are eluted at the element edge because of periodic rainfalls.

Based on the concentrations of chloride ions determined in the tested concretes with the method described in point 3.4, the coefficients of chloride extraction in the tested concretes were determined and their values are shown in Table 2. Figure 9 presents distribution of chloride ion concentrations determined in the tested layers of concrete n accordance with point 3.2 and calculated from the Equations (3) and (4) with the approximation method on the basis of the lowest value of mean square error.

Based on chloride ion concentrations at the test element edge, which were determined by the approximation method of computational curves expressed by the Equations (3) and (4), a simplified theoretical linear change in chloride ion concentrations for two time intervals from 1 to 10 days and from 11 to 21 days of extraction ($t_1 = 24$ h, $t_2 = 240$ h, $t_3 = 504$ h) was taken for the computational values. Then, concentration at the element edge was calculated after the extraction time of 24; 96; 120; 144; 192; 219; 240; 288; 384; 504 h using a corresponding linear equation describing these changes (cf. Figure 10a). Similar changes in concentration at the element edge were observed for both types of concrete. A significant drop in concentration of chlorides was observed in the first-time interval, which is suggested by a wide angle of the line deviation. A drop was minor in the second stage of tests. A similar method was initially proposed by the author in the work [34] where good results were obtained for concrete with Portland cement.

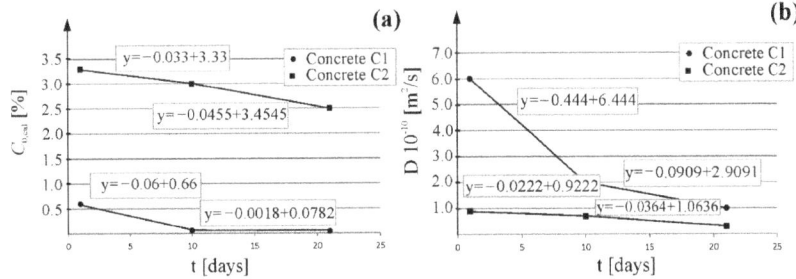

Figure 10. Theoretical linear change: (**a**) of the boundary chloride concentration, (**b**) of the coefficient of chloride extraction during extraction process.

Correspondingly, using previously determined values (D_e (m^2/s)) of the extraction coefficient at three different times (t_1 = 24 h, t_2 = 240 h, t_1 = 504 h), the theoretical linear change in that coefficient was assumed for two time intervals from 1 to 10 days and from 10 to 21 days of extraction. Linear functions expressing a change in extraction coefficient are illustrated in Figure 10b. A significant drop in the extraction coefficient was observed in the first-time interval for concrete C1, which is suggested by a wide angle of the line deviation. A minor drop in the extraction coefficient was observed in both time intervals for concrete C2, which could be associated with a slowdown in chloride extraction from concrete. However, the difference in values of that coefficient was so small, that it was within the measuring error tolerance. To model extraction, extraction coefficients (D_e (m^2/s)) were calculated after the following extraction times: 24; 96; 120; 144; 192; 219; 240; 288; 384; 504 h using corresponding linear equations describing these changes (cf. Figure 10b).

Table 2 presents calculated concentrations at the element edge and the extraction coefficient, determined for 11 times of extraction from both types of concrete.

4.4. Predicting Duration of Extraction Using the Coefficient of Chloride Migration and Extraction

Based on the above values of concentration and the element edge, coefficients of chloride extraction and the Equation (2), the development of extraction of chloride ions over time was modelled by plotting concentration curves of these ions at the thickness of 2 cm corresponding to concrete cover thickness. Figure 8a illustrates a group of curves presenting the distribution of chloride ion concentration at the concrete cover depth of 2 cm, plotted for the following times: 24, 96, 120, 144, 192, 219, 240, 288, 384, 504 h in concrete C1. The distribution of curves indicated that after 7 days of extraction, the concentration at the steel surface was already $C_{(x = 20)}$ = 0.1%, which should be a safe value with no corrosion risk according to the standard [2]. However, this value can present a corrosion risk according to stricter American standards [3,4]. This assumption was confirmed by results from the corrosion tests performed after 10 days of extractions. The determined densities of corrosion current C1.1(i_{corr} = 7.40 µA); C1.2(i_{corr} = 19.40 µA) were characteristic for steel with moderate corrosion activity according to the criterion described in the papers [29,30]. Another 11 days of chloride extraction did not produce such spectacular effects, but concentrations of chloride ions at the rebar surface achieved $C_{(x=20)}$ = 0.01% from the model (and $C_{(x = 20)}$ = 0.02% from the tests). Both of these values were lower than the boundary values specified in the standards [3,4] for compressed structures. Densities of corrosion current C1.1(i_{corr} = 3.96 µA); C1.2(i_{corr} = 1.11 µA) obtained from the tests performed after 21 days of extraction, indicated low corrosion activity according to the criterion described in the papers [29,30]. This process was effective with reference to chloride extraction from the structure to the value lower than the critical value required by the standard [2] which, unfortunately, did not guarantee safety of the structure. It should be also noted that the critical concentration C_{crit} = 0.4% at the rebar surface was achieved after 21 days of the migration process. Moreover, after the same time of extraction concentration was reduced at the rebar surface by 95%, which could suggest that extraction rate was slower than chloride ion migration to concrete within the electric field.

Figure 9b illustrates a group of curves presenting the distribution of chloride ion concentration at the concrete cover depth of 2 cm, plotted for the following times: 24, 96, 120, 144, 192, 219, 240, 288, 384, 504 h in concrete C2. The distribution of curves indicated that only after 10 days of extraction, the concentration at the steel surface was $C_{(x = 20)}$ = 0.3% (and $C_{(x = 20)}$ = 0.2% determined from the tests), which is a value posing a corrosion risk according to the standard [4,5]. This assumption was confirmed by results from the corrosion tests performed after 10 days of extractions. The determined densities of corrosion current C2.1(i_{corr} = 4.29 µA); C2.2(i_{corr} = 2.91 µA) were characteristic of steel with moderate corrosion activity according to the criterion described in the papers [29,30]. Another 11 days of chloride extraction did not produce such spectacular effects, but concentrations of chloride ions at the rebar surface achieved $C_{(x = 20)}$ = 0.2% from the model (and $C_{(x=20)} = 0.1\%$ from the tests). Both of these values are specified in the standards [4,5] as values, at which corrosion can develop. Densities of corrosion current (C2.1(i_{corr} = 3.56 µA); C2.2(i_{corr} = 3.07 µA) obtained from the tests performed after 21 days of extraction, indicated low corrosion activity according to the criterion described in the papers [29,30]. This process was effective but the critical concentration C_{crit} = 0.4% at the rebar surface was achieved after 21 days of the migration process. Moreover, after the same time of extraction, the concentration was reduced at the rebar surface by 75%, which could suggest that the extraction rate was slower than chloride ion migration to concrete within the electric field.

5. Conclusions and Recommendations

The performed tests demonstrated that cement CEM III/A 42.5 N-LH/HSR/NA improves protective properties of concrete against corrosion of reinforcing steel as corrosion current values were lower at higher concentrations of chloride ions. On the other hand, chloride extraction from the structure built from this type of concrete should be longer to provide as effective extraction as for concrete with cement CEM I 42.5 R. This effect can be explained by the fact that the majority of chlorides in concrete C2 were probably bounded in the cement matrix, so the real concentration of chloride ions in pore water was lower than in this type of concrete. The effect of binding chloride ions in concrete with cement CEM III/A 42.5 N-LH/HSR/NA was also confirmed in the paper [25], in which values of diffusion coefficients of chloride ions were determined, taking into account the process of binding chloride ions.

Furthermore, the determined coefficient of extraction in concrete must be considered with caution as the duration of the effective chloride extraction can be erroneously evaluated by comparing these coefficients without taking into account the distribution of concentration values at the concrete cover depth. The values of extraction coefficient in concrete C1 equal to ($D_e = (1-6) \cdot 10^{-10}$ (m^2/s)) were about 7 times higher than the values determined for concrete C2 ($D_e = (0.3-0.9) \cdot 10^{-10}$ (m^2/s)), which could suggest extraction in concrete C2 was 7 times longer than in concrete C1. Moreover, the analysed model from Figure 9b indicates that after 26 days of extraction, the concentration of chloride ions at the steel surface was C_{crit} = 0.01% and was the same as for concrete C1 after a 21-day extraction. However, it is known that too long extraction could deteriorate the mechanical properties of concrete and steel in the treated structure and increase the costs of the applied method.

The recommendations of this study:

The previously proposed method for assessing the effectiveness of electrochemical extraction of chlorides (ECE) from concrete can be successfully used to compare the effect of the cement used in concrete on the speed of the extraction process.

In another application, a rapid test on steel resistance to chloride ions can be performed using the drilled cylindrical specimen of concrete with fragments of reinforcement. This test uses the electric field to accelerate migration of chloride ions to concrete and then the densities of corrosion current are measured at the regular time intervals with the LPR method.

A very important point here is that extraction is a longer process than migration of chloride ions into concrete. Hence, modelling the extraction process with the migration coefficient is incorrect and can lead to a too short duration of this process. As a result, this method may be ineffective.

The limitations of this study:

The method still needs to be tested on a larger number of different concretes.

Funding: The research was financed by Silesian University of Technology (Poland) within the grant no. 03/020/RGH_20/0096 and in part within the project BK-222/RB-2/2022 (03/020/BK_22/0128).

Institutional Review Board Statement: Not applicable.

Informed Consent Statement: Not applicable.

Data Availability Statement: The data presented in this study are available on request from the corresponding author.

Conflicts of Interest: The author declares no conflict of interest.

Appendix A

Table A1. Comparison of results from analyzing polarization curves obtained for two selected specimens C1.1 and C1.2 measuring before (M0) and after 7 (M1) and 14 (M2) and 21 (E0) days chloride migration and after 10 (E1) and after 11 (E2) days chloride extraction.

Measure No.	Time Days	E_{corr} mV	b_a mV	b_c mV	R_p kΩ	R_pA kΩcm^2	I_{corr} µA/cm^2	V_r µm/Year
C1.1-M0	0	−205	169	78	4.90	111.6	0.21	0.2
C1.2-M0	0	−151	204	82	6.32	143.44	0.18	0.2
C1.1-M1	7	−540	369	88	1.30	29.58	1.04	1.1
C1.2-M1	7	−686	343	119	0.24	5.36	7.16	7.9
C1.1-M2	14	−661	374	101	0.05	1.18	29.25	32.1
C1.2-M2	14	−662	382	104	0.06	1.43	24.82	27.3
C1.1-E0	21	−751	230	91	0.03	0.77	38.94	42.8
C1.2-E0	21	−688	480	162	0.06	1.38	37.98	41.8
C1.1-E1	10	−751	461	81	0.18	4.04	7.40	8.1
C1.2-E1	10	−666	420	323	0.18	4.09	19.40	21.3
C1.1-E2	21	−752	118	141	0.31	7.04	3.96	4.4
C1.2-E2	21	−540	415	93	1.31	29.74	1.11	1.2

Table A2. Comparison of results from analyzing polarization curves obtained for two selected specimens C2.1 and C2.2 and measuring before (M0) and after 7 (M1) and 14 (M2) and 21 (E0) days chloride migration and after 10 (E1) and after 11 (E2) days chloride extraction.

Measure No.	Time Days	E_{corr} mV	b_a mV/dec	b_c mV/dec	R_p kΩ	R_pA kΩ cm^2	i_{corr} µA/cm^2	V_r µm/Year
C2.1-M0	0	−191	513	54	3.39	76.95	0.28	0.3
C2.2-M0	0	−221	414	69	4.93	111.91	0.23	0.3
C2.1-M1	7	−585	332	115	0.26	5.90	6.28	6.9
C2.2-M1	7	−416	255	104	1.35	30.65	1.05	1.2
C2.1-M2	14	−556	784	149	0.25	5.56	9.78	10.8
C2.2-M2	14	−617	451	133	0.44	10.03	4.45	4.9
C2.1-E0	21	−580	317	148	0.18	3.97	11.03	12.1
C2.2-E0	21	−559	289	158	0.43	9.67	4.59	5.0
C2.1-E1	10	−324	343	104	0.36	8.08	4.29	4.7
C2.2-E1	10	−338	272	131	0.58	13.21	2.91	3.2
C2.1-E2	21	−332	203	120	0.41	9.19	3.56	3.9
C2.2-E2	21	−337	290	140	0.59	13.35	3.07	3.4

References

1. Quraishi, M.; Nayak, D.; Kumar, R.; Kumar, V. Corrosion of Reinforced Steel in Concrete and Its Control: An overview. *J. Steel Struct. Constr.* **2017**, *3*, 1000124. [CrossRef]
2. PN-EN 206+A1:2016-12; Concrete Specification, Performance, Production and Conformity. PKN/KT 274: Warszawa, Poland, 2016.
3. BS 8110: Part 1; Structural Use of Concrete–Code of Practice for Design and Construction. Br Stand Institute: London, UK, 1985; p. 38.
4. ACI 201.2R-08; Guide to Durable Concrete. American Concrete Institute: Farmington Hills, MI, USA, 2008.
5. Revisions to: Guide for the Design and Construction of Fixed Offshore Concrete Structures. *ACI J. Proc.* **1984**, *81*, 632–639. [CrossRef]
6. ACI 222R-01; Protection of Metals in Concrete against Corrosion. ACI Committee: Farmington Hills, MI, USA, 2001; pp. 1–41.
7. Hausmann, D.A. Steel corrosion in concrete–How does it occur? *Mater. Prot.* **1967**, *6*, 19–23.

8. Yonezawa, T.; Ashworth, V.; Procter, R.P.M. Pore Solution Composition and Chloride Effects on the Corrosion of Steel in Concrete. *Corrosion* **1988**, *44*, 489–499. [CrossRef]
9. Kayyali, O.A.; Haque, M.N. The Cl^-/OH^- ratio in chloride-contaminated concrete—A most important criterion. *Mag. Concr. Res.* **1995**, *47*, 235–242. [CrossRef]
10. Castellote, M.; Andrade, C.; Alonso, C. Accelerated simultaneous determination of the chloride depassivation threshold and of the non-stationary diffusion coefficient values. *Corros. Sci.* **2002**, *44*, 2409–2424. [CrossRef]
11. Ann, K.Y.; Song, H.-W. Chloride threshold level for corrosion of steel in concrete. *Corros. Sci.* **2007**, *49*, 4113–4133. [CrossRef]
12. Morris, W.; Vico, A.; Vazquez, M.; de Sanchez, S. Corrosion of reinforcing steel evaluated by means of concrete resistivity measurements. *Corros. Sci.* **2002**, *44*, 81–99. [CrossRef]
13. Yu, B.; Liu, J.; Chen, Z. Probabilistic evaluation method for corrosion risk of steel reinforcement based on concrete resistivity. *Constr. Build. Mater.* **2017**, *138*, 101–113. [CrossRef]
14. Bamforth, P.B. The derivation of input data for modelling chloride ingress from eight-year UK coastal exposure trials. *Mag. Concr. Res.* **1999**, *51*, 87–96. [CrossRef]
15. EN 14038-2; Electrochemical Realkalization and Chloride Extraction Treatments for Reinforced Concrete—Part 2: Chloride Extraction. European Standard: Brussels, Belgium, 2020.
16. Chang, C.; Yeih, W.; Chang, J.; Huang, R. Effects of stirrups on electrochemical chloride removal efficiency. *Constr. Build. Mater.* **2014**, *68*, 692–700. [CrossRef]
17. Yeih, W.; Chang, J.J.; Hung, C.C. Selecting an adequate procedure for the electrochemical chloride removal. *Cem. Concr. Res.* **2006**, *36*, 562–570. [CrossRef]
18. Ihekwaba, N.M.; Hope, B.B.; Hansson, C.M. Structural shape effect on rehabilitation of vertical concrete structures by ECE technique. *Cem. Concr. Res.* **1996**, *26*, 165–175. [CrossRef]
19. Szweda, Z.; Jaśniok, T.; Jaśniok, M. Evaluation of the effectiveness of electrochemical chloride extraction from concrete on the basis of testing reinforcement polarization and chloride concentration. *Ochr. Przed Korozja* **2018**, *61*, 3–9. [CrossRef]
20. Toumi, A.; Francois, R.; Alvarado, O. Experimental and numerical study of electrochemical chloride removal from brick and concrete specimens. *Cem. Concr. Res.* **2007**, *37*, 54–62. [CrossRef]
21. Andrade, C.; Diez, J.; Alamán, A.; Alonso, C. Mathematical modelling of electrochemical chloride extraction from concrete. *Cem. Concr. Res.* **1995**, *25*, 727–740. [CrossRef]
22. Chang, J. Bond degradation due to the desalination process. *Constr. Build. Mater.* **2003**, *17*, 281–287. [CrossRef]
23. Carmona, J.; Climent, M.; Antón, C.; De Vera, G.; Garcés, P. Shape Effect of Electrochemical Chloride Extraction in Structural Reinforced Concrete Elements Using a New Cement-Based Anodic System. *Materials* **2015**, *8*, 2901–2917. [CrossRef]
24. Bouteiller, V.; Tissier, Y.; Marie-Victoire, E.; Chaussadent, T.; Joiret, S. The application of electrochemical chloride extraction to reinforced concrete—A review. *Constr. Build. Mater.* **2022**, *351*, 128931. [CrossRef]
25. Perkowski, Z.; Szweda, Z. The "Skin Effect" Assessment of Chloride Ingress into Concrete Based on the Identification of Effective and Apparent Diffusivity. *Appl. Sci.* **2022**, *12*, 1730. [CrossRef]
26. Herrera, J.O.; Escadeillas, G.; Arliguie, G. Electro-chemical chloride extraction: Influence of C3A of the cement on treatment efficiency. *Cem. Concr. Res.* **2006**, *36*, 1939–1946. [CrossRef]
27. Zofia, S.; Adam, Z. Theoretical Model and Experimental Tests on Chloride Diffusion and Migration Processes in Concrete. *Procedia Eng.* **2013**, *57*, 1121–1130. [CrossRef]
28. Stern, M. Closure to "Discussion of 'Electrochemical Polarization: 1. A Theoretical Analysis of the Shape of Polarization Curves' [M. Stern and A. L. Geary (pp. 56–63, Vol. 104)]". *J. Electrochem. Soc.* **1957**, *104*, 751–752. [CrossRef]
29. Raczkiewicz, W. Use of polypropylene fibres to increase the resistance of reinforcement to chloride corrosion in concretes. *Sci. Eng. Compos. Mater.* **2021**, *28*, 555–567. [CrossRef]
30. Raczkiewicz, W.; Koteš, P.; Konečný, P. Influence of the Type of Cement and the Addition of an Air-Entraining Agent on the Effectiveness of Concrete Cover in the Protection of Reinforcement against Corrosion. *Materials* **2021**, *14*, 4657. [CrossRef]
31. Green, W.; Lyon, S.; Scantlebury, J. Electrochemical changes in chloride-contaminated reinforced concrete following cathodic polarisation. *Corros. Sci.* **1993**, *35*, 1627–1631. [CrossRef]
32. Elsener, B.; Molina, M.; Böhni, H. The electrochemical removal of chlorides from reinforced concrete. *Corros. Sci.* **1993**, *35*, 1563–1570. [CrossRef]
33. Kim, K.B.; Hwang, J.P.; Ann, K.Y. Influence of cementitious binder on chloride removal under electrochemical treatment in concrete. *Constr. Build. Mater.* **2016**, *104*, 191–197. [CrossRef]
34. Szweda, Z. Estimating coefficient of chloride extraction from concrete. *Ochr. Przed Koroz* **2019**, *62*, 393–398. [CrossRef]

Disclaimer/Publisher's Note: The statements, opinions and data contained in all publications are solely those of the individual author(s) and contributor(s) and not of MDPI and/or the editor(s). MDPI and/or the editor(s) disclaim responsibility for any injury to people or property resulting from any ideas, methods, instructions or products referred to in the content.

MDPI
St. Alban-Anlage 66
4052 Basel
Switzerland
www.mdpi.com

Materials Editorial Office
E-mail: materials@mdpi.com
www.mdpi.com/journal/materials

Disclaimer/Publisher's Note: The statements, opinions and data contained in all publications are solely those of the individual author(s) and contributor(s) and not of MDPI and/or the editor(s). MDPI and/or the editor(s) disclaim responsibility for any injury to people or property resulting from any ideas, methods, instructions or products referred to in the content.

www.ingramcontent.com/pod-product-compliance
Lightning Source LLC
LaVergne TN
LVHW070213100526
838202LV00015B/2042